How to Kill an
ELEPHANT

Eighteen Months to Save the Planet

ROBERT PINS

authorHOUSE®

AuthorHouse™ UK
1663 Liberty Drive
Bloomington, IN 47403 USA
www.authorhouse.co.uk
Phone: 0800.197.4150

Published by AuthorHouse 11/23/2018

ISBN: 978-1-5462-9655-3 (sc)
ISBN: 978-1-5462-9656-0 (hc)
ISBN: 978-1-5462-9654-6 (e)

Library of Congress Control Number: 2018909732

Print information available on the last page.

Any people depicted in stock imagery provided by Getty Images are models,
and such images are being used for illustrative purposes only.
Certain stock imagery © Getty Images.

Cover image credit: Delamere Forest, 2012 (c) Charlotte Fox
https://charlottefoxphotography.com

This book is printed on acid-free paper.

Contents

Eighteen Months to Save the Planet
Introduction: The Executive Summary

How to Kill an Elephant should start with an apology.

World leader after world leader stands before the world's media and offers solutions to all that ails us. Given that these leaders' solutions are plausible, sincere, deeply impassioned, and above all else, fundamentally flawed, the time is right for a reply.

This is my reply: the time for politeness is past. I am hard-hitting and devoid of rose-tinted glasses. And rest assured, all princesses are dead.

I aim to open the eyes of my reader, to lift your thoughts to a different reality that plays out in your daily lives. You will identify with some sections of *How to Kill an Elephant* because you will have been exposed to the mind-numbing realities of the 'it's more than my job is worth' type of assertion that passes for sentient thought. Scanning the newspaper headlines, I frequently think, *At last, someone has actually worked it out,* before realising that whomever it is has merely breathed across the surface, afraid to penetrate even one cell below it.

How to Kill an Elephant should start with an apology, I agree. I am ready to accept yours.

Now, having set myself up to fail, faster than a speeding meteor, I will attempt to provide you with some structure to the book and its contents.

I have compiled a lifetime of un-thought practices which serve to ensure the free passage of our species as we each seek to earn a living in the race for survival. I have taken simple, everyday examples of actions and practices that serve to ensure that the job gets done and have dissected these to expose their inner workings. But it is not enough to expose them, because to achieve my task, I have to teach you how to do it for yourself.

So it is that I will return time and again to previously addressed topics, not to strip an onion, as the analogy goes, but to build an onion, layer by layer, hopefully according to exhibition standards. These un-thought practices exposures could best be described as my revenge—the revenge achieved by exposing all the mindless inequities forced upon my life and by calling their originators to task.

I will appeal to different aspects of your lives to attract your attention. It could be as simple as cost; the wasting of my hard-earned taxes; or the hours spent working for others so that they can waste my effort and, consequently, my life. It could be the cost of a packet of crisps or the cost of admission to the cinema. It could be the cost in lives of feeding the birds. (Which lives am I referring to, birds' lives, human lives, or the lives of something else?) It could be a cost in terms of STUs, not sexually transmitted, but deadly nonetheless. (A full explanation will be given later.) Perhaps it is a cost in a future lost while taking the time to reflect on a present and a past lost as well. All right, I know, any idiot can moan, and if truth be known, I could represent Great Britain in the Olympics if such an event were added. (Although, no doubt, my style of moaning wouldn't measure up to the style expected by the judges. See, I can even moan about moaning.) Rest assured, everything we do has a cost, a cost ultimately borne by earth.

After the exposure of the mindless processes that dominate our waking hours comes the search for the reasons for these processes. It is not enough to ridicule the actions and intent without gaining an explanation for—and understanding of—just why things are going wrong.

There is a search for understanding, a search for leadership, and a search for the controlling mechanism. And, given that ours is the only planet known to harbour life, a search for biology seems appropriate as well. Our species has the power of dominion over the beasts and fowl of land and sea, not through any God-given right but through the ingenuity of the human brain, a brain that is rapidly outstripping the capacity of our planet. The human brain offers so much promise. Indeed, it could even guarantee us a future, if only we were in control of it. I have found the control mechanism for the brain.

And elephants. How could I have nearly forgotten elephants? Consider *How to Kill an Elephant* to be a comprehensive guidebook to the identification of elephants, their feeding habits, their breeding methods, how to train and house them, and most importantly why we need to kill them. Inoculations for exotic diseases from far-flung places will not be needed because most of these elephants have never seen warmer climes. This is a new genus of elephant that you didn't even know existed and therefore that you singularly ignore. These pachyderms have a momentum and mind of their own and resist all attempts to control them. They are often sired (apparently) by that most damaging of agents: good intentions.

These elephants are not content with trampling on the field margins because their ambitions are without constraint.

Rest assured, everything we do has a cost, a cost ultimately borne by earth. While others earnestly engage in the search for other habitable planets (didn't some chap by the name of Noah try something similar once before?), let's call it Earth 2.0, the thought keeps coming back to me that the wise money would be on hedging one's bets by trying to fix Earth 1.0.

But in my comprehensive guide to the identification of elephants, *wisdom, money,* and *betting* all qualify as pachyderms. Pachyderm is the now defunct grouping of animals with thick skins into a related order where the thickness of the skin supposedly confers a common biological heritage. In my interpretation of the world, *Homo sapiens*—that is 'wise man'— money, and betting are all innately related and qualify perfectly as thick-skinned animals worthy of joining the order.

Just a thought—not really an afterthought, I must confess: the latter half of 'Rest assured, everything we do has a cost, a cost ultimately borne by earth' actually translates to 'a cost ultimately borne by humankind'. Would that be an inconvenient truth? Don't think for one instant that this is yet another book about climate change and our need to reduce our carbon footprint because that could be construed as a good intention. I can hear the patter of *Elephantidae* tiptoeing through the undergrowth. (Tiptoeing? Well, they do have five toes per foot, but they also have the ability to harden or soften the pads of their feet, enabling them to have a stealth mode. It's just that 'in stealth mode through the undergrowth' doesn't have as pleasant a ring to it. Actually, scratch the whole argument because the structure of the foot means that elephants are effectively always walking on the tips of their toes. Anyway, *How to Kill an Elephant* is not about climate change.)

By now I have either whetted your appetite or triggered a rush towards the exit. You may consider that the book's content is above your pay grade and that therefore you may leave it to 'those in charge' to sort out the issues. I have posited the question of whom the survivors would blame in the event of a nuclear apocalypse and have concluded that we would blame 'those in charge', when in reality the fault lies with you and me for allowing them to take charge in the first place. I could apologise for being so long-winded, verbose if you would prefer, but my aim is not to offer an alternative point of view. My aim is to offer an alternative way of thinking.

Perhaps an illustration of the pitfalls of an alternative view versus an alternative way of thinking would help, something currently filling the media screens that you can relate to. Harvey Weinstein and others have recently been exposed for their exploitative attitudes towards aspiring actresses. The floodgates have been opened, and the reactions have been displayed on the award ceremony red carpets and in the Twitter feeds. The position of power was abused by these individuals, and the redress is perfectly attuned to rebalancing the equation. What these men have (allegedly, pending court cases) done is *wrong*. What we have seen is the pendulum swinging from male domination to the other extreme, female domination. What we will begin to see in the near future is the abuse of power by women in the industry as they seek to dominate men. What we need is an alternative way of thinking, not an alternative way of doing the same thing. And while we are at it, I predict that in the new woman-controlled movie environment, the allure of the female body will still be used to bring the paying customers within range.

While written as non-fiction, *How to Kill an Elephant* is not unlike a whodunit novel that eventually reveals the real villains. But don't be jumping to any early conclusions as you decide on my apparent prejudices. Consider them to be red herrings, designed to throw you off the scent before the unveiling in the drawing room leads to the arrest.

We may even learn who killed the elephant from among the list of potential culprits, whereupon those of us who are innocent can continue on the path to living happily ever after.

Note that this volume you are reading is not the expurgated version. I have on occasion vented my frustration with some wholly appropriate rude words.

We live not in a real world but in a wholly artificial one created by … hmm, whom shall we blame? To create a really good illusion, it helps if those enveloped within it are also willing participants actively engaged in the supporting mechanism. For children it can be very simple, as they still believe in magic and have yet to learn the term *prestidigitation*. For adults, magic tricks astound and baffle us in equal measure, but beyond the bewilderment lies the knowledge that magic doesn't exist and that the mind is being tricked. For now, I will throw just one name into the ring for the creator of our illusion. Rest assured, there are many names to discover within this book, but I'll start with just one: advertising. It creates the illusion of need and desire that helps to drive our seemingly endless appetite for products in the acquisition of which to invest our waking activities.

I confess that *How to Kill an Elephant* is not an easy book to read. Many of the topics covered are denatured—that is to say, cooked to change their shape, taste, and appearance in order to give you a better understanding. Beef Wellington is still cow for example. For those of us unfamiliar with the kitchen, could we track the product from either end, from beef Wellington to a cow, or from a cow to beef Wellington? To add or strip a small number of ingredients and processes is to transform our understanding of the finished product and its origin. While we can all reach for the recipe book, there remains an element of magic that we don't fully understand, namely the process of cooking. A very simple explanation of cooking is that the proteins and other long molecules are damaged (denatured) by heat, and it is this that produces the different flavours. How willingly we embrace the process without ever understanding it. The old TV advert used to ask, 'How do you want your eggs, fried or boiled?' Or do you want them scrambled, poached, or in an omelette, or for that matter soft or hard boiled, sunny side up, or over easy? The point I am trying to prove is that our appetite can stretch far beyond the original idea and that different styles of denaturing can offer a variety of different tastes and textures to excite the palate (with or without soldiers? That would be bread fingers for dipping in the soft boiled egg for the demilitarised uninitiated). I will try to give an understanding of the 'magic'.

I will not apologise for the constant repetition of conclusions. This is deliberate. Every hour of our waking days, non-solutions to our most pressing problems are drip-fed into our environment. I have felt the need to respond with my own drip-feeding lest the message be diluted.

I tend to write on my tablet in random sections, often late at night when my brain seems to clear, and I only give up when the words stop coming. Then I mow the lawn or feed the boiler or collect logs, using the time to think about and polish my earlier thoughts. You should try it sometime. I'll even show you where to stack the logs. I react to all sorts of external stimuli, such as passing comments, or TV news or newspapers, and invariably attempt to rip the pro forma to shreds. And after a while, it becomes effortless—second nature to me. You will understand the attention to pro forma as the book's theme develops. By the very randomness of the media process, the topics can be widely dispersed, although the treatment becomes more and more focussed. My aim is to treat each topic as a single fibre. Over time, you will realise that all these fibres have begun to form an intertwined rope that binds all the disparate thoughts into a central tenet.

You need to divest yourself of personal involvement and understand the distinction between us and them, the latter being the faceless masses. And before you consider my comments about the 'faceless masses' as being tantamount to racism, or a similar anti-social attitude, know that the book's ambition is to give *all* of humankind a future by changing its present. This book endeavours to place the human race into a petri dish for closer examination. A degree of dispassion is needed to observe correctly.

I should also say that I am not fixated on speeding, although I don't speed in 30 mph zones. I use speeding as an example of how something so simple is really very complicated. It isn't complicated at all; it is actually very simple. And it is a wonderful example that can be readily understood by all drivers, if only they would vibrate a brain cell. While I concentrate on speeding vehicles and the law, I could just as easily have concentrated on the banning of pistols in the UK or on 'Thou shalt not kill'. If you would prefer, 'don't trade in ivory' equates to 'don't exceed 30 mph'.

Finally, I have endeavoured to add a little humour to the text, if only to ease and reward the reader's task.

No more procrastination.

Lies to Children

When we are young with limited vocabularies and attention spans, we often receive short, convenient answers that solve the immediate problem. These answers are called 'lies to children'. We can all think of examples, although the answer to one about where babies come from, 'under the gooseberry bush', is surprisingly accurate when considering the slang. The problem is that these lies become a shorthand guide to life and knowledge. We are bombarded with so much data that our brains have to develop a shorthand method of filtering out what is of importance, and we compartmentalise our lives accordingly. So it is that we travel through our lives firmly grasping these staple lies without devoting any thought to them. And we develop others to fill the non-existent cracks in our thought processes.

Some people have a rude awakening but fail to extrapolate beyond the immediate problem. Remember when your mother took you to the doctor for some 'magic medicine' that would make you better, or tell you to always ask a policeman if you got lost? Develop a chronic illness or complain of child sexual harassment in Rochdale, and you will understand the folly of these lies.

In these two simple examples, we have blown open two immense social topics that can be widened to cover the medical profession and the police. But this isn't enough, because it questions the National Health Service (NHS) and the criminal justice system. Yet if we question these, we must also question government and how they function. Rochdale involved the systematic sexual abuse and exploitation of predominantly white girls at the hands of predominantly Muslim men of Pakistani origin, so we must also question social attitudes.

We are exposed to millions of snippets of information throughout our lives. Some are formally presented to us at schools and in textbooks, whereas the vast majority are just out there on the news, in newspapers, on the Internet, and in television programmes. We learn from colleagues, friends, and family. This information-gathering process shapes and steers us throughout our lives. Very often we are too busy thinking about what we do and how we do it that, paradoxically, we don't in fact think about what we do and how we do it. As a simple example, spend some time watching labourers move material with a wheelbarrow. See how often they fill the barrow before turning it in the direction they want to travel. Turning an empty barrow is easier than turning a full one, but the job is to move the material, and the material will be moved.

You may take pity on a beggar in the street and slip a couple of quid to help. But in the process, you reward the beggar and remove any incentive to actually change his or her life. Yet you think this helps. Now I'm not talking about some blind boy with a handicap in India here. I'm talking about your city centre beggar who's surrounded by our entire welfare state and the product of our free education system. If you stop and think, every individual born and raised in Britain for the last seventy years has benefited from all the state has to offer in regard to health, welfare, and education, a programme of works set up to eliminate the need for begging.

Most of us buy a poppy in the annual Poppy Day collection period leading up to Armistice Day and Remembrance Sunday, and we are bombarded by the choice of 'retail source'; we can buy them at work, in the street, at pubs, in shops, and even at the door to our house. If you've bought one at work and are wearing it, you can happily walk past all other point-of-sale outlets without buying another, since you have already contributed. Most of us pay tax. Even if we don't earn any or enough money to pay income tax, we still pay VAT or duty on products we buy. So I am quite happy to walk past a beggar without taking my hands out of my pockets on the basis that I have already contributed. Why do it again? I can also ignore the clever marketing ploy of 'Have a nice day,' because I am and don't feel the need to sit in dirty clothes at my pitch all day. This assumes that I can ignore the beggar on the basis that somebody else (the state) has already made provisions for him or her, so I don't need to. (We will talk about cheese wells later.) But the state doesn't have any money. They have my and your money collected via taxes, and we charge them with the responsibility to ensure nobody has to beg to survive in Britain today.

Let's reverse the situation and turn the clock back, say, 150 years. Begging, poverty, illiteracy, and ill health are common, and massive deprivation exists throughout the nation. Let's do something about it. Let's engineer a safety net, a catchall to ensure that all citizens can achieve a minimum standard of health, education, wealth, and happiness. Let's engage the greatest brains available to devise an equitable arrangement to make it so and, in the process, eliminate begging from our streets. But one of the earliest problems you will encounter will be that these greatest brains will all approach the problem with different ideas and perspectives. Leave it to the debating committee and one of two things will happen: nothing (and nothing is not an option) because you cannot reach agreement, or something because different factions will be overrun or ignored, or the brains with the greatest vision and strength of personality will drive through their own ambitions to the exclusion of others. Now this process could almost be used as a model for politics and politicians; if you haven't got the answer, press on anyway, because it coincides with your beliefs (well, actually, I've been offered a promotion/plum job if I move my grouping across to agree). Straightaway, we have encountered a major issue, because our new system for providing for the poor and destitute begins life by providing for the greatest brains and all that such a thing entails.

So this is why we end up with compulsory religious education in a secular state. No doubt we need to thank the bishops for this, as they were looking after their own interests. Sorry, our Christian interests. The real questions will be, does it work and has it worked?

My childhood interest was biology, and while it has never featured in my career, it has featured large as a hobby. For years I have photographed and lately filmed any aspect of nature and learnt about the habitats we occupy at home and abroad.

While many regard *Homo sapiens* to be beyond nature, I will try to demonstrate that nothing could be further from the truth. For all our intellect and ingenuity, we are still bound by our DNA and follow biological models, although we never accept the idea in our actions an understanding of this premise is essential to our continuing survival.

Why Is a Christmas Tree a Tree and Not a Bush?

The growing tip, or apical bud, of a pine tree releases a chemical that travels down the stem, inhibiting the growth of all side shoots. Damage the apical bud and two or more replacement buds will become established, each dominating the buds below it. This is one reason deer and squirrels are considered pests in forestry plantations. Eat the top bud and your tree loses its shape and becomes a stunted bush.

Why is a gooseberry bush a bush and not a tree? (That would be *real* gooseberry bushes.) After all, they both contain lignin, the substance of trees—wood.

By extrapolation, the apical bud does not dominate the side shoots and they all have an equal chance of growth potential. Now the bush shape is determined by sunlight, wind direction, animal damage (watch out for the thorns), and competition from other plants.

But you will never find a gooseberry tree.

While staying in the Picos de Europa in northern Spain, I chanced upon a leaflet directing me to an ancient holm oak. As I was staying just a few miles away, I took the turn off the main road and took the loop in search of this specimen worthy of such comment. I could not find it, not at the first attempt anyway. At the second attempt I found the plaque advertising its presence and looked in wonder at such a tiny tree. This set me to thinking. And then I realised that all the other holm oaks are bushes and I had seen thousands of them. Here the damage is done by the goats, both herded and feral, that crop the growing tips, and in times gone by, by deer. So here we see a strategy for survival. This tree can survive quite happily as a shrub and will flower and set acorns to secure future generations, but occasionally through luck, or an absence of damage, it will achieve its full potential and win an award (plaque).

This is a natural process, a strategy for survival—and one that works. Travel to the rainforests of Asia and seek out the apex predators, the poster boys of the wildlife protection efforts. A healthy forest or ecosystem supports its apex predators, we are told, and by definition the presence or absence of tigers gives an instant snapshot of the value of the habitat. But why only one tiger? Why not three different types of large predator? Think apical bud domination. The presence of tigers within an ecosystem inhibits and prevents alternative apex predators from surviving or developing. They don't do it with chemicals but with tooth, claw, and muscle, and they dominate. The tiger therefore becomes an evolutionary block to other species that would occupy the same niche.

Leopards are other big cats that survive in these forests, but they survive because they occupy a different niche from the tiger. Given the opportunity, the removal of tigers, and enough time to evolve, leopards would develop to achieve the full apex domination that the tiger enjoys.

That is not to say that a leopard is an imperfect tiger. The leopard is a perfect leopard, but it is held in its niche by the presence of tigers, which in turn determines what a leopard will be.

Think Google, Microsoft, Amazon, Apple, and so forth, each an apex 'predator' or, should we say, 'dominator'. Each has developed and grown to dominate its own niche within the business environment. Each, through its actions, inhibits the growth and development of its own competitors while at the same time creating a demand for successful mimicry. Does every leopard seek to become a tiger?

But the rainforest is more than tigers, more than mammals; it is full of birds, insects, reptiles, fungi, moulds, bacteria, and so forth. Business is more than Google; it is full of stationers, furniture makers, coffee makers, et al. who use Google to facilitate their own requirements as they in turn provide or facilitate for Google.

We all marvel at the delicate beauty and grace of butterflies. Where is the apex butterfly? It doesn't exist, so the conclusion is that in plant terms, butterflies are bushes.

But not all 'trees' have to be large. Consider the bracken *Pteridium aquilinum*; this invasive fern grows to 1.5 metres or so but crowds out light and prevents that light from reaching the vegetation beneath it, thus dominating its environment. The humble zebra mussel (*Dreissena polymorpha*) has invaded European waters from the Caspian Sea. It tolerates low levels of oxygen and can exist in such huge numbers that it can seriously reduce the oxygen levels in freshwater rivers, effectively denuding them of the normal fauna. Trees themselves form forests, dense masses of canopy stretching for miles in all directions, until halted by geographical features such as rivers, cliffs, or bare rock. It is only at the fringes that full sunlight can reach ground level and that other ecosystems have a chance to survive. Shade-tolerant plants can thrive to form undergrowth, and seeds and young plants react vigorously to chance opportunities such as a forest giant dying and falling. But that is not to say that an impoverished ecosystem exists. Rather, an abundance of life is supported on the trees and by the waste products of trees, namely dead leaves and decaying wood. Just don't expect to find grass.

We describe mature woodland as a climax environment, one that has reached its natural zenith, its climax, and is in equilibrium with the climate, an environment that has arrived and is not journeying to something else. This of course ignores the evolutionary pressures that exist within the entirety of such an environment, that is the drive for an advantage, an edge to secure the survival of each individual over its neighbour.

Where does humankind sit in all this?

Take *Homo* and assume for the purpose of the discussion that hominids have been on earth for two million years. *Homo sapiens* has been around for one hundred thousand years or so, but truly modern humankind has existed for perhaps sixty thousand years. In simple terms, that's two million years of being a bush and sixty thousand years of being a tree, an apex dominator. Maybe I am being too generous here and perhaps the invention of agriculture around ten thousand years ago is the true event that took us to local domination.

Curiously enough, trees are believed to have been the catalyst for the human ape, or rather the lack of them—trees, that is. A commonly held idea is that as the climate of East Africa dried out due to changing weather systems the tree canopy began to thin and gaps appeared between isolated pockets of trees. Our tree-living ancestors took to the ground to cross the gaps between trees. Learning to exploit this changed and new habitat provided the impetus for walking upright. Many fortuitous gene alterations later, we have hominids.

In the simple foregoing paragraph I have slipped in some remarkable claims:

- Climate change resulted in fewer trees and initiated the drive to adapt or die.
- Climate change destroyed the equilibrium that had taken our ancestors up to their respective points of development.
- Climate change forced the development of new actions and overcame resistance to change.

Scientists are collecting evidence to show that deforestation is producing climate change in West Africa and the Amazon. Trees, through the rapid transit of moisture from root to leaf surface and subsequent 'sweating' (transpiration), introduce vast quantities of water vapour into the atmosphere, which produces rainfall and sets the cycle going again. Remove a sufficient area of trees and the 'normal' rainfall patterns are disrupted and drought can occur in the 'rainforest', an oxymoron but nevertheless the apparent truth.

Cast your thoughts back to the end of the last ice age. As the ice melted, cold-tolerant plants migrated north, trailing the newly exposed ground. We call it tundra today, and it still exists in northern climes. Within a few thousand years, as temperatures rose, this tundra was replaced with woodland in successive waves. First with the windblown seeds that only generate in full sun, such as those of birch, and then the slower march of oak and beech, we arrived at what we would call climax forest, the Great Wood that covered vast areas of temperate Europe with cold-adapted conifers to the northern edge.

But if the modern example of forest destruction causing climate change is accepted, then by definition the development and expansion of the Great Wood also caused climate change.

So here we have a paradox. Climate change is good because it drives evolutionary change, the likes of which ultimately led to the origins of human beings. Modern-day climate change caused by the activities of humankind is bad and must be curtailed. Imagine for one instant that some great body, call it God, Mother Nature, or some superior alien life form, had the power to prevent climate change and could halt the present-day trend. Let's wave the magic wand and return all the fossil carbon to the ground and revert to the status quo. But while we are crossing our fingers in the hope that such an entity exists, we should be thankful that the same entity didn't exist a couple of million years ago, because without climate change, we wouldn't exist.

The counterargument to this is that the climate change that dried out East African forests was natural and what we are doing is not, because the latter is man-made.

Think back to the Great Wood full of trees, not bushes. Think back to the advent of agriculture, when humankind ceased to be a bush and became a tree. Now tell me that humanity's activity is not a natural process and that it is a 'bad thing' without at the same time condemning the Great Wood for changing the climate of Europe.

Climate change then could be directly responsible for the evolution of our species and our ability to change the climate today (if you buy into the argument that climate change is man-made that rages today).

Remember the millennium bug and the panic induced in computer systems operators around the globe that led to an entire industry of 'experts' setting upon the task of ensuring that computers didn't crash at midnight on 31 December 1999? Early newspaper reports suggested that planes would fall from the skies, defence systems would fold, and life as we know it would cease to exist. Governments and boardrooms worldwide set up task forces and programmes to prevent this computer Armageddon. I must apologise to younger readers, who will be asking, 'What the hell is he talking about?'

Well, as it turns out, nothing. Hospitals continued to prescribe medicines, garages continued to supply petrol at the pumps, pubs continued to supply beer, and not a single plane fell out of the sky. We certainly dodged a bullet there.

Am I seriously comparing climate change, the single greatest challenge facing humankind, with the millennium bug?

Consider the word 'momentum'.

Before delving too deeply into this, I will throw in a couple of personal anecdotes. Fifteen years ago I was referred to a surgeon about my sore feet. *Plantar fasciitis* (policeman's heel) was the diagnosis: inflammation, and tearing of the fleshy pad on the sole of the foot from its anchorage points. My surgeon promptly operated and temporarily solved the problem for me. It was decided to operate on my left foot because it was the worst of the two, with the second operation to follow as soon as I had healed (should that be 'heeled'?) Many years later I suffered a recurrence of this painful condition, but this time I ended up with a physiotherapist. She, bless her, spent five minutes looking at my lower leg and suggested the root cause of the problem was over-tight Achilles tendons. Fifteen minutes of calf muscle stretching removed 90 per cent of the discomfort within a day. I asked her if I had needed the operation, and she replied, 'Probably not.' In a similar vein, I have arthritis in both shoulders and had an arthroscopy on my left shoulder to relieve the pain. I was originally booked in for my right shoulder, but during the wait my left shoulder became the most painful, so we switched sides. As soon as my left shoulder healed, I would have the right done. I had extensive physiotherapy on my shoulders after the first operation and haven't been back. I have had two operations, each one of a matched pair, but in both instances I never felt the need for the second operation, because the physio had done the job for me.

My point is, ask a surgeon and he will operate; ask a physiotherapist and be offered therapy.

So, ask a climate change scientist if climate change exists and don't be surprised at the answer.

Back to momentum. The millennium bug originated because somewhere in the early mists of computer programming, the programs were written in such a way that they included a date, a date that began with the numeral 1, as in 1979 or 1993, and nobody knew what would happen to the computer code if it failed to recognise a number beginning with the numeral 2, such as 2000. This was because nobody had thought in the early days that programs would extend into the twenty-first century. I personally knew of a programmer employed by a pub chain who was paid £50,000 for a six-month contract to ensure Y2K compliance. (It even had its own catchy title.)

Throw in some newspaper articles and television reports, and before you knew it we had a slow-speed stampede of worry feeding off its own momentum that crashed like a tsunami against the rock of indifference that was 1 January 2000.

Now, of course, climate change is different, if only for its scale. Scientists, bless them, have learnt that the key to securing funding for whatever pet project they have in mind is to mention climate change. How can I secure funding to study bats, dolphins, jellyfish, mountain pastures, Antarctic ice formation, polar bear populations, lichen growth patterns, or [insert name of your personal interest here]?

Think it's hard to secure funding? Try securing funding to prove that climate change doesn't exist and you will find out how irresistible the avalanche of climate change is.

But it goes beyond that; climate change is now an industry in its own right, a topic that I want to explore using different examples.

During the 1970s Margaret Thatcher came to power in the UK and introduced many sweeping changes to government policies. (Many regard her as divisive, but love her or loathe her, it doesn't matter.) Among these policies was the selling off of state assets to the public, be it private individuals or city organisations.

This saw the sale of British Telecommunications, British Gas, and electricity and water boards, to name a few. Thatcher's early years were characterised by high unemployment figures. People on radio phone-ins often referred to themselves as one of 'Maggie's three million', blaming her for their lack of a job.

During the 1980s my company employed a number of 'secretarial staff', some of whom had come from water and electricity companies. The tales they told about their employment experience were mind-blowing.

Example 1: Three women worked in one small office designed for one person. They produced tape measures to ensure that each individual had an equal and demarcated space between them. They would fight for the chance to type a letter and would then spend all day typing and retyping the same perfect letter because they had something to do. One woman spent her days sharpening and blunting boxes of pencils. Three women were employed to do twenty minutes of work a day between them.

Example 2: One woman worked for Anglian Water in a large typing pool of thirty women. If she was lucky, she typed one or two letters a day.

With privatisation, these non-jobs rapidly disappeared and the ranks of the unemployed grew. But Thatcher didn't sack the employees; she simply created an environment where such blatant waste of resources was no longer tolerated. These companies now worked for the benefit of the 'evil' shareholders and chased 'evil' profit at the expense of jobs.

So how did things work before Thatcher's privatisation? In simple terms, each company had a salary scale structure, a scale that increased with an employee's responsibility. A manager employing twenty people had a more responsible job and therefore a higher salary than a manager employing ten people. A senior manager employing junior managers earned more if his or her junior managers employed more people. A CEO earned more because his or hers was a more responsible job given that his or her company was such a major employer.

The main message here is, be careful how you measure things, and remember that people will always cheat to achieve their goals. In simple terms, introduce a measure and you will get the result you are looking for.

Ask a surgeon and …

Ask a climate change scientist …

Ask a school to improve the education standard in schools …

Ask a computer programmer if your system needs updating for Y2K …

Now I realise that these previous examples demonstrate how managers in charge can corrupt the system.

Example 3: In 1988–89, my company built an office block and we found ourselves in need of a carpenter at the same time as a local deep coal mine was shutting down. To assuage the impact of major job losses, a 'jobs club' had been set up to find employment for the redundant coal miners. They provided us with a carpenter. At the end of this carpenter's first days, it was apparent that his skill base was woefully inadequate for our needs, so we terminated his employment with us. He did open up to my chargehand about his working experience at the pit. He told us that he had worked with bricklayers who would lay a single concrete block down the pit in an eight-hour shift. All the shift workers started at the same time, but face workers took priority when it came to the lift access to the pit. Construction workers had to wait for an hour or more before travelling underground. When underground, construction workers were not allowed to use the train system, as this was reserved for production workers only.

The construction workers therefore had to walk what could be a distance of miles to the construction site, where, 'Lordy, lordy, I've left my trowel on the surface. I'll have to go back for it.' A further wait to return to the surface followed the walk, and this in turn led to another wait on the surface before being allowed to return underground to begin the long walk again. Knock up some mortar, lay one concrete block in a wall, and then return via the waiting system to the surface so that you can clock off at the end of your shift.

Example 4. In January of 2015 I employed a man in his early 20s to sink a borehole in my garden. His previous recent job experience included working for the local authority on highway maintenance. The council had introduced a limited tracking device to their fleet of vehicles that recorded the start and finish time of the day's work by recording when the vehicle was first started up and when it finished for the day. Trouble is, it didn't record the vehicle's location, so the old hands began work every day by firing up their council vehicle and leaving the engine running on the drive to their house, where they enjoyed a leisurely paid for breakfast.

The main message here is: be careful how you measure things, and remember that people will always cheat to achieve their goals. In simple terms, introduce a measure and you will get the result you are looking for. In this instance I would suggest that the clocking of the starting and stopping of the engine was to ensure that the workers were active for the full paid shift and didn't use the vehicles outside working hours. (If only we knew when they were working?) But if you stop and think for a minute, you realise that it is the management who is once again at fault. Prioritising the pit lifts and underground train system created an environment for construction workers to legitimately avoid doing any work. Measuring the length of the working day by recording the starting up and turning off of the vehicle engines fails to measure the productivity of the worker.

Ask the minister for education how our schools and education system is working and you will probably receive an answer which includes 'a record number of passes at x level or above in x number of subjects' or 'we are spending a record amount of money on the education budget' or 'a record number of schools are assessed by Ofsted as outstanding' or 'a record number of children are studying in classrooms of less than thirty pupils'. I think you get the picture.

Ask the prime minister how he assesses the performance of his education minister and you will probably receive an answer which includes, 'Under his [or her] stewardship schoolchildren have achieved a record number of passes at x level or above in x number of subjects' or … I think you get the picture.

Ask an employer of school-leavers what they think about our schools and education system and you will probably receive an answer along the lines of, 'School-leavers today are woefully short of our requirements for employment' or 'They demand significant resources on our part to train them in the most basic requirements for employment' or … I think you get the picture.

Ask a recently employed school-leaver what they think about employment and our schools and education system and you will probably receive an answer along the lines of, 'This isn't what I was prepared for' or 'It's been such a steep learning curve' or 'I had no idea that you had to pay tax on your earnings. Why didn't they teach me this at school?'

Ask a teacher what they think about our schools and education system and you may receive the answer I got, along the lines of, 'I wonder what we are teaching our kids today? We get initiative after initiative down from the Department of Education which we keep having to incorporate into our teaching methods' or 'We are so target led, we have to achieve the exam results demanded from us' or 'I am sick to death of evaluations. Everything revolves around evaluations of students. And when I've finished doing them, I have to evaluate my evaluations' or 'We are frowned upon for teaching facts. These are all available on the Internet anyway, I know, but it would be nice to teach them something to think about' or … I think you get the picture.

So why do we educate our children? What purpose does it serve?

Does it serve employers?

Does it serve parents?

Does it serve students?

Does it serve politicians?

Does it serve UK plc?

Does it keep children off the streets during the day?

The main message here is, be careful how you measure things, and remember that people will always cheat to achieve their goals. In simple terms, introduce a measure and you will get the result you are looking for.

What we have here is an education system that is pushed along by its own momentum, a momentum that overcomes and resists all attempts to steer it and fails to achieve its original aims. What were its original aims? Can anyone working in this field still remember? Whom does it serve?

Let me introduce you to Hannibal. No, not Lecter—the other one, with the elephants.

Hannibal

The following is a recently translated text recovered from a tomb near the site of Carthage. It appears to be the shorthand notes taken at arm's length by Hannibal's autobiographical ghostwriter, a chap called Hali To'sis.

'How to Kill an Elephant', or 'Hannibal's Thought Matrix

First of all catch and then train your elephant.

No, first of all train your elephant catchers.

Build your elephant training facility.

(I seem to remember reading about one in North Africa.)

Second, train new elephant catchers to replace the ones who got it wrong in the first place.

Rebuild your elephant training facility (to replace the old one, etc.).

No, the very first thing to do is ask why you need an elephant in the first instance.

No, the very, very first thing to do is to ask whether the elephant is really the solution to the problem and whether other alternatives are better employed.

So the very, very, very first thing to do is set up a committee to consider the role of the elephant on the battlefield. Thinking about this, you will need to decide on the committee chairman, the secretary, and other committee representatives first.

Then revise the committee membership to include warfare experts.

Then revise the committee membership to include animal behaviourists.

Then revise the committee membership to include animal behaviourists with knowledge broader than the herding instincts in goats.

Then set a date for the committee to hold its inaugural meeting.

Then set a date for the committee to hold its inaugural meeting that doesn't clash with the chairman's wedding anniversary.

Hold your inaugural meeting and do commission feasibility studies for catching elephants.

Hold your second meeting and do commission feasibility studies for training elephants.

Hold your third meeting and advise that Hannibal has made it quite clear that he just wants you to JFDI, OPOD (just fucking do it, on pain of death).

Catch some elephants.

Select some new volunteers and try to catch some more elephants.

Select some conscripts and try again.

Catch some elephants.

Rebuild your holding pen.

(The one I read about was built of stone.)

Catch some elephants.

Train your elephants.

Select some new volunteer elephant trainers.

Select some new conscript trainers and try again.

Parade your new trained elephants in front of Hannibal.

Urge the city council to authorise funds to rebuild the market and continue the slum clearance.

Retrain your elephants to cope with crowds.

Load your elephants on board a ship to cross the Mediterranean to Spain.

Design a new six-tonne-capacity loading ramp for your ship.

Design a new access ramp to the ship's hold.

Design a new ship.

Build a new fleet of ships.

Evaluate what skills you have learnt in the transportation of elephants and how these skills could be transferred to other projects, such as hydroponics (you never know when your fields could be damaged by a sudden ingress of salt, for example) or the place of paper folding in warfare.

Sail to Spain.

Try in vain to find a cure for seasickness in elephants.

Arrive in Spain and disembark.

Design a new offloading ramp to the dock.

Build a new offloading ramp.

Disembark.

Re-evaluate what skills you have learnt in the transportation of elephants and whether this has had any knock-on effects on hydroponics or paper folding.

Assemble your army and march north towards Roman territory.

Overface a small contingent of Romans and watch as they withdraw in disarray.

Evaluate the first eyewitness accounts of elephants running amok when injured and causing more casualties to your own troops, and dismiss these as teething problems.

Follow the retreating Romans.

Cross the River Ebro.

Discover that elephants are really good swimmers.

Rebuild the Roman bridge over the Ebro so that your army can cross.

Have your NES interview a Roman prisoner to try to find out what the sign 'Max IV T GVW' on the approach to the bridge means.

Clash with the Romans again.

Evaluate the second eyewitness reports of injured and frightened elephants running amok and causing casualties to your own troops, dismiss these as 'a failure to embrace the new technology', and sack a divisional commander to underline this fact.

After the third eyewitness reports come in about your own troops being more frightened of the elephants than the Romans are, issue whistles to the elephant drivers, which they are to blow to warn your own troops of potential 'amokery'. Don't bother to evaluate.

Cross the Alps in hot pursuit.

Note with curiosity that your elephants have turned white. Put this down to the cold and altitude while crossing the Alps.

Decide to have some defeated generals' wives knit some thermal blankets for your elephants' return trip. Have these blankets decorated with battle scenes.

Meet some more Romans in battle.

Order the execution of all the whistle-blowers for lowering the morale of the troops.

Rescind this order when told, 'There will be nobody left to drive the elephants.'

Promote the man who told you this to captain for using his initiative, and then have him executed for countermanding your order.

Have the elephant drivers line up and count to every twelfth man. Have numbers 11 and 13 kill number 12. The reason for this will not be apparent for two millennia, but rest assured that the BBC and Discovery Channel scriptwriters and narrators will struggle with dodecimation. They can't cope with decimation: 'Front line troops were decimated with over 35 per cent casualties,' for Christ's sake.[1]

Enquire why the elephant drivers have a blue NHS emblem on their uniforms and flags. Discover that your senior regimental sergeant major commissioned them for the National Elephant Service. Decide against (*a*) telling him there is no *H* at the beginning of elephant, (*b*) having him executed for bad pronunciation and spelling, as he's a distant relative on your wife's side and you can do without that argument this decade, and (*c*) you already have a National Execution Service, (NES) albeit with a red badge.

Instruct your remaining divisional commanders that for best effect, the elephants should be used in tight formation to punch through the Roman lines and that your own troops should follow in close formation to exploit the opportunities created.

Meet *all* the Romans in battle.

---

[1] For those of you who don't know, after the two Roman legions were defeated outside Rome by the slave revolt under Spartacus, the Roman commanders ordered that each tenth man was to be killed by the other nine by way of punishment and to 'put some backbone into them'.

Realise that it was a mistake not to train your white elephants not to panic when exposed to flames.

Realise that your white elephants are retreating headlong into your tightly packed forces held in tight formation to exploit any breakthroughs.

Realise that your white elephants are causing considerably more harm than good.

Wish you'd thought of how to kill an elephant.

Lose the battle.

Lose the war.

Flee the scene and return home to Carthage.

Receive an approach from W. E. Fillourpockets and Partners, Solicitors LLP.

Sue your employers for failing to offer sufficient training in how to exploit the use of elephants in battle and failing to advise about the potential pitfalls.

Claim damages for embarrassment, emotional distress at having been forced to watch the destruction of your own army, lack of further employment opportunities, and loss of office.

Remain on full pay on garden leave for two years while the tribunal deliberates.

Win your tribunal case—and costs.

Write your memoirs to warn others that they too will be shafted by the system.

Boast about your achievements.

Invest a small amount of money into humane methods of dispatching elephants.

When the Romans finally defeated the Carthaginians, it was not unknown for them to prevent the re-emergence of rivals by ruining the soil by spreading salt on it so that the city could never be reborn. History is silent as to whether the Carthaginians used hydroponics or not.

The translation has some glaring omissions, and it is fair to say that citing this text as historically accurate will not win you any prizes or exam grades.

I lied.

When Hannibal took his elephants into battle, his troops already knew how to kill an elephant. Each driver carried a bag with a hammer and a bolster (chisel, not long pillow) and would deliver a dislocating blow to the elephant's neck vertebrae.

So even two millennia ago people had the sense to know how to kill an elephant before releasing it onto the battlefield. It is such a pity that our own rulers never considered how to kill the irresistible forces they have unleashed upon us. How do you stop the NHS, social security, the welfare state? Too radical for you? Consider this: the original aims of all of these enterprises, where we are now relative to these original aims, and what purpose they currently serve. Our modern-day elephants have their own inertia and are as irresistible as bull elephants in charge against unarmed civilians. You might as well attempt to halt the movement of *Apollo 11* ten seconds after lift-off, by hand, on your own, than stop them.

I had thought it would better illustrate my point if indeed the Hannibal section were illustrated with pictures. My test readers suggested that the mental images created were clear enough and pictures were not required. Now that I think about it, however, I realise that this little tale has already received the Hollywood blockbuster treatment many times. You only have to consider humankind creating an unstoppable machine to solve a problem, only to have it bite back. Consider the *Terminator* and *Matrix* franchises as perfect examples of failing to fit an off switch.

Did Adam's Descendants
Learn Anything?

I spend a lot of time in northern Spain, specifically the Picos de Europa. This limestone mass sits roughly in the middle of the Cantabrian Mountain range, which bounds the northern edge of Spain facing the Bay of Biscay. I am drawn repeatedly to this area by the stunning scenery and wildlife. The Picos could drop neatly inside the M25 London orbital, and yet it and the immediate surrounding areas cover a large range of habitat from sea level to eight thousand feet. The majority of the agriculture is best described as pastoral, and transhumance (seasonal migration) is practised. Fields range in size from a few dozen square yards to open ranges on the mountainsides. Tractors are often no bigger than large lawnmowers, and scythes are regularly used in the production of hay. Villages are small and have largely escaped the mass construction boom that has afflicted so much of the rest of Spain. Communities are small and tight-knit. Indeed, some have only been accessible by road for about thirty years, and Bulnes still has no road access. (A tunnel railway has been provided. Does that count as isolated?) Most houses have vegetable plots, and shepherds and cow drivers live in the villages or winter stables maintaining their herds. Individual farmers produce cheese matured in caves within the limestone rocks. Water is drawn straight from the streams and rivers. I have found community-wide treatment plants for sewerage. Hydroelectric power plants were introduced from the 1920s onwards and serve the local area. Tourism is another major wealth-creation element in the region.

Accommodation ranges from state-run paradors to guesthouses. El Cable lifts tourists to six thousand feet as a direct substitute for the old mineral line to provide employment when mining was banned. Crime is lower in Asturias than anywhere else in Spain, which has a crime rate lower than in the UK. Forestry and agricultural workers think nothing of leaving their car engines running and their car doors open while getting their morning coffee fix at local shops and bars.

Even in the small towns it is possible to eat in a bar in the central crossroads location and still see cows on one side of the quadrant. The biodiversity is spectacular: three times as many butterfly species as the whole of the UK, thousands of plants including endemics, and the place is even wild enough to have a bear population nudging its boundaries. I have to think back to time spent in my youth in rural Derbyshire to remember swallows charging up the roadway in search of insects.

The nearest thing I have found to a natural paradise, Picos de Europa is only one thousand miles away from where I live.

And yet.

On numerous occasions while filming plants, I have been subjected to dismissive actions (I don't speak Spanish) by locals who cannot understand the interest in weeds. I was photographing a snake swallowing a large toad when a local happened onto the same track and, seeing my interest, instantly searched for a stick to kill the snake with. He thought better of it.

And then there is Teresa. Teresa went to the Picos as an ecology student on a field trip and never really left. Having graduated, she returned and earns her living by writing books on the wildlife and giving guided tours to mainly rich Americans and other groups, such as the Royal Horticultural Society—and me, once. She lives in a local village through the year, returning to the UK for the winter. Her neighbours know her purpose in life, and yet she returned home one day to find a note from a neighbour advising that they had spotted a snake in a flowerpot by her front door and had killed it for her. This was a snake that she was fully aware of and observed in her leisure time.

Her view, based on over twenty years of experience, reinforced my own, namely that the locals neither have a clue nor care about their environment outside their own specific needs. They worry about overgrazing and moving their livestock to better grass. They worry about maintaining their rights over their patch of the mountain and keeping their neighbours off it. In the evening, you regularly see one bunch of cows being driven down the mountain while another group is being driven in the opposite direction.

Communal dustbins are emptied on an almost daily basis. Large objects or loads that will not fit in the dustbins are taken down the road and tipped off the edge somewhere. These range from demolition material to beds and sofas. But hey, everybody does it, and every village has its own midden. In themselves, the middens are eyesores, but at the same time they provide an opportunity for wildlife. I bought a ceramic-faced teapot stand from a local tourist shop. Entitled 'Nature's Paradise', it features an attractive motif. So, from the vast diversity of natural wonders, what icon had the locals selected to represent their 'Natural Paradise'? And the winner is domestic fruit. Says it all really. They have not got a clue.

What if All Sticks Were White?

It is human nature to exploit nature. (Indeed the Japanese believe that humans are part of nature as well, and therefore anything humans do is part of the natural order of things. As for Catholics ...)

Humans perceive a need and satisfy that need to achieve their goal. They do not care how they achieve it or at what cost.

Within the UK we are regularly fed well-meaning propaganda by so-called environmental agencies. The RSPB (Royal Society for the Protection of Birds) want us to feed birds. In the spring of 2015, BBC *Countryfile* featured the training programmes set up by the Game Conservancy Council to encourage farmland wild birds. These included set-aside margins full of weed species and supplementary feeding to support bird numbers at the end of the winter. Farmland birds are in significant decline with many populations crashing. And?

A theory has been proposed that the Romans came to Britain for grain. Iron Age farmers were generating an estimated crop of some 11 cwt per acre that provided a surplus and was a major source of income/bounty to justify the occupation. Fast-forward to 1914 and the start of both the First World War and the German blockade of trade to the UK's shores. In 1914 the average estimated yield of grain production was 11 cwt per acre, and action was taken to boost production. Nowadays, with modern varieties, petrochemicals, and fertilisers, yields are significantly enhanced to the tune of 3.5–4.0 tonnes per acre. (Thinking about it, when Paris decided to introduce a proper piped sewerage system on a par with London's, plans were met with significant opposition by local farmers. They relied on the wagonloads of 'night soil' to fertilise their crops.)

Simplistically, the decline in bird populations is blamed on the use of 'winter wheat'. In my youth, harvested fields were left empty until sowing time the following spring. Nowadays harvested fields are ploughed and resown in a few short weeks, thus denying the wild bird population the spilt grain and weed seeds to eat. Working on the basis that the amount of spilt grain is a percentage of the yield, and the idea that despite our most efficient harvesting methods this percentage has not changed, the available spilt grain (birdfeed) would also have increased significantly since 1914 until the advent of winter wheat. So the question is, what happened to the farmland bird population when yields and spillage increased post-1914? I would postulate that bird populations boomed during this period and that the subsequent crash is simply a return to pre–agrarian revolution levels. But hey-ho, we can correct this by artificial feeding.

As a teenager I used to help out at Bennett's Bird Seed Merchants, which used to be situated on the corner of Sketchley Lane and Rugby Road in Hinckley. I would mix the various seeds by weight to produce Canary No. 1, Budgie Tonic, etc. Linseed, buckwheat, white millet, panicum millet, Niger millet, blue maw, red and black rape, and the like came in 1 cwt hessian sacks from all over the world to end up in budgie cages throughout the UK. Bennett's actually began as a seed-growing operation during the war. Import restrictions caused by the bally Germans forced canary breeders and other cage bird owners to grow their own feeds on allotments.

Can you see where I am coming from here? What is the difference between 'Who's a pretty boy?' in a cage bird fed with seed imported from abroad and farmland birds being fed with seed imported from abroad (or grown in UK fields, given that we've displaced our wildlife to produce the crop)?

If bird populations are impacted by agricultural practices, why are the RSPB and GCC (Game Conservancy Council) actively exporting the destruction of habitat and changing agricultural practices worldwide for birdseed production so that *our* birds do not suffer?

What is the bird population of the UK? When was the golden age of birds? 1950? 1850? Was it before 1914? But this dismisses the post–ice age circa twelve thousand years ago. What about the Great Wood, which existed before Bronze Age and Iron Age communities began the great clearance? How about hedgerow-nesting birds before the Enclosures Acts created hundreds of thousands of miles of field hedges?

This leads to a simple conclusion. None of the UK's habitat could be described as natural; all has been impacted by humankind. Effectively the UK is nothing more than a garden controlled and managed to a greater or lesser extent for our ends. The RSPB and others are little more than advocates for lawns over flower beds, shrubs over trees, rose bushes over garden ponds, etc. They are not actually advocates for birds, unless those birds are ours, and then only the ones these organisations can earn from.

How then do we define a 'healthy ecosystem'?

Travel to the jungles of India and the answer would be 'A healthy ecosystem supports its apex predators'. Tigers and leopards become the poster boys of the process although they represent only a tiny fraction of the biosphere.

I am reminded of a discussion I had with my commercial estate agent. I asked how the markets were doing, and he spoke about a general improvement in most fields except shops, the situation of which was dire. I asked if he knew when the golden age of shopkeeping was. He didn't, but I do. The answer is now. Since 1980 the total footage of shops in the UK has trebled, yes, trebled. Yet all you hear about in the papers and on the television is the decline of the high street and corner shop. Presumably the white stick seller's trade is booming.

This then begs the question: how do we measure something that we do not understand? Is a healthy ecosystem full of pretty birds twittering away in plain view of our ignorance? I use the term, ignorance, because most UK individuals wouldn't know a bullfinch if it ate all the buds off their fruit trees, or the difference between a blue, great, or coal tit, or a tree or house sparrow, or a sparrowhawk or kestrel, or a bee orchid or storksbill. Nor do they care.

To address these concerns, we have 'ecosystem services' that are imposed on communities by organisations. A prime candidate in the UK would be the Environment Agency, supported by Natural England and the planning process. Let us introduce our blinkered interpretation of an enriched environment to the masses. Legislate for national parks, impose planning restrictions on selected species (great crested newts and bats, for example), increase planning densities to retain more countryside (sorry, garden land), and boost the requirement for public open space within residential schemes.

This leads (conveniently) to another soapbox I will get up on.

In the drive to protect the green belt, housing has become more and more compressed, the planning process insisting on tighter and tighter densities through the planning process.

The obvious consequence of this is the reduction in individual garden size; a brick box on a postage stamp springs to mind. At the same time, public open space has been developed to offset these densities. These are normally identified by the 'No Ball Games Allowed' signs to protect the local residents. Section 106 Agreements (legalised bribes) are used to demand money for the provision of playground equipment. This is social engineering at its most basic and, I believe, accidental, because no one has the foresight to consider the consequences. What do you do with your 'postage stamp'? A garden shed would fill it; a greenhouse, similarly. A game of football would be impossible. A tree or large shrub is out of the question. What skills can the children grow up with? Where do they learn how to use a hammer or a screwdriver (hammers used to be called a Birmingham screwdriver when I was a kid, the perception being that city dwellers hadn't had a clue, even in the 1960s), to build a rabbit hutch or an aviary, or to grow plants or view wildlife, any of the thousand and one things I used to do living on the edge of a village with a big garden?

I have spoken to numerous young tradesmen, electricians, carpenters, tilers, et al. and asked them about their skill base. Time and again they advise, 'Mates ring me up for help to erect shelves.' And when they turn up expecting to cloak an entire wall, instead they find a B&Q pre-made item 900 mm long with all the fittings that they could install in their sleep. But it is way beyond their friend's capability.

Yet why do we protect the precious green belt?

Recently it was reported that Surrey has more acreage under golf courses than it has under housing. I found this difficult to believe until I took the train to London. Go for a drive in the countryside; you'll find that all roads lead to houses. That's why they are there, the roads and the houses. You need a road to service a house, and in one sense you need the house to service the road. Any non-motorway journey delivers a constant succession of houses; you cannot escape from them. Towns and cities cause interminable delays, and the countryside roads become the gaps between delays. Take the slow train to Euston from Rugby, however, and leave the houses behind. Northampton, which is a pain to drive through, barely mentions a comment on the journey, and cross-country to Milton Keynes is empty of houses. It's only on the approach from Watford that densities start building up. I do believe that Surrey has more golf courses than housing.

This begs the next question: what are we protecting the countryside from, this 'green and precious land'? The conclusion is that we are protecting England from the English, and if that is the case, then whom are we helping, generations unborn? Is this social engineering or social stupidity?

Many years ago I read an article in the *BBC Nature* magazine. An area of chalk down in southern England had been taken out of short-term cereal production and was being returned to 'natural chalk grassland'. You know the stuff; it's been there since the climax oak and beech forest was cleared during the Iron Age. An Oxford University project had determined that to encourage the greatest diversity of downland flowers, the grass sward had to be kept short, and a heavy grazing regime needed to be maintained for management.

One month later, the entomology department of an Oxford University college project studying the same site responded by stating that for the greatest diversity of insect populations on chalk downland, the longer the grass sward, the better.

Unfortunately for us, most ecology initiatives are similarly blinkered in their objectives. The RSPB buys a wood and stuffs it full of nest boxes to boost the tit population. Tits eat caterpillars that eat trees. More tits equals fewer caterpillars, which equals fewer moths and butterflies, which could also mean fewer bats. But bats are not birds, are they, so they do not care. Fewer caterpillars means more tree canopy, which means less light reaching the ground and fewer woodland flowers, but these are not birds, are they, so they do not care.

Recently, on BBC *Countryfile*, Leicester and Rutland Trust for Nature Conservation were filmed stripping reeds from the edges of Rutland

Water. Done by rotation over several years, this prevented the reeds from being colonised by birch trees. It was also done to encourage the further establishment of bitterns (*Botaurus stellaris*). This secretive bird is an apex species within the realm of ecosystems (sorry, birdwatchers) and deserves to be encouraged.

Fair enough. Why were the dried reed stems piled up and burnt? Hollow stems provide refuge for overwintering insects. Not anymore; they have just been barbequed. These insects themselves would have been food for swallows and swifts, and for the dragonflies over the water's margins, all of which provide food for hobby falcons (*Falco subbuteo*)—you know, the rare summer visitor that is also an apex species. Piles of vegetation rotting down provide another habitat for worms and other invertebrates (bird food) and can even serve as hibernacula, basking and nesting sites for snakes. The reeds that have not been burnt, that is.

So we have the blind leading the blind. And not only that, but they are also enjoying the destruction of habitat and opportunity in the pursuit of habitat and opportunity for the one headline species.

Within Nassim Nicholas Taleb's book *The Black Swan*, reference is made to Greek philosophers and a discussion about belief in God. In short, during a shipwreck, some of the crew made it to safety and survived. All of these survivors, not surprisingly, believed in God. One philosopher opined that this was proof that God exists, as all the non-believers had drowned. The reply was, 'Ask the ones who believed in God who drowned.' This can be summed up in a simple phrase: the dead witnesses were unable to make their voices heard and had become 'silent witnesses'.

The success of the Leicester and Rutland Trust for Nature Conservation's venture will be measured by how many bitterns are counted at Rutland Water. Nobody will count the swallows or swifts or dragonflies or snakes (the silent witnesses), but curiously enough they will count the hobbies.

In March 2015, a report by York University was picked up by the news media. Spending £30,000 per year to provide specialist cancer drugs to provide a reasonable extension of life is too high. A more reasonable figure was put at £13,000 per annum, because cancer spending diverted resources away from other areas that were now disadvantaged.

What? Really? It took a University of York study to work that one out? Every other deserving case that couldn't receive treatment because of financial constraints becomes a silent witness.

Many years ago, my structural concrete products manufacturing company employed a chartered structural engineer as part of our BSI quality assurance scheme. Peter was the 'checking engineer' for Blaby District Building Control. He also lectured engineering to De Montfort University students.

His problem was, he lectured architects and engineers. Engineers understand maths, whereas architects do not. Peter had to describe in words and pictures engineering concepts, and in so doing he realised that much of the maths he had offered to engineers was inappropriate for the problem. He coined the phrase 'cookery book engineers'. If they did not have the cookery book for the recipe, then they could not work out how to calculate the solution to the problem. But they would also quite happily use the wrong recipe to fit the question. In fact, they fell woefully short of the requirements to be engineers. But as they are only qualified structural engineers, it's not as if anyone's life depends on it, is it?

So, what cookery book are the 'ecosystem engineers' using in their deliberations, given that, in my experience, nobody has a clue?

Which brings us back nicely to climate change. This two-word mantra opens doors to funding for research and grants for infrastructure (wind farms etc.) and can be used as an unassailable political device to force unpopular or unpalatable notions into law. It reminds me of that other two-word mantra that seemed to disappear rapidly up its own anus: millennium bug.

So is climate change real? Does it matter when an entire industry of academics and intellectuals are earning a living from it? But of course for every pound spent on the new 'emperor's clothes', there is a one-pound shortfall in all the non-qualifying schemes, *the silent witnesses*.

Climate change prevention is now an industry in its own right and has the potential to do immense damage beyond its remit by deflecting resources away from things that might actually matter.

When my sister-in-law died, over £400 was raised in donations for Save the Children, a charity she'd supported. The warm glow in the belly lasts for about three milliseconds when you realise that this doesn't even pay a day's salary for the chief executive. Save the Children and all other non-governmental organisations (NGOs) are an industry in their own right, all pushing a plausible demand for help.

Daytime telly is full of adverts for refugees and poor children displaced by world conflicts such as that in Syria. War is terrible and should always be avoided at all costs. But what is the cost? The displacement of refugees, death, and destruction all mitigate to the desire to bring an end to the conflict. And yet by providing support and succour to the displaced, are we not creating an environment whereby war becomes palatable? The only ways a solution will be achieved in Syria is by one side winning and vanquishing its foe, or the realisation that the price is too high to continue and that a peaceful solution is paramount. It is human nature to want to render assistance, but it is also human nature to fight. But do we ever question whether our help is helping? By providing succour to the Syrians, are we prolonging the conflict as readily as those who supply armaments and 'training'?

When my daughter was at Lincoln University, we parents regularly visited and had to run the gauntlet of the homeless beggar whose pitch was under the bridge at Brayford Water. Turns out he wasn't homeless; he lived with his girlfriend and received about £15,000 a year in benefits. Begging was simply his career choice. As it is for the CEO of Save the Children—begging, that is.

What cookery book are the charities working from, other than the one that promotes their own existence? Is it a case of being cruel to be kind, given that kindness extends the conflict and the cruelty by making these things more palatable and sustainable?

In the UK it could be argued that our single most important form of help is the welfare state.

The welfare state was conceived as a safety net to catch the ill and otherwise unfortunates within society. I had a great-uncle who served as a councillor in Derbyshire in the 1930s. During his house clearance, my family discovered the council minutes that he had kept. Our early interest was in the stamps, but the minutes proved to be much more valuable. Annually, a schedule of 'charity cases' was produced. This schedule listed a variety of individuals who were variously described as street sweepers, along with the pension that they received. They were categorised as either 'idiot' or 'cripple'. The list was contained on one side of foolscap paper. Imagine the list today: 'too fat to work', 'too lazy to work', 'too stupid to work (unemployable)', 'too high on drugs to work', 'too drunk to work', 'too bipolar to work', 'too busy working the black economy to work', 'too ....' Somehow I don't think it would fit on one piece of foolscap.

If it is now considered in some quarters counterproductive to offer aid overseas, then why do we persist in doing it at home? If all of these modern-day individuals were given the choice of entering work or taking a one-way trip to Mars, I suspect the spaceship wouldn't need to be very big.

During the early 1980s, the factory where I worked was plagued with breakdowns. They were expensive in lost production, and we had to factor in downtime when determining delivery dates, which often cost us orders. The solution was very simple: get in front of the problem by having planned breakdowns. That is to say, introduce a programme of machine servicing at our convenience, which largely prevented the breakdowns from occurring in the first place. Our fitters went from offering a 'fire service' to a maintenance service, and we benefited greatly.

'What's that got to do with the price of eggs?' I hear you ask, assuming you have not lost the will to live yet.

I remember watching a documentary years ago that examined the death rates for two conjoined towns in Lancashire. They sat in the same valley and enjoyed the same air and water, and yet life expectancy was markedly different, to the tune of twenty years. They even performed the same employment in that both were 'cloth towns'.

Painstaking analysis by the researcher finally latched on to the difference. In Victorian times, one town's mills were run by Quakers the others were not. Big deal?

As it happened, it was. The Quakers allowed expectant mothers to have time off work both before and after birth, and they were encouraged to enjoy a healthy diet. The dark satanic mill owners in the other town didn't. Women worked until they went into labour, and returned after childbirth to finish their shifts. Quaker babies were fed when needed; non-Quaker mothers snatched quick breaks from the mill when they could. This directed the researcher to look at social environmental conditions. What he discovered could actually be measured. One of the hospitals in the Liverpool area had taken around twenty-seven measurements of all live births in the district. This database was available to study. The major factor in determining life expectancy was birth weight in comparison with the weight of the afterbirth. In simple terms, any baby born lighter than the afterbirth was a runt and his or her life expectancy was reduced by twenty years on average. The researcher could trace 'runts' born in the 1910s and find the death certificates in the 1960s. Relating this back to the two mill towns, the conclusion was that 'runts' gave birth to 'runts'. You may find the use of the term 'runts' offensive. I use the term because the researcher had stumbled upon a simple truth known for centuries by animal husbandry, namely that the runt of the litter ne'er does well.

The reasoning is that failing to achieve optimum performance levels for such things as the cardiovascular system compromises a foetus deprived of adequate nutrients during its early development. The fate of these babies was sealed before they were born, but so was the fate of their own children. A farmer will never breed from a runt.

One of the central planks of the welfare state is the National Health Service and its 'cradle to the grave' ambition. The fire service deals with breakdowns after the event, when the damage has already been done. Where is the planned maintenance that prevents the breakdown in the first place? Where is the dietary supplement for the expectant mother? I won't even bother to ask why some expectant mothers don't stop smoking or drinking.

Homo sapiens has, I believe, a major evolutionary design fault, namely a self-destruct gene. We all drive too fast and too close to the car in front. We all drink too much alcohol, eat too much food, eat too much sugar, fat, salt, processed food, etc. We even rush off to war with a song on our lips and a rosy expectation as to the outcome. Now some of this is deliberate; the male has not formed the proper risk–threat perception section of the brain until the age of 25. I believe that this puts us on a par with immature male lions. These are driven from the pride and have to roam the wilderness for a few years, pushing their horizons, but most importantly not breeding with their siblings. It is no accident that young men fight the battles while old men direct them. Who would want a non-risk-averse general telling you what to do?

Interestingly enough, age 25 is the threshold where your car insurance premiums drop. It's no accident. (Forgive the pun, intended.)

When my second son entered the Lutterworth Grammar School, or whatever it pretends to be now, I was asked to enter into a covenant with the school. I forget the precise details, but in essence my son and I had to commit to abide by the school's rules—no swearing, spitting, fighting, farting, insolence, etc.—and comply with any edicts it issued. I think this covenant had fallen off the radar by the time my daughter enrolled. My point is, where is the covenant between the NHS and its customers?

We all know it is dangerous to smoke and chew tobacco. The latest report from Australia suggests that two-thirds of all smokers will die from smoking-related illnesses. The very first report on the dangers of smoking was published nearly ninety years ago. (And if you want to delve a little deeper, neonicotinoid insecticides are implicated in the collapse of honey bee numbers and significant reductions in insect populations and the birds that feed off them. They are chemically similar to nicotine.)

We all know it is dangerous to imbibe large amounts of alcohol. We all know that drinking and driving increases our chances of seeing the inside of a casualty unit or failing to see the inside of a wooden box.

We all know that being fat elevates a large number of risk factors, including those for heart disease, diabetes, and cancer.

We all know that we should brush our teeth twice a day and reduce our exposure to sugary drinks and sweets.

We all know that unprotected sex puts us at risk of sexually transmitted diseases, some without a cure.

We all now know that high exposure to sunlight ages the skin and increases the risk of skin cancer.

And yet?

Hey-ho, that's what the NHS is there for.

I ask again, where is the covenant between the NHS and its customers?

Why is such a thing important? The NHS is a finite resource funded by wealth generated by the economy. Decisions are already being made every minute of every day as to who lives and who dies, who gets the treatment and who doesn't. And things can only get worse.

Imagine the universal condemnation you would receive if you declared that all alcoholics, druggies, obese people, smokers, et al. were to be denied treatment by the NHS for what are essentially self-inflicted injuries.

And yet, I repeat, the NHS on a daily basis is already denying people potentially life-saving treatment. But they are classified as unlucky. Sometimes the term 'postcode lottery' is used. These people are the silent witnesses.

How many people are denied access to a medical professional in a timely manner because doctors are too busy dealing with the self-inflicted injuries? Try to get an emergency ambulance at pub chucking-out time on a Friday night.

How many minor complaints become major issues because there is a lack of resources? But it is not a lack of resources; it is a failure to develop a comprehensive plan of action. Government agencies suggest that adults should take an interest in their children's education.

Lesson one, how to stay alive: don't smoke, don't drink excessively, don't get fat, eat healthy foods, brush your teeth, etc. (Dental caries is implicated in heart disease.) Enter into a covenant with life, with the NHS, with the welfare state. 'Nanny state' will be the cry from media far and wide, but guess what: fuck 'em; you cannot have it both ways. What you could do, however, is *think*.

Learn at school the values of life and healthy living, and the obligations you have to the welfare state and not its obligation to you. Do it apolitically. But hang on, everyone with half a brain cell already knows what you should and shouldn't do, and yet they don't apply this knowledge. As we used to say at work, 'If you are going to get nothing for it, do nothing.'

So we arrive at a paradox. Our education system doesn't prepare us for life, and the welfare state doesn't in fact provide any welfare. The welfare state traps people in poverty and ill health, the very things it is supposed to end. It's screwed up. But then welfare is an industry in itself.

Give a purpose to education and in turn give a purpose to life.

Another universal tenet is that people are redeemable and always deserve a second chance. Does that include Adolf Hitler, or Joseph Stalin, or Pol Pot, or Fred West, or Harold Shipman? These are extreme examples, I know, but then again there's Jihadi John. And what about the drunk driver who's just killed a family of four? That's the same drunk driver that has already been banned from driving for causing serious injury while drunk driving.

The car thief who stole my brother's car was given a six-month suspended sentence and went on to kill a farmer at Wolvey—while stealing his car.

But we don't need to consider the redeemability of those that harm others in isolation. The NHS offers redeemability to those that harm themselves as well. 'I know I am an alcoholic and have destroyed my liver by drinking and ignoring all previous warnings and people's attempts to help me, but I deserve a second chance.' Prior to the transplant era, there was no second chance. Is this what Christian Barnard worked so hard to achieve with his pioneering heart transplant work? Is this what funds were raised for by charities? 'Give us enough money and we will cure cancer.' And while they are doing that, we will invent a new way to abuse our bodies and negate their work. In the meantime, of course, one can earn a good living by working for a charity.

This brings me to a universal truth that is universally ignored.

In the nineteenth century in the United States, there existed a species of bird, the passenger pigeon, *Ectopistes migratorius*. Flocks would darken the sky and take days to fly past (the 'passing' pigeon). Estimates suggest that some 6 billion individuals existed in 1850. Martha, the last surviving passenger pigeon, died in a zoo in 1914. Essentially the species had been wiped out by humankind's activities in about 50 years. Let's do the maths: $6,000,000,000 \div 50 \times 365$ days equals a third of a million birds a day. That's 330,000 birds a day, every day for 50 years. This is somewhat simplistic, as obviously the birds continued to breed and replace some losses. Passenger pigeons had a certain evolutionary design fault, namely stupidity. When they heard an alarm call from one of their number, they flew towards it, straight to the scene of the bloodbath, so to speak. (It's where expressions such as 'stool pigeon' come from.) Hunters would trap an individual in distress and wait for its colleagues to show up at the scene of their own death.

The relevance?

Six billion is an awfully big number, nine noughts, beyond our comprehension. Now, 330,000 is a bit more manageable or even imaginable. Cast your mind back to Boxing Day in 2004. Remember the tsunami that spread death and destruction across the Indian Ocean? Early estimates feared that as many as a third of a million people had died. Have you watched *Independence Day*? It features superior aliens wreaking mass destruction upon cities worldwide. One wonders how long humankind could survive such a disaster. Well, in simple number terms, a third of a million people, equivalent to a city the size of greater Leicester or Coventry being destroyed every day for 18,250 days, or 50 years. This again is ignoring the breeding potential.

Do you want to know what the universal truth that is universally ignored is? There are too many people. If you control the population and reduce demand on the earth's resources, then you free up space for other species to continue to exist. Even climate change would stop. Anything else is pissing in the wind. Everything else could be considered as a sop to make us feel happier while the same old process continues unabated, while we rest easier in our beds secure in the knowledge that our views have been heard or have resulted in a new industry in its own right.

But hey-ho, pissing in the wind is an industry in its own right.

So what does it matter if we are on the path to a major extinction event? Vast arrays of species that share our planet are at risk of disappearing forever. But are they? If the people who point those glass tube thingies into the night sky are to be believed, earth's path is already predetermined and ends with our local star running out of fuel and expanding out to include our orbit within its own mass. And while this may be the extinction event to end all extinction events on earth, it will not be the first to have come close. Defined as an event whereby species die out at a faster rate than they are created at least five major events have been recorded in the fossil record and as many as twenty have occurred depending on your definition of major. Interestingly enough, scientists can only really examine the fossil record of marine species given the readiness with which their carcasses become stratified in seabed sludge on its journey to becoming rock. In that regard we tend to ignore the microbial diversity that is both greater and more difficult to observe and for that matter, the animals and birds that fill our lives with their more observable presence.

Think back to the tiger, the jungle's apex predator. As the apex predator, it acts like an apical bud on a tree. Its presence prevents the development of side shoots that could challenge its dominance as 'the lead shoot'. Remove the lead shoot and you get a new leader. In that sense, the tiger is an evolutionary block to the development of alternative species that are even now waiting in the wings to grab their chance at the limelight.

Perhaps an alternative to Google already exists? The successor already exists and is biding its time before it strikes.

But on that basis, the greatest apical bud is humankind.

I will remind you again about the apical bud. The topmost bud of a tree produces a hormone which inhibits the development of side shoots. This is why a tree is a tree and not a bush. Damage the apical bud on a Christmas tree and you end up with two or more leaders and a misshapen and subsequently rejected tree. So are human beings the villain of the piece, or are they simply doing what comes naturally to them, as unthinkingly as a tiger?

Ultimately the question is irrelevant. When earth ceases to be, does it matter how many species are alive on earth or not?

All it comes down to is what sort of cultivated garden do we want to live in, a rose bed or lawn, a tree or shrub, and filled with diversity or filled with survivors? What quality of life do we enjoy? Shall we put it to a vote? Cookery books to the fore, please.

The Picos nearly fits in to the idea of humankind's stewardship of the natural world, apart from the fact that the people of the area do not know what they have in terms of their own environment. But why does the Picos exist?

Remember Hannibal—no, not Lecter; the other one, with the elephants. He crossed the Alps to give the Romans what for. Thing is, he led an army of pastoralists who were motivated enough to risk life and limb by going up against the other world superpower of the day. This army was motivated by the Romans who, in imposing frontiers, had blocked the transhumance routes and were damaging the Carthaginians' way of life (unlike the ensuing battle, which damaged their *lifes*). The point is, the Picos and the way of life there has been going on for thousands of years because it works. It has evolved in to what it is. It is its own climax environment, a climax environment that includes humankind. You cannot plough the ground; the soil is too thin, the slopes too steep, the rocks exposed or too near the surface. And the limestone supports massive diversity. The Picos was created by nature and humankind, but not by nature and humankind with a cookery book. True, government as a good idea arranged the introduction of tourism, and it seems to be beneficial, but only time will tell. (When I first went to the Picos in May 1991, I remarked to the manageress of my hotel that I would like to return in the summer. 'We are fully booked,' she replied. 'This is where the Spanish come on holiday.') Since 2008, I have made another fourteen or so trips, and time after time I have stayed in hotels where I was the only guest. (Fellow potential guests couldn't have known it was me attending, so it wasn't my fault.) The recession in Spain has caused massive hardship. Only now is tourism picking up again.

But in itself, tourism is simply an attempt to impose a flower bed over a lawn, or at worst it will block the transhumance routes and bring an end to the very thing that the tourists once went to the Picos to enjoy.

When I first returned in 2008, I was shocked to see the damage wrought by state attempts to develop the area by way of a massive road improvement scheme. Flower-strewn verges had been cut back to bare rock, and all the bends had been straightened. Old sections of road were abandoned under a thin coating of soil that makes for impromptu pull-offs while searching for something to photograph. *Rampant vandalism* was the thought that sprang to mind. And the good news is, we, that is the UK, paid for it through our net contribution to the EU budget, Spain being a net recipient. If you don't believe me, then take a trip to Spain yourself and see the EU funding signs that herald every road development even today. Only God will know what cookery book the EU is using. And you can guess my views on God.

By 2011, however, I was encouraged to see the recolonisation of some of these impromptu lay-bys by what I personally regard as an apex species, orchids. I have even filmed orchids growing out of the old tarmac between the crash barrier and the new road. The equilibrium between nature and humankind was re-establishing itself; but in fact it wasn't. Orchids thrive on the margins. Orchid seed has no food store and is no bigger than a grain of pollen, and these seeds can only germinate in the presence of fungi, from which they can steal nutrients. Orchids were doing what comes naturally to them, surviving. That's why they still exist.

Remember the cornfield on the chalk downs. In its cornfield form, it represented its own climax environment when assisted by humankind. But it wasn't sustainable; the cost in fertiliser was too great to support its yield, and the intensive agriculture was brought to a halt and the site 'donated' to science.

My thoughts turn to cities. These must be seen for what they are: climax environments for people. Where else can so many people be found in such a small area? I nearly referred to the cornfield as a monoculture, but this would deny the existence of bacteria, mould, fungi, weeds, birds, mice, insects, etc. that co-inhabit this space. By the same token, you cannot describe a city as a monoculture with its parks and rats and so forth. So where do our planners tinker with the margins to improve our city life? More open spaces. One person's park is another person's dog toilet; one person's quiet place for contemplation is another person's football ground.

I designed an industrial unit in 1999 and had to provide a *minimum* of thirty-seven car parking spaces to prevent my co-workers from parking on neighbouring side streets. In 2008, my tenant asked me to design an extension to this unit, and I duly obliged. The unit footprint increased from 18,000 square feet to 40,000 square feet, an increase of 220 per cent. Imagine my surprise when I was told that I needed to provide a *maximum* of thirty-seven car parking spaces, 'because if people cannot park their cars, they will travel by public transport instead'.

Travel down Queens Road in Hinckley, or any other one of thousands of similar roads throughout the UK. Its housing stock is Victorian terracing, and it was built before the advent of the motor car. See how easy it is to drive down this road when cars are parked virtually end to end down both sides of a narrow road? This is what one hundred years of reducing the demand for cars looks like. And yet some bright spark in central government has successfully driven (forgive the pun) this policy onto the planning rules. Try to drive around the Harrowbrook Industrial Estate in Hinckley. Every path and street corner is half buried beneath parked vehicles. Try the same on modern housing estates that have parking provision limitations imposed through the same planning process.

Plagued with insect pests in Australian sugar cane crops, another bright spark came up with the idea of introducing the cane toad (*Rhinella marina*) from its native Central America and South America to solve the problem. (Bureau of Sugar Experiment Stations brought in cane toads in 1935 from Hawaii.) This large, extremely poisonous toad has munched its way through significant native fauna, and poisons any native species that attempts to eat it. Thankfully we are cleverer than to do such a thing now. We would never introduce an alien species into such an environment without considering all possible and potential outcomes, and even then we would shy away from action for fear of what we don't know yet. Multiple safeguards and protocols would need to be realised before such an event could occur. I am reminded of an experiment with an alien salmon species introduced into the Eastern Seaboard of the United States. All eggs were irradiated before the fry were released, so that when the experiment ended after three to four years, the returning salmon could be harvested for food production. All salmon that evaded the nets spawned unsuccessfully, as they were all sterile. All introduced alien adult salmon then died.

Yet look what we have done. In just nine years, from 1999 to 2008, the cooks in charge of car parking spaces decided to switch from parsley to oregano without any control or understanding of the potential consequences.

Does the divorce rate increase because there is a lack of garden sheds?

I had an aunt who tried to buy a one-bedroom house in a block of four on the Barratt estate in Burbage in the 1980s. One of her potential neighbours, a young woman, slept in the house for one night and then returned home to her parents, as she was too afraid to sleep on her own. Others on the estate could walk into the show houses and buy a house with a £75 initial reserve deposit on the spur of the moment. Barratt would sell you the house and arrange your solicitor and your mortgage for you. If you hadn't had a big enough deposit, Barratt would arrange a second or even third top-up mortgage for you. By making it so easy, Barratt almost single-handedly brought the then housing recession to an end and sold 22,000 houses in the year, thus priming the system for other builders to take advantage of. How many buyers would have trawled the high street for a mortgage or solicitor? If the young woman had had the wherewithal to do that, she would probably have had the balls to sleep there on her own. Given the minimal threshold to owning a house and the singular lack of commitment to the process, how many doomed marriages were brokered by Barratt? Conversely, how many marriages were launched by Barratt? We can never know whether this was a doomed experiment in social engineering or not, but it solved a construction industry crisis at least.

And yes, I really am comparing planning initiatives to the introduction of poisonous alien toads into Australia. Why? Because these little experiments that can change from one administration to another cannot evaluate a fraction of the potential implications that they can throw up. They are not even sterilised before exposure.

A research group has recently published an idea together with supporting data. Other groups together with supporting data have instantly dismissed this. Whether they are right or wrong doesn't matter. It's the idea that is so interesting.

Data from the Middle Ages confirms the 'civilisation' of our society by considering the murder rate per head of population and logging the murder rates major decline. That is to say, you are much less likely to be murdered today in our enlightened age than you were back then.

Team A suggests that this is down to the 'hanging judges' who systematically hung criminals for even minor crimes (stealing a rabbit, living with gypsies for a month, murder) and thus removed the 'criminality genes' from the population.

Team B suggests that we are effectively more 'civilised' and are better educated, more reasoning, and more trusting in the redress of the law without feeling the need to seek redress of our own. Or is it more fearful of the consequences?

Team A or team B, it doesn't matter. What does matter is that for the duration of the study period, successive generations have been subjected to a social engineering experiment with the gallows in one corner versus education, reason, and fear of the consequences in the other. Although it does strike me that without the gallows there were no consequences. And yet at no time in the duration of the study period did anybody consider that they were engaged in a social experiment. The law saw a problem and provided a solution, permanently, a solution that retained its own momentum until the 1960s, when capital punishment for murder was abolished.

Was it right or wrong? We will never know. But the lucky ones lived with the consequences. The unlucky ones didn't.

Thinking about it, though, is team A correct? Remove all thought processes from the equation and use the brown rat, *Rattus norvegicus*, as your test guinea pig (I'm sure the rats won't mind being called guinea pigs). Introduce traps into their environment and you will remove all the nosy, inquisitive, bold, and incautious rats and be left with careful, nervous, cautious, shy rats that may well pass these characteristics down through their genes. The surviving rats are likely to become less and less susceptible to trapping, even though they have no memory of 'being dead'. The dead rats are of course silent witnesses.

By a remarkable coincidence, a natural experiment is thought to have been conducted in the mangrove forests of the Indian subcontinent known as the Sundarbans. Here Western trophy hunters have never hunted the tiger population, which appear to have no natural fear of humans and simply view human beings as part of the menu. You probably have a higher chance of being killed and eaten by tigers here than anywhere else on earth.

In neither of these examples does civilisation or redress to the law feature in the thought processes, although education could certainly be a factor for the tigers. No, scratch that, the bold aggressive tigers would all have been shot (silent witnesses) by trophy hunters, leaving the naturally shy, cautious ones to remain to populate the region.

Back to the human model. We do introduce new controls, such as education and the fear of the consequences. Alternatively, we hanged all the stupid ones (those who got caught) and are now left with a genetic super-race of murderers who have learnt how to hide their victims. I wonder whether the study teams considered the thousands of 'missing people presumed alive' in the UK today as actually 'missing people with no known grave', victims of a five-hundred-year social experiment? Apologies to anyone with a missing loved one.

One of the consequences of living in 'monocultures' for the primary mono species is the increased risk of disease and the devastating impact of virulent spread. Think of the potato famine in Victorian Ireland and the resultant mass starvation of the principal beneficiary of the monoculture caused by the removal of the staple food source by potato blight, *Phytophthora infestans*.

Think of the cholera outbreaks in Victorian London or the Crimea War, the 'Spanish flu' epidemic that killed perhaps fifty million people at the end of the First World War.

Two examples from nature spring to mind (without forgetting that potato blight, cholera, and flu are all natural, of course).

- Locusts, *Melanoplus spretus*. We are aware of the potential for crop destruction that locust hordes can wreak on any ecosystem, natural or agricultural. But did you know that some of the largest swarms ever recorded were in North America? But the United States doesn't have locusts, does it? Back in the nineteenth century, the US did have locusts, and then they mysteriously disappeared from the face of the earth, although samples can still be found of specimens that flew too high into the Rocky Mountains and ended up trapped in glacial ice deposits. The current theory is that locusts bred in a few sandy valleys in the Rocky Mountains, and when cattle rearing descended on these valleys, the patter of thousands of bovine feet crushed and destroyed the locust eggs before they could hatch.
- Monarch butterflies, *Danaus plexippus*. These colourful butterflies cover the whole of North America from Mexico to Canada, and at the year's end migrate thousands of miles to overwinter in a few select valleys in Mexico, where they exist in their millions. They produce a true visual spectacle but also present a spectacular opportunity for the species to be destroyed by a natural occurrence such as a forest fire or virulent pathogen.

This naturally leads to the elevated exposure risks of humankind's climax environment, cities. We've all seen the news reports from Western Africa and the ten-thousand-plus deaths of Ebola victims. But Ebola and its ancestors have been around for a long time; studies suggest for at least twenty million years. Even in Africa earliest cases have been tracked back over fifty years in people, so why the sudden upsurge in fatalities? The answer can simply be suggested to be population density. Earlier outbreaks were in small rural areas, where the outbreaks were more easily contained. Now the disease has spread to major urban centres, where the potential for transmission is exponential.

The same model has been proposed for HIV/AIDS, which has arguably been tracked back to the 1920s. Interestingly enough, 'charity work' exacerbated one of the earliest outbreaks of Ebola in humans. Pregnant women were being treated in a missionary health centre, and all women were routinely injected against disease with *one* new needle a day shared between all. It didn't cause Ebola to occur, but it did ensure that an isolated death became an outbreak.

Researchers are continuing to conduct studies to try to find the pathogen's resistant host, a vector that can spread the disease without suffering undue consequences itself, the carrier.

Now imagine two scenarios:

1. Ebola has become established in a major world city with millions of potential victims, say, Tokyo, Mexico City, New York, London, or Shanghai. Oh, and let's remember that these are all major transport hubs connecting all the major cities in the world.

2. We stopped spending vast resources in pursuit of pet projects supporting personal ambitions in the name of climate change and actually devoted these resources to finding a cure for something that is likely to have a much greater impact on person's well-being, such as Ebola and the like. (Although, as we will discover later, this might not provide a valid solution either.)

But let me suggest a different scenario. Success in controlling the basic illnesses associated with high densities of people has allowed for the ever higher densities of people. Cholera, typhus, and tuberculosis spring to mind. The present theory about HIV/AIDS rationalises that the disease has been around for a long time but surmises that it only became an issue when population densities reached a sufficient level to allow multiple transmission between significant numbers of individuals. It was changing human activity that supported the emergence of AIDS into the wider population. Bird flu has the potential to cross the species divide and create global pandemics simply because enough high-density people live in proximity to enough high-density birds to make this a very real scenario. What then is the real issue here, the population densities of an ever increasing human population or the emergence/increasing virulence of disease? While we expect our governments to provide a solution to the symptom (Ebola), our real problem is population density, which is something that our governments ignore.

Curtains Made of Iron

There was a chap in South Wales. Who he was is not important; it's what he did that matters. He wanted to become rich by building a coal mine, and he even had the foresight to construct a coastal dock so that he could ship his product to market before roads or canals were up to scratch. This suggests that he was already rich and that he wanted to become richer.

Having established his infrastructure, his next task was to employ workers to extract the coal and process it to market. Trouble was, in those days people worked for the money in their pocket, and if by Tuesday night they had enough money for the rest of the week, they didn't work for the rest of the week. In those days everything was done piecemeal. You'd clear an acre or so of field and wouldn't return to continue the job until your money ran out.

This chap tried to entice his workers to remain through the week by offering more money and even a pension, but still he struggled. Each worker had only the most basic of needs: food in the belly, a hovel roof over the head, the clothes on his back, a homemade chair, a plate and spoon, and a mug to drink from. What would one do with all the excess money? You couldn't buy furniture or ornaments or fancy clothes because nobody made any, except for the super-rich, and this was mainly bespoke.

But then the more skilled among the workers began to make tables and better-quality clothes, thereby beginning to fill the void formed by this surplus of money. Before long, the chap's workers began to enjoy the finer things in life—a cutlery set, a table to eat off of—and consumerism was born. Who could imagine that this would lead to the TV set in your lounge, the microwave oven in your kitchen, designer goods in your wardrobe, or for that matter your wardrobe? Without this process beginning, there would be no supermarkets, no football league, no cars in the drive, and no holiday flights abroad. The list is endless.

Some of you will be annoyed by the absence of names. Who was the man with the coal mine?

It doesn't matter. You will never meet him. His name isn't important, and his actions resulted in consequences he never intended. It was effectively an accident. If you want some names, I will give you two:

1. Winston Churchill, on 5 March 1946, gave a speech highlighting the partition of Europe by the Soviet Union. He coined the term 'Iron Curtain'—powerful stuff from a man who should be respected.
2. Josef Goebbels, on 5 February 1945, published a newspaper article in which he coined the term 'Iron Curtain'—powerful stuff from a man who is considerably less respected.

My question is, which is more important, what was said or who said it?

Most of us know the story about the emperor's new clothes, an outfit designed by a con man out of magical material that only the cleverest people could see. No one would admit that they couldn't see it for fear of being branded an idiot, in spite of the fact that the emperor was naked.

It took the innocence of a child to break the pomposity of the spell by declaring, 'Er, the emperor is wearing no clothes.' (Have I mentioned designer clothes?)

So if I upset you by not naming names, forgive me and get over it. It doesn't give my ideas any more or any less authority.

This brings me to another unexpected consequence. Modern planning restrictions impose higher densities on housing developments, resulting in smaller houses on smaller plots. Modern houses don't have a separate dining room but rather a kitchen/diner or lounge/diner.

It's the 'modern style', we are told by estate agents who are embarrassed by the size of the products they have to sell. This in turn has led to the demise of 'formal dining', where families interacted around a common meal, an essential anchor in the cohesiveness of family life. Nowadays most meals are served on a plate on the lap in front of the TV or monitor and eaten with the fingers. So as a potential result of this 'planning initiative', we see the breakdown of the nuclear family unit and an increase in single-parent families and single-person households that demand an ever increasing number of houses to satisfy the new demographic—houses that need to be built in ever higher densities to reduce the pressure on 'our green and pleasant land'.

And as additional bonuses come the destruction of the high-class pottery industry of the Staffordshire Potteries together with the erosion of personal skill bases due to the absence of garden sheds.

Do you remember the garden festivals of the 1980s, government initiatives designed to stimulate the regeneration of run-down areas throughout the UK? Liverpool, Glasgow, and Stoke-on-Trent each had one, if I remember correctly, and now we know why, in Stoke-on-Trent's case at least.

While on the subject of unintended consequences, 'finger food' tends to be a processed convenience product—you know, the stuff that is less healthy for you than whole foods you prepare yourself. I remember serving a traditional Sunday lunch to some teenage American guests and being staggered by their attempts to eat roast beef with a spoon. Is it reasonable to blame so many consequences on a supposedly benevolent planning policy? Decide for yourself. Fancy a poisonous toad for tea?

Would You Live in a Caravan?

Many of us have holidayed in caravans at some stage in our lives. They provide cheap accommodation and are mobile 'B & B's. For me they always had one major disadvantage, namely the multifunctional aspect of the accommodation. The double bed is made from the lounge area, with the addition of the dining table doubling as a mattress base. You either have a bed or a lounge or a dining area; you can never have any two at the same time. The other disadvantage is the tightness of space; everything squeezed in to achieve a width and length conducive to towing. Not for you? You prefer your home?

Look at your home through estate agent speak. You have one reception room, three bedrooms, a kitchen diner perhaps, a garage, and delightful gardens to one or more aspects. What does 'reception room' mean? Put simply, these are rooms where you 'receive' guests, visitors to your house. Normally this doubles as your lounge, but it also trebles as your entertainment area (your TV) and quadruples to your music room (your stereo system). Did I mention that it's also your children's play area, your wife's hobby room (knitting, dressmaking, cutting-out area), and of course your library (bookshelf/bookcase)? And as for the garage, what, you keep your car in the garage?

So you couldn't live in a caravan?

The old Edwardian films screened occasionally on TV documentaries always intrigue me. The overriding impact is of vast quantities of people milling around in the shot. Sunday school parades with hundreds of kids marching behind banners watched by crowds several deep, worthy of a royal visit, or parks full of hundreds of elegantly dressed individuals promenading on the paths. The conclusion for me has always been that people 'lived' on the streets while sleeping and eating at home.

Entertainment at home was limited by the lack of space in the two-up two-down back-to-back terraces that slept enlarged families of several children.

Thankfully most of these housing units have long been consigned to the rubble pile, with modern housing stock replacing these 'slums'.

We seem to have learnt so much, don't we? We have a fixation on the number of bedrooms and the value this ascribes. Elsewhere in the world houses are judged by the square footage. I remember a case back in the 1980s or 1990s of a housing development in London that went to court and surfaced in a construction magazine. A developer had begun development of a site and had sold a number of dwellings by taking deposits from and entering into contracts with buyers. In the meantime the local authority put in a request for the developer to install a roundabout on the perimeter of the site.

The planning consent was already granted, and the local authority had no authority to impose this demand on the developer. Through negotiation, the developer agreed to fund the roundabout in exchange for a variation to the planning consent. This was agreed, and a number of the units went from two- to three-bedroom units and the price of each increased by around £70,000. One of these units had already been sold. The developer cancelled the contract with the buyer and demanded an extra £70,000.

Not unreasonably, the buyer objected and subsequently won his court case to have the original contract honoured.

Fair enough, but what's the relevance? The footprint of the three-bedroom house was identical to that of the two-bedroom house. One L-shaped bedroom was dissected in two by the addition of around eight feet of plasterboard wall; an extra door was supplied' and an extra light switch was added, yours for the bargain price of £70,000.

Why not change the planning rules to allow the development of proper housing stock that serves the population better than the present caravans foisted upon us? But I hear you cry, 'We wouldn't be able to afford it!' In the not too distant past, builders would talk about thirds. The cost of a house was one-third land, one-third build cost, and one-third profit and administration. Release the stranglehold on land supply and chop a chunk of the first third out of the equation. In 2007 my company selected a builder to obtain planning consent for a development of around two hundred dwellings. The cost of the traffic, environmental, ecological, noise, and landscaping reports, etc., ran to £300,000—this for an outline planning consent we haven't finished yet.

That's £1,500 per house just to get the initial consent. All of these reports have a shelf life of a couple of years and need to be revisited before the build operation can begin, after the 'validation' exercise of agreeing to the planning conditions that have been imposed.

On a smaller scale I designed an industrial unit in 1999. I sat down around a table with my tenants, the planning officer, and the environmental health officer and agreed on hours of operation. We built the unit and then, several tenants later, approached the same council officials for a variation of the working hours. Now, in 2010, I had to obtain traffic and noise reports to substantiate my request, at a cost of £6,000.

Both reports confirmed that no adverse effects would occur, but my request was rejected on the grounds that 'there was no material difference to satisfy the removal of the original condition'. These were the conditions that were set so scientifically by a chat around a table with a cup of tea and a biscuit.

Six thousand pounds is small beer, I know, but consider this. My son is a chartered landscape architect working for a firm that designs landscapes and submits the designs through the planning process. In 2007 they worked for one major supermarket chain that was spending £1,000,000 a week on consultancy fees for pre-planning reports. Ever wondered about the cost of your cornflakes? But it's only a supermarket, I don't care. The same supermarket that was buying up inner-city locations was building flats above the supermarkets and had become, by accident, a significant player in the UK housing market.

My own two-hundred-house application produced a report dossier of information that forms an A4 pile over a foot high, a third of a metre of paper, all of which has to be read and assessed by planning officers, be it Highways, Ecology, Environmental, or other. (Thinking about it, that's one thousand pounds per millimetre of paper.) This includes all council employees paid for out of your and my rates. I can't believe the planning application fee covers it.

So the conclusion is, we now have a planning system corrupt with stupidity hurtling along under its own momentum without any real control that you and I pay for. We pay for it in high house prices, high rates bills, inadequate housing, social engineering experiments that lead God knows where, and time. It takes about twenty years to bring a new road to fruition, for example. That's twenty years of waiting for the new bypass to relieve the noise, danger, and pollution from the centre of your community.

That's the same community that your council is working so hard for.

Planning and all of its associated support services is now a massive industry in its own right. What purpose does it serve? Why was it set up in the first place. How far from its original design goals has it strayed? Can anyone even remember what those goals were? How do you kill an elephant?

The multifunctionality of the modern-day caravan house has recently been supplanted by the liberal dosing of 'handheld' computer devices that defeat the constraints of space by providing an escape. The personal computer is in decline, overtaken by the tablet and ever more sophisticated mobile phones, which have opened the horizon of the smallest box room bedroom and overcrowded lounge into a World Wide Web of social intercourse and escapism from the reality of the housing stock. Imagine, Edwardians just went for a stroll around the park. It is often commented that younger generations will happily sit en masse in public places engrossed in their handheld devices while ignoring their immediate companions. Perhaps now we know why.

Let's look at the planning process from a different viewpoint, the people it serves.

My father lives in a house bought in 1971 that was built by a local architect for his own use. The house enjoys a large plot of land, a plot that is now worth considerably more than the subsidence-damaged building sitting on it. Now that my father is in his twilight years, it was decided to obtain planning consent for future development. The house in 1971 had one neighbour on a bigger plot and then open fields to two aspects. My father now has twenty-two neighbours, eighteen of whom have written letters of complaint to try to have the planning application rejected. How selfish and short-sighted can you get? Virtually every dwelling in the UK can be described as being 'in somebody else's backyard'. Every one of these new house owners had no hesitation in buying their new house built in my father's backyard, but their backyard is now precious. But of course that's different. Let's wind the clocks back to 1971 and issue a 'magic wand' to ward off evil neighbours' houses from being built. Where would you moaning fuckers live now?

One of the local 'super quarries' struggled to overcome objections from its neighbouring houses, but eventually, at the cost of extra landscaping and a new access road, achieved consent. The quarry produces 'granite', a hard stone used throughout the construction industry that isn't actually granite, but never mind.

But when we say construction industry what do we actually mean?

Quarried stone is used in:

- concrete for foundations and house floors
- concrete for structural products such as precast house floors and beams
- concrete for making and supporting road edges
- concrete blocks for building walls
- sub-bases for roads
- tarmac for making roads
- ballast for railway lines
- concrete for drainage pipes and sewer systems
- concrete frames for high -rise buildings and office floors
- concrete for factory foundations and aprons
- concrete for airport runways.

I think you get the picture.

Just to add the cherry to the icing on the cake, it was also used as the basic stone material that many of the objectors' stone-built houses were made of.

Let's turn it round then.

We all want to live in a house, served with a road ('But I don't drive'), or on a bus route (happy now?), that can take us to a train station to take us to the airport, so we can enjoy a relaxing holiday and spend our hard-earned money enjoying a drink that was delivered to the hotel by truck, money that was earned at our factory/office/shop that was built using hard stone from the local quarry that we want to shut down because of the noise and dust.

Where do you people get off! If you don't want to live next to a quarry, don't live there.

I've even had comments from Environmental Health because residents have complained about the noise from the school playground next door, saying, 'It was the summer holidays when we bought the place.'

Or how about, 'It's so important that my husband gets a good night's sleep. He has to leave for London at 5.30'? Does he push his car off the drive and round the corner before starting up for fear of disturbing his neighbours?

Does he, bollocks.

For those of my readers who are not familiar with this English colloquial term; 'Bollocks' is used as a slightly less polite form of 'nonsense'. It can be used in a positively supporting role, as in 'the dog's bollocks'. It comes from Middle English origins and means testicles. Rest assured that you can always interpret bollocks within *How to Kill an Elephant* in the pejorative sense. It is best verbalised with a sense of venom for best effect.

Then we get our politicians talking about 'divesting power to the local communities', 'giving them the power to control what happens in their area'. One possible solution would be to issue a crossbow and training to the local neighbourhood so that they can deal with antisocial London workers, quietly.

Anyone for Skinny Milk?

One proposal for controlling obesity in children is to ban advertisements for fatty foods from television screened before the 9 p.m. watershed. I presume the ambition is to stop the little darlings from begging their parents for the next fix of junk food.

For this strategy to work, we have to assume that the parents are also immune from the suggestive powers of advertising. Walking around my local supermarket during school time, I see numerous parents placing 'junk food' into their trolleys. There are no children present to ask for it, so the purchasing decision has already been made without undue influence from the kids.

Another idea is to introduce a 'fat tax', as if our taxes weren't fat enough already.

Many years ago a colleague of mine asked his milkman a question, but before I give you the details, I will drop in another history lesson.

As you walk or drive around inclined paths and roads, you often come across yellow council bins labelled 'Grit' containing salt for de-icing surfaces. So why if it contains salt does it say 'grit' on the outside? Some of you clever ones will be saying, 'Because it's got "rock" salt in it,' a good attempt, but not the truth. Up until the 1950s, councils did in fact chuck grit onto the roads and paths for traction. And then someone came up with the idea of using rock salt instead. This coarsely granular material not only provides traction but also has the physical ability to melt snow and ice—an improvement.

The grit was quarry waste, basically rubbish with no apparent commercial value, a by-product of the stone-crushing processes. Quarries began to fill up with this waste, and then a concrete company, Cowley Concrete, in Abingdon came up with a use for it. Ron Amey, the quarry owner, was intrigued to find out where this waste was going. He had to threaten Cowley Concrete with no further supply before he learnt that Cowley Concrete were mixing the 'grit' with cement and making concrete blocks for wall construction. Prior to this invention, all house walls were built of brick or stone. Eventually Ron Amey bought the company and created a network of block plants around the country under the trade name of ARC (Amey Roadstone Corporation).

Now, back to the milkman. See if you can guess where this one is going.

The question my colleague asked was, 'What do you do with all the fat that is extracted from milk to make it skimmed or semi-skimmed?'

'Good question, that. It used to be a right problem, but not anymore' was the reply.

So assuming they don't build houses with it, where would you dispose of a homogenous lake of *edible* milk fat?

Fat, sugar, and salt are all 'good' within the food industry because they have taste. Fat retains the flavour experience in the mouth, and sugar and salt provide taste. That's the taste of sugar and salt anyway, not the food itself. Consider this: make a ham sandwich out of cheap, tasteless ingredients on cheap, flavourless bread. Now dose it up with fat (mayo) and salt and throw in some red onion. What have you got? An onion sandwich that's bad for your health.

Or add mustard, or curry, or Thousand Island, or any other flavour enhancers thrown in for our benefit to disguise the otherwise plastic ingredients. The time was when spices were sought out as a flavour disguiser to cover the taste of rotten meat, which was the result of a shortage of adequate preserving techniques. Now their function appears to provide flavour for otherwise tasteless ingredients.

I used to buy sandwiches for the trip home from the supermarket and soon realised that it was a bad thing that they kept reducing the prices under the banner of 'good news'. How can you keep reducing the prices while maintaining the quality? Wages will be set at the national minimum, and transports cost are rising, so the only thing left is debasement.

Thirty years ago my company supplied several truckloads of concrete product to a national building products outlet, not a builder's merchant, rather a crossover aimed more at the general public, the DIMers (do-it-meselfers). Their immediate supply needs meant that I had circumnavigated their standard buying conditions, but orders would dry up unless I signed up.

There were numerous reasons why I didn't, but I will concentrate on one for now. 'We as a company do not accept price increases. If you experience difficulties, we are always available to talk, but we will not accept any price increase.'

So, given that my cement, steel reinforcement, aggregate, energy, and distribution costs were all outside of my control, how could I possibly agree? The only option left open to me was to debase my product, a structural product that held up people's houses. Every now and again I forget and buy something from one of these DIM stores, but I am soon reminded why I'd stopped. For one thing, their incandescent light bulbs had a life expectancy of days, not weeks. For another, if you over-tighten a screw in a light fitting by just a fraction of a turn, you strip all the threads and the screw falls out.

Not everything sold at the DIMers' store is debased; the cement manufacturers, for example, will put their prices up as they need to, because they have the muscle to do so.

The same can be said of staple products at the supermarket such as milk and potatoes. It's manufactured goods where the problem lies. That's not to say these staples are immune from price constraints. Ask any of the dairy farmers driven out of business.

My maternal grandfather worked for Lyons as a development chef through and beyond the Second World War. In simple terms his job was to design and bring new cake products through to the market. In practice his job was to achieve the desired product quality and then work out how to make it ever cheaper without compromising the standard below an acceptable level.

It is a source of great disappointment to me that in this age of great technology, when every effort could be used to achieve near perfection in food, all the technological expertise is devoted to driving the price down and down. We are not far from cutting out the middleman of the digestive system before our lovingly prepared food reaches the sewer.

Why spend millions of pounds on adverts to promote products unless you know the advertising works? So by deduction it must work; otherwise, the money wouldn't be spent, would it? I used to watch Sky News in the morning while getting dressed and eating breakfast and would take an interest in the weather forecast in particular. This was always announced before, 'Let's take a short break; stay with us', and then the adverts. I soon realised that I didn't know what the weather for the day was going to be because I disengaged mentally from the TV during the break and failed to reengage when the programme returned.

In days gone by I have searched for telephone numbers for suppliers, only to realise days later that I was using their page-a-day diary on my desk complete with all their contact details. I've even allowed my desk jotter pad or the corporate calendar hanging on the wall to merge into the background. Am I alone in this? Have I got very early stage dementia, or does everybody reach the saturation point where you have to disengage to retain your sanity?

Do the advertisers ever realise that their constant activity merges into the wallpaper?

How many times does the same advert have to be screened over and over again, day in day out, before somebody cries 'enough'? Sky channels are the worst. The bastards have now learnt to synchronise their advert breaks so you can't skip channels to avoid them. (Thank God for the PBS channel that didn't get the memo and is out of sync.) But here's an idea: instead of screening forty adverts an hour at £2,000 a pop, why not screen five adverts an hour for £16,000 each?

You produce the same revenue and retain the viewer interest during the thirty-second break. One can't even have a piddle or brew the tea without missing the programme one actually wants to watch. (To any advertising/ TV execs out there, I apologise for not having a clue as to how much a single advert costs.) (Then again, will you apologise to me for serving up endless snippets of highly polished dross?)

But then another thought springs to mind: if our thought process are broken every fifteen minutes for three to four minutes at a time, how much long-term damage does it do to us? Are we training generations of individuals to only concentrate in short bursts, incapable of sustained thought? Are we effectively training our mental abilities to be reduced? This seems to me to be a good research project, if only you can tie it in to global warming.

But why single out the telly? Try browsing the Internet, listening to music on YouTube, or reading a magazine, or for that matter picking up your post off the doormat, or watching a sporting fixture, or passing a billboard, or passing a bus stop, or buying a car park ticket, or watching a bus go by, or reading the sticker in the back window of the car in front, or checking out the 'This roundabout is sponsored by …', or reading a newspaper, or removing the flier from under your windscreen wiper, or listening to commercial radio. I think, provided you can concentrate for long enough, that you get the picture.

So, Mister Advertising Executive, can you remember what the original aims of advertising were? And are you achieving them? Does it ever occur to you that on a scale of 1 to 10, where 10 represents saturation point, your industry has reached 100? But you don't care, because advertising is an industry in its own right, with its own unstoppable momentum that now controls you even though you think you are in charge.

During the 1980s I was a structural products estimator. During quiet periods I would let my mind wander with calculator in hand. What I used to do was convert the price of my lunch into figures expressed as pounds per tonne. Try it yourself and see what you can discover. My prices are all out of date, but the principle will remain. Plain biscuits were £800 per tonne, biscuits with fruit were £1,000 per tonne, and biscuits with chocolate were £1,200 per tonne. The really interesting one for me was Walker's Crisps, which came in at £16,000 per tonne, when I knew at the time that potatoes were £80 per tonne, and I also knew that wheat was a similar price. That represents a 200 fold markup. Now I realise that I didn't understand the full processes involved in buying, peeling, slicing, frying, sprinkling with flavour, packing, selling, and distributing potato crisps in a bag to shops, but I still questioned where all the money went given that most of the same activities occur at the biscuit factory and there was a predictable step in price between plain and fruit or chocolate.

The conclusion I came to—and I might have been wrong; I will let you decide—is that the difference was spent on marketing or, should I say, advertising. Even in those days Walker's spent millions on TV campaigns (I seem to remember newspaper reports read at the time). I reasoned that every time you bought a packet of crisps, you were actually paying for *Coronation Street* and the rest. My wife used to eat a pack of crisps a day. I tried to persuade her to have a small jacket potato instead, because that way I could pay for the microwave oven in short order. I had laser surgery on one eye at about the same time. Allowing for my eye tissue to weigh the same as water, the cost equated to £25,000,000 per gram, enough to make your eyes water. Sorry about this trip down memory lane. You will begin to realise how much of an anorak I am.

So, if my assumptions are correct, we are actually paying advertisers to ruin our viewing experience and destroy our brain function. Yep, that sounds like *Corrie* to me.

Imagine a scenario where advertising didn't exist beyond a scale of 1 to 10, where 10 represents the saturation point, and consider the cost of your weekly grocery shop. 'But what about *Corrie*?' I hear you ask. Well, there's the thing. At about the time of my calculator experiments, the actors were being paid about £20,000 per episode (newspaper reports again, so it must be true). Why? Because they were worth it, of course. Now if the revenue coming in dropped and they could only be paid £1,000 per week, how many would still turn up? The answer might well be none, but turn it round. If they had never been paid £20,000 in the first place, who among them wouldn't jump at the chance to earn £1,000 a week? It's not as if there is any heavy lifting involved, is there?

Let's consider the imaginary grocery shop without the cost of advertising.

During the 1980s I began my interest in photography. I bought a 35-millimetre SLR camera and started to photograph attractive things in nature: flowers, butterflies, scenery, and the like. I soon learnt to avoid Kodak film because the photos had a very unflattering green that is not much use when working with plants. I found out later that Kodak was very good with 'flesh tones' and was pitched at photos of people long after I had switched to Fujifilm, which captured greens very well. My employers soon knew of my interest and I was regularly despatched to photograph and record work processes and products, with the instruction to produce multiple copies for distribution. By that time my local camera shop had shut down and I ended up using SupaSnaps for convenience purposes for printing and replacing films. Now SupaSnaps sold its own brand of film along with Fujifilm and Kodak. The price was around £5 for Kodak and Fujifilm, and around £3 for SupaSnaps' own brand. It was obvious that SupaSnaps didn't make their own film and bought it in from a major manufacturer, but which one? I think at the time the only alternatives to Kodak and Fuji were Konica and Ilford. By a process of testing and a hint from 'somebody in the know', I discovered very little difference between Fuji and SupaSnaps and concluded that they were one and the same, so I could now save £2 or so per film, although the range of film speeds was limited. Why would Fuji sell films at £2 below their normal price range? I reasoned that Fuji had reasoned that they had sold all the film they could sell to customers willing to pay £5 but could afford to sell a limited amount of film for less because Fujifilm sales covered their overhead and extra sales increased their profitability. But Fujifilm were walking a tightrope. They couldn't afford to sell all their film for £3, because that would damage their profitability, but they could afford to sell some, because this enhanced their profitability, provided their involvement was kept secret from the public. And SupaSnaps didn't advertise their products but 'popped up' on the high street and won business through speed and convenience and cheapness. They were always busy producing people's holiday snaps, for example. (It's what we used to do before digital cameras and, later, smartphones arrived.)

So the supermarket shop?

Fill your shop floor with obscure products of brands you've never heard of made by companies you've never heard of which support no advertising and all the associated cost multiples that involves, including the making of *Coronation Street*.

Leave the products on the pallets and reduce staffing levels to give the illusion of a cheap set-up, and pursue your ambitions relentlessly.

But you don't have to imagine it, because this model is presently carving its way through the 'Big Four', Asda, Tesco, Sainsbury's, and Morrisons, to their alarm and dismay. True, the Big Four all dabble in their own brands made by somebody else, following the Fuji–SupaSnaps model, but the likes of Aldi, Lidl, and Netto have it engrained in their DNA. Advertisers, beware; is your bubble about to take a hit?

That being said, however, you cannot fail to have been annoyed by the inane product comparison adverts pushed out there by Aldi. So has the DNA suffered a mutation? These naturally staged 'quality comparison' adverts with 'real people'? I don't think so. I've yet to hear, 'Fuckin' hell, that's dirt cheap, that is.' Perhaps that bit got edited out? And if that bit has been edited out, what does that say about selective quotes from real people? 'This product is crap' will also see the cutting-room floor.

So what are these obscure products in brands you've never heard of?

Tomato sauce, baked beans, spaghetti, chocolate, smoked salmon, etc. Hardly obscure products, are they? For good measure, sell only one variety of each. Offering choice is expensive; it increases your storage and distribution costs and the area of shop floor required. You will notice that these 'German discount supermarkets', as they are labelled, have smaller stores and still sell the same basic range of goods, but without the multiple choices and, more importantly, the costs to the customer. Now this leaves the Big Four with a fundamental problem. To compete they need to do a number of things: reduce prices, reduce costs, reduce choice, and reduce shop size, and this at the end of a major store-building programme. Now it is true that Tesco have shelved a number of planned stores or even built new stores, but this only begins to scratch the surface.

So what have the Big Four presented to the great British public?

A price war between them to fight for market share and, I nearly forgot, massive advertising campaigns. In the process, they have also 'leaned' still further on their suppliers by demanding monetary assistance for marketing and price reductions for the products. The *Financial Times* and other commentators are already discussing the disappearance of multiple small firms from the UK supply chain. This will lead to less choice for the consumer, and supermarkets will not be able to fill their shop floors.

So, to use the Great Wood analogy, the Big Four climax shopping environment has just been introduced to Iron Age man with axes and a plough. In simpler terms, the Big Four pro forma has been supplanted by another.

There is a more sinister impact to these market 'corrections'; each supermarket will fight tooth and nail (back to tigers again) to protect what they have, namely market share, shareholder confidence, and their salaries. You can understand their position perfectly; market share equates to market clout, and the more you sell, the more you can bully your suppliers to keep your prices down. It's a merry-go-round.

So now where does that leave the suppliers, up a creek without a paddle and nowhere to turn, except to squeeze something else out of the product: debasement? So we end up with horsemeat in our lasagnes, as a result of criminal activity we are told, although I personally have not heard of any prosecutions. Right or wrong, horsemeat is there because of pricing pressures; somebody somewhere in the chain of production has cut a corner to remain in business. But that isn't what worries me particularly. Some people choose to eat horsemeat, and there are no health implications, only moral ones for horse lovers (live ones, that is).

Try the following experiment at home. You need a George electric grill, although a normal grill would suffice, and some beefsteak from two sources. Buy one steak from your local supermarket and the other, equivalent-sized one from a quality family butcher. Cook them one at a time. The George grill slopes to the front, and all juices superfluous to the meat runs down the grill and is collected in a plastic tray. When I did the test several years ago, before I started buying my meat exclusively from the family butcher, I had to empty the tray three times with the supermarket meat and only half filled it with the local butcher's meat. To be fair, I was cooking for a family of five and the George was filled to capacity, but such was the case in both instances. OK, you accept that the supermarkets add water to the meat to make it more succulent and cheaper. (If rump steak is £25 a kilo, that's £25,000 a tonne for beef. Your local water board charges a couple of quid a cubic metre (one tonne), so do the maths, as they say.) But that still isn't what worries me. Weigh yourself before you get in the bath, and then weigh yourself after getting out. You will not notice any real change. So how do you entice all of this water to enter the meat and stay there? After all, you would reject any trays with water in them at the point of buying. (Although you probably thought that the little absorbent pad under the meat was to collect any blood. Still, you don't get them at the family butcher's.) You need a medium, some product that will enter the meat and facilitate the uptake of water, such as nitrates and nitrites. Ever wondered what the difference is between pork and bacon? It's not a different breed of pig. Bacon and ham are treated with salt and nitrates to 'cure' pork and produce the bacon and ham. In recent years studies have suggested a direct correlation between the consumption of bacon and the incidence of pancreatic cancer. Eat a rasher of bacon a day and your risk of developing pancreatic cancer increases by 18 per cent. Double your daily consumption and you double your risk factor to 36 per cent.

Now my maths is quite good, but every reader should be able to see that if this was plotted on a graph, you would draw a straight line at 45 degrees. One problem with all of these studies is that it very difficult to establish a direct causal link between something you ate or did years before and the development of a disease, but in this instance the researchers felt confident enough to go to print. What they didn't feel confident to announce was the idea that bacon is effectively poisoning you. The more you eat, the more you are likely to suffer. After all, the graph line is straight and plotted at 45 degrees. Now that isn't to say that everybody who eats bacon will develop pancreatic cancer, but the study does suggest that people who develop pancreatic cancer are significantly more likely to have eaten bacon. It's almost as if the researchers stumbled across an inconvenient truth that had to be reported, but also that this was done quietly and slipped under everybody's radar. So, poised for the next big supermarket war, I note that my local Big Four proudly announces, 'Satisfaction guaranteed or your money back'. What I have yet to see in the small print is 'We reserve the right to poison you.'

I have talked before about the two Greek philosophers and the shipwreck survivors all believing in God, only for the comment about the silent witnesses. Some of you will be really annoyed because I haven't named names or provided dates and places. Throughout your later schooling years, you are encouraged to name your sources. Why is that, I wonder?

Our factory made structural building components and we were compelled to sign up to and implement BSI quality assurance, later replaced with ISO 9000 to comply with the EU. To comply we had to introduce a paper trail of documented responsibility through all aspects of our operation. Our operatives would take ten measurements from each production run, and the immediate supervisor would take a minimum of three. Then our quality control inspector would check each run before releasing product to the storage areas for distribution. Our loading teams were also instructed to visually inspect the quality of the finished goods during the handling processes and were authorised to quarantine any suspect goods. All of this was duly documented on the paper trail for recording and later inspection.

So why did I receive complaints about poor product delivered to my customers' sites?

There is a tale about a city ruler (I think it might come from *One Thousand and One Nights*. Forgive me; I wasn't taking notes. I was only a kid at the time) who decided that no one should starve in his city. He devised a plan whereby all the rich city merchants would fill one of the city wells with milk (I know, genius, a milk-filled well in Arabia. I told you we would cover cheese wells later) so that all the poor and destitute could seek nourishment from it. When the day came, the first street urchin lowered the bucket and withdrew pure water. Yes, you've guessed it: every rich city merchant thought, *No one will notice if I put water in it. It will just be slightly diluted.*

Back to my poor-quality product. There was not much of it. I'm not bad-mouthing my employers: they have kept me clothed, fed, and sheltered for fifty years.

Everybody cut corners on the basis that somebody else had already checked the product or would check it, 'so it doesn't matter if I don't'.

If you quote an idea or thought, do you have to ascribe it to a source so that your readers can simply accept it in good faith and not have to think about it themselves?

The second Greek, the one with the smart-alec reply, was he the genius, or had he heard a man called Elvis down at the kebab shop offer his thoughts on the subject? Had he actually interviewed one of the shipwreck widows and learnt that her poor departed partner had been a devout believer in a god? Had he copied the line from a really important text that has been lost to the sands of time? We will never know, so why ascribe such importance to the source? Ascribe the importance to what was said.

Still annoyed? I bet some of you know the names. Answers on a postcard, please, to my publisher. The closing date for entry is 31 December 1999. Prizes will not be awarded.

Oh, all right then, it was Diagoras.

Happy now?

But this was a story presented by a Roman, Marcus Tullius Cicero.

Now that I think about it, does every unascribed quote make the originator a silent witness?

Alternatively, does every ascribed quote become one of the 'lies to children'? Accept the shorthand of the declaration within, either challenging it or thinking about it, because somebody else has done the thinking for you.

It's 18 March 2015, Budget Day, and the news media are full of the chancellor's initiatives. One of the 'giveaways' is a new scheme for first-time buyers, the endangered species of the housing market. The basic issue is one of affordability: house prices are rising faster than the time you need to save up the deposit, and the banks, post-recession, have all tightened the lending criteria. Now, qualifying first-timers can get a cash boost from the government (us, the taxpayer, given that the government has no money, only ours) whereby for every £200 saved, you will receive an extra £50, up to a maximum of £3,000.

This isn't the first initiative from this or the previous governments to aid the beleaguered first-timers; we have had a number of variations to stamp duty as well. For those of you yet to enter the house-buying experience, I will explain what stamp duty is. In essence it is a tax payable by the purchaser of a dwelling calculated on the value of the transaction. In simple terms, the more expensive the property, the higher the tax.

Like with income tax, there is a threshold where tax begins to be paid, and it is typically banded so that as the price enters a new threshold, the percentage of tax due also increases from 1 per cent, 2 per cent, 3 per cent, etc. For example, in days gone by, houses up to £120,000 were free of tax and therefore nobody would attempt to sell a house for £122,000, as the £2,000 increase was further boosted by the 1 per cent tax on the whole amount. You tended to end up with shop-style prices of £119,995 or £179,995 if the 2 per cent threshold was set at £180,000. The figures vary with time, so accuracy is not needed; it's enough for you to understand the principle.

Consider what would happen to the asking price of an £119,995 house if the threshold were to be increased to £130,000. You have removed the artificial ceiling that restricted the price to less than £120,000, and the new ceiling becomes less than £130,000. Now repeat the procedure at the thresholds between 1 per cent and 2 per cent, or 2 per cent and 3 per cent, and think about the effect on house prices.

Previous initiatives have included so-called stamp duty holidays where first-time buyers would have no tax to pay, so for them the thresholds had gone completely. So now you have a deadline to secure the purchase of your house, in competition with other first-timers in the area. In 2014 the Scottish Parliament also varied the thresholds for stamp duty north of the border, and for that matter major changes were introduced in England and Wales in the Autumn Statement.

The net effect of all of these changes has been to reduce the tax take by the UK governments and to increase the prices of all the houses available to first-time buyers. If you can afford a £140,000 house and pay the tax at, say, £1,400, then you can afford to pay £141,400 for the same property without the tax. But don't take my word for it. The Royal Institute of Chartered Surveyors have published data to that effect. Whatever the tax saving is, it's rapidly transferred to the asking price. This latest savings initiative has just added up to £3,000 to the asking price of the properties, so what help has been achieved?

At a stroke, the chancellor and the previous office holders, plus their contemporaries north of the border, have increased the price of all properties in the UK. And you call this help?

Can you believe for one instant that our benign masters, whether it's Conservatives/Liberal Democrats or adherents of the Labour Party or Scottish National Party, are blind to this fact? Are they really so stupid as to seriously offer these initiatives as a solution to the housing issue? The solution to the housing issue is to build more houses; it's as simple as supply and demand. If there is a shortage, the price goes up. If there's a surplus, the price goes down. Artificially boosting cash when there is a shortage simply increases the prices and at best solves nothing.

What makes these initiatives even worse is that many of them have been time limited. Beat the clock or lose the help. This has induced near panic in the market before now.

Back in April 1989, the then chancellor announced that the second personal allowance of MIRAS (mortgage interest relief at source) for unmarried couples would be phased out by 31 August of that year. House sales boomed, prices shot up, and builders ran out of materials and labour until September, when everything stopped and the housing market dropped into recession until 1991. MIRAS gave tax relief on mortgage payments up to £30,000 for married couples, but £60,000 for unmarried couples. It has long since been phased out completely.

So, we wouldn't be stupid enough to introduce cane toads into Australia now, would we? We are much too clever to do that now. We know we shouldn't introduce something without understanding the consequences, don't we?

I wish.

But in this instance we do, in fact, know the consequences. Please note: I am not making any party political points here; they are all as bad as each other.

Still not convinced? At the height of the recession, the local authority at Stamford announced that they wanted all new housing sites to include 90 per cent social housing, leaving 10 per cent for sale on the open market. This initiative was to provide subsidised housing to the young who couldn't afford to live and work anywhere near where they grew up because they had been priced out of the market. Now Stamford is a very attractive place with it medieval alleyways and stone-built houses, and it attracts people with money who push up the prices of the housing stock because they can afford it. So what you need to do is build more houses so that the price pressures are reversed. In the process you will destroy Stamford for what it is and will turn it into a less desirable place to live, at which point you will need less housing and the prices will drop. What you don't need to do is limit the number of new houses coming onto the open market, because this will drive up the price of all the existing housing stock, making it less easy for young people to afford to buy a house anywhere near where they grew up. But that isn't fair on the young generation, is it? That would be the younger generation who grew up surrounded by the delights of Stamford, who watched their parents drive the house prices up around them. So whose fault is it anyway? Is this not the natural order of things? When a forest reaches its climax state, is not the access to space and light dominated by the mature trees? Don't the younger saplings have to fight to achieve dominance for themselves? Don't the trees themselves have mechanisms for the dispersal of their seeds and fruits? Living in a forest has obvious advantages for those 'make it'. The advantages for those struggling in the shade are less so, but ultimately we know it works. What we have here then are local government policies that will exacerbate the problem, not solve it. Did anyone go to school and learn anything? Is there a social experiment here that cannot accept that rich people like to live in rich communities and don't like to have to trip over poor people all the time? For good measure, national government policies have sought to increase the minimum requirement for social housing from 20 per cent to 40 per cent on all new developments since 2007.

It's the day after the Budget Day. I've broken the habit of a long time and bought a newspaper, the *Daily Mail*. (Other papers are available, but I'd already seen some of the content in my local café.)

Before I examine the contents with you, I thought I would share a news report with you that I picked up in the 1980s. My local newspaper ran an article championed with this headline: 'Forgers Foiled in Local Bank'. So let's flesh the story out, shall we?

A man had gone into his local building society with his wage packet (cash in those days, not computer transfers) and had tried to pay in some money. The eagle-eyed teller had spotted some discrepancies and had both refused to accept and confiscated two £20 notes because they were forgeries, leaving the man £40 out of pocket. At what stage in this process was any 'forger foiled'?

The *Daily Mail*'s front page is emblazoned with the latest government initiative designed to provide a boost for first-time buyers. Nearly all of page 5 is devoted to an article about a potential £3,000 fund for first-time buyers, with a case study of a featured couple under the premise of speeding up lives by aiding the generation of house-buying deposits.

I do agree with the front-page sentiment, but only if you interpret 'leg up' to mean 'trip up', as in, 'I legged him over.'

So given the stunning quality of our 'opinion peddlers', what chance do we have of changing anything in a logical, meaningful way, using policies that might actually help rather than hinder? To be fair, if you happen to have a house for sale suitable for a first-time buyer to buy, you have just received a £3,000 windfall. (Don't for one moment think I am trying to peddle opinion.)

The Footpad and Highwayman Code

When I passed my driving test back in the 1970s—yes, I really am that old—I was taught that a 30 mph speed limit sign set the maximum speed for an area, assuming it was safe to travel at that speed. At no time since has the Highway Code been altered to make this maximum speed limit advisory.

Since the late 1970s I have seen the introduction of various devices to ensure the 'observation' of this limit. These have included:

- painting lines on the side of the road to make the road seem narrower than before
- hatching the centre of wide roads to make them seem narrower than before
- filling the centre of the road with white hatching on red tarmac to emphasize the fact that the road appears narrower
- installing speed bumps or 'sleeping policemen'
- installing speed tables (fat sleeping policemen—must be all the doughnuts)
- constructing on-road parking bays to make the road look narrower and to act as chicanes
- constructing chicanes, locally restricting the two-way traffic to one lane only
- instituting 20 mph zones, 'because nobody does 30 anyway'
- using telematics, flashing electronic signs telling you to slow down
- using telematics, flashing electronic signs telling you what speed you are doing

- installing movable speed cameras
- installing fixed speed cameras
- installing average speed cameras
- posting speed enforcement area signs
- posting speed camera signs
- erecting short sections of wooden fence at village boundaries to give the impression that you are entering a private space
- installing rumble strips
- creating mini roundabouts
- placing white triangular patches at the roadside on the approach to the 30 mph sign
- posting rectangular signs with three, two, and one black diagonal stripes counting down to the 30 sign
- airing television public service notices advising of the likelihood of breaking bones at 40 mph as opposed to 30 mph impacts.

And let's not forget the revision to the driving test with its two-part theory and driving elements.

Forgive me if I have missed any. I think this is enough to give you the picture.

That's twenty-two initiatives designed to slow motorists down. Why? Every single one of these ideas has cost money to implement, and every single one has been ignored. It reminds me of the castle scene in *Monty Python and the Holy Grail* whereby the castle is rebuilt several times in the swamp to replace the ones that have already sunk, even though any castle built on the site continues to sink.

So you don't think the cameras have been ignored? Then why are drivers still paying fines and receiving points?

Either you are serious about controlling the speed of vehicles in residential areas or you are not. Which is it? (I did meet with the highway engineers shortly after they had covered a local village with red tarmac graffiti. They told me it had been a great success and had slowed the average speed by four miles per hour. The same village is no longer red. The shock value lasted a matter of days.) Usual comments are: 'If I'm watching my speed all the time, I can't watch the road, can I?' It is not without the wit of car manufacturers to offer a 'speed ceiling optional' button on every modern car. Push once for thirty, twice for forty, etc.; it ain't rocket science. Alternatively, you could always learn to drive.

So are you serious about imposing the speed limit or not? If you're serious, enforce it. If not, stop wasting money on initiative after pointless initiative.

When Four Multiplied
by Four Equals Nothing

We have discussed the way that *Homo sapiens* seems to have a self-destruct gene and apparent death wish, albeit in slow motion. We have all been educated and know in broad terms that smoking, drinking, obesity, etc. are injurious to health. To put it in simple perspective, let's look at dental care. We all know we need to avoid sugary snacks and drinks and to brush our teeth twice a day with fluoride toothpaste. Let's see what the 2013 Children's Dental Health Survey for England, Wales, and Northern Ireland has to say on the subject. In round figures, half of 15-year-olds had decay and a third of 5-year-olds and half of 8-year-olds had signs of decay in their milk teeth. There were signs of improvement over the ten-year period between the current and the previous survey, with an overall reduction in cavities based on ten thousand surveyed.

So are we serious about our children's teeth or not?

Many of us have been exposed to little risk assessment forms: these simple little matrices pose a risk down one column, against consequences plotted on the rows. You allocate a number, say, from 1 to 4, against the chance or frequency of it happening, and then repeat with 1 to 4 against the consequences. Once you have your numbers simply multiply them together to give a score from 1 to 16.

Using this simple grid, you quickly identify the risk factors. A dangerous occurrence (4) happening frequently (4) would yield a score of 16 and require immediate action to prevent it from happening. Let's take a reasonably understandable example: you would never ask a street sweeper to sweep the hard shoulder of an open motorway. The chance of being hit is high (4) and the consequences lethal (4), scoring 16. A concern about health and safety would therefore prevent you from carrying out this task without addressing the issues. In this case, the only safe way to proceed would be to shut at least one lane of the motorway, screen your workers with barriers (not cones), and impose a reduced speed limit of, say, 50 mph. The same street sweeper could be employed to sweep the pavement in a residential area, because the chances of being hit would be reduced (1) and the consequences of a slower speed impact reduced (2), scoring an acceptable 2 overall. Issue your man with hi-vis clothing and with training to be traffic-aware and you've covered most bases.

Seems perfectly logical. People up and down the country have been trained in these simple 'risk grids', and our workforce is safer because of it. It works.

Now working at the motorway edge requires training, so you would ensure that all your employees were trained in safe working practices. If your employee ignored your instructions, you would sanction them and ultimately terminate their employment, because not only does their life depend on it but also your livelihood does. If you get it wrong, someone can die. That's why we have Health and Safety. I am not going to attack the overzealous media reports of 'conkers' being banned in schools. When you kiss your loved one goodbye in the morning to go to work, it's a perfectly reasonable assumption that you will return alive and well. The health and safety culture developed in the UK has much to be proud of.

Now if the same health and safety training manager were to consider a different question using the same techniques, we would arrive at a similar conclusion.

So what is the different question?

Take a risk factor and matrix it within the grid to arrive at the risk compilation factor. Expect scores from 1 to 16. For scores of 1 to 4, take no action; 5 to 8, offer training; 9 to 12, offer training with the threat of sanctions; and 13 to 16, introduce a radical training regime with sanctions.

Review every three months, and increase the sanctions as necessary.

So are we serious about health and safety? If you are, enforce it. If not, stop wasting money on initiative after initiative.

So what is the risk factor? Well, there's lots really. It could be obesity or smoking or alcohol consumption or drug abuse. I think you get the picture.

We either want to be serious about health and safety or we don't.

And if you're not going to be serious about your own health, why should I be?

Let's have a classic response from a smoker: 'The amount of money I've paid in fag duty, I've paid for my stay in hospital, so get stuffed.'

Fair enough: when you buy a lottery ticket for £2 in the hope to win the £80,000,000 EuroMillions rollover, you accept that for your ticket to win, 39,999,999 tickets will not. So the question for the smoker is, if he's going to use up all his tax contribution for his end-of-life services, who's going to pay for his wife's or child's or grandchildren's needs? If he can be a selfish bastard, then so can I, by refusing to pay my taxes into the kitty.

Is it possible to have an educated, adult debate about the NHS? Just to remind you, that's the National *Health* Service.

Any 'attack' on this treasured institution is instantly rebutted by party political attacks.

Imagine taking your car to the garage to have it serviced and then having it break down the next day. Imagine how you'd react if the garage said, 'Yes, I could see the car was at risk, but it's not on my service list, so I did nothing about it.' Would you give that garage any repeat business?

Imagine that you'd broken down on a French toll motorway where it costs you €200 to get towed off the motorway (because breakdowns are preventable and the French authorities effectively fine you for not taking precautions—even though you'd had your car serviced the day before). Would you give that garage any repeat business?

Imagine that because you broke down on the toll, you failed to reach your hotel and got billed even though you weren't there, and had to pay for emergency accommodation as well as your garage repair bill.

Am I going over the top?

Imagine going to the doctor for treatment for a skin rash and coming out with some ointment.

Imagine ending up in hospital the next day with a heart attack and ringing up your doctor to report your 'breakdown', only to be told, 'Yes, I could see you were at risk, but it wasn't on your list, so I didn't do anything about it.'

Seems far-fetched and couldn't happen, could it?

But visit your doctor with a skin rash while an overweight smoker and you come out with the ointment and nothing else. This brings me to the risk matrix and the introduction of preventative 'maintenance'. (If you asked the doctor why, they hadn't addressed your other obvious health issues, he or she would probably reply, 'I haven't got time to address all my patients' needs' or 'You try it and see how much abuse you get.' Is this because the system is based on offering a 'fire service' and has not got the resources to offer preventative maintenance? Is the reason it hasn't got the resources to offer preventative maintenance, because it's spending all its resources on 'fire service'?

If you're not serious about enforcing the 30 mph speed limit, stop tossing about with it and take the signs down.

Where is the contract between the NHS and its customers?

If the provision of healthcare doesn't motivate you, then how about the cost? In 2015 we will spend £140,000,000,000 on the NHS. That's a fourteen and ten noughts, an incomprehensible amount.

That's 3,750,000 passenger pigeons every day for 18,250 days, or 50 years. This ignores the breeding potential (of the budget amount).

There used to be a joke—obviously it still exists; otherwise I couldn't repeat it. 'So you want to live to be a hundred? Don't drink, don't smoke, and only ever have sex with one woman in your life.' 'And will that make me live to be a hundred?'

'No, but it will seem like it.'

Please tell me when drinking, smoking, and promiscuous sex became essentials for enjoying life?

A significant proportion of the population hasn't had a good week unless they get 'blathered' at the weekend, that is to say drunk to the point of falling down at least.

That's fine by me; get drunk to the point of unconsciousness if you like. But why do I have to pay for your actions through my taxes? Why does somebody else have to die because their ambulance is busy dragging you out the gutter?

In 1988 the Control of Substances Hazardous to Health (COSHH) regulations were launched into the UK workplace. These regulations required all employers to assess and control all substances that might represent a hazard to their employees' health. You may remember this as the time when your cleaners stopped using bleach to clean the toilets and had to use 'safer' products instead. Another favourite target was your 'Tippex' correction fluid, or Liquid Paper, for correcting errors on typed papers, because of the toxic thinners used. In the factory all products would be assessed by those responsible against manufacturers' data sheets, and the obligation was to remove hazardous substances and replace them with safer alternatives.

I had the responsibility of producing our COSHH data sheets for our customers to use.

A few years later I chanced upon a TV interview between the then president of the EU, who was asked by a journalist if he could foresee a time when Europe would follow the US example of banning smoking in the workplace. His reply went along the lines of, 'Ah, yes, the legislation is already in place.' There was no further explanation. This set me to thinking. I realised the following things:

1. Cigarette smoke had been firmly established as being hazardous to health.
2. All employers had a duty of care to protect their workforce from hazardous substances.
3. All venues where employees would be exposed to hazardous smoke were covered by the legislation.

So at a stroke, I realised that compliance with this EU directive meant that all pubs, restaurants, theatres, and other places where people worked also had to become smoke-free. Yet, this inconvenient truth sat around ignored until July 2007, when the smoking ban came into force.

I waited in vain for a trades union to bring a test case against an employer, but I decided that too many of them smoked to want to challenge it and at the same time had no interest in protecting the very lives of their members and would sooner preserve the rights of their members to smoke.

Assuming I was right and that COSHH could have been used to control smoking, this illustrates perfectly the disconnect with the obligations of the health and safety manager.

Something else that seems to have slipped past the world without comment is the protection racket that is the sunscreen industry. This is the current-day version of the suntan industry that for many years championed the 'healthy look of tanned skin'. Tanning oil and products with sun protection factor 5 or 10 were marketed with magazine pictures of scantily clad females basking in the sun displaying their perfect brown hues. How strange that an industry that unknowingly promoted the rapid growth in skin cancer is now the very one that will save us from it. Why in this compensation culture of ours has no one had a pop at them? Without their multimillion-pound marketing campaigns, how many fewer melanoma sufferers would there have been? Even today the peer pressure to colour up with a tan is immense, in direct contrast with earlier generations, where colour was considered common and indicative of outdoor workers.

You may be lucky enough to dodge the skin cancer bullet, but you will not be able to avoid the ageing effects of sun damage to the skin. Never mind: there is an even bigger industry out there providing the 'antidote' with anti-wrinkle creams. And please, what the hell is a serum other than something made from blood?

Immunei bollocksii

We have free fruit for schoolchildren. I grew up with free school milk in primary and junior school. We have advertising campaigns pushing 'five a day'—five portions of fresh fruit or vegetables in the daily diet.

Shopping centres and village centres are seeing the introduction of heart defibrillators.

We have health awareness programmes on the telly.

We have screening for various cancers—bowel, cervical, and breast, to name a few.

We have opticians screening for glaucoma (diabetes induced) and other eye diseases such as macular degeneration (exposure to high levels of UV is considered a risk factor/cause).

We have government lifestyle adverts under the banner of 'fit for life'.

In contrast we have had massive TV advert campaigns pushing the virtue of yoghurt packed with 'healthy bacteria' to make you 'feel good inside'. Has anyone else noticed that the health claims for these products have vanished from the tag lines? The EU said to prove your claims or stop saying it. To date, of the thousands of claims made by advertisers, only a handful have made it through. The best the advertisers can come up with now is 'Proven to reach the gut alive'. Now as a bit of 'self-justification' here, the smart ones among you will buy the idea that it costs too much for the manufacturer to prove the claims, but the claim must have been based on some data and therefore the earlier claims still hold true. No? Then why is the discredited immunei bollocksii still on the shelf? The really smart ones among you will realise that you have been the victims of an advertising con trick and you should leave the product on the shelf. Have we moved on at all from the quack tonic salesman's pitch featured in western films such as *The Outlaw Josie Wales*?

One product that failed to make it through the claims process is glucosamine, promoted for its 'joint lubrication' properties and even recommended to me by a doctor and a physiotherapist.

Strange how the immunei bollocksii yogurt product rapidly filled the supermarket shelves and our fridges. Have we learnt something here? We want health, but only if it comes out of a bottle and doesn't require any effort on our part. The advent of the slimming pill is on the horizon. The active ingredients will bind with dietary fat and render it indigestible. Or you could eat less of it. I know where the smart money would place its bets (chemically induced bulimia without the need to vomit).

So we arrive at the paradox.

Health versus lifestyle, health initiatives versus lifestyle, health service versus lifestyle, take your pick. Are we serious about our health or not? If you are, enforce health initiatives If not, stop wasting my money on it.

Supermarkets are the merchants that have carved out the oligopolies of the grocery world, aided and abetted by the smart people in advertising who control the customer and reinforce the undoubted clout the supermarkets have, which they swing to control what the manufacturers can and cannot provide, namely the quality and price. I saw a newspaper report which claimed that Walmart had done more to control inflation in the United States in a decade than all the US government policies had achieved.

Supermarkets control access to the customer; advertisers control what the supermarkets want by creating demand; supermarkets control the manufacturer by dictating the price pressures; manufacturers control the advertisers; and the customer ends up with the product of this process, a cheap weekly shop, but at what cost?

Back in 1825 plant collectors introduced Japanese knotweed (*Fallopia japonica*) into Britain. This not unattractive plant has red stems and a vigorous growth habit. For some reason it does not produce seed in the UK and all reproduction is done vegetatively; that is to say, a physical part of the plant has to be transferred from site to site. Unbeknown to the introducers of this plant, the vegetative powers of knotweed are quite remarkable; a plant can develop from a piece of plant, stem, or root just 6 mm long (that's a quarter of an inch). But it gets better. Its roots can penetrate to a depth of 5 metres (16 feet), and mature plants can send these same roots horizontally up to 7 metres (23 feet). Not only that, but also these same roots have the ability to penetrate tarmac and concrete foundations and walls of buildings, resulting in extreme cases with houses being condemned as unsafe for habitation. As a final trick, even though you have 'killed' the knotweed with systemic weedkillers, it can remain dormant for up to four years, before resurrecting itself from its chemically induced slumber. It would be fair to say that there is effectively no safe exposure limit for Japanese knotweed. In the UK, the Environment Agency estimates that we spend £150,000,000 a year in trying to eradicate it. It is a shame we cannot recover the costs from those well-meaning individuals who introduced it into the UK. (And as a final irony, there is talk of introducing a Japanese aphid into the UK to deal with the knotweed, approved in March 2010.)

Since 2007 the 'consumer world' has suffered a major recession brought about by the villains of the piece, 'evil bankers' who shamelessly pursued short-term profit over substance. One of the principle triggers reported at the time was the banks' appetite for subprime mortgages, that is to say, mortgages provided to customers who had no realistic chance of repaying them. Closer examination revealed entire catalogues of globally intertwined bad deals that management simply couldn't get a handle on. Here in a nutshell is the consequence of working to a single target, in this case profit, but it could just as easily be exam pass rates, for example. And while I say 'consumer world', let's not forget the silent witnesses, so-called Third World countries striving to join the big guys' club whose fledgling economies took a hammer blow.

We have since learnt of PPI. (Payment Protection Insurance) Anyone with a pulse couldn't have not learnt of it because of the compensation culture circus that rings my (and probably yours) telephone, fixed and mobile, several times a month. Then of course there are all the television adverts and newspaper adverts including details of the time limit for claiming, which is years away.

Then there's the LIBOR scandal involving corruption in the currency markets. But don't worry, our financial rulers are on the case with billions of pounds in compensation cases and fines, some of which monies are siphoned off for charitable causes such as Afghanistan veterans.

I am sure there will be some I've missed, but I think you get the picture.

In 1903 scientists heated oil to high temperatures with a nickel catalyst and succeeded in changing liquid oil into a hard, greasy, grey lard-like substance while searching for a cheap substitute for candle wax. Universally known as trans fats or hydrogenated vegetable oils, these things soon found their way into food production. Food scientists discovered the substance could be used as a substitute for the much more expensive butter, and it is taste-free with a very long shelf life without the need for refrigeration. In the human body it acts like a Trojan horse. It is welcomed as a vegetable oil but behaves as a saturated fat and distorts cholesterol levels, encourages obesity, and has a major impact on heart disease risk. These products obviously created an advantage in cost, whether through easing the manufacturing process or by replacing more expensive ingredients.

In 2003 Denmark banned the material from the human food chain because of its danger to health. Denmark even went so far as to declare that there 'is no safe exposure limit'. Switzerland banned the product in April 2008. The UK opted for a 'voluntary arrangement for manufacturers to set their own house in order'. Just to prove that this issue matters, Denmark's heart disease rates have dropped by 40 per cent. It just goes to show that we are clever enough not to introduce Japanese knotweed into the food chain/garden in this day and age.

In recent years our nice, caring manufacturers (they certainly care about profit) have even boasted about 'contains no hydrogenated vegetable oils' (HVOs) as though they are doing us a favour: they have successfully 'extracted' it from their product, unable to imagine where it came from. Two products that spring to mind that contained HVOs are Mars Bars and Hobnob Biscuits, if my memory serves.

It's surprising in this compensation-driven world in which we live that no one has tried to recover the costs from those well-meaning individuals who introduced HVOs into our diets. Unlike the bankers who are facing all sorts of punitive fines and rule changes, no individual within the food industry seems to have been brought to book, unlike construction industry bosses who face 'corporate manslaughter' charges for failing to prevent fatal accidents to their workforce through health and safety legislature.

'So what did you do at work today, Daddy?'

'I forced through the use of HVOs to save our company money.'

'Why did you do that, Daddy?'

'Because I had to cut costs to keep my business with my largest single customer.'

So what is the cost of our cheap weekly shop: premature death, incapacity through stroke, and a lack of fitness and lifestyle choices due to hypertension? I am no doctor, but I'm willing to bet that a doctor could come up with a multiplicity of effects caused or compounded by plaque build-up in arteries.

Now I hear you cry, 'That's a pretty extreme case. Nobody knew HVOs were so injurious to health.'

Can you also claim that nobody knows about the injurious effects of extraneous salt, refined sugar, and fats in our processed food? Would you like to bet?

Would you also claim that nobody knows that processed foods, which are short of fibre, are less healthy than wholegrain or unprocessed foods?

Let us arrest the ringleaders and put them on trial. Let's use a venue on neutral ground, say, Nuremberg, and let's imagine what the pleas would be. I'm not talking about whether those who introduced HVOs are guilty or not; I'm talking about the reasoning and justification they will use:

- 'I was following orders.'
- 'You can't believe all the scaremongering; you'd never get anything done.'
- 'Everybody else was doing it.'
- 'The people knew what we were doing. They bought our goods. They didn't have to.'
- 'Nobody forced them to buy our goods. They can all decide for themselves.'
- 'We can't possibly react to these different fads and phases. One minute it's bad for you; the next minute it's not.'
- 'We do sell sugar-free foods.'

In reply to cross-examination, you might get: 'Well, of course if we reduce the sugar, we increase the fat and vice versa. You have to have flavour.'

Do you think I'm being extreme, using the example of the Nazi Nuremberg trials at the end of the Second World War? Estimates suggest that as many as 50,000,000 people perished during this conflict in all areas of the globe. What we should say, however, is not deaths but premature deaths—deaths that occurred as a direct consequence of the conflict, where without the conflict life would have continued. You could pick an entirely arbitrary figure out of the air and say that 50,000,000 lives were reduced on average by 25 years. That is 1,250,000,000 years of life removed from the population pool. While many consider that it all began in 1939, this dismisses the Chinese–Japanese conflict that began years earlier. So let's take an arbitrary start date of 1937 and make it an 8-year conflict, resulting in an average of 156,250,000 years of life per year of conflict. And as we've been quite free with our assumptions, let's round the figure to 150,000,000 per year. Not all the world 'enjoys' our superior diet, but taking North America and Europe together with richer sections of society throughout the world, it would be reasonable to suggest that 1,000,000,000 people do. So let's do the maths. If the average life expectancy of these 1,000,000,000 people was reduced by just 6 years and 8 months, then that equates to 150,000,000 years of life. Seems far-fetched? Predictions are already being made that this current generation is likely to see a reduction in its life expectancy. And this is against a background of health screening, inoculations, free at the point of use healthcare, and a reduction in the smoking population. But this 150,000,000 years of premature death is per generation, so we are effectively looking at one year of World War II being repeated every 20 years or so.

The latest large-scale analysis of smoking from Australia suggest that two-thirds of smokers will die from smoking-related diseases and have a ten-year reduced life expectancy. All of the previously described health initiatives will have had a beneficial impact on life expectancy (supposedly), so if life expectancy is actually going to drop, then these initiatives are obviously working against a headwind. Have we found the headwind?

Does Hitler or Stalin deserve a second chance?

This puts food suppliers in the same cart as the evil bankers ignoring the consequences of their actions in pursuit of short-term profit.

The obesity epidemic began in the sixties and has spread around the 'First World'. As a child at a large village junior school, I remember there was only one fat child in the entire school.

My parents used to do the weekly shop at the Tesco supermarket in the Coalville precinct, and later at the Hinckley Asda. I remember the Leicester Asda store opening at Fosse Park and the identical mirror-image opening at Coventry Walsgrave. Then Asda took over a store at Oadby and built another one at Thurmaston and one at Nuneaton. This whole process took over forty years. The point I am making is about the rise of the oligopoly that is the supermarket and the decline of the alternative independent stores with limited buying power. Reinforce the supermarket power with fast-food outlets and you have an irresistible stranglehold on the public's buying habits.

And you have an obesity epidemic?

But the authorities will protect us from harm, won't they? In 1904 the use of lead-based paint was linked to childhood lead poisoning. (White lead paint contained up to 50 per cent lead carbonate/lead sulphate to provide the white pigment.) Some countries in Europe (France, Belgium, and Austria) banned the use of interior lead-based paint in 1909. In 1922 the League of Nations attempted to ban the use of lead. Lead poisoning causes brain damage and is irreversible if exposure is maintained.

Lead poisoning has been suggested as a possible cause of the decline of the Roman Empire, as lead oxide was used to sterilise wine. This was also used to sterilise West Country cider. Lucky sufferers would travel to the spa at Bath, where records indicated that 'treatment' was successful. Patients were immersed for several hours up to the neck in the hot spring baths once or twice a week for months until 'cured'. It took NASA to discover the reason why centuries later.

NASA put people in warm water for long periods to simulate weightlessness but tested their subjects to see what other effects it had on the body. They discovered that after several hours of immersion, their subjects excreted lead out of their skin. As the Romans built the baths at Bath and elsewhere, had they inadvertently stumbled on the cure for their own affliction?

So when was lead-based paint banned in Britain? Try 1992, under European Union legislation, and even now it can be used for historical buildings and artwork as long as children have no access to it.

One of the reasons bat populations have suffered throughout Western Europe is down to pesticide treatment of structural timber in roof spaces. A common treatment was Lindane, which was banned in parts of the world many years ago. In tests, a bat was placed in a glass tank with a piece of Lindane-treated wood treated two years earlier. The bat, a mammal, died within minutes. You would sleep underneath Lindane-treated timber for your eight hours.

One of my industrial tenants produces timber roof trusses and has a treatment plant which is now rarely used. The manager told me that the main threat to roof timbers is a beetle whose larval stage bores through untreated wood, causing enough damage to weaken the roof structure. The beetle is confined to an area around Aldershot (Savernake Forest), and the males are flightless and unable to spread. The species is now endangered, and consideration is being given to removing treated timber roofs and replacing them with untreated wood on army houses under an EU directive.

As another thought, are we witnessing the decline of the First World Empire? As a society we are getting fatter with a reducing life expectancy, we are pressing the accelerator on the self-destruct gene, and we are being systematically dumbed down by our media. Where are our advantages working for us?

Suntan Lotion

If we lump suntan products into the 'cosmetics' basket, then we have gone full circle: encourage people to buy your products to get a healthy tan, realise the dangers of UV skin damage, and produce sunscreen as an antidote, but as a major bonus invent anti-wrinkle/anti-ageing products to disguise the effects of sun damage—this at a time when people naturally hid from the sun (although not Englishmen, who took mad dogs out for 'walkies' under the midday sun) and when pale was considered beautiful.

Another question I have yet to get an answer to is, how does beauty come out of a bottle? Walk around Boots et al. and you come to the beauty products division. It's normally near the serums. How does that work? 'Put this on your top lip and become beautiful'; 'paint this on your eyebrows and become beautiful.' OK, I understand that you could have a nose job, chin job, and cheek and lip implants, and have your crow's feet stretched to look disfigured, but how do you become beautiful out of a bottle? If the answer is disguise, then why not buy a mask and be done with it?

I have talked before about the 'non-jobs' that existed at the water and electricity boards and gave an example at the local coal mine. In all instances I laid the fault at management's door, a fault due to either simple greed or stupidity. What I haven't commented on was the involvement of the unions. The coal mine had one of the dominant unions in the UK, and under the aegis of president for life Arthur Scargill, the NUM (National Union of Mineworkers) had bought down the Conservative government of Ted Heath in the early 1970s through strike action, which had led to power rationing and the three-day week. This lack of coal at the power stations was also supported by the power generation unions, which had forced power cuts on the populace.

This caused immense damage to the economy. The realisation of the power within their grasp rushed to the unions' heads. With the 'genie of power' steadfastly refusing to be put back into the bottle, they forced more and more strikes on the UK workforce, culminating in the Winter of Discontent with uncollected household refuse piling up in the streets and the end of the Labour government of James Callaghan.

The public, similarly discontented, elected Margaret Thatcher's Conservative Party into power, and she set about loosening the union stranglehold and effectively broke the union power.

I used to work during the summer holidays and would eat lunch at the pub next door to the factory. During the Ford motor plant strike that lasted for several weeks (twelvish, and there was more than one), the landlady's son (a Ford worker) did casual work behind the bar to earn some money. He told me that ballots were passed by a show of hands and strike action was called even though a clear majority had voted against. When they finally returned to work, having won a few-pounds-a-week wage concession, he declared that he would need to work for twelve years for this extra money to cover his losses during the strike. TV news reports showed the picket line at a British Leyland car plant being interviewed. I remember the reporter commenting about the legal need for peaceful protest. The pickets stood behind a barricade covered with barbed wire. One of the protestors said, 'This is a peaceful picket line, and we are within the law.' Another added, 'Yeah, we're peaceful all right, but if anyone tries to break through …'

The sentence wasn't finished, but the men were all holding pickaxe handles, the apparent weapon of choice for peaceful picket lines.

During the national miners' strike under Thatcher, people died at the hands of strikers who dropped concrete blocks off road bridges and through car windscreens. It was against this background many years later that the mine management failed to control its non-jobs. Needless to say, if the management had been successful, the mine may have stayed open longer than it did. We will never know. The old argument went something along the lines of: 'We, the unions and the management, both want the same thing, the success of the company. What we are arguing about is the share of the reward.' There are still militant unions in existence today, and they tend to favour the older, traditional occupations such as railway operators and bus and tanker drivers. Union leadership tends en masse to try to control the Labour Party, with varying degrees of success. Today's union leaders can also normally be characterised by their inflated salaries, company cars, and even grace and favour houses. They certainly understand the argument about the 'share of the reward'.

Don't get me wrong, I'm not trying to be party political here. I am simply trying to show that other methods of 'control' have been attempted in the past.

In the unions we have an example of the good intent being altered to serve the benefit of the new masters, the union bigwigs themselves.

Somewhere around the year 2000, my company was approached by the bank, who needed to ascertain that we were real people and not some 'front' for criminal activity. All of us directors were requested to produce a valid passport to satisfy the new initiative designed to flush out criminal activity.

The bank representative duly arrived at our offices. I 'entertained' her while waiting for the other two signatories to arrive from around the factory site. She began to explain that she was looking for money launderers, drug smugglers, and arms dealers. After leaving the room, she soon returned, having made a photocopy of my passport. I asked her how many money launderers she had found. The reply was 'none'. The second signatory arrived, and she returned with a photocopy. I asked how many drug smugglers she had found, and the reply was 'none'. This was repeated after the third passport. I asked how many arms dealers she had found, to which she said 'none'. Having decided I was having a go at her, she said, 'Why do you keep asking me about money launderers and drug smugglers? This is a serious job, and I am personally responsible for vouching for you.'

I hoped that her employers were paying her enough for the responsibility, which she decided was another dig at her. I had to explain to her what personal liability meant. If she failed in her duties, she could be sacked by her employer, taken to court by the Crown Prosecution Service, and fined up to £20,000 or jailed. In a nutshell she could lose her wealth (wage and fine), health (worrying about her court case), and happiness (unless of course you want to be an unemployed poor convict). I noted with interest the new government initiative whereby all social workers, police, teachers, et al. can now be personally responsible for failing to take action to protect a child in their care. I personally would require a massive pay rise to accept this responsibility.

So back to the discussion about 'share of the reward'. Management is tasked daily with putting its health, wealth, and happiness on the line for any one of hundreds of actionable occurrences. Businesses are required to have a 'director of death' to step up to the plate in the event that one of their employees dies, for example as a result of a motor accident while engaged in company business. Responsibilities will include proving that the employee was competent to drive and adequately trained and had sufficient time to carry out the allocated workload.

Businesses are required to have a 'director for cash' who will establish a policy for receiving cash payments from customers so that the firm is not engaged in money laundering.

These are just two examples, without even considering such vast areas as health and safety.

Owner-directors also have their own personal assets committed in the enterprise. So how much reward would you want to warrant putting your head in the noose every day and night? All the average worker sees is the big car and expensive suits and none of the responsibility.

This brings me to taxation and the socialist clamour for giving to the poor and taking from the rich. The examples from the seventies give an indication of what happens when you let the unions have control. But I have another, more up-to-date example. Northern Rhodesia succumbed to black majority rule, fell into the clutches of the duly elected Robert Mugabe, and changed its name to Zimbabwe. Zimbabwe was an agricultural powerhouse within Africa and a major exporter of food products.

The trouble was, white colonialists owned all the farms, and that didn't sit well with the dispossessed blacks. You can understand perfectly why.

Various land acts went before parliament. The early result was the seizure of land from white owners. Now I am not saying it was wrong to redistribute the land. What I am saying is that the way it was done destroyed the very thing the Zimbabweans were trying to secure. The indigenous population seized the land but were not capable of making it work, which meant the subsequent failure of agriculture in Zimbabwe.

The country went from breadbasket to basket case in swift order, with widespread famine, and went from exporter to importer of food—food they could not afford to buy, having lost their export revenue.

If the Zimbabweans had retained white control of the land by keeping the white landowners on as well-paid managers, they could have maintained the production of wealth and then had a more equitable distribution of wealth, rather than the destruction of wealth.

So let's tax the rich out of existence and see what we end up with to support the poor (France has successfully driven its richest people from its borders. Why bother to invest there?)

Chickens are mass-produced in now carefully monitored-for-welfare sheds where densities are strictly controlled. Typically they exist for thirty-five to forty-two days from egg to death on super-rich diets that produce explosive growth rates.

In the 1970s the World Health Organisation (WHO) declared chicken meat to be an excellent food. Now the same organisation has changed its mind. These super-chickens beloved of supermarkets reach their dispatch weight of three to four pounds, but the meat contains one pint of oil.

Asda spit-roast and sell these ready-to-eat birds for £3 to £4. With an eye to global warming, they responsibly collect the spilt oil and convert this to diesel for the truck fleet. They boast about this as a virtue. Personally I would sooner eat chicken and not oil with chicken. The birds grow so fast that their skeletons don't even form properly, and they are so heavy that they can't even stand up properly. They used to appear on the supermarket shelf with hock burn, the hocks having been damaged by the ammonia in the bedding. I haven't looked for a while, but fresher bedding would remove the symptoms while leaving the skeletal problem. I call these chickens 'berries'. They look like chickens, but squash them and all the juice comes out.

Betty Didn't Buy Some Butter

The Adam Smith Institute acting as a watchdog announces tax-free day. This is the date in England when people stop working for the government and start working for themselves. The change in date is a very simple calculated measure of the current regime's spending policy. In 2013 and 2014, the date was towards the end of May. The calculation includes direct taxes such as income tax and National Insurance, as well as indirect taxes such as VAT (value-added tax) and corporation tax. In both years, which are broadly indicative of recent annual trends, we worked around 40 per cent of our time for the government. If you paid no taxes at all and presently worked a forty-hour week, then you would achieve the same net pay by working just twenty-four hours a week. The counterargument to this is that our government is working for us. So, if my calculations are correct, we work two days a week to give to the government so that they can give it back to us, minus the handling fee of course.

So what price is your weekly shop?

A 40 per cent reduction in Danish heart disease is probably greater than all the 'successes' achieved by all the health screening and other government initiatives combined, all of which we pay for, when all we needed to do was eat proper food with butter and no HVOs.

So here is an example of a totally misplaced and mistargeted approach to a problem not unlike the stamp duty fiasco—what a screw-up! Set up a charity to drive down heart disease and then raise large amounts of money including all the attached overhead and chief executives to pay. If the issue was to increase survival rates from heart attacks wouldn't it be better to reduce the number of heart attacks in the first place? How do the charities measure success? What was the original ambition of the British Heart Foundation, to halt the heart disease epidemic? The focus is entirely in the wrong direction, namely research the disease and how to cure it, whereas we could lobby intently for the food industry to stop killing us. There is even a Plan C, whereby you take personal responsibility for your own health and change your lifestyle and diet. This, incidentally, is also the most powerful form of lobbying that cannot be ignored. In its simplest form, any product left unsold on the shelf won't be reordered and the manufacturers must revise their recipes or go out of business.

It's nice to know that your charitable contributions to the British Heart Foundation are spent on such a vital task, correcting the impact of the British Food Industry. It's such a blessing that your weekly shop leaves you with spare change to give away to such good causes.

Treat the symptom, not the cause. Our lifestyle is killing us.

Voluntary arrangement? Look at the 30 mph debate. Effectively our governments for forty years have, to my knowledge, failed to solve the problem. What we have is a 'voluntary arrangement' whereby you volunteer to pay the fine if you get caught. (I'm not sure that the fatalities of speeding incidents would consider that they too had volunteered.)

Many years ago my elder brother was speeding through his estate when he remarked, 'Suppose I'll have to start slowing down soon.' In reply to my question why, he said, 'Because my kids will be playing out in the streets soon.' I got a grunt from him when I pointed out that everybody else's kids were already out in the street. I still get left behind by cars that speed through 30 mph zones only to turn in to a private drive. These will be the same people complaining about vehicles speeding through their estate/village/town.

Pareto Rule, or the Law of Diminishing Returns

Pareto states that by measuring result against effort and plotting it on a graph, you arrive at a stylised S-shaped curve with a gentle slope at the start and finish. In multiple examples covering many disciplines, the Pareto rule confirms that 80 per cent of the result is achieved with 20 per cent of the effort. Conversely, 80 per cent of the effort achieves the final 20 per cent of the result. If you accept that the 40 per cent reduction in heart disease in Denmark is the greater part of the result, then all the remaining activity will only achieve a further 8 per cent of return (20 per cent of the 40 per cent achieved to date). But this means 52 per cent is unaffected by research and new developments. But what figure is (*a*) acceptable and (*b*) achievable? Do we really expect to solve 100 per cent of heart disease? Or world poverty, or world hunger, or world disease, or armed conflict, or …? If we considered the question from a different perspective, how much information about healthy lifestyle do you need to continue to ignore while waiting for the pill or treatment to make you magically better? There is already a guaranteed solution to avoiding premature death as a result of smoking tobacco. It's called 'don't smoke'—either your own or somebody else's smoke.

Last year I attended two tennis tournaments, Queens and the ATP World Tour Finals at the O2. At Queens I watched with curiosity how people crammed in to restaurants and bars to 'watch' the tennis on the TV, when the real thing was happening yards away, and thought, *How ironic.*

The real eye opener was at the O2 Arena, where the audience is in the dark during match play. Here the blue glow or shiny reflections of hundreds of mobile telephone screens caught my eye. Numerous people devoted the majority of their attention to the taking of 'selfies' with the court in the background. These were then beamed around the universe to, no doubt, envious friends via Facebook or whatever, and the responses all dealt with in real time. It struck me that the event was only real because it was shared. I have seen tables of diners in restaurants who interact as much with the Internet as with each other, and the conclusion I have come to is that many live in a strange 'virtual world' where they seem to have difficulty in selecting what's real or not.

I have asked the question as to whether you could live in a caravan or not. This modern generation of tech-savvy networked individuals seems to have resolved this issue. Not for them do the confines of their accommodation apply. You can dine on your own surrounded by disembodied friends having inane conversations about God knows what using handheld gadgets that act as music systems, libraries, encyclopaedias, map atlases, game modules, and 'reception rooms' where you meet and greet your friends. You can pull up short video clips of grumpy cats and 'fails' and car crashes without the need to venture into the real world. It doesn't matter if it's real or not; it's on the screen and transporting you off to somewhere else, somewhere real.

Watch *The Life of Pi* and realise that you don't need real tigers anymore; CGI fills your reality. Thinking back to the birth of consumerism in Wales, the individuals, by working five days a week, began to turn their backs on the real world of experiencing the wind, rain, and birdsong as they laboured away underground. They would emerge from their shifts to trudge home in the dark, isolated from the natural world, only to return the next day to more of the same. What they needed was time to appreciate the natural world around them, time which their employment and desire for consumer goods denied them. Fast-forward to now, when we can experience the real on our TVs and computer monitors that never leave our sides, and be connected to what? *The Matrix* begins to look scarily real, doesn't it?

Think I'm going overboard? Commute to work in your air-conditioned cars with the radio DJ for company, or sit with your earphones in a steel tube with windows hurtling underground from station to station in the people mine, to surface briefly before plunging into your air-conditioned office or shop until the end of your shift, when you can retrace your steps, dreaming of your retirement to the countryside and the end of your own rat race when you realise that you don't know what's real or not and how disconnected from life you now feel. Watch one of the BBC's Natural History programmes and see how real they are. I've filmed in the Picos every photogenic scene, plant, insect, animal, and reptile I can find to sit still long enough. I refuse to sit in a hide for hours at a time, preferring to capture it all in the raw or not at all. The BBC Natural History unit fills it programmes with cross-season long time- lapse sequences, helicopter/drone aerial views, and studio-staged sequences. Then they boast about their achievements in the Diary section, explaining how they filmed the original spring woodland sequence in the wild before painstakingly rebuilding the scene in the studio, complete with potted plants, to capture a tracked time-lapse sequence. My questions are:

- Why did you bother? You might as well have used CGI.
- What interpretation of the term 'natural' are you using?
- By 'history', do you mean that real wildlife is now to be found in the history books?

It occurs to me that if you travelled to one of BBC Natural History's featured locations, say, the geysers in Yellowstone National Park, you would come away utterly disappointed by the reality of what you found on the ground. In fact you'd be much better off staying at home and watching it on the telly.

Still not convinced? Dip your toe into the Christmas advertising reality of perfume adverts. WTF?

While waltzing spectacularly against some city landscape magically devoid of people, Keira Knightly tempts us with the allure of body scent; 'Come buy this bottle and smell like me.' Which scent of KK is that? Sweaty after playing football in *Bend It Like Beckham*, or after six weeks aboard a ship without female bathing quarters in a Caribbean climate?

How about Charlize Theron doing a disappearing Indian rope trick into your attic space without as much as a snake charmer? What does that smell like? And she's thrown your pearls on the floor, the bitch. Doesn't she know how much they cost?

So you think I'm being unfair by targeting poor Keira? Let's consider in more detail:

I don't know her; I've never met her and probably never will; her 'presence' in my life is illusionary; she appears in films wearing costumes and pretending to be someone else, or at premieres wearing costumes supplied by Emperor Clothing Company pretending to be somebody else; she features in gossip magazines; and she is as real as the tiger in *The Life of Pi*, yet I am expected to fill my loved one's stocking with her featured scent because I want my wife to smell like her.

Ford Illusions

For those of you lucky enough not to live in caravans, we have the executive house.

I have looked at thousands of house plans while estimating structural components and always tried to form a view about the design. I've also looked at numerous houses while going through the process of house buying. What strikes me is how the house owner is sadly let down by the whole process. Architects and architectural designers by and large do not live in large houses and as a consequence do not understand them. You cannot bolt two Ford Escorts together and call it a limousine; equally, you cannot bolt two Ford Transits together and call it a limousine. (I'm not talking about the extra-long pimp mobiles of the Hollywood boulevards.) I visited Lord Litchfield's house Shugborough Hall in a small private tour group and stood in awe at the staterooms, the salon, the dining room, and the ballroom and thought, *How do you* live *in this?* The tour guide informed us that because we were a small group, we could have a peek at Lord Litchfield's personal living quarters, which we duly did. Then the penny dropped. You don't live in the staterooms; you live in a cosy homely subsection and roll out the big rooms for the big functions. So why are modern-day large houses full of huge rooms of non-human scale designed to impress visitors while at the same time being uncomfortable to live in (i.e. two Ford Transits bolted together).

The illusion of a big house.

Take a big news story of major importance, add a TV news programme, and feature it. There has been a tragedy: interview the neighbours/relatives/

school friends/head teacher/local councillor/police spokesman/MP. Learn from these near eyewitnesses that it has been tragic. Film some grieving relatives/friends/neighbours: see that it has been tragic.

Despatch some reporters to the home town/incident scene/police press conference. Interview some experts and add some in-depth analysis. Ask some searching questions and receive purely speculative answers about this tragedy.

Hire a helicopter to give an aerial view of the scene so you can display the latest sensational pictures of the tragedy. Sell some space, justify your existence and importance, prove what a good journalist/news channel you are.

Does this seem familiar?

I mentioned before that I don't buy newspapers, but I will happily browse through someone else's given the chance. Take a news story in the press that has run for several days and read it carefully, but do it backwards, that is to say, read the story from the latest to the earliest reports. As the story tracks nearer to the origin, you will normally notice that the 'facts of the case' continually contradict the previous 'facts of the case', until you arrive at the original story which bears little resemblance to the final story apart from the basic fact. Some will argue that this indicates investigative reporting at its best. But on what day was it at its best, day 1 or day 6? If day 6 contradicts days 1–5, who's to say day 7 won't be different again? In reality, where criminal cases are involved, the official truth (in the eyes of the judicial system) will not emerge until the verdict is in, and this may be several months after the event. So by what definition is the news, the news?

Let us take a purely hypothetical event, something that is not intended to be confused with any real occurrence, and consider the treatment that the story warrants.

One of Mrs Twiggywinkle's babies has failed to come home.

Tragedy Strikes Lower Field

Distressed mother beside herself with worry.

Police not ruling out foul play.

Friend of the family says, 'It's so unlike little Spikey to miss his bedtime; he is normally such a well-behaved kid.'

Local community leader asking for volunteers to search for brave Spikey.

'Spikey will be missed. He was such a pleasant hedgehog, such an attentive student.'

'I miss Spikey. We used to play together all the time.'

'This is the scene of Spikey's last-known sighting. Sorry, can you hear me over the noise of the helicopter?'

'Why did you feel the need to volunteer?'

'Well, you just do, don't you? That's what the communities all about.'

'Over to the studio now. Ron Weasel, of course we don't know what's happened to poor Spikey yet. How long could a young hedgehog survive on his own?'

'Well, that depends. If he's managed to find shelter and has got some food, he could survive for some considerable time, dependent on the weather of course.'

'Tell me, Ron, what mental state do you think he will be in by now? It's been thirty minutes since he was reported missing.'

'Well, I'm not a psychologist, you know, but as long as he stays calm, he should be all right.'

'Over to Fiona Fox in the field. Fiona, what impact will this have on the local community?'

'I've talked to the local community leaders, and they tell me that they are hoping for a positive outcome.'

'Fiona, what sort of community is Lower Field?'

'Well, it's your typical community really. Everybody knows everybody else. It's really tight-knit.'

'That's Fiona Fox there, reporting from Lower Field. We will of course bring you any updates as they arise later in the programme.'

'The day's other news now. I'm sorry, this just in. Unconfirmed reports are surfacing that Spikey, the missing hedgehog, has in fact be found alive and well after going missing while playing hide-and-seek.'

'We will bring you more analysis when we have it.'

Maybe this tragedy is a bit near to home; missing children cases are devastating in their impact. What I am trying to demonstrate is how utterly banal the whole media circus is, the whole performance full of utter trite based around the *fact*, in this case that Spikey the hedgehog hasn't been seen for thirty minutes.

Given that you have nothing to say, guess what, say nothing. But silence doesn't fill the airways, does it?

My offices used to be two doors down from Nigel Lawson, the previous chancellor under Margaret Thatcher. Yes, that's right, Nigella's dad. I left to go into town at lunch, and while driving back I noticed that the church had brought a crane in to remove the weathercock from the top of the high steeple. On arriving at the lane to our offices, I was gobsmacked to see the fleet of media vehicles which had responded to the rumour/tip-off that Nigel planned to resign. (I only had an hour for lunch, and nothing was there when I left.) The crane was nothing to do with the church; it was a transmitter dish on a stick boom raised higher than the top of the high steeple so as to catch the early evening news.

Just a thought, here's a news story you might want to run: BRITISH FOOD MANUFACTURERS KILLING CUSTOMERS BY THE TENS OF THOUSANDS. What, no appetite for it?

Or, HEALTH AND SAFETY IGNORING HEALTH.

Or, WELFARE STATE FAILING TO PROVIDE WELFARE.

But you can't bang on and on about these stories; the public has only got some much appetite for them. 'Push it too hard and they disengage and we can't get the message across.'

So what part of your endless trite dross do you consider we engage in now?

Consider smoking, such a robust example of persuading and educating people to cut down or stop the process of killing themselves. Very few smokers are alive today who didn't know it was dangerous when they took it up. But talk to the government or NHS and they'll tell you the antismoking campaign is a major success story. Latest figures suggest that only 20 per cent of the UK population still smokes. What's more, the number of younger smokers is dropping fastest. 'The message is finally getting across that smoking is bad for you. All of our initiatives have led up to this.' Go to any NHS building and see the 'No Smoking – This NHS site is a designated no smoking area. You may not smoke on this site' signs at the car park entrance. The message is normally reinforced at the building entrances, you know the ones, next to the smokers waiting to enter the building. If you are not going to enforce the ban, then why bother to put the signs up? If you're going to get nothing for it, do nothing. As for the reduction in young smokers, the theory is that they are too busy interacting/playing on their computers or on social media to smoke. Turns out we didn't need anti-smoking campaigns at all. What we needed was Facebook and Twitter. The same argument has been extended for the reduction in young people's binge drinking, another brilliant result of a long-term government education campaign.

So if we can't actually get news from our newsvendors, where can we get news from? Ever wondered why, remote control in hand, the same 'news stories' are presented on Sky, BBC, and ITV news channels virtually simultaneously? News editors across the land sit down together or have conference calls and decide what the 'news' for the day is. If they don't, who does? Somebody is coordinating it. If it's coordinated, then by definition it's controlled, and if it's controlled, what does that say about the freedom of the press? 'We have the freedom to report the same news items every day.' Where do we turn for honest, educated information free of trite, banal bollocks? Has anyone in the industry got half a brain cell? I understand you can't work for the BBC without a degree; they have a graduate-only policy. So what did their journalists graduate in?

The answer to this conundrum probably lies in standardisation; time is of the essence in more ways than one. In any line of work where repetition is involved, you devise a standard, a pro forma, a standard method of working, so that you do not have to invent something new every time you are faced with the same task. So you end up with an enquiry form, a quotation form, a complaint form, or something similar that solves the greater part of each task, with maybe a separate box entitled 'Other'. Fill in the standard form and you have the bare bones of the process sorted, swiftly and efficiently as well as boring and repetitiously. Business likes that, boring and repetitious; it makes things cheap and transferable between offices and locations. The head office can see what the regional office has done and vice versa. It works. Transfer this to the television news and you have a format that satisfies your time pressures. Produce it in time and produce enough time, that is 'fill the space'. Of course one of the issues in recent years has been the invention of the twenty-four-hour news channel, which presents the programme makers with an awful lot of space to fill.

So we have now established that there is a pro forma for each news event which triggers standard responses and standard news items. The paradox: how can it be new *and* standard?

In my youth I used to keep and breed budgerigars. I avidly read the weekly publication *Cage and Aviary Birds* for tips and information. In every edition they would feature an article centred on a champion breeder and his or her particular methods of choosing and breeding stock together, with advice on preparing and staging the birds at shows. Feeding regimes and nesting times and exercise allowances were also thrown in. The point was, eventually I had read virtually every possible permutation for keeping budgerigars. The publication ceased to hold any interest for me, so I cancelled my subscription. There was nothing new for my money.

So we have a standard method of reporting (I'm not even sure if that is the right term to use) an event, which churns out nothing new (i.e. no news).

There are un/confirmed reports of a train/plane/multiple vehicle crash/incident at _____, un/confirmed reports of casualties, some of which are believed to be non/life-threatening/fatal, with emergency services in attendance/trying to reach the scene. Numbers are/are not known at this stage. We will keep you informed when we have more. It is believed that Brits are/are not involved, but consular assistance has/has not been offered.

Reports are coming in of the discovery of a body/two/three/four body/ies, discovered at a house/car fire/explosion believed to be deliberate/accidental/suspicious/not suspicious.

Police are/are not looking for anyone else in connection with the incident.

Then for the creative bit, the padding: Our reporters are at/near the scene/victim's house/hospital/school/train station/airport where the journey began/was due to end.

We have managed to interview the victim's/killer's mother/father/son/daughter/aunt/neighbour/friend/pet cat.

We have a travel/military/former police/psychologist/criminal profiler/aviation/train/terrorism consultant who was the nearest available to a vacant studio and the best/cheapest we could get given the time. The expert made sense the last time we had him/her on hand to ask for his/her in-depth analysis. (Let's hope the person is awake and has been briefed.)

Newspapers suffer from the same time pressures with printing and distribution deadlines and a requirement to fill the space. You wouldn't pay the usual money for a paper only half as thick, would you? So the choice is, print some basic facts and pad the rest with supposition, or print what you know and leave half the pages blank?

But of course there is always the Internet, where stories can be updated when they are ready and time pressures are lessened. But are they? It's no good producing an in-depth report a week after the event when everyone (i.e. your customers) has moved on to the next bit of tittle-tattle.

And for good measure, recycle the 'news' and report on last night's *Coronation Street* or *EastEnders*. Flesh it out with stories of personal tragedy of an actress's aunt/sister/childhood friend/marriage/parent/friend, with the flimsiest of links to the TV programme to hook viewers in. The *Huffington Post* regularly relaunches old stories by repeating them weeks after the original, just in case you have forgotten/didn't read them originally/needed to be reminded/thought the news content was a bit thin. (They even rerun three-week-old weather warnings, for fuck's sake.)

One conclusion we could draw is that we have too many news outlets chasing too few stories with too much time to fill while competing 'tooth and claw' with each other, while often copying the competitors. A bit of a climax forest situation, really, where the forest giants crowd out the shrubs and seedlings below while seeking to spread sideways into their neighbour's canopy.

But the whole operation can be argued to be suffering from one fundamental issue: momentum. How do you kill an elephant?

So what's the conclusion then? And why does it matter?

We are killing the planet, the air we breathe, our neighbours (beasties), and ourselves in the endless pursuit of 'making a living', essentially off the backs of everyone else.

Why? Destroy the illusion of self-importance we engender in our everyday tasks and consider the reality.

How many advertising executives, food scientists, government departments, healthcare professionals, journalists, shop workers, charities, and research workers (if you think about it, the list is endless) do we need?

How many non-jobs/non-occupations are there, all paid for out of your and my pocket? We even pay those bastard nuisance telephone operators to call us. If they didn't earn from it, they wouldn't do it.

'But we'd have mass unemployment,' I hear you cry. No we wouldn't. Instead of filling up the newspaper with useless padding, print fewer pages. Instead of working forty hours a week, work twenty. Use the money you earn to feed, clothe, and entertain yourselves without supporting all the bullshit.

Fancy a Tipple, or Twelve?

Here's a thought: make it illegal to serve drunk people in public houses. That way people can only get so drunk in public and therefore will be able to get home before collapsing (on their beds instead of out in public). I know there's a problem with this, namely the time difference between drinking and the drug being absorbed into the bloodstream.

This time lag presents a challenge to controlling the problem and always gives an excuse for the liquor industry as to why they can't be held responsible. There is one flaw in this argument: it is already illegal to serve a drunk in a licenced premises, and major penalties exist in terms of fines and lost licences.

There is a hard core of binge drinkers who start the night by 'pre-charging', that is drinking cheap supermarket alcohol, before going out to enjoy themselves at the more expensive public venues.

Gross Domestic Product

Governments often talk about the UK's productivity and its impact on the UK economy. The present economic recovery is still considered unbalanced. One of the reasons given is that productivity per worker is below what it was pre-recession. This will have a direct impact on UK plc's competitiveness when measured against other countries. So to ask the question then: how do you measure productivity?

I have made a previous comment about the problems of measuring and the issues of focussing on single measures: think bankers and short-term profit, or exam grades and grade inflation for example.

So how shall we measure it then, productivity? Let's take the gross domestic product and divide it by the number of workers and arrive at an average value per head. This will do.

Let's consider all the 30 mph initiatives that have failed.

Let's consider the anti-smoking initiatives that have failed.

Let's consider all the un-news journalists who fail.

Let's consider all the advertising industries that fail.

Let's consider all the fit4life initiatives that fail.

Let's consider all the charities that fail.

Let's consider all the telephone marketing campaigns that fail.

Let's consider all the planning policies that fail.

Let's consider all the school initiatives that fail.

Let's consider the welfare state that has failed.

Quantify the failure of all of these initiatives, and activities, and all the time, effort and resource devoted to them, and then consider what the true productivity of UK plc is.

Let's consider what the true cost benefit of all these activities is.

The *Matrix* and *Terminator* films have one basic central plot line, the domination of humankind by machines. These are machines that we invented and failed to fit a stop switch to. These are machines that have then evolved to control, enslave, and kill us to various extents. Scary stuff.

Where is the stop switch for the motor car?

I don't mean the ignition key or engine start code; I mean the stop switch, the one that stops the vehicle from controlling, enslaving, and killing us.

'What are you talking about?' I hear you ask.

Let's start with killing. I don't mean crashes and collisions; I mean pollution, carbon monoxide, and nitrous oxide, all of which cause asthma and premature death.

Enslaving? Cars are not cheap to run, and you spend a significant proportion of your working week earning money to pay for your car. You are a part-time slave to your car.

Controlling? 'I drive; therefore, I will. I can commute twenty miles each way to work, so I will.' The commute adds an hour or more each day to your workload. No, it takes one hour each day from your life.

In medieval times, markets could only be held with the consent of a royal charter, and these were jealously guarded. One of the principle limits on markets was the fact that no two markets could be within six and two-thirds of a mile from each other. It was reasoned that you could walk twenty miles a day, so this allowed you a third of the day to walk to market, a third of the day to shop, and a third of the day to walk home. Your market day was regulated in those days. Compare that with the freedom of the motor car (ignoring the killing, enslaving, and controlling, of course. Haven't we moved on so very far?).

I sat in my lounge, which has a view to the rear edge of my front lawn as it merges into woodland, and during the course of around two hours I noted the presence of several different species. During the two hours, occupying the same small area, I saw three fox cubs, two adult rabbits, three magpies, a muntjac deer, and an adult fox. This struck me as odd, because I would have thought that the rabbits and muntjac would have detected the scent of the foxes and run away. I also noted that my pony paddock to the rear of the house was also well populated with rabbits, even though a buzzard was nesting only 15 metres from the edge of the small paddock. It occurred to me that prey animals are afraid of nothing in particular (foxes, buzzards, and so forth) but rather everything. Their nervous disposition leads them to run from everything that poses a threat, but not the threat of a threat, as if they do not have the intelligence to process this idea. The likelihood is that a rabbit only realises that a fox is dangerous immediately before becoming its food source.

When I was a child my elder brother was deeply interested in becoming a gamekeeper. He and I spent many hours roaming the local fields and woods looking for 'prey'. The logic at the time, I don't know if it still is, was that birds couldn't count and the way to encourage them within range of a hide was to use a simple subterfuge whereby two would enter a hide and then one would leave.

Birds would watch the person leaving and then relax, unaware of the second person in the hide.

I have also observed small wall lizards in Spain and have noted their inability to cope with a threat. If you approach a sunny wall, the lizards will disappear to cover before carefully re-emerging, appearing to be nervous. If a second threat appears, such as a car driving by, they disappear back into the wall's crevices, only to return as soon as the car noise has faded away. Now, however, the lizards are relaxed in your presence. It's almost as though the second threat has erased the memory of the first threat, which has ceased to exist. You can move and film the lizards within limits without upsetting them and can even tease them with long grass stems without them reacting.

Humans, of course, have a much more powerful mind and would learn where threats come from and how to avoid them. We wouldn't knowingly set up camp next to a lion pride or a known grizzly bear haunt, because our learnt knowledge would seek to protect us from harm.

My reasoning about animals has led me to conclude that they are often running on an autopilot basis: don't think; just react. A builder was relaying a roof for me and explained that he was going to an African game ranch on a hunting safari. But he wouldn't be using a gun; he'd be using a bow and arrow. He told me that you had to be within close range, but if you loosed an arrow at a Thompson's gazelle while it was watching you, you would miss. Thompson's gazelles react so quickly that they will have moved away before the arrow has reached its aiming point. This spoke volumes about the evolutionary arms race between the hunter and the hunted. It also suggested something else: the gazelles react without conscious thought, a reaction I will call a tropism. Now my schoolboy biology taught me that tropisms are automatic responses to external stimuli that require no thought. So, for example, plants grow towards the sun, while their roots grow down—that type of thing.

Having had these thoughts, I was out driving down a single-track lane approaching another vehicle heading towards me on a two-lane section, when the realisation dawned on the other driver that I wasn't getting out of her way. She stopped her car beyond the two-lane section and had to pull onto the verge—a full car length onto the single carriageway. Why, visibility was excellent and the road was straight, and yet her thought process in dealing with an approaching car *and* a narrowing roadway seemed to be too much, one threat more than could be coped with. I began to think that people also use tropisms and simply react to the problem instead of anticipating it.

Sometimes the driver of a car will overtake a large puddle by crossing into your carriageway or steer round an overhanging branch directly in front of you. The driver would sooner have a head-on collision than run through water or scratch his or her paint.

Let's expand beyond driving and consider humankind as a species:

- We react to an external stimulus without thinking.
- We are unable to cope with more than one threat at a time.
- We are not the logical-thought-processing superior animal we think we are.
- We follow very basic patterns of action.

These things are deeply embedded within our DNA.

If we understand this concept, then we can arrive at a number of conclusions, as follows:

- We can be easily led.

Let's have a call to arms. Why did the Second World War follow so soon after the First World War—remember this one, 'the war to end all wars'? Let's have a riot, be it in London or Los Angeles. Let's go on strike. Let's all gang up and bully a certain individual. Let's all drive at 85 mph on the M1. Let's select the gay/Asian/white/black/disabled/Eastern European anomaly for 'differential treatment'.

Conversely, let's all brush our teeth twice a day. Let's all drive at 30 mph in built-up areas. Let's all eat sensibly and stay thin. Let's all follow government health initiatives.

- We can be easily controlled.
  Let's go and buy the latest fashions. Let's lust after the latest telephone/tablet/gadget. Let's all eat junk food. Let's all watch the latest blockbuster film.

We can be considered a herd animal, but herds are not without leaders: stallions in zebra groups, matriarchs in elephant groups. The mega herd is made up of lots of these small groups that all appear to have a common aim and thus appear to be moving as one mass. In reality, each group is moving, taking hints from its neighbours, but essentially acting within a small group. They collectively respond to external stimuli and collectively move en masse, but as individuals or small groups. The mega 'herd' in this sense is an illusion. The herd doesn't move en masse; individual members simply remain within a few yards of their neighbours, and in this way the whole herd moves as though under the control of only one brain, when in reality such is not the case. Think about a dense shoal of fish or flock of starlings and realise that those on the inside cannot even see where they are going. They just follow, instantly.

So what conclusions can we reach?

It's thought that 'the herds' can detect the presence of rain and the promise of fresh grass from many miles away. Elephants have recently been shown to react to ultra-low-frequency sounds produced in nature by thunderstorms and can possibly hear/detect thunder, perhaps one hundred and fifty miles away. Imagine being in London and hearing the thunder in Sheffield! In times of drought, elephant herds will travel this distance for the promise of fresh vegetation and water.

This behaviour is unlikely to be instinctive and must be learned from the elders within the group. We can see the benefits of peer pressure in modifying the behaviour of the herd, but given the absence of a language between elephants that we can understand, I would assume that the family group simply follows its leader, with subordinates ready to assume control if the leader were to die.

How does the whole human herd move?

Let us suggest that we are like the starlings: those on the outside see the threat or promise (i.e. a sparrowhawk or roosting site) and move away or towards it, while those on the inside simply react. But continuing with the starlings, there are additional complications to the model (there always are): birds on the outside are exposed to the risk; birds on the inside gain all the benefits without having any of the risk. Dominant birds sleep deeper into the roost and are less likely to be picked off by owls and other predators. I don't know if anyone has established whether dominant birds fly in the centre or on the outside. Tracking an individual bird would be an interesting task. (See if you can get a grant?) Dominant birds have been shown to feed in the middle of the smaller feeding flocks, and dominant male antelope hold their rutting territories in the heart of the breeding area, away from predators. Starlings don't spend all their time in these dense winter flocks; the roost splits into smaller groups that spread far and wide to feed, and then return to the congregation towards dusk, some having travelled for twenty miles. Starlings therefore obviously have two strategies for survival (call them herding patterns. Sir David Attenborough will love that, a herd of starlings) for winter, and then other patterns will take over for the breeding season.

As a complete aside, I have always found one or two intact starling eggs on the ground during each breeding season. As kids, we used to think that the egg-laying birds had been 'caught short' and failed to reach the nest in time. Researchers conducting a major study of starlings nesting on the salt marshes of Holland were amazed to find researcher-marked eggs in the wrong nest boxes. The birds are sneaky enough to transfer their own eggs into other starling nests so that those other birds can raise the chicks instead.

So if we are to truly influence the human herd, we have to establish the pattern of movement, or rather the patterns. If crowds can spontaneously riot, why can't we spontaneously brush our teeth?

Jamie Oliver has launched various heartfelt campaigns at the grass-roots level to change people's attitudes towards food and cooking both at home and abroad. Jamie's approach is one of targeted education: get the school cook on board and you are pushing at an open door, then generate some momentum as the head cook pulls the junior cooks along. Oliver has also tried with small nuclei of parents in towns, attempting to work through their neighbours and friends. Now that the TV cameras have gone, we can but guess as to whether he has succeeded or not, but it's fair to say that no national revolution has taken place. If Jamie Oliver could tap in to the right herd instinct, his job would be done.

How do you move the entire herd?

How does the whole herd move? The leading edge is followed by the main body, going where others have already gone, and then comes the rearguard, the stragglers, the old and young. Leaders take the risk and blaze the trail. The main body has the benefit of a well-trodden path but the downside of reduced grazing, while the rearguard has a reduced diet but travels through sated predators. (Alternatively, the leaders see which way the herd is going and decide to be at the front, making them the leaders, but this means there are no leaders.) The real stragglers are most vulnerable with reduced food and lost protection of the herd. Within the lifetime of the individuals, the position they occupy will vary. They are at the rear when very young, at the front while at their prime, and in the main body when compromised by pregnancy or breeding exertion (males) or middle age. The old, the ill, and the injured will slip to the rear.

This assumes an almost antlike existence without any real control or cognitive thought process, whereas we think we know that elephant herds are led by a cognitive leader, an experienced matriarch.

So where does that leave us? Are we antlike or cognitive? My thoughts are that we are both, displaying both tropismic and cognitive reactions to circumstances and reacting accordingly. This way we follow the herd by following the car through a built-up area at whatever speed it is doing, but conversely we decide not to brush our teeth twice a day. We munch through junk food with our neighbours but ignore campaigns about healthy eating. We play the fool in class and ignore the need to get a 'good education', whatever that means. So why not dispose of the cognitive advertising campaigns and initiatives and design tropismic alternatives that might just work?

Or alternatively, is it that which all the dross is targeted at?

It has become commonplace in my experience to see small A3-sized advertising posters above urinals in public conveniences. The ones at Fosse Park are all information- and 'good cause'–oriented. They are changed every couple of weeks or so and have featured appeals for Syria, clean water for Africa, the rescue of overworked donkeys in poor countries, that you check your bowel habits for cancer and talk to your doctor, and the like. The one that really infuriated me for its abject stupidity was the one calling for campaign support to make cancer an election issue. Genius, absolute fucking genius. Apparently cancer will soon affect 50 per cent of us at some time in our lives, and we should introduce a campaign group to force this issue to become a political football, requiring pledges from political parties to spend, spend, spend on cancer.

So, given the fact that there are finite resources with which to fund the NHS, extra money for cancer prevention and treatment removes money for everything else. So, you fuckwits, what do you want to see shelved so that cancer treatment can become top priority? Dementia, diabetes, heart disease, multiple sclerosis, motor neurone disease, asthma, vaccinations, and so forth must all take a back seat to cancer treatment. Boast of your success at money captured for your project, but all the others, which will suffer, become silent witnesses.

A growing number of the world's super-rich are now pledging their immense fortunes to charity or so-called philanthropic projects. The Bill and Melinda Gates Foundation, Warren Buffett, and others are pledging billions. Perhaps these ultra-rich are feeling guilty because of their success and want to redress the balance by helping others. Why are the super-rich, super-rich? A lot of it is luck, being in the right place at the right time. Most often the wealth has been generated from being in at the start of soon-to-be world dominant market players, Microsoft, Apple, and Facebook for example. These companies have generated wealth through the single-minded pursuit of their ambitions and by being successful. Theoretically, each of these companies could have been overtaken by any one of a dozen players in their early days (the also-rans), but through luck, hard work, innovation, and dedication, they succeeded as others fell by the wayside. One could argue that their ideas and managers were superior to those of the ones that failed. (Success also breeds momentum, which means that these companies new ideas rarely suffer from the constraints of underfunding or the need for organic growth.)

So having amassed this huge amount of wealth by being successful, you give it away to charity.

But these charities would have no need for money if they were in themselves successful. Their management puts them firmly in the category of the also-rans that never saw the winner's enclosure. So why reward failure for failing to thrive? If charities are going to work properly, the management structure needs to be as successful as the donors' businesses. In simple terms, think about all the lottery winners who have been 'set up for life' only to find themselves poor a few years later because they mismanaged their own funds—designer lifestyles and all the trimmings.

But if you stop and think, the UK lottery has five winners: the shopkeepers who sold the ticket; Camelot, which runs the lottery; the government, which takes a tax slice; the punters who win; and of course the lottery-funded grants and charitable recipients. So who controls the last group, the lottery grant winners, who suddenly find themselves in possession of large windfalls of cash whereby they can achieve their own dreams? Who audits the grant spending and monitors the management? If the cause were worthy and the management adequate, then the additional funds would not be needed. Don't get me wrong, anybody receiving a massive money injection will find a use for it, but that's the problem, 'finding a use'. Don't forget, need is endless. It's a bottomless pit. The more you have, the more you want, and your desire for more is very easy to justify. The next time you hear a lottery draw summary of all the billions raised for 'good causes', think about what that actually means, and think about all the penniless lottery and pools winners of years gone by. Ask yourself what makes charity recipients any better at managing windfalls than now destitute lottery winners.

Let's think about the charities a little more. The Bill and Melinda Gates Foundation has announced the aim of ending the scourge of malaria in Africa. A worthwhile pursuit? Research has been funded for the development of new drugs; staffs have been trained; and insecticide-impregnated nets have been issued to communities in need.

Take a sweet pea seed and grow it. However long you look at it, it's won't grow unless it has water—not too much, but just the right amount. The seed still won't grow without an appropriate temperature, neither too cold nor too hot. Now that your seed has germinated, it will soon use up its food reserves without light to power photosynthesis and generate energy for life and growth. So it needs access to light and an increasing amount of water, the latter of which it secures through its developing root system. It also needs minerals and nutrients from its support structure, in this case soil (if only the Carthaginians had persisted with hydroponics) which has the right pH for the plant and the right texture. Too hard and the roots will struggle; too light and the plant may struggle for anchorage. The plant needs protection from predators and extremes of weather, wind, and heat, and it needs these elements to be in an optimum range to produce a strong healthy plant that can resist disease, moulds, and fungi. It also needs support. The plant is a trailing climber equipped with grasping tendrils that search out other vegetation for support, which again needs to be not too vigorous in its own right for fear of swamping the sweet pea. Then of course you need an insect pollinator to ensure the cross-fertilisation of the flowers, failing which the plant can secure the next generation by self-pollination.

So here in simplistic terms are the requirements for growing a sweet pea plant. If any one of the requirements fails, at best the plant is compromised. At worst it dies. So let's deal with one element of the sweet pea in isolation and treat the plant with a drug to prevent pea blight, a well-known disease that devastates sweet peas that I have just made up. Will this treatment guarantee the success of the plant? The simple answer is no. All the other requirements for space, light, warmth, support, minerals, water, and so forth remain.

So how does removing the scourge of malaria solve the problem in Africa? The people there still need, warmth, food, space, shelter, water, and protection from fungi, parasites, and moulds, and other diseases. So let's assume that the Gates Foundation has got its head screwed on and is providing more support than a malaria 'cure' alone by providing access to clean water, comprehensive health, and education to boost local agriculture and the like. But what are we actually achieving? The Gates Foundation can now be described as the best-funded 'gardening operation' in Africa, where the crop is people, not sweet peas. But while we are creating the perfect garden for people to thrive in, we are also by definition removing the weeds and all other factors that inhibit our garden crop from growing. So what have we achieved in this Garden of Eden that is perfect for humankind to survive in, other than the destruction or marginalisation of everything else that used to survive there? Is the Gates Foundation's legacy to be one of curing the world of malaria or being the single biggest force in the destruction of natural habitat in Africa, since the influence of the white man began centuries before? Billions of dollars to provide a flower bed over a rose garden. Haven't we learnt so much? At what stage will the modern Adam and Eve bite the apple and lose their innocence and realise that the world they occupy has changed irrevocably and is all the poorer for it? Does this make Mr and Mrs Gates God?

Here's an interesting example of how a social issue has gone from pariah to acceptance. For decades homosexuality has been outlawed in Britain. We can all think of notable examples of persecuted individuals, such as Oscar Wilde and Alan Turing, who really suffered the consequences. In just a few short years, the entire population has seen a volte-face in attitude and acceptance. It now seems inconceivable to many younger individuals that homosexuality was once a reviled crime.

Rich or poor, well-educated or poorly educated. Why does it matter? Take it to its extreme conclusion and you end up with such a division between the haves and the have-nots.

Once again, Hollywood has beaten me to it. Watch *Elysium* (if you can), which depicts a utopian world free of disease and the poor, suspended in orbit above the earth populated by human detritus. (Interestingly enough, the name Elysium is from the Elysian Fields, an ancient Greek vision of the afterlife. By analogy, its occupants haven't died but have still gone to heaven—but, as ever, only those who really deserve it.) Socialists seek to relevel the equilibrium by stripping money from the rich and returning it to the poor. As fast as they strip the rich, the rich seek out ever new ways of regenerating their wealth or hiding it. But the equilibrium will maintain itself simply because the desire to gain an advantage is paramount within our species. The very rich will build their 'Elysiums' on earth, and the harder the drive of socialists to remove their wealth, the greater their desire to protect it. At the other end of the scale, the recipients of this stripped wealth will grasp it readily as their means of gaining advantage. The desired 'equality' will never be achieved. It is not within our nature to accept such a concept. Indeed socialism is flawed in its basic premise. Taken to its ultimate conclusion, everybody becomes universally rich, or should that be universally poor. Our species simply cannot function with that concept. If life is based around the premise of survival of the fittest, then the mechanism that provides for universal equality disappeared once life itself was created. If survival of the fittest had not come in to play, then life itself would have stayed at the primeval soup stage and evolution would have failed to exist. Inequality is a function of life; get over it. We arrive at an interesting concept whereby our 'socialist principles' seem to be in direct competition with nature, not just human nature but all nature. I suppose the idea is that we will use our intelligence to countermand our base instincts, a process akin to swimming up a waterfall, Niagara perhaps. There may be hope for us yet.

Think about the pensions debate which came into focus in the seventies under Margaret Thatcher: as the baby boomer generation reaches pensionable age, there will not be enough younger workers paying taxes to support the pension burden. If you want to have a financially comfortable old age, then you should make provision for your own pension and should regard the state pension as fit for subsistence only.

In my own working lifetime, I have seen both the old stagers retiring from our company with no pension provision of their own and others leaving years later with final salary schemes. The message has largely gotten across, but still the government has seen the need to introduce 'compulsory' schemes whereby workers have to positively opt out rather than opt in.

Having seen moves towards a solution for this 'ticking time bomb' of unfunded pension demands, we have developed another problem of the modern age, an epidemic of obesity. Now we will have an imbalance between the taxpayers and the unhealthy tax-takers who are unfit to work but still demand resources. Think I'm being too harsh? 'Human rights' dictate that obesity can be treated as a disability. Sit in any hospital waiting room, as I have done over the years, and notice the advent of the extra-wide chairs that never existed years ago. How to solve this ticking time bomb? Why does it matter anyway? Politicians of every hue occasionally spout, 'We are a rich nation and can afford it.' Ask that question in Greece and see what reaction you get. Hospital care is severely compromised, pharmacies are closing down because the government will not pay them for the drugs they have supplied, and trained medical staffs are leaving for employment elsewhere in the EU. Why? We can ignore the arguments about the Euro and Greek failure to achieve financial parity. The simplest and most obvious cause is the failure of the Greek authorities to collect taxes, reinforced by the failure to eliminate the widespread fraud of state-funded workers not working. (Whereas we have too many state-funded workers working to achieve nothing at all. Fancy a sleeping policeman or twenty thousand pointless humps in the road. Just a guess.) Using the simplest possible description, there is more money going out than coming in. So how is our model any different? 'We are a rich country and we can afford it.' Please explain to me how being a trillion pounds in debt makes us a rich country?

Please also explain to me why I should pay my taxes into the pot to support those among us who self-engineer their own exclusion from the work rota?

But we have all sorts of initiatives in place to counteract obesity, don't we? Do they work? Do they, bollocks. It would be bad enough if obesity simply prevented workers from working, but it doesn't. Obesity fills the diabetes, heart, and cancer clinics, to name a few, but these people are now protected under the umbrella of disability, so I am spouting heresy. Or is it common sense?

I realise that large swathes of the population will be feeling left out, so let's strike out the mention of obesity for fear of my being called 'fattist' (like a racist, but with more food) and substitute illegal drug use or binge drinking or smoking.

So, dear government, when are you going to get off your smug arses and do something positive about the obesity problem? When are you profit-junkie food manufacturers going to start supplying healthy foods only to the profit-junkie supermarkets? When are you peddlers of killer junk fast food going to change your recipes and provide real healthy food? When are you journalists going to put down your nanny state commentaries? When are you waste-of-space advertising/marketing executives going to promote something worthwhile and really beneficial for a change? And if you stop and think about it for just one minute, are you happy to know that all the taxes generated by your activities are going into the pot to alleviate the symptoms generated by your activities? I ask the question again: how do you kill an elephant?

Back to Greece for a moment, which, as I write, in April 2015, is seeking to renegotiate the terms of its bailout from the EU (again). The Greek people got themselves into this mess through decades of misspending and tax avoidance, so how is it my problem to bail them out? If the Greek people themselves won't sort themselves out, why should I? The Greek people have enjoyed the 'fruits' of their corruption for years, but now that the bill has arrived, they don't want to pay it. What is the difference between a destitute former lottery winner and a poor Greek unemployed ex–civil servant who never went to work even though he or she took the salary? The lottery winner is treated with scorn for 'throwing away a golden opportunity', and the Greek is bailed out. Greeks have a major issue with the dominant Germans and their resistance to bailing them out. Germans have a major issue with the Greek prodigal son who has thrown it all away and wants some more. But there is history here. The Germans have had to tighten their belts and work hard to assimilate the former Eastern Germany after reunification. The Germans didn't look for help elsewhere; they sucked it up, knuckled down, and got on with it. What does this tell us about the herd instinct in Greece and Germany? More importantly, what does it tell us about the political agendas of the EU, where failure is not an option?

At Last a Use for Elephants, Nearly.

I was Christmas shopping last year and came across an 'ethical shop' in a Leicester arcade. One of the ethical items offered for sale was small bags of elly poo—yes, really, dried dung for use as a garden fertiliser. Obviously I had to ask. The answer came back: 'The income from the sale of this product places a value on the wild elephants bordering the villagers farms and encourages them to respect and nurture these creatures.'

For those of you who don't know, elephant dung supports a whole ecology of its own, from salt-seeking butterflies through to dung beetles and the birds that feed off them. So, my little ethical genius, how does bagging up elephant dung and shipping it halfway round the world benefit the local wildlife, elephants excluded? Another customer for the white stick shop, I presume. Do you want petunias or French marigolds in your garden?

So why the negativity? Where's my plan for action? I haven't got one, but I have tried to prove throughout this text that no one else has either. So what's the solution?

Various theories have been expounded about the demise of the Roman Empire. It's too far back for us to know for sure. In a peculiar way I am going to suggest that the more successful and sure of itself the Roman Empire became, the nearer to collapse it moved. The more control it exerted, the less it actually achieved. Whatever measure of success we attribute to ourselves is false and corrupt, not in a criminal sense, but inaccurate to downright wrong. When the Roman Empire was built, it was built, not designed, debated, organised, 'committeefied', or measured.

It just happened. Now I know that you will all be shouting at the book in your hands, declaring that the Romans were designers, debaters, organisers, and great measurers of everything, just like the Chinese are now. But while we sit content with our lot, smugly satisfied with our carefully controlled success, consider that the Chinese are well on their way to constructing the second-largest economic power in the world in a little under thirty years. If they want something, they build it—now, not after twenty years of planning debates and processes. They don't worry about global warming or health and safety. Those things will come later, as their empire begins its decline, when they actually have something to lose. This premise suggests that success breeds failure. Success leads to greater controls, and controls need measurement. Measurement leads to corruption as we serve the target and not the problem. We are blinded by our targets and thus we fail. Ticked any boxes for your employer lately?

Do you remember Chernobyl, the greatest nuclear peacetime disaster yet? We had radioactive sheep in Scotland and prophesies about the end of civilisation as we knew it. Ukraine has paid a heavy price, with significant populations forcibly removed from contaminated areas and large swathes of no-go areas. Surprisingly, early signs indicated that small mammals were surviving and breeding. To begin with, this was put down to their short lifespans and insufficient time for gene damage to occur to produce cancers. Later studies seemed to indicate that fats in the bodies were offering protection from the radiation. Without the interference caused by humankind, the area now has significant breeding populations of wolves, deer, bear, and bison, all large mammals with extended lifespans. What I have found particularly interesting are the studies of birds. There are large and varied populations of resident birds successfully breeding in the 'danger zone'. Swallows entering the area after their long migration set about to breeding but soon display white patches on feathers and other physical defects. The birds fail to breed successfully and they die. The danger area has become a black hole for swallows; they enter but do not leave. Other resident insectivorous birds do not appear to suffer the same fate, so we know it's not the diet. We know that human cells are attacked by free radicals, which cause damage to our cell DNA, which can lead to cancerous cells forming. We also know that antioxidants neutralise these free radicals and offer protection. That is why it is recommended that we eat our 'five a day' of fresh fruits and vegetables, to boost our supply of antioxidants, vitamins.

The suggestion is, for the migratory swallows, that they use up all their antioxidants during the flight from Africa and are unable to defend themselves against the radiation damage. Resident birds maintain their antioxidant levels and survive to breed successfully. While we are not exposed to the same extent as the animals around Chernobyl, we are nonetheless exposed to free radicals throughout our hopefully extended lifespan. This news should drive you to your 'five a day'. I have heard it said that if you could take a magic pill that could protect against many forms of cancer, the majority of people would want it. The trick to this question is in the reply: the magic pill already exists. It's called five portions of fresh fruits and vegetables a day. I haven't introduced anything new into the debate as yet. I am simply repeating what I have seen elsewhere. But I have also seen another report that hasn't been linked in to the debate. Fruit serves a purpose for the plant. The plant invests significant energy and resources into its own production, but altruism is not in the plant's vocabulary. Animals will eat the fruit and through their mobility provide a dispersal mechanism for the plant seeds, thus furthering the future of the plant. But spoiled fruit will be rejected by the animals, so at the point where the fruit is going off, the plant maximises the presence of antioxidants to prolong its 'shelf life' as long as possible. For the greatest beneficial effect, eat fruit that is showing signs of distress, on the cusp of spoiling. So what do we find in the supermarkets, tray after tray of perfect, uniform fruit. Perhaps you should make it six or seven a day. Perhaps the fruit and veg are pristine because they are stuffed to the gills with vitamins, or bred for a long shelf life by removing 'spoiling components', or have been stored in optimum artificial conditions to retain their appearance. Do we know? (You will not be surprised by the texture of the grossly enlarged strawberries that have appeared in recent years, seeming to have not a little amount of lignin in them.) Are plant breeders effectively denaturing fruit and vegetables of their natural defence mechanisms and, replacing them with non-vitamin replacements?

So who, if anyone, leads the herd? In the earlier example, Jamie Oliver tried passionately to essentially extend people's lives. Of course Jamie is what we would know as a 'celebrity chef'. This is a chef whose reputation and influence elevates him into the stratosphere. His books have sold millions, his TV programmes have been viewed by millions, and yet for all that, he has failed to move the herd. Is this—sorry, I can't resist it—because he is a 'recipe book engineer' of social eating habits? Is the reason for failure to have the desired effect because we know, deep down, that he is a celebrity first and a chef second? How do you become a celebrity chef? What are the criteria for entry?

I am quite sure that a trip to the entry for Jamie Oliver on Wikipedia will list in chronological order his history and body of works to date, including his 'breakthrough' role and key moments in his career as seen by the biographer. I wonder if Jamie ever realised his life was as simple as the Wiki pages will describe?

Conversely, if Jamie were to produce a 'magic' yoghurt or similar product that would cure us of our dietary health problems, then his work would have been done by now. Remember, given the choice of taking a pill to bind with dietary fat or eating less fat, we can guess where the smart money will be.

Yet the media world is dominated by the cult of celebrity. Entire TV series filling the airways are devoted to *Bushtucker Trials* and *Big Brother* and all the celebrity pointless/*MasterChef*/charity telethons/*Mastermind*/*Britain's Got Talent*/*Strictly Come Dancing*/*A Question of Sport*/Sports Celebrity of the Year/*Weakest Link* programmes. Newspaper are filled with lurid tales of who is shagging whom; wearing the wrong clothes; appearing without make-up or with unshaven armpits; putting on weight or losing weight; plagued by cellulite; divorcing, marrying, breeding; and so forth—everything we need, really, to ensure our day's knowledge is complete. And in the same way that the sunscreen industry is now providing the cure to the problem that they created, the newspapers are filling column inches with tales of illegal hacking of celebrity phone messages so as to ensure that it won't happen again.

Until the next time anyway.

Yet the power of celebrity is certainly engaged in the minds of marketers worldwide. How can we possibly sell perfume without an Indian rope trick and just the right celeb disappearing up the rope? The celebrities don't even need to be real; think Sergei, the animated meerkat, or Brian, the robot, for car insurance websites. I can't believe I just wrote that, and hope you spotted it as well. Have you been paying attention? The celebrities don't even have to be real. They can't possibly spend their lives pouting and posing. When the cameras have gone away, they become flesh and blood, the act and illusion put on hold. Even Sergei spends his off-air time in a drawer somewhere.

Have I just hit on the solution? The advertising industry is lazy.

Use an established figure to endorse/market your product. Design a pro forma. Cue seductive/action-oriented/dynamic/popular music jingle; add comely/trustworthy/ sexy/rugged/dynamic/stuffed/robot celebrity (Nicolas Cage would cover all bases) looking happy/ smug/content/ satisfied/alluring/seductive and speaking to the camera/voice-over from the naturally appearing script.

'Yes, I think I can get KK, Charlize, Beckham, Roberts, Portman, or someone similar, provided you can stump up the budget. Of course we'll need a proper director and studio producer, provided you can stump up the budget. And just to remind you, we are on a percentage fee.'

That way you can advertise the celebrity and not the product. This makes your life so much easier. And as for the fee? You're worth it.

In fact, why don't you just CGI it?

But you know it works, because the sales figures prove it, so why don't people brush their teeth or stay slim? Where is the magic switch that connects with the populace and moves the herd? As a sweeping generalisation, our screen heroes are slim and attractive. There are obviously exceptions—Frank Cannon, PI, springs to mind—but we spend money to watch attractive people performing. KK wouldn't be in the Caribbean if she were fat and ugly, unless she was playing a fat and ugly character of course (probably a villain or piece of realism), and then she wouldn't be at the top of the billing. Visually attractive people, male and female, fill our screens and red carpet premieres and magazine shoots and music venues. (Tests have shown that audiences assess 'attractive performers' as having a better singing voice over less attractive singers, even though the ugly one might have the better voice in a blind test.)

And they all have perfect white teeth.

Why do we engage and want to fill our lives with them and the illusion of them and yet won't put in the effort to be like them? I realise that there are always exceptions to the rule, and that a significant proportion of the population buy cosmetic teeth treatment, hair dyes, hairstyles, designer goods, and clothes and try to hang out at celebrity haunts, where some of the magic dust might just fall and stick to them. You might even be mistaken for one of the celebrities. Gone are the days when schoolkids wanted to be train drivers or police officers or nurses. A present-day commonplace request is to be a celebrity. Have you noted how modern-day parents don't actually want to be parents but instead wish to be their children's best friends? Discipline and learning seems to have gone out the window, as kids now rule the roost. Kids are firmly in charge, and adults serve only to make the kids' fairy-tale lives complete. They are all little princes and princesses these days. Is this why, as they approach adulthood and are faced with the realities of working for a living (some of them anyway), they have to find a reality changer to make their pathetic lives palatable? Is this what creates the urge to take drugs, binge drink, run up store cards, gamble (for fun, so the adverts say), get a tattoo, get Botox injections, and get cosmetic surgery? How fitting really that one of the common themes of celebrity is the rise to amass fabulous wealth so that you can achieve all the lifestyle trappings that frequently include the ultimately fatal drug overdose. So if the pursuit of celebrity is escape from reality, why the need to escape from the reality of celebrity?

An Introduction to Princesses

I always considered that there was not a school in the country good enough for my children to attend, but I contented myself with the knowledge that school is where you go to learn and home is where you get your education. I am pleased to report that my efforts have been rewarded. I am delighted by my young adults' attitudes and successes in life and by their undoubted ability to think, the most priceless skill of all.

It seems the modern generation has been bought up on a diet of get-rich-quick schemes (*X Factor*, *Britain's Got Talent*, *Big Brother*, *The Apprentice*, etc.) that don't require any work and don't usually make you rich either, in spite of all the phone-in costs. These schemes do make somebody rich, but that's another story. Kids don't play football these days; they role-play as their 'heroes': it's not little Johnny kicking the ball; it's a 4-year-old Ronaldo, complete with shirt and hairstyle.

True, when I was at school, I and my friends wanted to be Geoff Hurst or Bobby Moore, but when the game stopped, so did the role-play. It's a bit harder to remove the haircut along with the shirt. Have you noticed how every unusual artefact dug up on *Time Team* once belonged to a princess or ruler, 'very high status' at least? The need to introduce some drama and excitement into the mundane dictates the presence of an ancient celebrity. Is it because we are all junkies for some content in our lives? We fail to engage in the life that we have, so we submerse ourselves into the life we imagine we want instead. So what's so wrong with the life we lead? We emerge from a minimum thirteen years of education with all the aspirations in the world and none of the tools to achieve our goals.

Young people these days are held hostage to the premise that all children are born equal and must all be given the same chance in life to excel. You cannot condemn a child to a life of manual labour; you must allow them the freedom to develop into rocket scientists, computer programmers, and a thousand and one other occupations that they are not capable of doing. In the olden days, 15-year-olds could leave school and join the workplace (I think my school phased 15- year-old leavers out in 1971–72) and start to earn a living. Ask any teacher 'hand on heart' to identify the pupils who will fail to prosper in school, and he or she will produce a list. Well, guess what, all children are not born equal, and some will fail while others prosper: fact. But once again we are back to the use of measurement and the problem that it engenders. Richard Branson was a self-confessed 'duffer' at school but went on to become a billionaire. He could claim to be a late developer, or he could simply say school wasn't for him. Either way he was failed by school but succeeded in life. Why not acknowledge that you cannot expect a square peg to fit into a round hole and introduce some non-socialist/liberal realism into the debate? Why? We equip our children with the skills to dream but give them nothing to achieve them with. Cue fantasy escapism of celebrity, or will it be cannabis, alcohol, bulimia, crime, or depression that fills their need? Then of course we arrive at the paradox: the highest level of education ever achieved in this country's history in terms of years and cost is woefully inadequate in teaching the skills of life. Think about it. Even the 'successful' ones go on to careers in media, marketing, politics, teaching, and so forth, where they serve what purpose? Use all your accumulated years of education to strike out the alternatives from the pro forma of your chosen career path. I think even I would be looking for an escape route. The unsuccessful ones have learnt that their labour has a minimum price, a threshold below which they won't lower themselves to get their hands dirty. So we import people to get their hands dirty instead, East Anglia has filled with eastern Europeans who work on the farms or in meat processing, or as domestic cleaners, fast-food servers, and the like.

Meanwhile, our low performers disappear to the dole office and commit to a life on benefits, because 'I ain't working for that much.' One of the contradictions of this system is that many of the 'imports' are better educated than our own welfare parasites.

Occasionally the TV news or a documentary will feature a story about schools in Africa. For many, schooling is a privilege and not a right. The thing that makes the strongest impression is the eager intentness and work ethic that these African students possess. Compare that with the 'problem' students in the UK (that's if you can get them to turn up at school in the first place).

So why do we devote so many resources to keeping these kids in school when the only useful purpose it seems to serve is to keep them off the streets during the day? Why not engage them in workshop facilities where they can learn the value of learning and develop into something worthwhile? Years ago, if your dad drove a JCB, he would take you to work on a Saturday morning and show you the basic skills for operating such a machine. Many kids of these parents could 'drive' a machine by the age of 13 or 14. If a parent took a 14-year-old on a building site in this day and age, a number of things would happen: police would be called, the parent would be sacked, and the employer would be fined. Conversely, what could you have achieved? Child enthused, skills learnt, parent–child relationship strengthened, responsibility acquired, confidence developed, expectations grounded. Instead of this, we get the 'Take your son or daughter to work day', whereby children are exposed to the idea that their parents might actually work for a living. I worked in a factory during the summer holidays from the age of 9 onwards. It grounded me and developed me as a person. We return to the idea that education is a purpose in itself, a means to a means, stuck so myopically to itself that it can no longer remember what purpose it is supposed to serve. It is pushed along by its own irresistible momentum to where? In my lifetime I have seen the school starting age drop to 4 and the leaving age rise from 15 to 18. That's another four years of education to achieve what, exactly? What's the next step, a university education for everyone and another three years to achieve what we couldn't achieve in fourteen? Are our lives really so technically demanding that we are failures if we do not achieve five or more GCSEs in C or above?

'Immigrants are good for the UK economy.' That's the official position pushed by politicians, but let's look at some of the reasons given why. I am not getting into the argument about its rights or wrongs, but simply citing a narrow economic case.

Generally the case is made that immigrants are young and educated and come here to work. This makes them taxpayers and net contributors to UK plc, but at the same time they have already been educated while living abroad. Now this is important because of the cost of education: put simply, using a Pole as an example, the cost of all the non-productive and expensive years spent at school have been met by the Polish government and Polish taxpayers. We therefore reap the benefits of ready-to-work Poles with none of the costs of getting them to that stage. While here, they work hard and pay their taxes and are net contributors, until they reach pension age, when they start to become tax-takers.

But of course now the model is to pay for your own pension for your own old age so the impact on society will be reduced.

So if this is good for UK plc, please explain to me how the opposite is not also true. We force our children to spend an extra four years at school, four years of high-cost unproductive time when they are net-takers. They ain't all going to be brain surgeons, are they? If educated immigrants are good for the UK economy, then by default all schoolchildren are bad for the UK economy. The reply to this assertion would be, 'Of course it's a price we have to pay.' According to my *Daily Mail* bought the day after Budget Day, our current expenditure on the education budget is £99,000,000,000 a year. If we simply divided this figure by 14, then each year of schooling could be argued to be costing us £7,000,000,000. When I was 18 and still at school, only about 10 per cent of the student body made it into the sixth form. Are we really so much cleverer now?

The old education process of streaming children and selecting the brightest pupils for further education is anathema to today's social liberals, who see it as class betrayal. Try studying in a class with students who have no interest in being there and you will understand what betrayal in the class means. If you honestly believe that our education system is so brilliant and appropriate for its purpose, answer this one simple question: why are our super-trained and super-qualified school-leavers always at the back of the queue when it comes to employment? The age differential is so noticeable that unemployment figures are separated out into people under 25 and the UK in general. I grew up in an age where people would say that your first proper job is your first proper qualification. This notion still holds true. Ah, but you see, what employers are looking for is experience over training. So why do we promote training and destroy any opportunity to gain experience? When are the people who control education going to get their heads out of their arses and actually think about what they are doing, beyond worrying about their salaries and pensions? How do we kill the elephant?

I realise that the foregoing will strike a chord with so-called right-wing thinkers and that I have spoken heresy to the left-wing thinkers. So let's think about this, shall we? Every child is born equal and deserves an equal opportunity to succeed in life, without artificial barriers of class or wealth or background or environment or religion or race or gender, and must be treated accordingly. Let's turn this around slightly and come from the other end. As an employer with a vacancy to fill, you will be faced with a requirement to select the most appropriate candidate for the job. Now every candidate you view was born equal and deserves an equal opportunity to succeed in life without artificial barriers of class or wealth or background or environment or religion or race or gender. So why do employers have a selection process? Why don't they simply pull a name out of the hat? So after fourteen years of intensive schooling, where all are held on a level playing field of opportunity, there is a reality check: you are not all equal. 'Yes, but they have all had the same chance' will be the reply.

Let us return to *Elysium*. This is the film where a utopian disease-free world full of rich people orbits above a detritus-filled earth. What would the entry-level requirement be to join this project? Shall we populate the new planet with billionaires and Nobel Prize winners and rocket scientists? Why would you go there; what would be the attraction? The cost would be immense, so the project would need to be supported by a vast hinterland of resources, in this case all the poor left on earth, so that the privileged few can sit 'kings of the castle'. Of course it would never happen in my lifetime, would it?

Where do the super-rich currently live or, more correctly, reside? Is it Monaco, or the Cayman Islands, or Honduras, or any other tax haven? The very rich populate their own enclaves, such as Kensington or Chelsea, while the rich settle for Stamford or Cheshire. They don't tend to live in the Welsh valleys in a near derelict area after the coal mining industry collapse.

Where shall we school the children of the super-rich? It won't be the local comprehensive; you can guess that much. Where shall we treat the ills and diseases of the super-rich? It won't be the local NHS hospital; you can guess that much.

There are always exceptions, of course; Warren Buffett still lives in the marital house he bought decades ago, and decades before he became a billionaire. It sort of begs the question: Why become a billionaire? When is enough, enough? I guess it's a sort of hobby for Buffett. At least he knows how to dispose of his wealth to benefit us all.

Consider the collapse of the Soviet Union, built on the premise that everyone contributes as they are able and takes as they need. It didn't actually work like that though, did it?

So how do you reconcile the debate between Left and Right, takers and givers? Ducks and geese fly in V formation and share the burden. It's hard at the front, so the others take turns and spread the load. All benefit. Watch the Tour de France and see how each team supports their own team leader to get him to the front in as good as condition as they can. Watch how the breakaway groups fail when they don't cooperate and get drawn back into the peloton. Vampire bats have been shown to share blood with unsuccessful roost members but are less likely to repeat this practice unless the favour is returned. Cooperation is not unknown in the natural world where the benefits are shared, but true altruism?

As a species we cooperate. How else can we build Stonehenge or cities or schools or hospitals or roads? Is it part of our herding instinct? Yet out of the cooperation, we all seek to gain an individual advantage: 'I will be the high priest'; 'I will live in the West End'; 'I will become headmaster'; 'I will live in the countryside beside the new road.' 'I will roost in the middle of the starling flock, as it's warmer and my neighbours are more likely to die than me.'

'I will be the best footballer on the pitch and earn more adulation and money.'

'I will support my fellow team members, as they help me to be the best footballer on the pitch and help me earn more adulation and money.'

'I will leave my football team if I think I can get a better chance at another team.'

'I will play as hard as possible with my new team to defeat my old team.'

'I will be the seventh-best footballer in the team, and I will support the best footballer on the pitch because his success is shared by all of us.'

'I will support my chosen football team because their success gives me a nice glow in my stomach and bragging rights at my local.'

Why is the West End of London more expensive to live in than the East End? In simple terms it's because the air quality used to be better there: the prevailing winds in Britain blow from the west to the east, carrying all the smoke from Victorian London away from the rich and dumping it on the poor.

It appears, however altruistic we attempt to be, that somebody else always ends up with the shitty end of the stick, and we all work to try to make sure that it isn't us. By definition, then, we also work to ensure that someone else does.

So we have an education system designed to give individual pupils the shitty end. No? Then why do we have a race at the end of every school period through the examination process? If the winners win, the losers by definition lose. But all children are born equal and deserve equal treatment, until they enter the examination room anyway. Is this a paradox or hypocrisy? An education system free for all to provide a universal schooling irrespective of race, creed, background, etc. that is simultaneously designed to separate out the wheat from the chaff. But in spite of what the socialists preach, we are not all born equal. Our DNA separates each from the other. Our diet, environment, habitat, access to medical care, and a million and one other variables conspire to separate us. (As an aside, I will in the company of women announce to a male companion that a man's brain is bigger than a woman's. Before I can complete the sentence, enraged women will have rounded on me. What they never let me complete is the rest of the statement: 'But women have smaller bodies, so on average they have more brain per kilogram of body weight than men.' The last time I did this, the women involved were training to be army doctors. My reason for writing this is to ask if you have jumped to any conclusions about my 'genetic supremacy arguments'. I have said we are all different; that's all.)

Is this because you pass your exams as the exit route to life? Is it because as you leave the education process, the school's job is done and the teachers and administrators don't give a shit what happens next, the next batch on the conveyor belt of dreams already demanding attention? (Something like a pair of blackbirds: build a nest, mate, lay eggs, incubate for fourteen days, feed the young to fledging, feed them for the next few days before abandoning them, and return to nest building, never to see or think about your offspring again.) After all, teachers have done their job and can prove that they have done it. Their pupils have achieved grades *x*, *y*, *z* in something really valuable or other. We are back to the problem of measurement again. Why not measure a school's performance by the number of former pupils in full-time employment, or the amount of taxes they pay after one, two, three years, etc.? Perhaps then our educational standards might actually serve a purpose.

I use a local café for lunch most days of the week and have done so for several years. Successive 'crops' of young people are employed and pass through onto higher education or full-time jobs. I study their development as first-time workers. This thirty-five-seater café serves as a microcosm of the world outside as I watch the young employees develop, some faster than others, into rounded individuals fit for life. As a counterbalance to this, I drive past my local grammar school/college/academy, whatever it calls itself this year, on my journey to and from the café, past the thirty-foot-long poster bragging about record exam results. Well, whoop-de-fucking-doo.

So Left or Right, does it matter? Wouldn't it be better just to be correct? But what do we mean by correct? Well, there's the rub. How do you measure it? Is my glass half empty or half full? Let's polarise the debate by looking to the extremes:

1) The rich can go to Elysium and look down from space; the poor can live in a refugee camp where all their needs are met by beneficial agencies.

2) The rich can stay within the castle walls with the drawbridge raised, while the poor (who built the castle) can starve outside.

3) The rich can quarantine themselves in walled enclaves, while the poor can die from Ebola.

4) The rich can refuse to pay their taxes and watch the poor starve through lack of resources.

5) The rich can secure the services of all the healthcare workers, while the poor can die for lack of assistance.

6) The rich can invest in secure hotel rooms, while the poor are exposed to suicide bombers in the marketplace.

7) The rich can ensure that they never have to grasp the shitty end of the stick, while the poor are left with no option.

Socialists would ensure that none of this would happen. They'd raise money for charitable appeals; change the law to reduce inequality; provide universal opportunity, universal education, and universal healthcare; and most importantly get somebody else to pay for it. Socialism must work. Otherwise, the following could never have happened.

1) People would live in Kensington and Chelsea, while the poor live in deprived inner-city areas living off welfare.

2) People would live in the First World while preventing the free migration of Third World people, having exploited the poor to become the First World.

3) First World governments have introduced health screening and cancelled flights to West Africa.

4) The rich can opt out of paying taxes by residing in Monaco and the like, thus reducing available tax income to spend looking after the deprived in society.

5) The First World has a monopoly on healthcare, while the poor can die from diarrhoea.

6) Politicians and journalists stay in 'Western hotels' in places like Kabul or Tripoli and have their food brought to them.

Speaks for itself really.

How do you socialists sleep at night with all this global injustice going on? Let me guess: 'Rome wasn't built in a day.' And it's fair to say, it never would have been built if you'd had your way.

Rich people accept that they have a place in society, normally at the head of the table as far away from the shitty end as possible, but they do accept that the most vulnerable should receive protection and assistance. But I'm going to suggest that they also have a view as to what's fair and what isn't.

The pendulum swings from one point on the curve to another.

When Margaret Thatcher came to power, she reduced the high rate of income tax from 60 per cent to 40 per cent, and the amount of tax taken by the UK government went up. How can you reduce the tax figure by a third and see an increase? Quite simply, the rich saw 40 per cent as fair and were prepared to pay the tax. Prior to this, the rich, who are normally in positions of responsibility/ownership with regard to businesses, found ways of avoiding the tax. My wife's ex-boss provided a 'tuck shop' for his workforce, sold hundreds of pounds worth of company sweets a week, and pocketed the proceeds. Others would let the company pay for home extensions, or fact-finding trips to foreign competitors in sunnier climes, taking their families along for company. While I don't think there will be any socialists left reading this by now, let me redress the balance somewhat. Plumbers, builders, gardeners, and chip shop owners regularly handle cash and only declare what they think they need to. Feeling left out? All the plumber, builder, and gardener employers happily paid them cash to avoid paying tax as well.

The message is, if you want to tax the workers, do it equitably. If you want to provide welfare to the poorest and least able in society, do it equitably. Our present social security and Personal Social Services budget (*Daily Mail* again) stands at £262,000,000,000 for the 2015–2016 tax year. That's all the income tax and three-quarters of the National Insurance spent.

So what is the measure of equitable? Back to measurements, my favourite subject.

I have said earlier that there is endless need but only finite resources, so by definition not everyone will be satisfied. As a nation we can afford to provide for the needs of many, but the question is, how much? Let us suggest a poll of our taxpaying population and ask them what percentage of their taxes they would like to see spent on welfare. Let us consider the options open to us: we can pay nothing, we can pay every penny that we currently do, we could increase the amount we pay, or we can adopt some point in between. Can you imagine any government actually asking this question of us? This is not something we can be trusted with after all. We need to be guided by the hand to the correct conclusion. (In 2015, a survey indicated that a majority (52 per cent) of the UK population actually agreed that capital punishment was wrong, for the first time ever.) Imagine that you live in a small island community with limited resources where the labour of many is needed to stay alive and warm and fed. For convenience, let's say this island community has one hundred souls in total and the only shortage of resource is the ability to extract it from the environment. That is, it has to be won by labour. Children are essential, because they represent the future and without them there is no future, so we will exclude them, within limits, from the equation.

When people fall ill or become pregnant, you all pull together and cover the shortfall; it might be you next time and you will be grateful for the help. When the elderly reach the point where they are no longer 100 per cent productive, you support them in their needs, as they have done for you and the elderly who went before them. Everybody's happy and it all works. Some times are easier than others, it's true, but it's manageable. Imagine that the day's labours are all gathered together and distributed at the end of each work shift and an equitable distribution is made. Every eye would be looking at the contribution made by each individual, with judgement made according to the abilities of each.

Some will have a better skill base and abilities in given areas, so an allowance will be made. What would you give the person who contributed nothing day after day for no reason? What would you give the person who contributed nothing day after day because they were drunk? What would you give the person who was unable to walk, unless no one was watching of course? What would you give to the person with twelve kids when you have deliberately limited your own to two? What percentage would you volunteer for 'charitable' uses? What percentage would you volunteer for piss-takers?

It's not so clear-cut, is it now? Or rather it is. The community elders would sanction the 'misfits' by removing food and shelter until they pulled their own weight. If this failed, my guess is that the misfits would be issued with a rowing boat and sent into exile. But then again, maybe not, being that such behaviour is encouraged by anonymity. Because everybody knows everybody, the respect between the individuals of the community would probably prevent this circumstance from arising. (It's in the anonymity of the crowd that rioting occurs; normal rules of social engagement disappear in seconds and the mob takes over.)

But our system is so much fairer than that, isn't it? It doesn't matter what you do, or rather what you don't do; you will always be provided for. But for every volunteer non-worker you support, you are denying help for an involuntary non-worker. There are thousands of households in the UK which shelter three generations of families where no one has held a proper job or paid into the kitty. That is virtually the lifespan of the welfare state.

For every non-productive individual, look for the silent witness: we can't afford to treat your illness, we are short of resources' we can't afford to repair your road, we are short of resources; we can't afford to provide a care package for your elderly parents, we are short of resource. No, we are not short of resources; the only thing we are not short of is net-takers. The reason this happy state of affairs persists is in the breakdown of communities and the anonymity of the scroungers. But of course this is only partly true, because entire communities of benefit-seekers exist.

Think about the Greek model, outgoings exceeding incomings. Think about our trillion-pound debt. When does a necessity become a luxury we can't afford?

So we are back to the herd problem. How do you persuade the net-takers to contribute? My view is simple: kill the elephant that is the welfare state, strip it down to its original aims and ambitions, and set it on its course again, but this time with a kill switch. Still not convinced? How can we import and find jobs for hundreds of thousands of workers and still have millions unemployed?

I was watching *The Simpsons* on TV a couple of years back and came across a severely depressed Lisa Simpson who uttered a profound one-liner that for me summed up the entire school process. She asked why the school was continuously preparing her for the life that she wouldn't lead.

Because that is the education doctrine: every child born equal to tread whatever path they seek irrespective of class, creed, religion, or ability. Such a noble sentiment, something that prepares our children for everything and nothing while simultaneously streaming and focussing each and every child towards a narrower and narrower range of options that ultimately and essentially have to fit the timetable. So we have the sentiment that no child is born to be a bricklayer, since this would interfere with the heart of the socialist/liberal philosophy.

'My dad is a bricklayer, and his brothers and his father were as well' is anathema to the concept of free will and 'we must save the child from social deprivation and protect his or her right to choose'. In an age gone by, life depended on having a skill or trade to earn a living from, and learning a trade from parents or neighbours was how it was done. The smith would train his sons and neighbours. The thatcher, the butcher, and the fletcher all took it upon themselves to protect their children's futures. Others would put their children in bonded servitude for years to achieve the ambition of securing a trade for their offspring. This process, while not written in our DNA, was written in our surnames instead. Our children are placed in bonded servitude for fourteen years to achieve what, their full potential? Honestly? 'As long as it fits the timetable'; 'We only have these resources available'; 'We don't do that at this school'; 'There wasn't enough interest, so we dropped it off the curriculum'; 'We've really struggled to get good staff in that area'; 'He's only a stand-in; it's not his trained subject.'

I cannot comment on his education, but I will comment on one of our early customers from the late 1960s and early 1970s. John Bloor was a plasterer by trade who slowly built his operation up from plastering other people's houses to building his own. Today his company is the largest privately owned house-building firm in the country, and for good measure he also resurrected Triumph Motorcycles and has the factory to prove it. So what's wrong with being a bricklayer or a plasterer or a carpenter or a ...? How many of our building firms employing thousands of people started out from such humble beginnings? A lot of the great car manufacturing companies started out as smiths or bicycle repairers/builders and used their metalworking skills to grow into new fields. Many of the older-style garages began life selling fuel and bending metal as blacksmith forges. The message is, it's not what you are; it's what you do with it that matters. 'How dare you condemn little John Bloor to a life as a plasterer? He must have the right to choose.' Bollocks. The clever ones find a way. They learn the pro forma of life and how to make it work for them.

Conversely, how dare you condemn little John Bloor to a life of mediocrity by locking him up for fourteen years in school and preventing him from gaining a trade to use as a launch pad to a better life?

Of course the educational empire has its elite, the children of the rich who study at the expensive private schools, where they get the best schooling that money can buy, followed by a shoo-in to Oxbridge, followed by a shoo-in to a glittering career. Notice I didn't say the best education.

But isn't this a simple example of the bricklayer message? A bricklayer knows how to lay bricks and can build walls that build buildings; his sons can also learn the same trade. A rich person knows how to make money that can pay for the 'best schools' to teach his or her kids how to make money. All we are arguing about is the scale, another measurement. It must be dreadful realising that you are the child of Lord Loadsofmoney and will be forced to attend seventeen years of Eton and Oxbridge and learn how to make money. Much better to be the child of a factory worker and being forced into fourteen years of the local comprehensive, where you learn what exactly?

But there is another message to learn from John Bloor, and it is a supremely simple one: he succeeded by always generating more than he spent. If he had spent more than he'd earned, there would be no Bloor Homes and no Triumph Motorcycles.

In years gone by, people used to talk about the class divide. Poor people were condemned to a life of poverty, whereas the rich enjoyed the privileges that their wealth provided. Nowadays we have social mobility. Poor people can have a 'proper education' and can go to university so that they can better themselves. And for every one that succeeds, how many others are swallowed up in the morass of fourteen years of schooling that they neither want nor need? I remember election pledges of years gone by when political parties were promising to reduce class sizes to less than twenty-five pupils. I also remember analysis of the time which proved that there was no statistical evidence of any deterioration in test results in classes of up to thirty. It was only when classes of thirty-one or more occurred that results suffered. Once again, this is clear evidence of politicians spouting rhetorical bollocks fully in the face of the facts. Once again I am left with only one question: how do you kill an elephant?

When you drive a car, you are exposed to potential and real hazards all along your journey, and you have to drive accordingly. When learning to drive, everything looms large in the hazard-perception mechanism and you find it hard to get past 20 mph. With practice, you filter out what's important and what isn't, you learn to brake only when necessary, and your speed improves. This applies to every task we do. We are bombarded with extraneous information that our brain filters out, hopefully leaving us with what's important. That way we can walk down the high street without stopping to look in every shop window on a quest for shoes. We go to the shoe shop instead, having filtered out the butcher and florist and chip shop as not required. I realise this may be an alien concept for the female of the species, walking past other shops on the way to the shoe shop without looking in. I actually believe that women are genetically preset to browse, picking fruit and berries or mushrooms, while men are genetically preset to hunt and pursue our quarry to the conclusion, without getting distracted by other 'spoor' on the way. That's upset you, hasn't it? But all I'm asking you to do is look at the behaviour of men and women and accept that there is a difference. (And no, this doesn't make one superior to the other; it just establishes that there is a difference.) In the vein of men's brains being bigger than women's, try saying that men and women are not equal; see what that kicks off. Bear in mind that whatever method of measurement you use, it will always favour one sex over the other. By definition that means we are not equal, simply prepared in different ways to deal with whatever life throws at our DNA. This is another interesting example of how the method of measurement corrupts the truth.

As children, walking through shops is like being in a wonderland of new experience, and the need to touch and hold everything is overwhelming. Children explore in order to develop the filters for use later in life. These filters enable us to function in later life. Imagine how long it would take to do your supermarket shopping if you gave your children free rein to control the pace of your progress. If nothing else, the aisles would be full of discarded product littering the floor. Let us call this learning and consider that we all develop these filters to ignore what can be ignored. Now throw in to the pot advertising executives and see what they preach: a never-ending assault on our visual and audio fields that we steadfastly filter out and ignore.

When at school, I decided that the term 'selflessness' didn't actually exist. I reasoned that when we are given a choice, we will always make the decision that suits us. So if I hold a gun to your head, you can decide whether to give me your wallet or not. If you were in the queue to live, you may decide to swap places with someone in the queue to die at the gas chamber.

The latter instance would be described as a selfless act, but in reality the 'volunteer' has decided that they couldn't live with themselves for one more day knowing that another has died and they could have prevented it. So the volunteer has actually been selfish, although another has benefited, so it is termed a selfless act. As a mother, would you rather go without food or let your children go hungry? The choice is yours, but the end result will be, 'Yes, I can deny my children food and still be able to live with myself' or 'No, I will go hungry rather than see my children suffer.'

Either way you will take the selfish option.

So why do I share this bit of philosophy with you now?

Because I'm going to be selfish by insisting that the money that I donate to help the less fortunate who need help should only go to the less fortunate who need help. Seems simple enough, doesn't it? But as you will see, it becomes virtually impossible to achieve.

1) I want to give money to the RSPB to protect birds.
2) I want to give money to the RSPCA to protect animals
3) I want to give money to save children.
4) I want to give money to cure cancer.
5) I want to give money to support people who have lost their jobs.

These will do for examples. Following are the results of contributing to each cause.

1) We have already considered the actions of the RSPB in light of protecting a select number of birds at the expense of other bird species, either at home or abroad. If I give them money, I have failed in my ambition.

2) The RSPCA has now become the NGO for the enforcement of animal welfare within the UK. Do I really want my money to fund lawyers to take out prosecutions against uneducated pet owners or fox hunts?

3) I have already highlighted my disgust that the CEO of Save the Children is on a mega salary. Do I want my money to pay for all the other inflated salaries of the organisation staff, and advertising and websites and everything else, or do I actually want to give a child money?

4) By giving money to research cancer treatments, I can ensure that the best medical minds and research facilities are turned to this disease. In the process I have created an internal market for excellence, whereby other, less well funded, but equally deserving diseases are left struggling for resources. Money also breeds money. Why work for the MS Society researchers when they can only pay £50,000 a year, when you can get £80,000 working for cancer research and they've got the better facilities?

5) I want to help the recently unemployed while they search for further employment. How can I ensure that my money doesn't go to people who simply won't work and have chosen this as their career path?

In all of these instances, the stated intention of each good cause bears no resemblance to my ambitions in giving it my money.

They are all elephants let loose into society with no off switch, crushing along with unstoppable momentum. I am willing to bet that if you sat in on one of their board meetings, you would not be able to differentiate between a charity and any other major industrial/commercial organisation. 'But that's a good thing,' they will cry. 'We can't have any amateurs here. We are all deadly serious about what we do'—and that includes maintaining their personal salaries and contacts with the rich and famous.

How can you tell the difference between a business and a charity? Each is trying to earn as much as it can get while marketing its respective products and fighting for market share. At least with M & S you get a pair of trousers.

But I've also slipped the dole into the discussion, so it isn't only charities I have a gripe with. Mr Government, please enter into a covenant with me that the money you take from me in taxes will go to the correct recipient. I do not need any help from you to spend my money, much less piss it up the wall. I am quite capable of doing that for myself. Any left-wing socialists reading this will now be spitting into their muesli. But I'm not attacking socialism; I am attacking waste on an industrial scale. Think about it; get rid of all the non-deserving claimants and you have much more left over to support the people who really need it. Carry on the way you are and, as I've said before, the real deserving are left as silent witnesses at the end of the queue with nothing to receive. Is this what socialism means to be? Surely you must support my argument if your real aim is to help the people who need it.

Think I'm being too cynical? Remember that 2014 was the year of the Ice Bucket Challenge, you know the one, get a cold shower or donate a forfeit to charity.

The small obscure charity that launched this soon-to-go-viral idea went from tens of thousands to millions in donations and soon upset the other charities who saw their own incomes drop and wrote off the originator's charity because they couldn't possibly use that amount of money. All's fair in love and war and charities.

Dear charity, hand on heart, please explain to me the purpose of your existence. Please confirm to me that your sole interest is in the single-minded pursuit of your stated aims.

Dear Government, please confirm to me that your sole interest in governing me is to ensure that my needs are met in the most cost-effective manner possible.

There is an old saying, 'charity begins at home'. I have often wondered what this means.

Mission Creep

To build an empire, you need a single-minded ruthless organisation with a clear understanding of the aim and ambition. You need to override your competition and be prepared to assimilate them into your operation. You need an advantage, an edge, something that tips the balance in your favour, and you need to maintain that edge over time. Britain's empire was founded on the navy and on our control of strategic ports worldwide that gave us a geographic advantage. Gibraltar, Malta, and Alexandria controlled the Mediterranean; the Falklands controlled the southern passage to the Pacific; and Mauritius controlled the passage around Africa to the Indian Ocean.

But you don't have to think that big. Woolworth's started from a market trader's store in London and expanded onto every major high street in the country. Colonel Sanders started with one fried chicken shop which was superseded by a road bypass, but he went on the road to build Kentucky Fried Chicken into a billion-dollar empire. Little Chef began with one roadside café offering food to the travelling motorist. Now half of these 'empires' have gone while others remain, so what's the difference? Woolworth's got overtaken by other stores and lost its edge to its competitors. Little Chef was overtaken by quicker and cheaper fast-food outlets and became disadvantaged.

But there is another way to build an empire, and this one has a major advantage built in. Think back to the secretarial staff working for the electricity board. Management had built an empire of people supported by an endless supply of money/income for commodities that were monopolies.

The magazine with the highest circulation in the country was *The Radio Times*, closely followed by *The TV Times*. It seems inconceivable today with a two-week programme guide on your Sky controller that these were the only publications allowed to publish programmes up to one week in advance.

Nobody else was allowed to publish more than a day in advance, until Thatcher's government changed the rules. What we have here are examples of artificial empires supported not by having a real advantage but by having an artificial one. These advantages were created by government rules and protected artificially because of them. In understanding this, we are just one small step away from understanding the biggest artificial empire of all, and that is the government itself.

You don't want to rely on the NHS for healthcare, so you can pay Nuffield or BUPA instead. The only trouble is that you still have to pay your share of the NHS as well. You don't want to send your child to state school, so you pay for private school instead, paying your share of the education budget as well. You want to provide your own pension for your old age, but you end up paying for the state pension as well.

The instant you suggest that you should privatise the NHS, you will be decried by socialists/liberals for trying to take a state asset of the people, for the people, and make it into a vehicle for profit for private individuals, when in reality it is a state asset providing profit (wages) for all of its employees. Where is the distinction?

To make these private businesses work, you would need a clear understanding and aim to drive towards your goal. It's true that not all of them would succeed, and new companies would need to take over the failing ones. But think about it: if your local Starbucks coffee shop started to fail because it had lost its way and was taken over by Costa, which hasn't lost its way, you'd still have the same number of coffee shops, but they would be run successfully for the ultimate benefit of the customers.

Unlike the present system, where you have 'failing hospitals' put into 'special measures', at least in a failing Starbucks you only get a bad cup of coffee. People don't die unnecessarily.

What is the socialist/liberal problem with profit? Is it because profit is seen as an additional burden imposed on the taxpayer? You must be fucking stupid if you think the present model either offers value for money or is sustainable. Has that got your attention? The present system is unsustainable, and I don't mean the management; I mean the whole thing, the NHS. Walmart claims to employ the highest number of employees in the history of the world, ever. How do you think our NHS management systems would compare with theirs? Yet the NHS employs 1,200,000 people. How many spare typists is that, do you think?

Now there's a thing: having decried the supermarkets for continually debasing food products, I am now offering them up as champions for the healthcare industry. When I was a kid, if you had your appendix or tonsils out, this was followed by a lengthy stay in a hospital bed.

When my son had his tonsils out, he came home that same day. Has the hospital care been debased, or have the patients' needs been properly identified? Kids survive the operation these days, so the conclusion would have to be that the earlier hospital stay was overdone. Alternatively, advances in medicine have moved the goalposts.

I have spent some time in hospital over recent years, both as a patient on the general and heart wards and as a visitor to friends and relatives. The conclusion I have come to—and others reinforce this whenever I say it—is, 'If you can't look after yourself, the hospital won't do it for you.' Now this is a damning statement to make when you consider that hospitals are in the business of care. I will give you some examples:

1) An elderly Asian gentleman lay wrapped from head to foot in a sheet in the bed next to me and refused all food supplied to him. Shut off in his little world complete with a language barrier, he was obviously unable to eat. After two days a relative visited and the patient wolfed down the baguette that the visitor had brought in for himself. The staff had failed to identify and satisfy the patient's needs.

2) My father-in-law was moved to hospital to 'sort his pills out' and ensure that he was on the correct medication for his numerous age-related conditions. Please explain to me how sealing his prescription in the closed hospital pharmacy from Friday evening until Monday morning was 'sorting his pills out'. He died two days after leaving hospital.

3) An elderly neighbouring patient with mild dementia was escorted by the sister to the toilet and was left with the instruction to pull

the cord to summon assistance when finished. While the patient was away, dinner was served. Despite many toilet cord pulls, Arthur was left abandoned. My wife and I reminded the sister that Arthur was stuck in the toilet, only to be faced down with, 'I'm doing dinner. I can't possibly do the toilet while serving food.' Another member of the 'team' arrived to remove Arthur's dinner, with the comment of, 'Not hungry today then, Arthur?' She would have taken his food if we hadn't stopped her. You can imagine the comments he would have received an hour later when Arthur said he was hungry. 'Don't be silly, love. You've just had dinner.'

Now I know a lot of you will be saying, 'I owe my [or my partner's, or my child's, or my parent's] life to the NHS, and I won't hear anything bad said against it,' but there is also a large slice of the population who will say, 'I lost my partner's [or child's or parent's] life because of the NHS.' Trust me when I say that neither side is lying.

Then there is the matter of condition management. In my instance I developed an electrical problem with my heart and periodically ended up admitted with tachycardia. The simplest way of describing this is to say that the top chambers of my heart beat at three times a second, while the bottom chambers beat about four times a minute. Once out of immediate danger, normally after about five days in hospital, I would be discharged to outpatients and the care of a consultant.

Then the game starts. The consultant is required to see you within six weeks, but it takes two weeks for your computerised records to reach his secretary for her to offer an appointment. Once you see your consultant, tests are ordered. These come in about six weeks, after allowing for the two weeks for the request to reach the test department. Results take another two weeks to reach the consultant, who has to see you in six weeks. At this point you have now gone twenty-four weeks—half a year since being discharged before your consultant considers your first test results. Further tests are then ordered, and a further sixteen weeks pass. This of course assumes that you don't get the secretary on the telephone. 'My consultant has to see you within six weeks, but I'm struggling to find an appointment because of the holiday. Can I ring you next week when I've had a chance to look?' When you do get the phone call at the end of the following week, you are told, 'Because *you* asked for a delay, my consultant can't actually get to see you before …' And now you are out to ten weeks before your 'six'-week appointment arrives. In my case this process took eighteen months and three hospital admittances to arrive at the need for an operation, which was a further eighteen-month waiting list away. Now this whole game is determined by minimum waiting lists to see your consultant so as to comply with government targets.

Measurement again, bless it. We have so much to be thankful for, don't we?

I cannot comment on the work done in intensive care, because thankfully I have not experienced it.

In a similar vein, we have all seen newspaper reports of tricks to circumnavigate the four-hour waiting time at A & E departments. Patients are kept on ambulance trolleys complete with ambulance staff before being booked in, only when a realistic chance of being seen within four hours is achieved. Other hospitals have a primary waiting area where you wait until called through into A & E, where the four-hour clock starts after you have already spent hours waiting to enter. Once again this illustrates the absolute folly of measurement. The system, to give the official truth and hide the real truth, corrupts whatever method of measure you apply. Ultimately the whole operation supports the lie, whether it's the junior staff, the consultant in overall charge of A & E, the management team, the Hospital Trust, the Department of Health, the health minister, or the prime minister. Can anyone within the entire team deny that they have knowledge of the target corruption that goes on?

So what purpose do the lies it serve? Is it plausible deniability that is the root cause? Does the lie start with the junior staff, or does it actually start with the prime minister? Whom does it serve the most?

If you are going to lie and corrupt the four-hour target, then what else are you prepared to corrupt? Is it only limited to A & E, or does it permeate the entire NHS? (It certainly corrupts outpatients.)

We arrive at another paradox: we need the bulk abilities of the NHS to secure the treatments for the population, and at the same time the bulk abilities of the NHS cause the corruption. If you had a small dedicated team treating a small group of people, you would not be able to afford the expensive equipment such as MRI scanners and CT scanners. You need the bulk resources, which at the same time corrupt the truth.

Once again we have the irresistible bulk and power of an elephant that we have no control of.

As Hannibal demonstrated, failure to control your elephants leads to loss of the battle and loss of the war. So what is at stake? The failure of the NHS and the failure of the UK economy, and this leads to the failure of the entire state apparatus, including education, welfare, and defence, and total defeat in the war.

A lot has been said about the banks being too big to fail and the need for the state (that's you and me bailing the banks out by injecting billions) to intervene. The banks are small beer when compared to the welfare state. Who bails that out? The answer is 'We do,' and we have done so since its creation, because it has been in continuous fail virtually since its creation. Essentially the whole edifice is based on the premise that each of us puts in according to our abilities and takes out according to our needs. This is, in simple terms, communism, the same discredited system that has failed throughout the world since its doctrines have been put into practice. We in the West have sat smugly by watching the collapse of the Soviet Union, ultimately the result of its financial position, while at the same time presiding over our own corrupt communist system. Explain to me, please: what is the difference? 'But we need to help. We cannot stand idly by while people suffer.' If the history of the world teaches us one thing and one thing alone, it is that interference always has unintended consequences. Think cane toads, think global warming, think Jesuit missionaries in South America spreading the gospel and measles, think feeding the birds in your garden at the expense of other birds throughout the world, think the invention of consumerism, think cutting the trees down so you can have a lawn, think eradicating malaria in the Garden of Eden. Think.

But this leaves us with no answers. Like the system we have gives us any answers?

Is our model for success an illusion built from smoke and mirrors? The 'Great Forest' of our cities is a lie built upon a corrupt delusion because it is not sustainable and Mother Nature is always in the wings waiting to deliver the coup de grâce, the end of *Homo sapiens*?

Why is everything corrupt? Why do we cheat ourselves? Whom are we actually serving if we lie to ourselves? What would happen if the A & E department actually told the truth and admitted that the average turnround time was six to seven hours (just a guess)? First of all, the ambulances would be released as soon as practicable, and either they would be ready to respond to more emergency calls faster or you would need fewer ambulances. (This could mean fewer 'silent witness' deaths where the ambulance arrived too late.) You would only need one waiting room with one set of receptionists, not two. Patients could be treated in order of medical priority instead of time priority. The patients themselves would see no difference; it would still take the same amount of time to pass through the system. The NHS Trust would lose its bragging rights to the non-existent performance that it claims. If the truth were known, then resources could be allocated to ensure that the targets could be met, bearing in mind that these resources/targets would also be corrupted. If the truth were known, then the government would lose its bragging rights to the performance that it claims (left or right, they both make the same claims). A stress-inducing artificial time pressure target would be removed from the medical staff. The management team would be paid to fix the problem instead of claiming success for failure, bearing in mind that they would corrupt the new data to solve the new dilemma. I ask again, whom are your lies serving?

If you are prepared to lie about the four-hour rule, what else are you lying about (in addition to outpatients)?

Do you lie about infection control, drug control, patient safety, patient care, cause of death, staff safety, staff training, etc.? Can we trust a single statement you make? After all, it's not as if anyone's life depends on it, is it?

When I think about it, the exact same process occurs with planning, where applications have to be processed within a given timescale. First of all, it can take up to a month to book your application into the system for the clock to start. Secondly, if the planners cannot process it in time, they ask for it to be withdrawn, while at the same time asking for further expensive reports. Then it's *you* who asked for the time extension, which 'fact' lets them off the hook. The cherry on the top: councils receive a bonus for performing on time (as opposed to being fined for failure).

A friend of mine is a textiles teacher of 14- to 18-year-olds at my former school. A few years back she received her new intake from the lower schools. All had been assessed as A–B grade students by their earlier schools.

When asked, none of the intake had taken any tests or exams, so my friend did one of her own, and rated them all as C–D grade. Now at the time all teachers could earn a performance bonus to their salary by providing good teaching standards. So my friend was now faced with a mediocre intake rated as A–B who, if they achieved any less, would make her a bad teacher, while at the same time all of the lower school teachers were 'excellent' teachers in line for a bonus. My friend complained to her head teacher (manager), who told her to bury her findings. Once again we have rewarded failure and corrupted the system. Why? What purpose does it serve?

Somebody out there produces an annual report on corruption and transparency in governments, and believe it or not, the UK scores quite high. It's a wonder there are not more white stick shops on the high street.

Am I a genius? Am I the only person in the world to work this out? Does everybody who measures lie by default, because they don't understand the measure? Is the measure the pro forma? 'I have a pro forma, so I can fill the form in and do not need to think; thinking has already been done for me.'

The thought has struck me about pro formas and the effective standardisation of our brilliant somnambulistic minds: we do in fact recognise difference and even celebrate originality. Think about the original progenitors of style, Constable, Stubbs, Picasso, Rodin, Moore, Matisse, Van Gogh, et al., and the almost messianic influence they hold over the art world and its patrons' purses. Although in a not too peculiar fashion, the mega bucks shelled out on an artwork is a reflection on the herding instinct, albeit a very rich herd. Is a couple of square feet of oil on canvas really worth millions when you could produce an exact copy for hundreds? Have you ever thought what the value of a personalised number plate would be worth if number plates or cars became obsolete and anyone could make and attach a number plate to their bedroom wall?

I have driven Volvos for thirty years. Yes, I admit it; it was me. I have watched the company pass from Volvo to Ford and now to the Chinese. When Volvo was sold to Ford for several billion pounds, the reasons given by Volvo were down to profit. Now Volvo used to brag that they had never made a loss in the history of the company and were dissatisfied because they could only achieve a profit of 4 per cent in the car division. At the time I thought that they had a second option to the path they took: they could put their prices up by 4 per cent and, at a stroke, double their profits. The route taken saw the mega deal pushed through, which no doubt built the reputation of the CEO and management team and no doubt produced mega bonuses at the same time for the seller and the lawyers and accountants who handled such a massive deal. Compare that deal on your CV with 'I put the prices up by 4 per cent and doubled the profits,' or in English, 'I did the job I was paid to do.'

A decade ago I was laying some roof tiles when my plasterers, who also did general building works, arrived to render the building. They watched me finish my job so I could get out of their way. A short while later one of them said to me, 'I've been thinking about all the nails you were using to hold the tiles down with, and then I thought, *If you can't do your own job properly, whose job can you do?*' I was simply fixing the tiles in accordance with the manufacturer's instructions. So is this the great lie? We all work to earn wages. We all work diligently to the very best of our ability to maximise our company's earnings and hope to secure our company's and our own futures.

Conversely, we all do as little as is reasonable and lie about our production and our targets on the basis that everybody else is doing it properly, so we don't have to. Are we back to the cheese-filled well in Arabia made from pure water syndrome? If you were teaching your child how to work with textiles, would you lie to them? If you were sending your child to your own A & E department, would you tell them that you will be out in four hours? If you were the minister for education, would you send your child to the local state school on the basis that this is the best education? Now the answer to this last question is yes, because otherwise that would be hypocritical and you are a politician first and a parent second.

Alternatively, 'I am a politician; therefore, I must comply with my own pro forma of politicians.'

Can anyone be honest? Is there no gene for honesty in our DNA? Is honesty another term for selfless?

Let's lose some weight. Which diet to select, which support programme to use, where to spend one's money? Let's buy low-fat, low-sugar foods and trawl the supermarket shelves for foods marketed at 'weight loss', or ditch the carbs and only eat protein. Let's read a magazine or buy a diet book or search the Internet for the latest fad. The diet industry in the UK is worth billions. Why? Our weight increases because we consume more calories or joules of energy than we use. There are other factors at work, such as age, hormones, and even the idea that your body has an ideal weight and disposes of excess calories at the digestion stage. But the simple fact remains that if you consume more calories than you 'burn', then these excess calories can be stored and increase your weight. So here's a simple idea for losing weight: eat exactly what you eat now, only less of it. You don't need low-sugar, low-fat, calorie-reduced food, or high-protein, high-fibre, or any of the multimillion diet schemes or organisations out there. Just eat less of what you do eat.

Do you think my penny's worth of wisdom will have any impact on the diet industry in the UK?

Now dieticians will take exception to my comments because they will also want to 'correct' the nutritional deficiencies in your diet. They will want to boost your vitamins and minerals while trying to correct your fats-to-sugar balance and regulate your good and bad cholesterol. In simple terms, your diet can be unhealthy and you can still be slim. But the original question has been sideswiped from 'how to lose weight' to 'how to have a better diet', the latter of which broadens their involvement and earnings capabilities.

Now I am not trying to become the latest slimming guru with thousands of disciples, so why do I bring this up?

I do a lot of menial tasks I cut and carry copious quantities of wood to heat my house, and I mow grass for up to five hours a week, all of which occupies my body but not my mind, so I chew the fat. I think you should try it sometime.

My neighbour who is doing an MA in sustainability asked for my help with a questionnaire she had produced to gather opinion from our little community. Unfortunately for her, I cannot answer questionnaires, because I can always find ambiguity in the questions and also because I always analyse why the question is being asked rather than thinking about what the answer is. In this case, having worked out what the answers could be and what my neighbour planned to do with them, I found it impossible to give an 'honest' answer for fear of steering her away from her ambitions.

What she did get was a sixteen-page (A4 size) reply while I disgorged what her questionnaire triggered. Sorry.

But this catalyst in turn led to *How to Kill an Elephant*, so in one sense its focus is on sustainability. But as I am sure you have worked out by now, it has expanded in scope somewhat.

What struck me was the fact that you can now take sustainability as a degree or higher course at university. That set me to thinking. All these trained 'sustainabilitists' are going to enter the jobs market, where they will use their accumulated knowledge to launch sustainability into the public consciousness. They are going to create an 'industry of sustainability', which we are going to pay for. We are going to pay for it by:

- increased energy costs to support artificial subsidies for renewable energy (for 'artificial', read empire)
- increased house-building costs to promote energy-efficient insulation and house design and to reduce the carbon footprint (what a shame we are trying to reduce housing costs to make them more affordable)
- increased taxes and levies to dissuade us from flying abroad for holidays
- increased fuel costs because of the need to promote greener components to our vehicle and heating systems (biodiesel—and don't mention rainforest destruction for palm oil plantations)
- increased costs of low-energy light bulbs over the cost of incandescent bulbs (£4 to £12 per bulb, as opposed to 50p).

My own development of two hundred houses required me to contribute £50,000 a year to subsidise the local bus companies to run buses through the estate during the construction phase (seven years projected at the time), even though the bus routes would be cancelled as soon as the construction phase ended. (This stipulation was subsequently dropped due to affordability issues post–2008 recession.)

Will this do for now? But while we are busy beavering away to reduce our carbon footprint and working harder to pay our taxes to the sustainabilitists' salary fund at the expense of all the silent witnesses who have seen their funding go elsewhere, we, that is UK plc, are hell-bent on securing 2 per cent to 3 per cent annual growth to keep our economy healthy. Now bear in mind that this growth is compounded, so if you achieve 2 per cent to 3 per cent growth for ten years, you don't get 20 per cent to 30 per cent; you get 22 per cent to 34 per cent growth. So if the government is successful in its economic ambitions, in ten years' time our carbon footprint will have grown by 22 per cent to 34 per cent, minus whatever the sustainabilitists have achieved. But give it another ten years and you are looking nearer to a 50 per cent increase. Can we really reduce our carbon footprint by 50 per cent over the next twenty years in order to stay still? But while you are at it, don't forget that the UK government is committed, along with the EU and USA, to significant reductions already announced to tackle climate change. Why do I get the idea that this target will be another achieved lie?

Also while you are at it, don't forget that the Indian subcontinent, Africa, and South America are all intent on growing their own economies to compete on the world stage, and I won't even mention _____. I will throw my penny's worth of wisdom into the debate and suggest that if you want to reduce your carbon footprint, use less energy.

Forget the diets: just eat less.

But as this cannot be reconciled with continuous growth worldwide, what the fuck are we playing at, and why are people making a living at it?

I've written earlier that this is the 'golden age of shopkeeping' given that the area of shop floor has increased by 300 per cent since 1980. I also understand, because 'they' tell me, that shopping is now a major recreational activity that I enjoy. (Give me a blowpipe and some fresh spoor anytime.) That said, what do we fill these shops with? How do you increase the product on sale by 300 per cent and retain any relevance? What I'm trying to say is, I am sick to death of traipsing around endless acres of shop floor looking at endless variations of endless cheap tat. How many different novelty wine racks can you give away at Christmas? How many different-styled bottles of oil with floating bits in do I want? [Place your own pet hatred here.]

Personally, on a scale of 1 to 10, where 10 is saturation point, I believe that our present-day shopping system is well past 50. Were we really so impoverished for choice before 1980? Have our needs really changed that much, or are we sucked in to some whirlpool of commercial depravity/ hooked on the new opium? But if shopaholicism is the symptom, what is the cause, and how do you reconcile that with reducing our impact on the world? Before you condemn me for wanting everybody dressed in the same colour clothes, bear in mind that the shops are all filled with *this* season's colours and choice is actually quite restricted. And unlike that which I won't mention, I will mention the Internet and all the choice that it engenders.

So what does all this choice provide us with? The illusion of variety, the ability to compare quality, the imperative to empty our wallets?

As the number of TV channels increases, what happens to the quality of the television programming? I have moaned before about the advertising interrupted by the programmes on commercial TV channels. Let us now look at the programmes instead. In fact, let's start with the programme, as in the index, the schedule.

Every day there are hundreds of channels screening for hundreds of minutes of television programmes. We have dedicated film channels, children's channels, God channels, documentary channels, news channels, lifestyle channels, music video channels, sports channels, traditional terrestrial channels, and classic repeat channels, and the vast majority are supported or sponsored by advertising. Every week Sky showcases around four new films on its Sky Premiere channels. That is to say four new films repeated hour after hour on two channels set one hour apart, because only around four new films suitable for a mass audience are produced each week. We are all familiar with it, so I will not delve too deeply, but what I will do is ask, what has happened to the quality of the programme? Essentially all this choice boils down to watching the old telly repeated over and over and over again at different times and on different channels, with the added bonus of iPlayer or box sets should we miss something. The more we get, the less we actually receive. The point I am trying to illustrate is that exactly the same thing has happened to our shops. You cannot fill endless shelves of space without constant repetition, so why attempt to? Just to throw another thought into the pot, not only has the footprint increased, but so have the opening hours. We never used to have supermarkets open for twenty-four hours, excluding Sundays, nor Sunday opening for six hours. All of this is done in the interest of 'choice', when in reality there is nothing extra to choose from. I do understand the need for shift workers to shop at unsocial hours. They are so busy filling the supply chain for the shops that they have to work nights. (I remember when a furniture chain sold bags of carrots for hundreds of pounds with a free three-piece suite thrown in to get around the Sunday trading laws: they lost the court case.)

My car broke down in France. I was stuck in Montluçon for a week and used my replacement hire car to travel to a large hypermarket. Imagine my disbelief when I arrived to find it shut for lunch. Yet the French don't starve or panic-buy to cover the inconvenience of a shop actually shutting.

If you stop and think about it, if all shops opened for a fixed period of five and a half days (including half-day closing midweek; remember that), then the same amount of produce would be sold. Extended hours once gave shops an edge, an advantage over their competitors, but if they are all open all the time, where is the advantage?

In reality the net result is disadvantage, with increased staffing levels and site overhead, yet everybody else does it, so we must follow. Time to review the pro forma perhaps and realise that you have driven your artic into a long cul-de-sac with nowhere to turn around.

As a further paradox to the TV debate, as the quality of the programme has deteriorated, the quality of the TV sets has exploded, with new system after new system replacing its defunct predecessors, sometimes within months. Would you like to watch your repeat in ultra, ultra, mega high-definition, or will you wait for the three-dimensional version?

Welcome to the Jungle

What is the city? Ignoring the hard structures, it is a densely packed cooperative of people who share the facilities and commit to observing the rules. Some do more than others, it has to be said, but (take a deep breath) that's why we have neighbourhood watch; and CCTV; and police; and traffic wardens; and dog wardens; and car alarms; and house alarms; and glass embedded in the tops of walls; and razor wire; and anti-climb paint; and litter pickers; and private security patrols; and door staff; and locks; and bike chains; and security lighting; and gates; and environmental health noise nuisance measurers; and 'No Ball Games Allowed' signs; and by-laws; and double yellow and double red lines; and 'no alcohol' areas; and trading standards officers; and speed cameras; and traffic light cameras; and bus lane cameras; and social workers; and lawyers; and private detectives; and ASBOs; and speed bumps; and pelican crossings; and bars on windows; and steel shutters on shops; and anti-ram raid bollards; and blue lights in pub toilets; and parking enforcement tickets; and fly-tipping laws; and registration numbers etched into car windows and stamped into bike frames; and SmartWater; and PIN codes for mobile phones and credit cards and computer passwords; and the Ten Commandments—so that we can all get along so well with each other.

Let's try that again, shall we? What is a city? Ignoring the hard structures, it is a densely packed uncooperative of people who are trying to get through life as easily and as selfishly as possible with the minimum of effort and by taking every advantage of others at every opportunity.

Think I'm being too hard? Hand on heart, who hasn't parked on double yellows or the in the private pub car park while nipping to the bank/shops/school/office, or dropped litter, or failed to pick your dog shit up, or listened to loud music, or …?

From a biological point of view, every single leaf in the forest is vying for sunlight and water and nutrients and doesn't care what it puts in the shade or deprives of resources, as long as it can get what it wants. People are no different. True, some plants are better, more ruthless, than others. Strangler figs race to the canopy by climbing existing trees and eventually crowding them out to the point where they kill the 'host' tree. Let us think of them as hardened criminals in our people forest/city. So where does that leave society, or should I say, what society?

Welcome to the society paradox. We all want everything that social cooperation provides, clean water, clean hard-surfaced roads and paths, public order, safety, disease controls, clear transport routes, and the like, while at the same time we all think that the rules of cooperation don't apply to us: *I've got a temporary dispensation pass.*

At the same time, society is preprogrammed to render assistance in whatever circumstance and however extreme. Consider the killers of drummer Lee Rigby, who was run down by a car and then attacked with machetes on a London street. The police attended and were attacked, and the murderers were shot. So what's the next thing that happens? Yes, you've guessed it: emergency medical aid is rushed to the scene, and the killers receive emergency treatment for their wounds. The same thing happened with the surviving Boston bomber; the treatment saved his life. Now I'm not trying to make a special case around terrorist-related incidents. I am simply using these extreme incidents that are well known in the news. If a driver was chased at 120 mph down a residential street and, having crashed, received life-threatening injuries, the emergency services would be on hand to render every assistance. Interestingly in this instance, the rapid-response paramedics and fire brigade would put their own and other road users' lives at risk by travelling at speed under blue lights and sirens to save the life of an individual who had just risked the lives of everyone else within range of his dangerous driving. Now life isn't as clear-cut as I make out, is it? Was the speeding driver rushing home because his partner had just told him that she'd slashed her wrists? Were the killers of Lee Rigby mentally ill and in need of treatment rather than punishment? Had the Boston bombers been tricked into leaving a 'harmless parcel' at the marathon finish line by someone else?

So we are preset to respond and treat first, and ask questions later. In fact there is legislature to that effect. Say I found my chief mechanical engineer on his knees cutting into the company safe at two o'clock in the morning. I can't actually sack him without giving him the opportunity to defend his actions; otherwise, he can take me to tribunal and win because I hadn't followed procedures. Imagine that his wife and kids were being held at home at gunpoint and would be shot unless he emptied the company safe. Stranger things have happened to bank managers. That is why we have trials and the full process of the law, to establish the facts before judgement is made. This way we don't allow the 'judgement' to bleed to death at the scene.

But what happens if the incident happens in slow motion? There's an interesting thought: a slow-motion high-speed crash that would probably be fatal.

How about a heroin addiction where you spend several years killing yourself, or destroying your liver through alcoholism, or eating to morbid obesity, or smoking to emphysema or lung cancer? At what point do we give up on people who do such things and remove all but palliative care? If you're diagnosed with a terminal illness and treatment is not appropriate, then all you get is palliative care. But all of these examples can be described as self-inflicted, and all of the sufferers will have had multiple warnings and ample opportunities to redeem themselves, because ultimately only they can redeem themselves. In *Futurama*, the creators have inserted 'suicide booths' in to the street furniture where you can select the ease and means of your own death without making a mess or being inconvenient to the rest of society. We just sell alcohol and tobacco and shit food in shops and illegal drugs on the street. The problem is that we haven't got the 'making a mess or being inconvenient to the rest of society' bit worked out yet. But why should it become my problem, and why should my taxes be committed to redeeming the unredeemable? Haven't I got enough demands on my money as it is? But while I concentrate on my money, the cost is so much more than this, because while we are messing around with these selfish arseholes who cannot say no, other people are dying from treatable illnesses that are not self-inflicted. So after 'the amount of money I've paid in fag tax, I've paid for my treatment', you can add to the CV, 'I don't care that I've inadvertently caused the deaths of others.' Really, this ought to be CD or CM if my Latin is right. 'The story of my life' becomes 'the story of deaths I caused', *curriculum mortis* if my guess is correct.

Robin Hood and His Merry Men

There has been a lot of talk in recent years about the use of violence to defend yourself and loved ones when confronted by a criminal within your own home. If my study of American TV programmes is correct, if you shoot somebody climbing in through your window, it's self-defence if the person falls into the building and murder if he or she falls outside the building. Now as very few of us in the UK have access to guns, this is unlikely to be adopted here.

I am going to suggest a simple solution which uses an old expression but also common sense.

When a criminal commits an offence, they become an outlaw and should, by their own voluntary actions, be considered to be precisely that, an outlaw. That is to say they have voluntarily stepped outside the law and should voluntarily surrender their rights to protection under the law.

When a patient refuses to modify their dangerous behaviour that is injurious to health, they have voluntarily stepped out of the healthcare system and should voluntarily surrender their rights to healthcare and cease to be a patient.

In *Saving Private Ryan*, the opening landing scenes end with the overrunning of the bunkers and the 'shock factor' of surrendering Germans being shot. So if I get this right, after firing thousands of bullets into the oncoming army, one can, when circumstances are less favourable, stick one's hands up in the air as though 'I'd had my fingers crossed and didn't mean to do you any harm' and expect to be treated with kid gloves.

I've seen a Brazilian TV documentary where the 'fly on the wall' camera crew were disgusted after filming a shootout between armed bank robbers and the police which resulted in fatalities on both sides, only to be turned away from filming so that the police could shoot dead the surviving bank robber who had surrendered. So, 'I've done my best to kill you all to make my escape, but you can't hurt me. It's not fair.'

How about, 'I have been smoking forty a day for thirty-five years and ignored every attempt to try to make me stop smoking, but now that I've got lung cancer, I want you to save me so that I can continue with the only pleasure left to me.'

I have known two relatives who lost not one but two legs each to amputation because of smoking, and both refused to give up the habit until they died. I bet the surgeon has sky-high job satisfaction and really looked forward to devoting his life to saving people while at medical college.

Ever since some nut went on the rampage in Hungerford, we have had bans on pump-action shotguns. And after the massacre of kids at Dunblane, all pistols have been banned and strict controls introduced for sports guns. You can't even train for pistol shooting for the Olympics in the UK, because you'd be breaking the law. Now these bans have been in place for many years, and yet gun crime hasn't stopped, because outlaws, being outside the law, still obtain and use guns as and when it suits them. Every time I hear of a shooting on the news, I think, *They ought to ban guns, oughtn't they? Then there wouldn't be any gun crime.*

The funny thing is, and it isn't funny, that outlaws, being outside the law, don't obey the gun laws any more than any other laws they find inconvenient. So our entire legislature only controls the law-abiding and only serves to fix the penalties for digression for the outlaws who get caught. It's enough to make you sleep soundly in your bed, isn't it? And while we are on the subject, guns don't kill people; people kill people. So if you really want to stop gun crime, ban people. If you really want to hurt people, you still can. Think people deliberately mown down by cars or multiple victim stab attacks or gas cylinder bombs in cars, etc.

Way back in this section I dropped the Ten Commandments into the discussion. I think these need further discourse. I should say that other religious groupings are available. I grew up in an age when religious education only catered for Christians, and Christianity is the only religion

I am vaguely familiar with. I will point out that Christians do not have a monopoly on supposedly Christian thoughts or acts. Imagine a world where rules were observed and society truly worked in cooperation. Think of all the redundant posts we would create: no police, no CCTV makers, installers, or operators, no traffic wardens, no yellow line painters (strange how we have to have red lines; it's like saying, 'We *really* mean it this time'), no self-inflicted injuries, no slow-motion suicides, no door staff, etc.

We've tried. If you believe the Bible, you believe that even God tried through Moses to impose order and control, he left out the bits about 'Thou shalt not drive in a bus lane' or 'Thou shalt not park here for longer than twenty minutes until after 5 p.m. or all day Sunday and bank holidays.'

I think even Jesus had a go. I had lost interest in school by then and wasn't paying much attention. So why do we try to impose order when it obviously doesn't work? What we end up with is some kind of acceptable buffered equilibrium where we tolerate only so much before 'Disgusted of Tunbridge Wells' writes in and the rules are adjusted. In English, 'We will have a concerted campaign against burglary by pulling our resources from traffic and drugs, which we will ignore in the meantime.'

Now people in Mexico might take exception to this idea, as the absolute corruption of the drug cartels has skewed the balance completely. But while an awful lot of additional people are murdered as a result of the cartels' activity, is the level of corruption greater in the drug cartels or the legal government? At least if a drug lord says he's going to have you killed, you die. Whereas when the government says that your education and healthcare needs will be met, they won't be.

> The reason our government is so corrupt is
> quite simple: they are in receipt of an endless
> supply of money that corrupts the government
> process as completely as the endless supply of
> money that corrupts Mexican drugs cartels.

Empire of the Sun

I went to the Ideal Home Exhibition in London a few years back for some ideas. OK, my wife dragged me there. If you go infrequently, you notice the changes in fashions and discover the new fad, the new double glazing that offers plenty of earning opportunity for franchises and self-start businesses. We've had the age of replacement windows and Jacuzzi spa pools and conservatories, and watched the prices come down as it gets harder and harder to sell these things.

So that season's fashionable flavour was renewable energy. The hall was bedecked with wind turbines, solar panels, and external heat source pumps. I enquired at stall after stall for prices. I had it in mind to fit a six-kilowatt energy source to my property to reduce my energy needs and was looking for the best alternative.

After several false starts, I found a technician who could give me a price for six kilowatts of solar panel and also a six-kilowatt heat source pump. I should explain that these heat source pumps work like a backwards fridge or air-conditioning unit. Instead of extracting heat from the inside of the room or fridge and pumping it outside, these extract heat from outside and pump it inside, and you can collect heat from the air outside or run water pipes through the ground to collect heat from the soil. In simple terms, for every two units of energy you use to power the unit, you typically produce six units of energy for consumption. I also had prices for a six-kilowatt wind turbine to erect in the garden.

Three very different methods of producing energy using three completely different technologies and machinery, but they all have one thing in common in that all of the energy is sourced from the sun. Actually, I lied. They had two things in common: the price; they all came in at £24,000. What a remarkable coincidence. I wonder why that is for three completely different technologies? Through the creation of artificial rules and subsidy, we have a new Empire of the Sun to rival the Japanese one of the Second World War, where our benign government, in the pursuit of reducing our carbon footprint, has artificially set the price of these commodities. The last couple of years has seen the EU accusing the Chinese of dumping overproduced solar panels into the European market and the threat of import tariffs. Yet we haven't seen a reduction in prices below the artificial threshold. I for one will buy the products when the price is reduced to the true market costs of manufacture installation, and profit. Only then will the stated aim of reducing our carbon footprint actually become a reality, when the 'free market' makes it affordable for everyone to actually use less fossil fuel energy. Our government has even sought to reduce the subsidies paid and has been taken to court and lost.

I told you, if you want to lose weight, eat less food. Don't go on a diet. It's cheaper to do it my way, and probably more effective.

Animal Husbandry
and the Keeping of Pets

Most children growing up in the sixties were exposed to pets of one form or another. This could be the ubiquitous cats and dogs, but guinea pigs and rabbits and gerbils and hamsters and budgies filled many a young mind and developed responsibility. These furry or feathered creatures depended on us for food, shelter, protection, and their veterinary needs and were a perfect microcosm of our future lives, where we would become breadwinners and have our own little people looking to us for their needs.

And then something happened. Suddenly 'imprisoning' these animals became unacceptable, and campaigns were launched by animal rights groups who would petition stores, raid mink farms and release very efficient killers into the wild to save their pelts (such a shame these people didn't stop to think about the damage to the fauna of the UK their selfless acts would and still does cause, stupid fuckers), and harass fox hunts and anglers. They basically took all the pleasure and normality out of pet keeping. One pet shop owner in Coventry I knew received a viable letter bomb, and other campaigns of intimidation have poured violent vitriol into the workers and shareholders of animal research institutes, including threats of injection with HIV-infected needles. School building exhibitions of bird and reptile shows were closed down by the actions of animal rights activists, and the whole industry driven underground and onto the Internet. Now I'm not saying that the mink and research animals were pets—far from it—but somewhere along the line, the distinction between these creatures and pets seemed to have gotten blurred.

(Something similar has happened with fireworks. Neighbours all complain and the pleasure has disappeared, so we are driven to attend organised shows, or else we don't bother.) I suppose part of the definition of 'pets' is that they provide pleasure, as opposed to financial or product reward. They are an indulgence. A breeder and exhibitor of pedigree dogs seeking to sell puppies for thousands of pounds would stretch the concept of pet keeping a bit, as would 'pet' mink for the fur trade or 'pet' cows kept for milk and meat. No, I think a true pet is one that is kept for the love of it without financial reward. And they are an indulgence.

But the thing about pets was you learnt how to care and provide for them. My second son appeared one day with some tadpoles he had caught in a bucket and wanted to keep them. There then followed a series of questions from me that led to him collecting an old plastic tank and cleaning it before filling it with chlorinated tap water and then changing it to pond water at my prompting. 'Is it ready, Dad?' I was asked. Then I asked what the tadpoles would eat, and supplied the answer of pond vegetation with algae. Our natural pond had brooklime, a marginal waterside plant with a small blue flower that rejoiced in the Latin name of *Veronica beccabunga*. My son disappeared and returned with said *Veronica*. Three days later his school also began to educate the class with a tank full of tadpoles. When asked what they eat, my son's hand shot up. He answered, '*Veronica beccabunga*,' to be met by a stony silence. *Not bad for a 6-year-old,* I thought. And it proved the importance of enthusiasm and interest when it comes to learning.

Of course keeping pets can be quite expensive, depending on the animals kept. I suppose you could run the range from goldfish in a bowl (often won at the local funfair) to a pony. Housing could include everything from cages with wheels to aviaries and stables. The costs all need to be met, but the reward is sufficient to justify the expense of housing, training, feeding, providing clothing such as rugs, and providing medical treatments such as wormers and flea powders, right through to operations and long-term chronic treatments. And of course there is the pleasure, even love, enjoyed from the experience.

I suppose the pinnacle of the pet-keeping world would be horses and ponies. Here you have the full range of housing, bedding, clothing, training, feeding, mucking out, transport, medical bills, and farrier and teeth care, and either owning or renting the space for exercise and grazing. Most of the population is not rich enough or motivated enough in the pleasures of horse riding and competing to justify the not inconsiderable expense.

What I didn't realise until many years after my pet keeping youth, was the fact that I am keeping other pets on a full-time basis. I could be talking about the fungi and bacteria living on and in my body, but I'm not. These pets are really expensive and cover the full range of costs. I will list them as:

1) housing
2) bedding
3) clothing
4) feeding
5) training
6) transport
7) heating
8) space for exercise
9) dental care
10) medical treatments
11) social services
12) policing.

Because, if you haven't worked it out yet, what else is a never employed but otherwise healthy person other than a pet which I feed and house and train and treat with medicines and provide with everything else on the list through the liberal application of benefits? If you think about it, such people are not even good pets, because you can't stroke them and you can't get any pleasure from them either. What purpose do they serve for society? How do they make the world a better place?

How has this class of pets escaped the attentions of the animal rights activists? Why is there no clamour to release them and prevent them from being kept? If you think I am being extreme, what would you define them as?

Crime and Punishment

Commit a crime and get caught, and you can expect to get punished for it. Why and how do we punish, and what do we seek to achieve?

Punishment can take the form of:
- public humiliation
- financial penalty through fines, costs, or compensation (this can also cover community service orders, where penalties can interfere with your earning abilities)
- issuing of 'proceeds of crime' orders to seize assets generated through criminal activity
- restriction of movement by curfew or banning orders or ASBOs (antisocial behaviour orders)
- restriction of movement by physical imprisonment, which can be subdivided into high, medium, or low (open) prisons, and adult or young offenders' institutes
- release under licence on parole, where the threat of return to prison for infringement hangs over you
- requirement to register and maintain knowledge of your home address with the police and the Sex Offenders Register.

In addition, much is made of rehabilitation of offenders. Establish the reason behind the criminal act and 'train' or educate the criminal so that he or she will avoid such a process again.

Having touched on the subject of criminals, I want to look further into this area. When criminals are caught, they receive sanctions which range in severity from a bollocking to a whole life sentence without the possibility of parole. Now the purpose of these sanctions are twofold: to act as a deterrent to prevent crime happening or to remove the criminal from open society so as to physically prevent him or her from reoffending. As an added bonus, we can educate and rehabilitate criminals as reformed characters to re-enter society. Now we all know this works because nobody reoffends, and for that matter the deterrence factor prevents crime from happening in the first place. Not. So if it doesn't work, why do we persist in it? Criminals, new and practiced, soon learn that the sanctions imposed are often only a minor inconvenience and are no real deterrent. Most custodial sentences are routinely slashed for 'good behaviour' (sorry, what I meant to say was 'to prevent overcrowding'), and on balance the rewards justify the penalties. When criminals emerge from prisons, they are stigmatised within the employment world and find it difficult to obtain well-paid jobs. In the UK, around one hundred thousand individuals are routinely held in incarceration, although in the USA fully 1 per cent of the population enjoy this condition. Now to my mind we have a perfect 30 mph road sign situation that is subjected to initiative after initiative that doesn't work—another perfect example of attempting to redeem people who will only be redeemed when they redeem themselves. So why is it my problem? Why am I paying for successive failure after failure? When I pay my taxes to pay for the criminal justice system, I expect those funds to be used wisely and effectively, not pissed up against the wall. By adopting the 'outlaw' scenario outlined earlier, these outlaws should be stripped of their rights to the protection of the law and removed from society. If you are going to impose a 30 mph speed limit on a road, enforce it. If you are going to punish criminals, punish them. If you are going to incarcerate criminals, incarcerate them. Anything else is pissing in the wind and there is no deterrent. You might as well invite them to commit crime. How far away from a Mexican crime oligopoly are we sleepwalking towards? 'But what about the cost to society?' I hear you cry.

Well, what is the cost? Think about the 'take a deep breath' on page 167 and begin to quantify all the cameras and yellow lines and police and door staff and decide whether this burden is justified. It's all been slowly drip-fed into our cityscape without anyone's ever quantifying the cost. The whole system is an exact replica of that 1952 cumulative song about swallowing flies, spiders, and the like until the reality of death sneaks in, written by Rose Bonne with music by Alan Mills. At least we are much cleverer than that now, aren't we?

On our present scale, we've gotten to swallowing a killer whale to catch a leopard seal. The old woman is still going to die though, and we can't even remember what she swallowed in the first or second or third place. What people do bleat about is the cost of the judiciary and prisons, which can be quantified, while at the same time ignoring all the other costs. We can't afford to lock any more people up; it costs more than a five-star hotel to keep them inside.

While filming in the Picos de Europa in northern Spain, I have spent many weeks in Asturias, which is the smallest province in Spain. Most of the Picos is in Asturias, but it does nudge into Cantabria and Castile and León. As you drive around the narrow hill roads past the locals, one of the things that strikes you is how they look, I mean really look, into your car in the hope of spotting a friend. It's not dissimilar to when I had to drive past the gamut of journalists to my offices when Nigel Lawson was resigning. These people live in a community and expect to see and acknowledge the passing of neighbours. There is one road that is just outside the Picos where I have been flagged down on two occasions to provide lifts up and down the valley, once by an elderly man on his way to the dentist and another time by an elderly man who stopped me so that his granddaughter could travel four villages up. Not all the hotels I stop at offer breakfast, and you have to walk to the local bar, which is packed with the land workers charging up with fags and coffee and gossip on their way to work. It is quite common for the vehicles to be left, engines running and even the doors open, at the street side. The owners expect them to still be there ten minutes later. I 'read' an article in a local paper where the chart confirmed that Asturias had the lowest recorded crime rate in Spain, and I can believe it. Talk to the elderly in the UK and they will tell you how they never used to lock their doors when they were kids and crime was relatively unheard of.

Somewhere in the last few generations, we have entered into our present easy-come, easy-go world where everything has to be bolted down and protected or else is lost.

My father-in-law was born and lived his entire life in one village, only leaving for active service during the Second World War. He could talk for hours about the old village and how he knew virtually everyone within the community.

As part of the planning process now, we have an initiative called Designing Crime Out, whereby the police will inspect housing estate drawings and tweak the design with an eye to reducing crime. Extra windows and lights overlooking 'quiet' areas are incorporated, and footpaths sealed or expanded to cut down on boltholes.

Perversely when designing industrial estates, the policy seems to be one of designing crime in by hiding these elements within tree belts and earth bunds for fear of disturbing the neighbours. The irony is that as these housing estates are built at such scale and speed, there is no sense of community, and respect or responsibility to your neighbours; and crime therefore factors in their development. The message that we are sending out loud and clear is that we tolerate crime and actively support it by our failure to find a solution.

Many years ago my managing director used to receive a 'comic' of investment tips and ideas for a rosy future. The premise was that if you got in on the ground floor of a fledgling business, then you could strike it rich for very little outlay. A one-dollar share in Coca-Cola from the 1920s would, provided you subscribed to the scrip issues and whatever else, would now be worth one million dollars. The tips they offered for a rosy future in this issue were based on security—anything to do with security: alarms, body armour, Kevlar, security patrols, etc. The reasoning offered was that we were entering an era where the polar distinction between the haves and the have-nots was going to widen, and basically the haves would spend a significant amount of their resources stopping the have-nots from stealing it. I envisaged walled estates with armed security patrols and machine-gun towers and searchlights, and in places the reality is not far removed. Kids are being sent to school with stab-proof vests, and gated communities and private security patrols exist. The question I would ask is, why treat the symptom instead of the cause?

I do not pretend to have the answer, but I will opine that the one we have isn't the right one either. So do we continue the downward spiral until we reach Mexican standards, or do we sort it now? In films like *Escape from New York*, the city has been turned into one huge prison camp and all the criminals are housed in one area where they rule themselves. I am not going to ask what has happened to our 'moral compass' as though religion had the solution. Rather I will ask, why, in this day and age, with all the enlightenment and education that we enjoy, can we not accept that crime is wrong and unacceptable? Every time we accept that a shoplifter shouldn't go to jail because it's only the thirtieth time they've been caught, we actively invite more crime. Every time we accept that a custodial sentence isn't appropriate for a serious assault, we are actively inviting more of the same. Every time we release an offender back into the community without their redemption, we are actively encouraging more crime. Either impose and enforce the 30 mph zone or don't, but don't spend billions on fudging the issues. The term we are looking for is 'zero tolerance'; anything less is tacit approval and an invitation for more of the same.

'But we can't afford it' will be the universal reply. So we can afford what we have, can we? Remove car alarms, house alarms, computer locking systems, SmartWater, CCTV systems, store detectives, and door staff. Reduce your insurance premiums and empty your A & E's of assault and stab victims. 'I carry a knife for my own protection' turns into a gunslinger match between egos, with many stab victims stabbed with their own knives. 'I live in such a rough neighbourhood that I need an attack dog to protect me and my child' translates into 'I live in a rough neighbourhood because it's populated with attack dogs.'

Anyone who can leave the community will do so, leaving the effective dregs and those others trapped by circumstance. No decent person with a choice would want to live there. 'It's covered in attack dogs, for fuck's sake.' The fuckwits' cure is part of the problem. Turn prisons into work camps which are self-financing. Work to eat and keep warm and sheltered, or don't. Why should it be my problem?

Learn a trade and respect and leave, or stay there and rot. Why should I care?

These people have no compunction or care for their victims or crimes. Why should they have the monopoly on indifference? Would you get upset if a cure for cancer was developed that would condemn its existence to history? Why not treat crime as a modern-day cancer and eradicate it, compartmentalise it, and remove it from society? If it wants to continue in isolation, let it. If not, take down the 30 mph signs and accept that a cull of society will continue unabated. Or if we don't enforce the law, then we should accept that the cull of society will continue because we let it. In the meantime, I will continue to pay my taxes and alarm company for the protection I don't receive.

Ask any victim of crime if they have received justice. The only real justice is not to have been a victim in the first place.

We have an entire criminal justice system industry that does not serve its purpose of either controlling criminals or providing justice. I no longer want to pay for it. Straightaway I know my ideas will fall afoul of the Human Rights Act, but so what? You voluntarily stepped out of its protection when you committed a crime. Because the criminal justice system is part of the problem, it cannot provide the solution.

How do you kill an elephant?

Let's retrace our steps a little and return to the Mexican drug cartels in control of Mexico. They achieve their corruption because of the seemingly endless supply of money that their trade generates. But the principal market is not Mexico; it's farther north in the US and Canada. So by buying illegal drugs in these countries, the price, in terms of murder, is exported and paid at home. But the north is not immune from the corruption that money can buy, and murder and crime expands here as well. Drugs flow north, money and guns flow south, and people die. Seems a bit like buying birdseed for our bird tables, doesn't it? Our birds live, but other birds have to die to produce the birdseed. Nowadays of course other links are mooted between fraud and drugs and people smuggling with the funding of terrorist organisations such as al-Qaeda. But I drew a comparison between the corruption of the cartels and the corruption of the legal government. Now the government's problem is the seemingly endless supply of money (taxes) that it receives. This wealth corrupts just as much as the drug money. It's just that the politicians never admit it.

By now you will have concluded that I am so right wing that I am off the scale. Not so. I don't buy into the argument. What's wrong with being right or wrong, realistic or deluded, shuffling through this life plugged in to the matrix or detaching oneself and waking up?

Now as you may be aware, 'matrix' is another name for the uterus or womb, which houses the placenta, the exchange membrane whereby food, minerals, and oxygen are diffused into the foetus and waste products removed without the mother and the baby's blood mixing. We plug into the matrix of education, of work, of benefits, and it all works because there is a gradient whereby resources pass from high concentrations to low. This is the problem with poor countries, where the gradient between rich and poor is slight.

How can you gain an education from a teacher who has been poorly educated? How can you have a proper health service where the resources are insufficient? You can be inoculated against polio but die from diarrhoea. But of course in these poor countries there are often very rich people who, the socialists will argue, 'owe their wealth' to the poor and it should be distributed.

You don't have to go to poor countries, simply poor deprived areas in the First World. In deprived inner cityscapes populated with attack dogs and junkies, the 'placenta' is malformed and the proper diffusion of resources cannot occur. How can you be properly educated if high truancy has been tolerated? How can you build a strong community when residents are frightened to leave their houses for fear of burglary, assault, or dog attack? How can your children play and develop properly when the parks are littered with used needles? How can you search for work when your address condemns you to mediocrity, assuming you can overcome the peer pressure from your unemployed neighbours? Now some will want a better life, will learn hard and work hard to escape from the deprivation, and will succeed and become rich (I'm talking comfortable, not millionaires). Some cannot learn or work hard through disability, and they deserve our help. Most will ignore the opportunities offered to them and will settle on the mediocre, the easy line of failing to learn or to work or to contribute. The rich learn by their actions and get richer, and the poor stay poor. This is where I see the class betrayal. It's not that the rich betray the poor by failing to provide more resources. It's that the poor betray the rich by squandering the resources and opportunities provided. Every single poor person who is born fit and healthy and survives in that condition has been blessed with all the resources needed to better their life if they would just seize the opportunity. They get free education and healthcare and shelter to launch themselves into the world, but they choose instead to squander it.

Everybody of working age born in Britain has been provided with free education until the age of 15 to 18. Why is that my problem? How many times must I empty my pockets of change for these undeserving cases? Why are poor people poor?

Now don't go jumping to any conclusions, will you? Wait for the great unveiling in the drawing room while we sort the villains from the innocent.

Why do companies go bust? I don't mean shutting down or going out of business; I mean failing catastrophically with mega debts. Assuming that they have not been on the receiving end of bad debts themselves, there has been a failure of management. This will be compounded by a lack of communication or a lack of interpretation, or it is a problem that lies within the organisation. In simple terms, the company has spent more in supplying goods or services than it has received in revenue, but this alone would not cause a mega failure. Somewhere within the system something has broken. This could take the form of a mistake, a crime, a lie, denial of the facts, or a failure to act on the information received. Sometimes it's just bad luck. Borrow heavily to expand weeks before a major recession and you will struggle, but this is also a mistake, exposing yourself to too much risk. Discover that your chief accountant has been embezzling revenue and lying in the books to cover it up—still a mistake, because you can always check the bank balance. Many people with debt problems 'bury their heads in the sand' and ignore the consequences until forced to confront them by the bailiffs turning up on their doorstep. Company directors can do the same, but this is still a mistake for failing to act. Don't believe your quarterly accounts and refuse to accept that you have a problem? Still a mistake of judgement. In all of these examples, I have broken them all down to being mistakes, because if you hadn't made the mistake, you would still have a viable business.

How do government departments work when they deal in corrupt data from one end of the year to the next? How can you spot the mistakes, the crime, the lies, and the denial of the facts when the facts are corrupted? And how can you act on the information received when it is corrupted? If the greatest measure of our education system is exam grades, and if they are the product of lies, how can you measure our education system? If the greatest measure of our education system is money spent on education and the money can't be accounted for properly, how can you measure our education system? Teachers cheat to meet their performance targets and enhance their earnings, and headmasters allow them to because it helps them meet their own performance targets and enhances their earnings as well. Why volunteer to be poor?

Hospital managers cheat to meet their performance targets and enhance their earnings, and hospital boards allow them to do so for the same reason. If the measure of success in hospitals is done on the four-hour A & E turnround and then they lie, how can you measure it? If the measure of success in hospitals is done on the six-week waiting time in outpatients and then they lie, how can you measure it? If the measure of success in hospitals is how few deaths occur in surgery and then they lie about it by refusing to treat high-risk patients and let them die instead, but not in the theatre, how can you measure it?

If the measure of success in the dole office is how many long-term unemployed people are taken off the list and then they lie about it by moving them onto the disability list instead, how can you measure it?

If the measure of success for highway engineers is the amount of money spent on reducing the speed of vehicles in built-up areas, it doesn't answer the question, which is, how many vehicles observe the limit and for how long?

If the measure of productivity of council workers is when they start and stop their vehicles for the day and then they lie about it, how do you measure it?

If the measure of success in the planning department is how many planning applications are processed within the targeted time frame and then they lie about it, then how do you measure it?

If the measure of success in the police, department is how many hours officers spend on patrol and then they lie about it by 'parking up' and doing paperwork away from the office, then how do you measure it?

If the measure of success in the police force is to crack down on either burglary, drugs, muggings, pickpockets, or shoplifting and then this is done at the expense of the other crimes, how do you measure it?

There is a medical expression used when tissue is damaged beyond repair; they talk about the tissue being corrupted. To save the body in such a case, amputation is normally required. Have we just found a way of killing an elephant, by letting it die from untreated wounds, letting it die from corruption of its tissues? But this does not help us. We have caught, trained, armoured, manned, transported, fed, mucked out, and launched our elephant into battle, and the last thing we want is for the elephant to die before it can be effective.

There are articles that appear infrequently in the newspapers that highlight the plight of an individual who has been let down by the police and their failure to act. This may take the form of 'victims' driven to commit suicide or victims murdered by 'disturbed' individuals after multiple attempts to recruit assistance from the police have failed to gain any traction. Victims may have been driven from their homes because of these failures or may have been forced to defend themselves, which in turn has made the victim the criminal in the eyes of the law.

The criminal system could be seen to be corrupt between stated intent and action.

This corruption could simply been seen as a lie.

And after the lie comes a warning.

And after the warning comes a caution.

And after the caution comes a conditional warning.

And after the conditional warning comes a conditional discharge.

And after the conditional discharge comes a fine.

And after the fine comes a curfew.

And after the curfew comes a tag.

And after the tag comes a community order with an unpaid work element.

And after the community order with an unpaid work element comes a suspended sentence.

And after the suspended sentence comes incarceration. But by then it doesn't matter, because having swallowed all of the above, the victim is dead.

Community Firemen

Franklin D. Roosevelt coined the phrase of 'lending your neighbour a hosepipe when their house is on fire'. When the UK stood alone against Nazi-occupied Europe, Roosevelt knew that the American people had no appetite for joining in the fighting. His solution was to lend the armaments and equipment to us so that we might get the job done. He spoke in simple understandable terms that would carry the people and broker no argument.

Your neighbour's house is on fire and he is in need of help. You willingly assist him by offering your hosepipe for his use. Your neighbour can now put the fire out and save his house from destruction.

Would you be so helpful if your neighbour had already borrowed your hosepipe once before and had lost it without recompense to you?

Would you be so helpful if you had advised your neighbour not to light bonfires close to his wooden house on numerous occasions before?

Would you be so helpful if you knew that your neighbour was operating an illegal drug-manufacturing facility using dangerously volatile chemicals?

Would you be so helpful if you knew that your neighbour had deliberately set fire to the house himself to make an insurance claim?

At what point do you decide, if ever, that your neighbour's house is irredeemable and just sit back and enjoy the spectacle?

Benefit: 'a helpful or good effect; something intended to help; to be helped by something; to help someone.'

Help: 'make easier; to make it possible or easier for someone to do something.'

The smarter ones among my readers will have realised that I am not referring to burning houses but to benefits and welfare. Once again I will look at a 'slow-motion fire' as opposed to a normal-speed fire.

In this instance let us suppose that the neighbour's house is failing due to a lack of maintenance. For example, the wooden cladding needs painting and some roof ridge tiles are loose. In itself there is no immediate problem, rather a slow-motion problem that can be readily addressed by the application of a paintbrush and a bucket of mortar. In spite of the fact that the neighbour is aware of the problems, nothing is done to maintain the house until the point of water ingress or structural failure. Depending on the neighbour's circumstances, you may be inclined to offer assistance. Elderly or disabled neighbours would be a prime example. In their way, they would reciprocate as best they could, for example by taking in parcel deliveries or keeping an eye on your property in your absence.

Benefit: 'a helpful or good effect; something intended to help; to be helped by something; to help someone.'

Help: 'make easier; to make it possible or easier for someone to do something.'

Let's go back to the neighbour's burning house and accept that he didn't actually start the fire himself. You don't just lend him your hosepipe; you also operate it to try to put the fire out. You brave the heat and smoke and toil to put the flames out, when you notice that your neighbour has gone missing. You ask another neighbour where he's gone and get the reply, 'He's gone fishing.'

Benefit: 'a helpful or good effect; something intended to help; to be helped by something; to help someone.'

Help: 'make easier, to make it possible or easier to do something.'

So in the above example you have provided a good effect (your labour) and something intended to help (your hosepipe). You have also made it possible and easier to do something (put the fire out). But in reality you haven't helped; you've done it all. And if the fire were to reoccur and you were not present, then neither the means nor the knowledge to put the fire out would be present. I want my hose back.

The only possible justification for continuing to put the fire out is that your own house might catch fire as well. Is this the sole premise for providing benefit, to control the risk of contagion and to protect you? The other option is to build your house on the other side of a firebreak and isolate yourself from the risk of contagion. This is called paying attention at school and learning the lessons of life so that you can live in a 'nice area' and have a 'nice job' and avoid the 'arse-onists'.

Nearly Back to Runts

If we consider the idea of providing housing as a short-term solution to families while they get established before moving on to their own properties, then using the above analogy we have not in fact offered help. What we have done is provided a permanent solution with the introduction of council housing, which, once you've successfully qualified for it, is yours for life—and your children's and their children's if you play the system right. The concept of a helping hand or a start in life has long since evaporated into a permanent entitlement to the grave and beyond. (Yes, I personally know of one house now on its fourth generation of occupiers.)

Now one of the problems with the matrix is the diffusion of materials from a high to a low concentration. The greater the difference, the greater the potential for the movement of material. In educational terms, starting with infants, the potential is great. As the child moves through the school, the upper limit has to be increased so as to maintain the differential, until eventually the differential ceases. By this I mean that the pupil knows as much as the teacher and the pupil needs to evolve into a student to find the differential for herself, or leave and learn from different teachers, peers, or employers. In academia, students can go on to a greater understanding than their professors and become professors in their own right, as an example.

Let's chuck in the issue of measurement; you know how I love it. If the teachers lie about the performance and ability of their pupils, then how can you trust that the matrix differential exists and your child is getting a proper education? You can't. But I am going to offer another suggestion here which devalues the whole process. The matrix has been corrupted and has been converted into a pro forma. Learn the pro forma and ignore the fact that this is not in fact an education.

Great Brains

We have a self-appointed guardian of the government calling itself the Taxpayers Alliance, which is described as a right-wing think tank or pressure group. You can search for it on the Internet and view various press releases and appeals for spending reductions in government budgets. One of the things they are asking for is a £12 billion reduction in spending. Firstly, £12 billion is a minute drop in the ocean. Do they believe that saving this amount of money will make any difference at all, especially as the 'Great Brains' will lie about it? It's interesting to me that this organisation is described as right wing, as though it's only the right wing that wants to reduce spending, so by extrapolation left-wingers must want to spend more. We are back to the introductory argument about the greatest brains in the country deciding what is good or not good for us. I suggested that in reality the argument boils down to what is good for the greatest brains, and I stand by that. So what constitutes a 'greatest brain'? What are the entry qualifications for these esteemed positions, and how can I become one of the brains? One could start with acknowledged great thinkers, Karl Marx or Albert Einstein or maybe Stephen Hawking, but straightaway one needs to ask what it is that they actually thought or think about, and if it helps us.

In my little village, which has a real sense of community, my wife is active in the Women's Institute Committee (WI) and has helped at various village extracurricular activities such as the Millennium Ball and Queens Jubilee celebrations. What you notice is that the same faces crop up at these activities, a hard core of public-spirited individuals who give their time and effort to ensure that these things happen to the benefit of all. While you would not necessarily describe them as the greatest brains in the community, they are in the sense that they are the brains who get involved. So in our microcosm of the village, our greatest brains are effectively qualified by turning up and proffering their services. The real intellects, if they exist, who don't turn up, don't have a voice and exclude themselves from the process.

Our government is split into different elements that combine to form our 'greatest brains', if for no other reason than that they turn up. We have elected MPs, appointed civil servants, appointed members of the upper house, appointed researchers, and appointed consultants or special advisers, all of whom are constrained by the constitution and the law lords, who ultimately interpret the meaning of the legislation issued by Parliament. I have ignored the devolved and local governments that operate on a similar principle and the EU, which operates as a law unto itself. Our MPs generally aggregate into factions of like-minded individuals. We call these 'political parties', and they collectively produce manifestos of policies designed to garner public support. MPs are obviously voted into power by the electorate, and these manifesto 'promises' are the 'sales literature' released to the media to attract votes. The most successful party normally forms the government, and the second-most successful party forms the official opposition, designed to act as a counterweight to the ruling party and to provide checks and balances through debate. While each group of like-minded individuals tends to agree on policies, some areas of debate can split these groups. It becomes possible through voting in the House to overturn contentious proposals that cannot command a majority of support. The civil service is designed to be apolitical, and they act as the clerks and administrators in the preparation of new legislature for the benefit of the ruling party. The appointed researchers and consultants advise the MPs and help to shape the policies. As you would expect, they are selected because they agree in broad principle with the ruling party. So right-wingers employ right-leaning consultants, left-wingers the opposite. The centre ground employs consultants who lean either way or have no strong opinion, depending on who is talking to them.

Now I realise that this is a very simplistic approach to the process that governs us, but I seek to highlight one important point. Where in this process do the greatest brains come in? Because you've been voted in by the majority of your constituents simply makes you popular, not a great brain. Because you've been voted in, you can appoint your own consultants, who will be appointed because they were available, they have a similar view on the proposed outcome, they have a track record of talking sense (they don't rock the boat), and they can be worked with. They also take a salary, often a big one, and therefore have a vested interest in saying yes and 'agreeing'. Where in this process do the greatest brains come in?

Theoretically our MPs are motivated by serving the public, but as the winner of our elections simply has the highest number of votes (a first past the post-electoral system), it is possible to be an MP by winning one more vote than the runner-up. For example, the winning MP could receive 20,001 votes for, the second-place candidate could receive 20,000 votes, and the other six candidates combined could receive 20,000 votes in total. So which public does the MP serve, the 20,001 who gave him the mandate or the 40,000 who didn't? Don't think that changing the electoral system to include proportional representation will change the system in any way either. Both systems are equally flawed. Collectively I would suggest that each MP serves the party he or she represents first, followed by his or her voters, followed by his or her constituents. Individually I would suggest that each MP serves himself first, second, and third, and only occasionally does an MP commit slow-motion suicide by letting her emotions get in the way by working to her conscience. (Think back to the selfless–selfish debate.) This last government under Cameron has fixed the government term of service to five years, so I would suggest that the first four and a half years will be spent serving the party, and then a gradual switch to serving the electorate will kick in upon the approach of the next election.

I have often received planning advice from planning officers to hold an application until after the local authority elections, because committee members are that busy chasing support that they won't support any applications with public letters of complaint for fear of losing votes. The same application will be passed unopposed a few weeks after the election. So whom does this process serve, the public or the government in all its guises? But we have to have government; otherwise we have anarchy, don't we?

In biological models the rules are a little simpler. Bee and ant colonies are controlled by chemical secretions from the queen in an apical bud manner. Estimates indicate that the greater animal biomass on the African savannah isn't the great herd of wildebeest, zebra, or gazelles but ants and termites, so if nothing else, this proves that the chemical secretion works. One could argue that the great dictatorships have worked on a similar model, using fear as the chemical control element. Mountain gorilla groups are controlled by the dominant male, who wins his position through physical strength. In human history, the man with the sharpest sword and strongest arm has replicated this on a larger scale. Elephants are a little different in that the oldest female, the matriarch, uses her memory and knowledge to guide, protect, and lead her family group. But there has to be a mechanism whereby larger family groups can split and prosper, simply because otherwise there would only ever have been one group of elephants. The human model could be the village elders in Papua New Guinea who hold the influence, but periodically new village groupings and new villages would appear.

Now all of these biological models have one thing in common, and that is size. If the ant colony exceeded a certain size, then the chemical signals would become too diffuse and the system would break down. This could be the trigger for the rearing of sexually active males and females to be undertaken with the resultant dispersal of the colony. This could have two causes: the queen is dead or sick and a new queen is needed, or they have been so successful as a colony that they have surplus resources to spend on rearing sexually active individuals. In human terms we could think of small sections leaving the main body of the grouping and wandering off over the horizon to fame or failure.

Our problem is that we don't have any horizons anymore on our crowed island; there is no vacant hinterland to colonise, no vacant living room. So our great herd bunches up and congregates in the monocultures we call cities, and we conform to the single leader, which I will call the 'Great Brain'. But how many of us know the instruments of control on a personal basis? How many see the legislature designed, enshrined, and delivered to the populace? What we catch is the faint whiff of a pheromone that indicates 'No Ball Games Allowed' or 'Clearway for 5 Miles', reinforced with an education system that says 'no spitting, no swearing, no farting, no insolence, etc.', and that underlines the fabric of our society, given that we all have our own 'personal dispensation card' that excuses us from compliance.

I have an interest in orchids and am particularly fascinated by the *Ophrys* genus in particular. These *Ophrys* can be found in rare sites throughout Britain but are typically at the northern extreme of their range. The nearer you get to the Mediterranean, the greater the species and numbers you will find. To our eyes, they mimic insects and have names like fly, bee, bumblebee, and sawfly orchid. They appear to have wings, compound eyes, and antennae, together with 'furry abdomens' (*Ophrys* is from the Greek for 'eyebrow'). They transfer their pollen from plant to plant by enlisting the help of insects that attempt to mate with these pseudo females: pseudo copulation. Some plants appear to have specific 'fertilisers' and have enlisted the aid of solitary bees, for example where the males emerge a few days before the females and attempt to mate with anything vaguely familiar. But what scientists are now discovering is that these flowers also produce chemical secretions that mimic those of the female bees. While they are mimics, they are not a perfect copy, and as soon as real females appear, mating success with the flowers diminishes rapidly. How the plants successfully copy the insects' pheromones is by trial and error. Dozens of chemical compounds have so far been isolated from a single species, one of which seems to do the job. As a fail-safe, these flowers will, in the absence of suitable insects, self-fertilise and effectively clone themselves.

Nice story, but what is the relevance? Our chemical receptors are so far removed from the queen (Great Brain) that we follow the mimics and copies of these chemical controls (laws) and are deceived into complying with something we don't understand (shagging the wrong flower). Let's be a little less poetic: The government decrees that it is not safe to exceed 30 mph in a built-up area. We all know that exceeding 30 mph is dangerous, but we follow the actions of our peers and other road users and do whatever speed we like. Occasionally a speed camera or police trap reinforces the law and the 'chemical control' is reasserted, we think. But it isn't, because we are so removed from the central control that the message that is asserted is incorrect: 'Bastards are always fleecing the motorists'; 'Haven't you got some proper criminals to catch?'; 'Why didn't you stop the car in front? It was going much faster than me?'

I keep banging on about the 30 mph speed limit because everyone is familiar with it. I could just as easily talk about shoplifting or illegal drug taking or murder. If education is the mechanism that binds us all into one society, and if education is sourced from millions of 'events'—some formal, such as schools and textbooks, and some informal, such as from employers, relatives, TV, and so forth—then there is no formal education and there is no overall control, and we therefore live in a sort of mutually beneficial symbiotic anarchy. Don't shit on my lawn and I won't shit on yours.

Occasionally the mutually beneficial element breaks down and central control is re-established by the boys in blue and the law courts, but this central control by its nature can only be demonstrative. You could not fine and issue penalty points to every speeder because then no one would have a licence. In a sense we are back to the great mass of starlings. We react to our neighbours inches away, but the majority cannot even see where they are going. Break the starlings down into the small feeding groups and you arrive at a parallel with human families at a function, a wedding for example. Everyone is free and an individual but is bound by the rules of behaviour acceptable at a family gathering, before the drinks start flowing anyway. But break the starlings down into breeding pairs and you have the equivalent of individual family units—mum, dad, and children—and the rules are different again.

Let us try to construct a hierarchy of Great Brains dependent on circumstance for the life of an individual from birth.

- Single-family unit: mum and dad.
- Group family meeting: grandparents, mum, and dad.
- Infant school: teacher, mum, and dad.
- Junior school: teacher, friends, mum, and dad.
- High school: friends, media, teacher, mum, and dad.
- Upper school: friends, media, teacher, mum, and dad.
- University: special friend, friends, media, tutors, mum, and dad.
- First job: partner, friends, colleagues, media, boss, mum, and dad.
- Parenthood: partner, fellow parent friends, colleagues, media, boss, grandparents.

This hierarchy will continue through to the death of the individual; unless there are special circumstances, the Great Brains of government are by and large an irrelevance. If you were the victim of a serious crime, or if a close relative was murdered for example, then the state spectacle of police and judiciary would loom large. One can argue that life-threatening illness and the newfound dependency on the medical profession and the lottery of funding availability might well politicise us. I think the biggest single event that the Great Brains can impose would be declaration of all-out war. Now we really would all be starlings in the great flock.

Yet there have been times when, and there are places in the world where, crime is reduced, where people are slim, and where they follow health and government advice, and there are also parts of everywhere that follow suit. There are people in Britain who drive at 30 mph, who follow an exercise regime, who are mindful about what they eat, and who worked hard at school. So what is the fundamental common denominator. What moves entire populations or just part of it? The word best-suited to express this is 'culture', but this isn't top-down culture; this is grass-roots-up culture. It doesn't matter how much money you throw at a problem, it will never succeed unless the culture is changed. This is where the welfare state has failed; the white knight charging in to the rescue has failed to move the common culture in the desirable direction. Not only has it failed to move it positively, but also it has successfully moved it negatively and made the problem worse, or as they say in the slums, worserer. Jamie Oliver was right in his attempts to move people to a better diet by attacking the problem at the grass roots; he just failed to generate enough momentum to breach the dam of indifference. But even though Jamie has failed, he should take comfort in the knowledge that the trillions of pounds spent over decades of welfare and education and healthcare has failed before him and continues to fail now. The problem is not how much money we spend, which is the question that separates our greatest brains; the problem is how we can change the culture. Our successive governments are busy shovelling money into the wheelbarrow, without turning the empty wheelbarrow around first. They believe that the job is how to spend the money under the mantra 'The more you spend, the better it will be.' Watch any electioneering prior to any election and all you see and hear is boasts about how much money will be spent, who will receive it, where we will get it from, when the books will be balanced, how much the politicians care (about getting re-elected), and how much they are looking after our interests. As we say in my household, bollocks.

Our successive governments have been shovelling money into wheelbarrows facing the wrong way. They see their task as moving the money, not making the task easier or solving the problem. They have lost sight of the issue and the reason, and their focus is entirely pointless and their effort pointless. But in place of their effort, you should read our money. Stop and think what sixty-five years of the welfare state has achieved. Where will it be in another sixty-five years?

Compare the impact of this money on our culture with the cultural change in attitude towards homosexuality. In sixty-five years it has gone from a crime punishable by prison or treatment by chemical castration to civil partnerships, gay marriage, gay bishops, and gay adoption. Yet the irony is, it's all happened for free, no mega spend, no government commissions, no law changes to make it happen, just law changes to acknowledge the injustice. (We didn't even need a 30 mph sign.) This change wasn't led from the top down but from the bottom up, and this is how real change has been effected. But this change in attitude isn't universal. IS (the Islamic State) is summarily executing gays, and Christian Uganda is jailing and threatening homosexuals with the death penalty.

Politicians will talk about the need to change culture and then promptly do nothing about it because they don't know how to, but they will gratefully claim the honours when change happens. The Great Brains are self-serving idiots who have precious little idea about reality. They assume that the poor, illiterate, ill 'lower classes' will magically transform into ambitious, educated, health-conscious, upwardly mobile individuals in response to a liberal dosing of money. Yet they must know after sixty-five years that this is not the case, so why persist with more of the same?

We have discovered an anomaly with culture. This is one skill that we can finally argue does set us apart from other animals, and yet it is a transitory skill that can languish for centuries before reappearing.

This seems a strange claim to make, but there is an idea that has been developed about our awareness of self: the Romans exercised to improve their physique and body image, and they also shagged anything that moved and, no doubt, some things that didn't. It has been suggested that after centuries of 'imposed' religious constraints of sexual proprietary and the need to worship God rather than the human body, we have as a society returned to near Roman standards. I say near because child and animal sex is still rightly taboo, but the Romans held no distinction between hetero- or homosexual acts; it was simply pleasure. Many individuals today will happily 'sculpt' their physiques in gyms. Although, not surprisingly, our 'enlightened' narcissists will commonly resort to steroids in pursuit of the perfect body (or disco muscles) without too much effort or thought for their health. Although it is true that the ancient Greeks would attempt to cheat by consuming testosterone-filled testicles (presumably after they'd been harvested).

I said earlier that architects cannot design large houses because they don't live in them and don't understand them. By exactly the same token, our educated Great Brains cannot understand the common culture because they don't live in it or understand it. They understand culture, all right, but this is the culture of parliamentary candidate selection processes where you have to impress the local party bigwigs and know how to eat a bacon sandwich without looking like a pillock. They do understand the culture of polls and political commentators and the need to never answer a straight question with a straight answer. Who in our modern society does understand culture, Simon Cowell? He understands how to part people from their money and increase his own bank balance, but for every star he creates, there is another overnight sensation who hasn't seen the light of dawn. If you remember the Internet-driven campaign which saw Cowell's Christmas shoo in for the number-one slot supplanted by Rage Against the Machine, then you realise that for every 'culture' there is a counter one. Do advertising executives create culture or follow it, or do their efforts translate into ephemeral ditties and buying fads that fade into the background? Does this confuse popular culture and the intellectual culture of museums and theatre shows with the culture of the herd, which is the real definition we are looking for?

There is a theory about energy expenditure and how animals balance effort and reward. The idea is that an animal calculates that the reward must justify the effort. It's no good expending two thousand calories to secure one thousand calories of food. Now while the idea of calculation sits well in the human mind, I doubt that this is the mechanism by which animals actually achieve their goals. A cheetah can only sprint so far before using up all its oxygen and therefore its available energy and has to stop. The energy release also generates heat that will cause chemical imbalances and death. While the cheetah faces the problem as a predator, its prey has the same limitations and can also overheat and die or get tired and fall victim. Humankind learnt this a long time ago. Bushmen in the Kalahari Desert know that they can chase and catch antelope, not because they can outsprint them but because they can outperform them over distance (an evolutionary advantage of sweating and being able to shed excessive heat). Half a world away, the Yaqui Indians of Mexico have also learnt this skill.

When I was a kid, the TV often featured live animal capture safaris where rhinoceroses or giraffes would be chased by Land Rover with a dickey seat on the bonnet and a man with a rope noose on the end of a long pole to snare them. If these men had asked the Kalahari Bushmen, they would have known that many of the captured animals would then die from heatstroke. Nowadays, strict time rules are observed and the process is safer, relatively. So what has this got to do with the welfare state? My wife attends a local gym in Market Harborough that has an annual collection of unwanted or donated food for the Harborough food bank. Poor people struggling to make ends meet can visit the charity and, after having qualified, remove food parcels for two to three days' worth of provisions. The more food they provide, the more they need to provide. There are now, nationally, hundreds of thousands of people reliant on this supplemental boost. The official reason given is the failure of the welfare state to provide for all of their needs and the tightening of the government's purse strings. Minimum wage and zero-hours contracts feature large as well. But while I believe that animals simply get tired of running or flying in pursuit or being pursued and are physically prevented from expending what they haven't got, I believe humans can make the distinction. Free food means, 'I can buy another packet of fags, play another game of Internet bingo, have another night's drinking down at the local pub, have another tattoo, and update my perfectly serviceable mobile phone to the latest model.' Now don't get me wrong, there will be genuine individuals who would not eat but for the food banks. The question is, what proportion of those who use the food banks are in this latter position? What begins as a genuine desire to help the needy rapidly becomes corrupted to little more than a convenient lifestyle choice for those on the receiving end. I seem to remember the story of a Victorian gentleman who, distressed by viewing poor drunks in the street, decided not to give them money, which they would spend on booze. Instead he decided to give them a voucher for free footwear or clothes. In this way he could give them genuine help. This man was held up as a beacon for charities and began the principle of targeted help and ensuring that help went to where it was needed. History is silent on whether or not the drunks then sold their new shoes to buy alcohol.

Only last week newspaper reports of Victorian poverty in schools were surfacing, with hungry and inappropriately clothed pupils turning up for school. (I think by 'inappropriately clothed' they meant T-shirts in winter, not skirt hems up to their crotches.) The BBC featured a report on a charity set up to provide nutritious breakfast in London schools, which had risen from a handful of schools to hundreds, with hundreds on the waiting list. I am so glad I contribute my taxes to the welfare state; people would struggle otherwise.

Let me ask a question: if we doubled the welfare budget in the UK, would it end poverty, hunger, and ill health and remove beggars from our streets? Let's turn this question around: since the inception of the welfare state, how many times have we already doubled its budget (bearing in mind that every time we did it, we did it to end poverty, hunger, and ill health and to remove beggars off our street)?

Now the instant you raise this argument for discussion, you will be attacked for demanding the return of the 'dreaded and universally loathed means testing'. Fair enough, but remember that means testing was designed and implemented by the Great Brains of that era as the correct solution for the problem. If these Great Brains were wrong, did they have the monopoly on Great Brain error? I think not.

It strikes me that ending poverty, hunger, and ill health, and removing beggars from our streets, is now an industry in its own right. If the demand weren't there, you would bust a gut to maintain your market share and recreate the demand because you earn your living from it as well. If you are left wing, champion the needs of the poor in society by taking from the rich and expect to get votes from the poor voters. If you are right wing, champion the need to control the needs of the poor in society to protect the rich while at the same time ring-fencing budgets to prove that you care about the poor. My first thought is, *If the poor don't care about the poor, why should I?* My second thought is, *How much temporary help to get back on their feet do they need?* My third thought is, *If they can't be helped back to their feet, why do we try?* My fourth thought is to commission the space liner to Mars.

If you are serious about ending poverty, hunger, and ill health and removing beggars from our streets, then do it. Otherwise, accept that no mechanism exists to achieve these ends.

The general election has been called for 7 May 2015. Slipped through my letter box this morning was my local Labour parliamentary candidate's sales brochure offering twenty thousand more nurses, eight thousand extra GPs, twenty-five hours of free childcare to working parents of 3- to 4-year-olds, and the idea of working with primary schools to provide childcare from 8 a.m. to 6 p.m. each day. Now other parliamentary candidates are available and other inane policies are also available, so don't think I'm picking on the socialist Great Brain wankers. I told you the purpose of education was to keep children off the streets, didn't I? The policies we have don't work and are discredited, so let's spend even more money doing it some more. Priceless.

Another thing that strikes me as odd are the xenophobic newspaper articles about bloody foreigners coming over here to capitalise on our welfare state. Why should we provide for these non-contributing parasites who descend on our country with the sole intention of sponging off the state? How many million Bulgarians and Romanians are we expecting? Such outrage, this week anyway. There are only so many newspapers you can sell with one story. Yet the same newspapers tolerate our own home-grown parasites who have been raised from the cradle without crossing a border. True, we periodically get exposés of feckless unemployed drunks who have fathered forty children, all of whom are living on welfare, but once again the public appetite is only so great.

Is this the real reason why we haven't found the solution, a lack of appetite? Or is it a lack of ability, in that our Great Brains simply don't know what to do or how to do it? Do the honourable thing then, chaps and chappesses, and step down. Resign, leave your posts, and bring in the next batch to repeat your mistakes. Let's recruit them from the same political parties, products of the same universities and education system, and press on regardless (while earning a good living from it).

Don't get me wrong, I am quite happy to help people to help themselves. I have given work to an ex-con and paid him wages for digging my garden, and I've found work for temporarily unemployed on the same basis, but where I draw the line is in the 'something for nothing' category. If I didn't pay my taxes, I would employ more people—it's a big garden—and give those who want to better themselves the chance to do so.

But let me tell you about the ex-con: he was an assistant store manager and discovered a fiddle whereby he could turn returned goods into cash even though the goods hadn't actually been bought in the first place. He was discovered and sentenced to six months in prison. What he had done was found a way of obtaining cash without earning it; as a result he had been punished by the state. What he should have done, of course, is not gone to work but settled for a life on benefits whereby he would have obtained cash without earning it and effectively been rewarded by the state.

My father is elderly, very infirm, virtually blind, and bedridden but pays for a care package to address his and my mother's needs. My brothers and I all offer as much support as we can, but the day-to-day comes through the care package. When my mother-in-law died around 2000, my wife and I responded to the dreaded phone call and attended the scene. I rang for the doctor but was surprised when the first people to turn up were the police, who I learnt now respond to all unexpected home deaths as a matter of course post–Dr Shipman and his killing spree. Fair enough; I'm not going to argue with that. They collected all prescribed medicine and removed it from the premises.

Now the care package: the carers attend to my father and dispense drugs four times a day, removing them from pharmacy-supplied dosette boxes, sealed blister packs of pre-made-up drugs for the carers' convenience. But these won't contain the warfarin tablets, as these can vary during the four weeks of the prescription supply, so the family has dosed up the dosette packs with these additional pills between the INR (international normalised ratio) blood test assessments. So far, so good, or so we thought. The care supervisor appeared on the scene and announced that her staff could not dispense any drugs that had been tampered with and reordered the entire prescription in case we had slipped some poison into the dispensers. Fair enough; she was protecting her staff and protecting the patient. Just how is that exactly? If her staff decided to suffer an episode of Munchausen's syndrome by proxy, then they could easily add a pill of their own. If a member of the family decided that they'd waited long enough for the inheritance, they could equally as easily slip something in my father's tea. In the event that something was slipped into his tea, are the carers completely eliminated from the police enquiries? In the event that an additional pill had been added to his dosage by the carer, is she immune from prosecution because she only opened pre-packed drugs from the pharmacy? Yes, but the paperwork's intact. 'Pathetic, brainless, stupid drivel' springs to mind. Now I am absolutely certain that every single reader of *How to Kill an Elephant* will be able to recount tales of similar pathetic, brainless, stupid drivel. Send your versions in to my publisher and I can write my next book with no effort required. I will simply hijack your experiences. Thanks.

Patio Furniture

My daughter works at Compton Verney Art Gallery, and we parents often visit for the new season's exhibition. After one such trip, we dropped in to the Harvester restaurant in Rugby on the way home. Now Harvester supply approximate calorie contents for all of their menu items, including estimates for the 'help yourself' salad bar. Useful information for the health and weight conscious, I thought, that is until we watched the clientele enter and leave or visit the salad bar. I can honestly say I have never seen so many gratuitously fat people in one place in my life. We're not talking bulges here; we're talking arse balconies you could put a patio set on. In some perverse way, it appears that these people used the idea of calorie-controlled food portions as an instruction to dive in, because the thinking about health had already been done for them and therefore they could experience a guilt-free dining experience. Perhaps I was just there on a bad day. I must go back some time to check. Education is such a powerful tool. I now know why we bother.

Now a curious thing about our Great Brains who represent and lead us so effectively is the vast range of knowledge and wisdom that they bring to the post. The spectrum of government interest is vast and covers virtually every imaginable facet of our lives. And when the unimaginable becomes reality, the public servants react to legislate and control, as we have seen with the Internet and other new technologies such as mobile phones. How is it possible to develop, hold, and action a view on everything that crops up? What is the Conservative Party line on three-person IVF, for example? And there we have the answer: what is the party line? Now it all pops resolutely into focus. It's called a pro forma. Why investigate, consider, research, study, understand, debate, and get a thorough understanding when somebody else has already done it for you? That's why we need 650 MPs; it's so the other 649 don't have to do any work, because someone else has done it for them. All the other 649 have to do is either agree or disagree. But you don't even have to decide that for yourself because you take instruction from your constituents—sorry, I mean voters. Sorry, I mean party whips. (You do *want* that promotion, don't you? 'The PM even smiled at me. At *me!*')

Around 2000, I read a statistical report summary on Britain's population and predicted future needs and planning requirements. This 'science' of demographics is used to predict hospital bed and school places, together with housing needs—that type of thing. Demographics obviously has a very influential position within the UK government; otherwise, we'd have a shortage of housing. What the report was predicting was the 'end of the baby boomer generation' and the theory that our population was going to stall by 2050. The prediction was that our population would then halve by 2100. This set me to thinking. I began to formulate a plan of action for my children and grandchildren to cope with this apocalyptic scenario. Why apocalyptic? After all, this would be some natural occurrence, not disease- or famine-triggered. Stop and think. In 2050 everything you know will have reached its peak, and the only way is down. By 2100 half the shops, factories, hospitals, schools, houses, roads, trains, planes, garages, football stadia, leisure centres, and so forth will be redundant and obsolete.

Imagine being the government in charge of that scenario, effectively presiding over the collapse of Britain as we know it, and at the same time attempting to juggle the collapse of revenue plotted against the increased need from an increasingly grey-hair-dominated population. It can't be done. I have it in mind that this is the primary driver behind the last Labour administration's recruitment drive we lesser mortals would call immigration. Oh, and did I mention that immigrants are more likely to vote socialist? Maybe I am doing the demographers a disservice. They got the housing needs right; they just forgot to allow for all the immigrants.

But while this vision seems to have disappeared, for now at least, it poses this problem of all governments at all times. The path we have so earnestly set out and are pursuing so vigorously is for constant, continuous, never-ending growth in a finite world with finite resources. Methinks this cannot be. Methinks we need some more sustainabilitists. How else will we reduce our carbon footprint?

So who would I suggest as the Great Brains? Who in my opinion would warrant this accolade and position of such responsibility? The simple answer is, there is no one person or body or corporation or committee or institution or government or world body capable of achieving great brain status, and there never has been. I have tried to demonstrate how, using an ant colony as a model, that the control structure can only function up to a certain limit, the range of the pheromone. As soon as you step beyond the pheromone, false pheromones or controls replace what we could term the 'official' control. That is why we cannot brush our teeth twice a day or drive at 30 mph, because this is outside the control mechanism. We are back to individual herd members reacting to what their neighbours do. They cannot see the horizon, let alone know where they are going or even when they have gotten there. They react to what is happening a few yards away in a tropismic way, without conscious thought. They JFDI OPOD. Joking. (Well, actually, I'm not joking. If the wildebeest gets it wrong and fails to act, it could well die.) That includes our Great Brains who, I have hopefully demonstrated, have no more of a clue than anybody else but, more dangerously, believe that they have. Our Great Brains, sit at the high table enjoying the best of the fruits while tossing us mere mortals onto the sacrificial slab, because we let them. Have you decided how you're going to vote yet? The conclusion for me is that when the human being ceased to be a bush and became a tree, we overstepped the control mechanism. That is why we cannot kill an elephant.

While this conclusion condemns the nonsense of government, remember that we don't actually come into contact with government; we come into contact with the disseminated product that has been corrupted over and over and over again as it has travelled down the line. That is why you won't get treated within four hours at A & E or be educated properly (because nothing knows what properly is), or remove beggars from the street or eradicate poverty or hunger or disease, because like all good control systems, the flow of over and over and over corrupted 'product' flows both ways. But in the process of trying to govern the ungovernable, we are destroying the planet as efficiently and as unthinkingly as a swarm of locusts in a cornfield. Anyone for a cane toad burger?

Let's go back several millennia to the building of Stonehenge and consider its construction. The population was small, and the organisation and communication was short between the Great Brain, which we'll call the architect. Everybody knew what they were doing, if not why they were doing it, and all felt motivated enough to put in the extracurricular effort to lug humungous great lumps of rock around the countryside. This was more than survival, more that food in the belly and warmth and shelter overhead; this was culture, art, something to stir the imagination and put fire in the belly—an achievement. Was it done at the point of a sword, under the threat of a catastrophe, or for the love of an idea, or had they all thought, *OK we'll have a bash. It would be a shame to waste all that effort with the planning permission*? We talk about solar eclipses and calendars and religious imperatives, but in truth we will probably never know why Stonehenge was built, for lots of reasons, not least of which is the fact that we don't know what we are doing now, so why should we know what we were doing then? Religion, we are taught, has been a big thing for thousands of years. It has shaped our culture, our history, and our psyche, and has brought us from being animals to being civilised. Er, scratch that; what a load of bollocks. The Great Brains have determined that religion is the control and motivator of humankind, the original governing force that supplants violence, generates cooperation, and lifts us to a higher plane. Yet why should it have had any greater influence over us then than do the Great Brains of today who think that they govern us now?

Having demolished the notion of control in this day and age, you must also acknowledge that it hasn't existed in our terms before either. Rip up your history books and think again.

That's why Tony Robinson keeps digging up princesses on *Time Team*, because he's looking for them. Why couldn't the high-born princess have been the jewellery maker's daughter on her way to make a sale? Why couldn't the stash of gold coin hidden from the invaders been hidden by the invaders after the crime? I remember a TV documentary discussing a Viking longboat found in a peat bog somewhere in Scandinavia. A near perfect boat had been 'sacrificed', complete with 'expensive' iron weapons, and was remarkably well preserved. All sort of reverent, romantic, highbrow speculation was proffered for the great significance of this find and the reasons behind it. I thought, and I can be spouting complete bollocks, that the boat had been hidden. The local Viking bigwig had dispatched his bully boys to the village to collect tribute, and the locals had knocked them on the head and hidden all remains, including the weapons, in the remotest bog they could find. 'Tax collectors? No, we ain't seen no tax collectors here. Did they get lost?' I wonder if I could get away with something like that now. Better still, 'Tax collectors? Yeah, we paid them ages ago. Did they get lost on the way home?' While the official archaeologists look for the princess and the intellectual justification, I just look for the base motivation, because by and large that's what we do. We don't set out to conduct an elaborate and elegant murder; we lose our temper and try to hide the consequences, except the 'clever ones'. We didn't invent the football league; we corrupted a game based around chasing a barrel or an animal carcass from one village centre to another. Nobody envisaged a Premier League or a World Cup; they just kicked a ball around a field and enjoyed it. The rest came later. I am reminded of a joke that sums up the whole process nicely: An 'expert' was conducting a tour around an art gallery and was waxing lyrical about a painting of three black men, one of whom had a white penis. 'Brilliant. Absolutely brilliant. There, you see, it doesn't matter what colour we are; underneath we are all the same. Such foresight.' At this point a man who introduced himself as the artist and thanked the expert for his appreciation interrupted him. The artist said, 'Actually they are three coal miners, and one of them goes home for lunch.' Now while I describe this as a joke, something to put a smile on your lips, it demonstrates eloquently the pomposity of the Great Brain of the art world. I've been to the Guggenheim in Bilbao and I can't get a handle on Picasso either. And who gives a shit about an unmade bed or a pile of bricks?

Jesus didn't invent Christianity; he gathered a few impressionable acquaintances who went on to corrupt what they'd heard and seen. Not that it mattered anyway, because the corruption just became more and more magnified with every telling. Too harsh for you? What would your precious princess Jesus make of the modern-day church? Don't forget that the corrupted command-and-control structure flows both ways and that this one has had two thousand years to mature.

Think back to the ant colony which has grown too big to be controlled by pheromones alone. I suggest that this weaker chemical control leads to a natural conclusion, the spare capacity, and weakened control of the colony allows it to invest in future generations, which means that sexually mature ants are raised.

Back to Salisbury Plain and we have spare capacity. Food must be in abundance, the population strong, and the people willing to put in some extracurricular effort. Instead of 'splitting the scene' and setting up new tribes, this energy has been devoted to rock humping and the construction of Stonehenge, a marvel of the ancient world. But of course at the time, it was the modern world and at the cutting edge of cooperation and motivation. This aggregation of people had to have a control mechanism greater than family bonds or even tribal bonds. It's like the ant colony had developed a polymodal queen, that is to say many queens all producing pheromones to control the super colony. In a sense that's precisely what we have on the savannah now. It's just that all the ant colonies are 'independent' and squabble over food and territory and the right to live. My house forms part of the village that forms part of the parish that forms part of the district that forms part of the county that forms part of the region that forms part of the country that forms part of the continent that forms part of the world. This polymodal queen becomes the Great Brains, as distinct from the Great Brain, and can move mountains, well, chuffing great lumps of rock anyway. Did our newfound polymodal queen on Salisbury Plain wield her control over the population, or did the population cede its control to the PMQ? (No reference to prime ministers' questions intended.)

Could it be that the people were just bored, with time and energy on their hands, when one of them starting 'kicking' a stone around a field and others joined in? Before you knew it, they'd erected a set of goalposts, blown up a pig's bladder because it was softer than a stone, and, fed up with arguments about whether the bladder was over the line or not, decided to put up a crossbar. Some miserable elder turned up and said, 'No bladder games allowed here,' and had some more goalposts erected to use up the spare energy and to 'fence off the pitch'. Next thing you know, someone came up with some rules on how to play, 'Because you have to have rules, don't you?' And over a period of time they've organised a Neolithic World summer tournament with pie stalls and 'Druids' dressed in black to officiate with a hollow twig to blow through to be heard over the crowd noise. They even took advantage of a passing Welsh rock seller who was trying to get some blue stones off his books. Local tribes can get a bit partisan, so the next thing they needed was to segregate the supporters and place bouncers at the pitch sides in hi-vis woad and tall pointy hard hats to ward off blows from falling pies. Then the bloody Gauls turned up to watch, and before you know it Johnny Foreigner was better at playing their game than the locals were. It's just as well they didn't show the Gauls the timber wickets at Woodhenge, isn't it? But of course nothing like this could ever happen without some great architect to steer it, could it?

Bound by Charter

A British Medical Association (BMA) poll has declared that GP services are facing a crisis. They have surveyed fifteen thousand UK GPs and discovered that a third plan to retire in the next five years, a quarter are considering working part time, and one in ten are planning to move abroad. BMA GP Dr Chaand Nagpaul spoke of politicians' 'absurd' promises about doctors.

I like teletext, which might make me old-fashioned, but nonetheless it serves a purpose. I like the fact that a whole news story can be condensed to eighty words or so. It leaves me with great freedom to analyse the statements quickly and simply. The information always leaves so much unsaid, and this is where the chase begins. The story of the BMA poll piqued my interest, coming as it did the day after the Labour Party sales brochure through my letter box, and my initial instinct was to take it at face value to rubbish the Labour Party claims, not that I needed any help. I thought, *Why has Dr Nagpaul taken it upon himself to release this information on this day to the BBC, and why has the BBC taken it upon itself to run this story on teletext?* Now we all know that the BBC is bound by its charter to be politically neutral, and this neutrality is maintained by referring to 'promises' 'by politicians'. This is entirely neutral in that it tars all with the same brush, and I agree that all politicians should be tarred with the same brush.

Of course by having this information published we have to accept that Dr Nagpaul is also making a political statement by putting pressure via the media on governments of all colour and expressing his bias to resolve the issues of GP job dissatisfaction. Some fifteen thousand GPs were surveyed, but that doesn't mean that fifteen thousand GPs responded. It means fifteen thousand GPs had the opportunity to, but only so many did. Now what we don't know is whether the BBC knows how many responded or not. Without this information, we have an 'informal show of hands' at the strike call at the Ford factory. We have no way of determining the strength of feeling demonstrated by the GPs. The figures given are:

1) a third retiring within five years;
2) over a quarter planning to work part time;
3) one in ten planning to work abroad.

Let's play mathematical devil's advocate.

Point 1 states 'one in three'. That doesn't mean five thousand, because we don't know how many GPs replied, so it could mean one of the three respondents.

Point 2 states 'over a quarter'. That doesn't mean more than 3,750, because we don't know how many GPs replied, but it doesn't mean one of the four respondents, because it's more than one in four and decimal parts of doctors are a bit messy. It doesn't say how much more, so we will use one of four respondents for simplicity.

Point 3 states 'one in ten' recipients. That doesn't mean 1,500, because we don't know how many GPs replied, so this could mean one of the ten respondents. Now to do the maths properly, we need to factor in a combination of the figures combined in points 1. 2, and 3. This could mean a total of 10,250 out of 15,000 doctors will retire, work part time, or emigrate. Or it could mean that we take the three fractions and combine them together with the lowest common denominator as a whole doctor, in which case we get 41 out of the 60 respondents. That gives us 20 of the 60 retiring, 15 of the 60 planning to work part time, and 6 planning to work abroad, with no messy decimals.

So are we going to lose the full-time services of 10,250 or 41 doctors, please? Is this difference statistically significant, do you think?

Either the BBC doesn't know what the figures are and therefore they have been tricked into supplying false information, or the BBC do know what the figures are and they consider the volumes to be sufficient to run the story. Or the BBC don't consider the volumes to be sufficient but have run the story anyway.

But the story has been run, so we have been led to the inevitable conclusion by the impartial BBC that politicians are making promises they cannot keep.

In the above, substitute Dr Nagpaul for the BBC. Now it is Dr Nagpaul who is leading us to the inevitable conclusion that politicians are making promises they cannot keep. And Dr Nagpaul works for the British Medical Association, which is set up to represent medical professionals' needs. And while we are at it, 'I'm *planning* to score the winning goal at the FA Cup Final' doesn't mean I'm actually going to do it, does it?

Now while I do not believe that the BMA or BBC have used sixty replies to their poll to go to press with, I do not believe it to be fifteen thousand respondents either. Given this, there is a vast difference in the range, so how can you trust a single part of this eighty-word statement? Stop and think and then ask how many other pages of teletext should be treated with equal disdain. But then ask yourself how many people have read and accepted these statements at face value. Josef Goebbels always preached the principle of the 'big lie', saying that some part of it will always be believed. Fortunately I am not bound by any charter to be impartial.

Of course, if I was concerned enough I could delve deeper into the BMA poll report and request or obtain the full statistical breakdown of information. But, I am only human, I accept, along with the other millions of readers of teletext and newspapers and listeners of the television and radio outlets, the information provided to me at face value. Well, obviously I don't, otherwise I couldn't have written the foregoing. Don't, whatever you do, think about the new term that has entered into the lexicon during the Trump/Clinton election campaign. I refer of course to *Fake News*.

Paradise? Or Luton Airport?

Still in the run up to the May '15 general election, I went to visit my wife's sister and her husband near Luton. One of the gripes my in-laws raised was the recent council tax demand that they had received which asked for a 15 per cent increase in the police budget. This money would be used to fund a big increase in officer numbers, with a resultant projected impact on crime. Now the typical increase is of the order of 2 per cent. Local democracy has sprung into life and this item is going forward to public referendum, tagged on to the May general election. The public can either support or reject this tax rise and live with the consequences either way. If the increase is rejected and crime increases, then the local police chief can shrug his shoulders and say, 'I tried. You wouldn't let me. It's not my problem. It's all your fault.' If the increase is approved, then the ranks of the police force will swell, crime will be vigorously tackled, and the public will get what they asked for.

What did the public ask for? Oh yes, extra police numbers with a bigger presence on the street. Think about all the additionally apprehended criminals and the UK national upper limit of around one hundred thousand incarcerated prisoners. If you double the arrests and double the number of sentences, you will also double the number of prisoners. But of course there is no commitment to double the number of prison spaces, so you will have to double the sentence reductions to reduce the prison population through shortened sentences on account of 'good behaviour'. But the good people of the Luton area have a choice (and the bad people, whom I bet will vote no). If they press through and the plan works, then a number of things will happen:

- The local police budget will grow significantly.
- The number of criminals caught locally will increase, with more being locked up.
- The number of prisoners locked up nationally will remain the same, so more will be released early.
- The number of Luton-area prisoners as a proportion of the UK prison population will increase.
- There will be fewer criminals on the streets of Luton area.
- There will be more released prisoners on the streets of all other areas.
- There will be less criminal activity in the Luton area because the criminals will target less heavily policed areas.
- Less heavily policed areas, seeing this influx of criminal activity, will demand an increase in their police activity and will sanction increased budgets. These areas will now send more criminals to prison and will have a higher proportion of the UK prison population locked up, which will see more prisoners released early to target other less heavily policed areas.
- The pressure on the judiciary will increase to impose fewer or shorter custodial sentences, and fewer criminals will be sent to prison or for less time.
- The penalty for committing crime will therefore reduce and the deterrent element will reduce, which means that criminals will commit more crime.
- Less time spent in prison means less time to offer training or counselling. Less chance of reforming the prisoners, and more chance of them reoffending.

What Luton will end up with is more police officers effectively doing the same amount of work, jailing one hundred thousand prisoners, if jailing prisoners is the single measure of performance. Alternatively, what Luton will end up with is a bigger police budget, if the single measure of performance is how much money is being spent, because the more money you spend, the better it must be. Part of this budget will find itself in the senior police officer's pocket, because he employs more officers and therefore it's a more important job, and 'I'm worth it.' What Luton could do with is an investigative expert, let's call it a detective, to work out what impact this venture will produce. I wonder where the police could get a detective from. But of course I'm being way too cynical here, because the function of the police is so much more than jailing criminals, isn't it? They also maintain speed limits, and stamp out petty vandalism and antisocial behaviour, and control all the drunks at night, and dodge the pie throwing at football matches, and make sure we pay our vehicle tax and insurance, and so forth. Why do they control all the drunks at night? Isn't this some form of antisocial behaviour? Surely the police task would be to prevent people from getting drunk in the first place, unless of course you are treating the symptom and not the cause. If the police came across a youth kitted out with aerosol paint near a graffiti site, they would confiscate the paint and issue a warning. If the police came across a youth with a pocket full of money going into a pub, they would wish him a nice evening. Four hours later, one youth could have painted his bike while the other is fighting in the street.

Now I have it: by increasing the police budget through the council tax, you reduce the disposable income of households so they have less money to get drunk with or to pay their speeding fines with. On the other hand they will also have less money to insure their cars with and pay the car tax, so what would be the balance? What I am trying to demonstrate here is the utter futility of the policing system, our expectations and theirs, and yet the constant demand to maintain business as usual without ever achieving a resolution, because a resolution isn't possible.

So why are we increasing the police budget in the Luton area without simultaneously increasing the prison budget for the Luton area? And while you think about this, think how many times the police budget has been increased in the past without increasing the prison budget. It's like asking for an extra hosepipe to help to put a fire out when you've only got one tap.

What happens when policing breaks down? And I'm not talking Mexico here; I'm talking Northern Ireland, part of our jurisdiction. Soldiers were on the streets to keep the peace and normal police activity was suspended; they wouldn't attend road traffic accidents for fear that they were an organised ambush, for example. The troubles were caused by 'social issues' neatly identified as religious, with a Catholic–Protestant divide, but that could equally be split by race, Irish –Scottish. I remember the start of the latest 'troubles' when I was a kid and the army being called in to deal with the civil unrest that exceeded the police capability. I say latest troubles not because I expect more but because this was unfinished business from an earlier time. What motivated the people at the bottom? Were they all hardened nationalists or republicans, or did they just have minor grievances they wanted to settle, with unexpected consequences? If you'd had your Catholic eye on a plum promotion to senior sergeant major and found out the Protestant general's wife had had her distant cousin appointed instead, you had a grievance. If your Protestant neighbour's dog shat on your lawn one too many times without the poo being cleaned up, you had a grievance. If the Catholics kept beating your Protestant football team, you had a grievance. If the local regimental old boys' club wanted to march past your door banging drums and flying flags to celebrate some victory from ancient times, you had a grievance. Just how many of the population wanted near civil war on their doorstep? How many simply wanted to settle a minor score and see a petty injustice righted, compared with how many who wanted to kill and bomb and punish and maim? Somewhere along the line, the Great Brains hijacked the minor discontent, corrupted the intent, and set the problem on a path to suit their own needs. In one sense this answers one of my earlier questions about the greatest brains and how I can become one. Find a cause; recognise, generate, and agitate; put yourself in the leadership committee; overthrow the existing regime; and become a Great Brain. Yep, sounds like Vladimir Lenin, Adolf Hitler, Gerry Adams, and Martin McGuinness all right. Other Great Brains are available from the Protestant side; I'm not being partisan.

Now during the troubles, the rule of law was substituted by the rule of the protection racket, whereby local grievances and crimes were controlled by the paramilitaries for a price. Rapists and drug dealers would receive punishment beatings or 'kneecapping', or worse, and local businesses would be approached for tribute or 'insurance'. They could even offset cash protection monies against their taxes. We just get council tax demands and insurance bills and shell out for alarm systems, CCTV networks, prisons, and parole officers.

Ants are really stupid individually and really clever collectively. I often watch jumping spiders patrolling my sunny desk or window searching for insect prey to catch. These small decorative spiders jump on their prey and catch them. They have among the best eyesight of all the spiders. They are curious as well, and if you place your finger within range, they will often come and check it out from a few millimetres away before moving on. But for all that, their eyesight is poor compared with ours, although it is arguably better than that of ants. Even if ants could see up to 20 metres away, the habitat they occupy would rarely give them the opportunity when you consider that they are millimetres above the ground and in a mountain range of terrain relative to their body size. Yet they are among the most efficient of animals, so how do they function? The simple answer is through trial and error, but if the error is death, then there wasn't really much of a trial, was there? As ants travel, they lay down a scent trail and cross the terrain at random. If two ants travel the trail, the scent is reinforced and more ants are likely to follow. And if one thousand ants do, then you end up with a super-scented highway of ants. If an ant were to meet a predator, say an ant lion larva in its pitfall trap, then the scent trail would stop and the reinforcement reduce as it evaporates. While another ant may suffer the same random fate as its predecessor, the trail to the ant lion will remain unreinforced. When food is found, the paths are rapidly reinforced until the food is exhausted, at which point the scents reduce and the ants return to random foraging. They very effectively scour the home range for all available food supplies and survive by following the most basic of stimuli. Scientists have built mazes and recorded the activity of ants. They have recorded the remarkably efficient way that the ants find and recover food. I think—I wasn't taking notes at the time—that ants have been used to calculate flight patterns for deep-space probes, again with remarkable success. Our supercomputers took longer to work it out than the ant colony, which also came up with more than one answer.

So what?

Nobody ever designed a city. One house became two, and two became twenty as the conditions and availability of resources reinforced the decision to stay and build there. Given enough advantage, a ford or bridge or good soil or availability of building materials or employment, the twenty can become twenty thousand. Along the way roads and drains and shops and factories and schools and libraries and churches are built to serve the people, but only because the people are there. You cannot build a factory without any people to work in it. People will say, 'What about Milton Keynes, which is a newly designed city?' This is where the distinction is. Milton Keynes is a city that is what a city would be if it were designed. It's been designed by Great Brains who have designed their thoughts and prejudices into what they think a city should be. I have designed industrial units for clients by making them do it themselves. The normal process is to give a brief to an architect, who then interprets the brief and returns it to the client. Unfortunately for the client, unless he works out what he actually needs, the brief is incomplete, and the architect's interpretation is also corrupted by his 'interpretation' of an incomplete brief. I made my clients design the building by placing the furniture, the machinery, and the processes into the building before drawing the walls around it. Then I future-proofed it as best I could to make it suitable for other tenants and had the structures built. Apply that logic to a legal system or an education system or a government and what you get is, 'We know it isn't perfect, but it's what we have. Of course if we could start with a blank sheet of paper …'

My paternal great-grandparents left the quarrying industry in Lancashire to search out new opportunities in the up-and-coming Derbyshire quarries. Others descended on the coal mines of South Wales because that was where the work was. When the coal mines shut, our government, our welfare state, picked up the bill for the unemployment by providing benefits. Benefits from the state didn't bring the workers to the pits; the benefits of labour (wages) did. But benefits from the state have kept the non-workers there. Why does mobility work in one direction in one century and cease to work in another?

Let's return to the super-efficient, super-stupid, super-clever ants that have been on this earth a lot longer than humankind and that will probably remain long after we have disappeared. They don't think; they just do. The action that is reinforced is the correct decision, and they don't even have to look for it. The original JFDIs? When the labourers descended on the Welsh coal pits, they looked for reinforcement and found it in jobs and wages, so they stayed and prospered. When the reinforcement (jobs and wages) evaporated as the pits closed, instead of leaving in search of fresh reinforcement (jobs and wages), the labourers stayed and settled for the reinforcement of benefit, simultaneously settling for deprivation and poverty. Now don't get me wrong. Who would want to volunteer to throw a house and home away and all the community links to search for work elsewhere? The simple answer is your grandparents and their parents, who did exactly that when they went to Wales in the first place. But of course this is where the other problem that we have created emerges. Their ancestors gave up their hovels and shacks with no furniture or other possessions beyond a knife, a spoon, a dish, a cooking pot, a razor blade, and the clothes on their backs. This consumerism world of ours, which provides us with all the paraphernalia of choice, actually limits our choice, so we mask it in welfare.

Keep Your Eyes Peeled

The pope is the leader of the Catholic Church worldwide and has to be seen as the queen ant of all queen ants within their world. He has his assistants, the cardinals, who have their bishops, who in turn disseminate the doctrine and control to the masses (sorry, I couldn't resist the pun). Commonly the pope is referred to as God's vicar and allegedly is divinely linked to God. Let's go back a few centuries to a Catholic Church in England before Henry the Eighth and his dissolution of the monasteries. Take your typical illiterate peasant sitting on his pew, head bowed, listening to the Latin cant coming from his priest's mouth, oblivious to the meaning of the sounds falling on his ever so attentive ears. Not for him the literary version; he has to settle for the comic, the illustrated version etched into the church windows. Just how much of the pope's wisdom actually arrived in the peasant's brain, at best, with some explanatory detail from the priest in English? The peasant would receive a liberal dosage of 'lies to children'. Yet we learn about royal plots to overturn the illegitimate incumbent on the throne to replace him with a Catholic and thereby remain in the pope's favour. Forgive me for asking, but what the fuck does the peasant care?

History books are written around the pro formas of present-day conventional thinking and make assumptions based on today's perception of knowledge and control. We go to war with a song on our lips and adventure in our hearts because of the need to win and follow our queen and country. No, we go to war for the adventure, but not in the way that you think. Many years ago one of my colleagues at work had initially come for his job interview because he didn't want to do 'double maths' in the sixth form; he still works for us thirty-seven years later—and never expected to get the job either. He saw the interview as an escape from the routine, a release from an otherwise double dose of tedium that got converted into thirty-seven years of the same and counting. That's not to say that our work is tedious, but long periods of repetition with flashes of interest is the norm with most interesting jobs.

My house is on the site of a Victorian hall and benefited from a large staff of gardeners, sixteen in total. Many of my neighbours had lived here for many years and occasionally met former staff from the 1910s, including one gardener who started as number 16 and worked his way up to number 9. He had wryly observed, 'The problem was, you had to do what the other fifteen didn't want to do.' How many march to war because they've had enough tedium for the day or week and fancy a change, or want to break up with a girlfriend without upsetting her, or want to prove something to parents who refuse to let go and let them grow up? But none of this features in the history books, because everything is given a higher purpose. Nobody ever went to war because they were bored.

Just occasionally something appears on the TV that breaks the mould. A recent documentary looked at prostitution and its impact on the economic success of cities like New York and New Orleans. Prostitution was a major driver in the growth of America's largest city and is tied in the present-day entertainment venues of Broadway with brothel locations. Having exercised the body, punters would continue with exercise for the mind. Even the Catholic Church got in on it; one pope was so impressed with a French Catholic brothel that he had one built at the Vatican. At the time men were not getting married until the average age of 28, and the lack of access to women led to concerns about homosexuality. The pope sanctioned his brothel as a defence against deviants, and as an added bonus the pope took a cut of the earnings. It's not very often you can legitimately get 'pope' and 'pimp' in the same sentence. I say this breaks the mould, but it's presented in a semi-salacious manner as though you're being let in to a secret not to be divulged—a bit of titillation.

It was in Victorian England that the modern police force was brought in to being by one of the then great brains, Robert Peel, home secretary and later PM, the Bob in bobbies or the Peel in peelers. He introduced what he considered to be the nine fundamental principles of policing. We have these on record, so we can consider what the police were designed to do, and thus we can see where we stand today in relation to the stated intentions. The idea central to the concept was that this was 'voluntarily policing' for the public and would not be imposed. It would be provided because the people wanted it. No one individual could opt out of observing the law, but the people as a whole could. The police were to be civilians in uniform and had the same rights as any other citizens, who were also expected to behave in a responsible manner. They were also always to recognise the need for strict adherence to police executive functions and never to authoritatively judge guilt or punish the guilty.

Now this idea about of the people, for the people is interesting, if for no other reason than who the people were. Did Peel mean the whole population, or did he mean the educated voters who, back then, were property owners with a property value threshold (because universal suffrage for men didn't come in for another ninety years)? While no one was to be above the law, being able to accurately and authoritatively communicate your needs to the police would give the educated a distinct advantage.

If you drive down the road at excessive speed and get caught, then you receive points on your licence. Drive and have an accident or three and you may well end up with prosecutions for undue care and attention. All of these activities will also bring you to the attention of your insurance provider, who will start to increase your premiums until ultimately withdrawing your cover altogether as the final sanction.

Conversely, if you drive without being caught for speeding and avoid accidents, your insurers will reward you with reduced premiums under your 'no claims bonus'. So the message is, drive sensibly and carefully and reap the benefit; drive like an idiot and pay the price. These sanctions, if observed (because some people will drive without insurance), ultimately reduce the cost of insurance for sensible drivers because there are fewer accidents for the insurers to pay for.

Insurers also look to allocate 'blame' for accidents and ascribe proportionate costs to the accident causer against a scale agreed on by the insurance association. It's been many years since I read one, but last I read, if an overtaking car hit you as you turned right, the accident was 20 per cent your fault and 80 per cent the overtaker's fault, for example, because you should have looked behind you before turning. Your costs would be met, but the difference could be reflected in how much of the combined bill would be paid by each insurer.

Now of course all of this is a private arrangement between the car owner/driver and the insurance company, with the exception of accidents caused by uninsured drivers, where the full cost is met by the insurers, who levy a flat rate fee on all insurance premiums as a government-induced policy. Now the newspapers are full of the injustice of this fee and they demand a crackdown by the police to reduce the injustice.

In round figures, every man, woman, and child in the UK contributes £2,000 into the insurance pot that is the NHS. This money is used to pay for all the qualifying ills and treatments. But of course every man, woman, and child does not pay the payment. Taxpayers in proportion to their earnings pay it, and of course we cannot expect children to pay.

Now I am not going to attack this principle of the taxpayers funding the NHS. What I am going to attack is the principle of universal healthcare determined by need. If you eat like an idiot with a crap diet, if you eat like an idiot to excess, if you drink alcohol like an idiot to excess, and/or if you smoke like an idiot, then you should lose your no claims bonus and either pay more or else have your insurance cover removed.

Consultants will predict the bankruptcy of the NHS and will follow up this statement by mentioning their own personal interest, which may be diabetes, obesity, liver issues, or something else. They will give the NHS twenty to thirty years before it fails. These consultants don't have any great insight into the financial strengths of such a large organisation, but they do have an understanding of their own immediate budgetary situation and how they have seen it grow and deteriorate. Now of course we can do nothing and continue to let the slow-motion suicides dominate the budget, or we can address the issues while there is still a chance of rescuing the patient. But which patient are we trying to save, the NHS or the sick person?

Many years ago we had a Scottish manager at our northern factory who would return to Scotland at the weekends. While walking at the loch side, he was fortunate to see an osprey dive into the water and catch a fish. Frantically trying to grab the camera from his backpack, he failed not only to photograph the event but also to see the event properly. The way to photograph an osprey fishing is to suspend a tank in the water complete with a captive fish, set up the camera to focus on the empty space above the trapped fish, and wait. A German photographer set his camera and strobe light to capture a kingfisher in a stream and was surprised to photograph a robin diving into the water to catch the fish. He left the camera preset and triggered by movement and initially had no idea what he had photographed until the images were examined. Conventional flashes are not fast enough to freeze the motion, so high-speed strobes are used instead. The robin was in fact three times faster than a kingfisher. A strobe speed of 1/30,000th of a second was used to capture the image. At the O2, the tennis players are on the court. They are already inside the 'glass tank', yet the spectators still miss the action because they're too busy trying to take photographs.

Great Works

During the medieval period, great building projects to construct cathedrals were undertaken. These cathedrals took many years and hundreds of thousands of labour hours to construct. The devotion of the Christian labourers of the workforce was considerable. Entire working lives could be devoted to one structure. The probable reality with these things is that only the educated, God-fearing rich had any interest in securing their future in the afterlife. The poor simply toiled away in exchange for securing their present.

True, the poor may have understood the basic tenets of religion, but they were so busy trying to earn a wage to stay alive that extracurricular soul-saving probably didn't feature. Their biggest worry was the life they had, not one in the future. It took the rich with time on their hands to really invest in their place in heaven. You can imagine the Gates cathedral if you try. How many peasants, skilled and unskilled, would labour to construct a castle, only to see the drawbridge raised to protect the rich when the shit hit the fan? Yet you never consider that the church or cathedral was simply a place of employment; it was always a labour to God.

Do Crocodiles Smile?

Does it strike you as odd how simple things evolve into something huge beyond all proportion to their origins? A group of bored individuals started kicking a leather-coated bladder around a field, and we ended up with a multibillion-pound worldwide industry (I won't call it a sport) that bears no relation to its beginnings. Its rules and regulations are handed down from international committees of hardened businessmen who would probably keel over clutching their chests if they ran five yards. How did this simple pastime become this comparative monstrosity? How does a group of car enthusiasts struggling to extract greater speeds out of their cars become the Formula One circus that straddles the world with venue and money? How does a group of 'sporty types' develop from an athletics meeting into the four-year cycle and multibillion-pound extravaganza that we know as the Olympics? Better still, how does Sepp Blatter, Bernie Ecclestone, and the Olympic Committee rise to take control? What makes these individuals and committees so unassailably in control, so dominant in their positions? Who made them queen ant?

BBC News features various promises and announcements from politicians of every hue and persuasion. But why are they news? Surely news is reporting something that has happened, not something that is going to happen. Otherwise you could have a story along the lines of this: 'Reports are not coming in of a Boeing/Airbus Industries plane that took off carrying some passengers and has crashed somewhere on its way to somewhere else. We have no reports on how many survivors/deaths but will bring you more news when we have the information, if we get any.'

Now we know as sure as eggs are eggs that a plane is going to crash at some time and at someplace because of the inescapable physics of heavier-than-air aircraft flying. They can't float harmlessly back to earth. (Although, I have heard an engineer talk of the possibility of aircraft sized parachutes. Something for the future, perhaps.)

What we also know is that whatever promises made by politicians are published, only a proportion of them will be enacted. We know this because only one political party will rule and the other will be in opposition. OK, you can have a coalition and multiple parties in opposition. In a coalition, once the horse-trading has finished, a hybrid set of policies and promises will continue on to fruition. We also know that of the promises that go forward to become policy, much will be corrupted in their path from top to bottom—between the ground surface and the pit face, if you like, and then the return journey from pit face to ground level—and the reality will be very different than the intent.

Say you are trying to buy a financial product, an ISA or endowment, or to invest in a pension scheme. All of these have to comply with strict rules and regulations set down by the FSA (Financial Standards Authority). Making predictions about future performance is not allowed unless a fixed rate of interest is offered. Rather, you are advised to look at past performance to see how the product has succeeded or not as compared to that of other service providers. But the entity offering the product is also duty-bound to report to you that 'past performance is no guarantee of future performance'. Now this whole process is important because if any financial services operator fails to advise you properly, then you can claim compensation from them, policed by the FSA, to redress any losses that you might otherwise have avoided. Government created the FSA as a device to police and regulate the industry and to ensure consumer confidence in the process. Buying financial products is very important because you are trying to secure your future or present prosperity. It could be your mortgage or pension or university fund that evaporates into thin air if you get it wrong.

Strange that the government should introduce such policing methods to ensure that either promises are not made or promises are kept, while at the same time it has no accountability for its own promises. But of course the electorate has the power to respond and discipline errant politicians and parties by voting the incumbents out of power and replacing them with another set of equally flawed promise makers five years later. A common statement made by non-voters is, 'You can't trust any of them. They're all as bad as each other.'

Why not prevent these promises from being made in the first place and place them into the collective bucket entitled 'wish list', because that is all they are? Did I tell you I'm planning to score the winning goal at the FA Cup Final? Where is the ASA in all this? That's the Advertising Standards Authority, set up by government to ensure that advertisers' claims and promises made about their products are true and accurate and not designed to be misleading in any way.

One can look on the election as a job interview process. You put on your best clothes, cut your hair and brush your shoes, rehearse your speech, hide your bad points, and push the qualities that you think your employer is looking for. You can even lie. Several of the people I have employed in the past I later found out had lied, but I console my decision making by realising that these were only the ones I found out about and they had used their initiative. One can look on the election as a fashion parade where we engage the candidates in parading before the public, strutting their stuff, hiding their bad sides while they earnestly portray themselves as MP material.

Less generously, we can place them in the cattle market, jostling for position, avoiding the man with the stick, milling round in unfamiliar circumstances away from their comfort zones.

We console ourselves that this is democracy in action. Yes, we know it's flawed, but it's what we've got. Adolf Hitler rose to power through the ballot box; he was legitimately elected to power as a minority government and used this to his advantage to establish greater and greater power.

The sales bumf that drops through our letter boxes or trips across the airwaves to our ears and eyes serves to legitimise the process. I trust the BBC. They are impartial and report the facts. Their political analysis explores the issues and breaks the analysis down through interpretation to make it accessible to us lesser mortals who only have a brain. (Other news media are available, in the interests of impartiality.) (I might have had my fingers crossed when I made the statement about trust.) Or does the BBC serve as the apparatus of lies and misinformation, the harbingers of political truth designed to anaesthetise and seduce us into the folly of our beliefs? Who props whom up? Do the impartial media make the government, or does the government make the media? Remember where I wrote that all the impartial media channels report the same news stories almost simultaneously but on different channels?

Now don't think I'm looking for some great conspiracy theory here. I am simply trying to relate the facts. There is some power, some mechanism, at work here, some magic pheromone that supports this process within our DNA. This is how Sepp Blatter runs FIFA, how Bernie Ecclestone runs Formula One, how a few bored kids running around a field end up creating a multibillion-pound industry, although in fact the kids didn't do anything of the sort. This is how we end up with God's vicar on earth, threatening and controlling the masses through mind control across millennia and unafraid to resort to physical action when required. But ultimately it's because we let it happen. We accept our flawed democracy, our flawed dictatorships, our post-Neolithic world summer tournament in Qatar, our massive pyramid built to honour the death of one man, our call to arms against the godless hordes who pray to the same God before entering battle against us—after we've finished praying, of course. Is this the great herd process at work? And yet we never formed herds. For millennia we lived in small family groups living off the natural resources, moving or defending territory as our immediate needs demanded. We never formed ant colonies with young born to be nurses or workers or soldiers or breeders, yet that is precisely what we have ended up with: the rich breeding the rich, the poor breeding the poor. (Yes, I know we have social mobility. Nothing stops you from 'improving your station' in life. It's just that very few of us do.) Inept government leads the masses. And I don't just mean political governments; I mean FIFA, IOC, FIA. And we accept it as though we react only to our neighbour a couple of metres away, never seeing the horizon, never knowing where we are going, blind to the opportunities. We comfort ourselves by appointing dieticians to lose weight, doctors to monitor our slow-motion suicides, environmentalists to continue environmental destruction to fit 'the plan', sustainabilitists to produce sustainability, makers of traffic laws to keep us safe on the roads, and governments to protect and govern us to our benefit.

Common sense is as rare as hen's teeth, but this is a truer statement than you'd expect because it's not rhetorical. We all know that hens don't have teeth, but they do. Have you ever stopped to wonder why your teeth fit your mouth, and why your mouth fits your teeth? Yes, I know, dentists often have to remove excess or make room for overlarge gnashers. Isn't it a happy coincidence that they generally do? Your smile would be somewhat disjointed if you had the dentition of a crocodile. While we accept that humans have human teeth, we also accept that hens don't have teeth, when in fact they do. When the embryo is developing, teeth start to form in the chick's beak. These baby teeth or prototeeth develop for a number of days before gradually fading away and disappearing. Treat them with the right chemical and they will continue to grow in the laboratory, although scientists have yet to mature these birds to the point of hatching. I will argue that the embryos have teeth because this is the 'happy coincidence' that ensures that their non-existent teeth fit their existent mouths. The growth of the teeth is a necessary prompt, scaffolding if you like, to make the mouth grow to the right size. But it also belies the origin of the chicken. Not for me the debate about which came first. The dinosaur came first, and birds evolved from dinosaurs, which we know died out sixty-five million years ago. Now chickens (birds) had already 'branched' away from dinosaurs before that particular apocalypse, so their history is even older. The experiment with hen's teeth in the laboratory is to try to retro-engineer a dinosaur by reactivating the dinosaur genes within the chicken's DNA.

But let us extrapolate the hen's teeth and the dinosaur's gene argument still further, because the dinosaur also contained genes and DNA passed down by its ancient ancestors, which also contained genes passed down from their ancient ancestors, all the way back to the origin of life. And so do we. Our ancestors may well have existed in small family groups 'swinging in the trees', but their ancestors had ancestors that included ants, which formed colonies that bred worker, nurse, and soldier ants and which accepted the control of a queen. And we do not seem to be able to break from this 'tradition'.

No Need to Worry, I Have a Band Aid

You don't need to know the answer. You need to know the right question, and then the answer comes easier.

We are faced with many difficulties in this day and age, which is not to say that any age has not been rife with them. The difference today is that we believe that we can see the problems facing humankind and believe that we can, people willing, do something about these problems. The problem we have is one of measurement, as always. We measure potential threats and attempt to resolve them. Now 'attempt' doesn't sound very positive, does it, so let's examine some of the solutions we have fixed upon.

Most of us have heard about the supervolcano that sits biding its geological time under Yellowstone National Park in the USA. By looking at the geological record, it has been determined that a geological hotspot, a plume of molten core, has periodically announced its presence with eruptions that dwarf anything thus far encountered by modern humankind. This hotspot is gradually drifting beneath the United States. The US Geological Society (USGS) is monitoring the activity of this natural phenomenon with a view to giving a warning. In round figures, this volcano has erupted roughly every seven hundred thousand years. The next eruption is 'overdue' in so much as it makes its own appointment. Some of us have seen the drama-documentaries that predict the end of much of North America, as we know it, buried under vast quantities of ash, and a global 'nuclear winter' caused by the volcanic gases and dust blocking the sun's rays. While I say monitoring, the reality is that it's being observed out of curiosity, because in the event that 'it's gonna blow', there is precious little that can be done to alleviate the impact.

Now it's obvious that there is nothing that can be done. If the eruption happens, nothing is capable of reversing the situation. We have to accept that what will be, will be.

What may be more of a surprise is the current belief among some geologists that the next major supervolcano eruption may well be in southern Italy, under the Bay of Naples. Alternatively, we could just as easily suppose that it could occur anywhere in the world with a history of volcanic activity, but this excludes those areas with a prehistory, that is to say fresh areas with no known history because it hasn't happened yet.

We worry about a lump of detritus impacting the earth from outer, although by then it would be inner, space. We dramatise the coordinated human response to deflect/destroy this body before it gets its chance. And theoretically the science is developing to make these ambitions feasible, possibly.

We worry about a global pandemic, bird flu, SARS, AIDS, Ebola, impacting the population. Our governments reassure and placate us with tales of antiviral drugs and quarantine and our medical facilities being adequate for the task.

We even have our own human-designed Armageddon that we call the nuclear deterrent, and MAD, mutually assured destruction, sitting in fixed and mobile launch silos. I say 'human-designed Armageddon' because the concept of Armageddon is a human invention, so in that sense all Armageddon is human.

We worry about the rise of superbugs that have learnt to evade our antibiotics and render us at their mercy. Routine operations are threatened in the not too distant future. No more hip replacements or stomach stapling because of the unacceptable risk of untreatable infection.

We worry about the tide of slow-motion suicides that crop our population, from the self-inflicted illegal-drug-takers to the self-inflicted legal-drug-takers (alcohol, nicotine) to the obesity epidemic.

We worry about our taxes and provision for the weak in society. We worry about the National Health Service collapsing due to a lack of funds. We worry about our children's education and our pensions and our old age.

We worry about global warming and unpredictable weather patterns and sea level rises and reducing our carbon footprint.

We worry about crime and protecting our assets and being secure.

This list will do; you should be depressed enough by now. This list consists of two types of threat, beyond human control and within human control. In reality, this list consists of one type of threat, outside human control, because things that we could control, we don't. It's not in our 'nature' to tackle each problem with a solution. It's in our nature to let somebody else do it. Let another herd member sort it; that's what they are there for. If I stay in the centre of the herd, I am protected from predators, I can follow my neighbour to food and water, I never have to see the horizon, and I never have to lead. 'I cannot influence my energy use, and if I did, it would make no difference, because the Chinese would expand to fill the void.' In China they could quite legitimately point to India. The Indians can point to Africa or South America. Even without the name blame game, we are still hell-bent on growing our economies ad infinitum until all that is left is the super-developed husk of what was.

What I have tried to illustrate is the utter futility of humankind's solution; we don't need sustainabilitists or dieticians or climate change scientists or UN commissions or international treaties, because these only treat the symptom and never the cause. Even education is damaging because it fuels the development and exploitation of ever decreasing resources and drives the competitive edge that maintains our position in this world. We have to build our castle so that we can pull up the drawbridge when the shit hits the fan. We didn't need to build cathedrals and create elaborate fairy tales of angels and demons to survive on earth; we needed food and water and warmth and shelter and security. But what we are so determined to achieve is the exact opposite, a shortage of food and water (drought, pollution, and population growth), warmth (fossil fuels and forests), shelter (hurricanes, sea level rises, population growth), and security (nuclear arsenals, pandemics, the predicted wars over resources).

Welcome to our world.

Following are some arguments I frequently consider:

1) But we have to open our shops on Sunday because everybody else does.

2) But we have to have continuous economic growth to maintain employment and prosperity.

3) But we have to educate to remain competitive.

4) But we have to have a health service to be healthy.
5) But we have to have welfare to protect the disadvantaged.
6) But we have to have the police to maintain security.
7) But we have to have a government to keep order.
8) But we have to maintain our culture.

Following are my responses to the aforementioned arguments:

1) Because everybody else does it, so must I. What sort of logic is that? If everybody else was queuing up for the suicide booth, would you consider it a good idea to join them?
2) Continuous economic growth simply makes the inevitable crash bigger. Take the demographics-based prediction that the population of the UK will peak in 2050 and halve by 2100. Do you wait for the collapse and do nothing, or do you prepare for it and soften the blow?
3) You have to remain competitive essentially because everybody else is doing it. So if everybody is doing it, where is the advantage?
4) The 'health' service seems to be hell-bent on preserving ill health.
5) Welfare keeps people being disadvantaged; it puts a ceiling on achievement.
6) We don't need police; we need the absence of crime. One is a symptom; the other is the cause—and we tolerate it.
7) The government is corrupt in intent, in action, and in reaction.
8) Culture? Do you mean business as usual, herd instinct, or queen pheromone? Maybe you mean the exploitation of effort in the pursuit of mutually assured destruction.

Why focus on climate change as if it were the single most important threat to humankind? Bollocks, climate change is a symptom of our relentless obsession with making a living, and more often than not this is done off the backs of others. Let's go back to the first bipedal apes wandering across the grasslands between the trees that used to provide all of the needs of shelter, security, and food. They found themselves in need of a new strategy, a new model for survival, because without one they would have withered away to extinction. They learnt to exploit the new opportunities. Their bodies changed, their brains and paws changed, tools and fire began to feature, and the great opportunist began to evolve and establish itself. Within each grouping there would have been a division of labour, if we follow the chimpanzees' approach (as opposed to the pygmy or bonobo chimpanzee), where the group is essentially dominated by violence from the dominant males (as opposed to sex often led by female bonobos). The males coordinate hunts but control the group's access to meat through their dominance; females, infants, and subservient males are unlikely to get a look in. The likelihood is that access to the prime feeding locations would also be yielded to the dominant males, who would use this nutrient boost to maintain their physique and, through this, their positions. Dominant females may also have featured; how can we be sure? The point I am trying to make is that a pecking order, a regime of rank, is likely to have existed, and still exists to this day in modern humans. That is why we are prepared to yield rank to our 'peers' and to follow the dictates of our place in society. At times this has meant feudal ranking, and at others it can simply affect our pay scale and lead us to follow the boss's instructions.

Somewhere between the savannah and the city, dominant males grew in stature and influence beyond their physical abilities. You might be the strongest male among fifty, but how do you maintain that among a group of five thousand? You cannot fight and win every challenge, because exhaustion and injury would wear you down. Every dominant red deer stag can only hold its crown for a short time before being supplanted by fresher males. So, something happened. The advantage became more complicated than muscle and strength, probably intellect. Allegiance between potential rivals of both sexes would establish a hierarchy and maintain control, and this would form the origin of politics. Out of this politics would come the first leaders who stopped working but took tribute in return for the security that they offered. Better to calculate the optimum time to plant crops than work tirelessly but inefficiently in the field. Prove that it works, and your reward would be greater tribute and the success of your tribe. Become a leader, supported by technicians, weather predictors, and sun watchers to mark the passing of the seasons. Establish trades, the pot-makers, arrowhead manufacturers, hunters, trackers, leatherworkers, storytellers, even singers. Learn to receive tribute from the pool of earned resources. Take a wage for your skill and trade it for food, firewood, security, and wisdom. Construct a community of divisional labour—smelters, metalworkers, cultivators, hunters, fighters, carpenters, and jewellers—and build a town. Within how many generations would the skills of centuries be lost to specialists? Whereas before all could hunt, now only the hunters know how. Whereas before all could chip stone, now only a handful can work iron. Establish a hierarchy of worth: an ironworker's wage is higher than a goatherd's, and a carpenter's higher than a hunter's. Put a value on intellect and knowledge. Corrupt the values, boost your own specialist skill, and demand more for your labour, £20 an hour when the going rate is £10. Generate wealth above your neighbours; ensure you can have the pick of the bunch, the best of the herd, and the better land for cropping; and raise more healthy kids. Create management where you earn for what you know rather than what you do. Generate the wage multiplier whereby your wealth isn't limited by your hours worked (40 hours at £10/hour) but becomes thousands per day because you earn off the backs of others by taking a cut.

This wealth corrupts.

In the 1930s a man created the first conglomerate building materials manufacturer in the UK. He was interviewed, and his thoughts reproduced many years later on the occasion of his death. He said, 'There's a lot of talk in the city about nepotism and how it is wrong. If you can't look after your own family, who the fuck can you look after?' Well, whose family do we look after now, our own? Kids are abandoned, marriages and partnerships are abandoned for something better (this week anyway), parents are in nursing homes or hospital beds, because 'I gave at the office and don't need to pay again.'

We live in a mad world where we don't even realise that we are following the herd, just our neighbour a few feet away. We live in a MAD world where our day-to-day activities are guaranteed to lead to our mutually assured destruction. We are all actively engaged in our own slow-motion Armageddon.

Conservation: stick a plaster on it.

Environmentalism: stick a plaster on it.

Global warming: stick a plaster on it.

Poverty: stick a plaster on it.

Ill health: stick a plaster on it.

How to become a billionaire in one million easy steps. Extract one thousand units from one million people. That doesn't reflect your activity; that is work that you have done.

Dominance gives you the pick of the territory, the breeding, and the food, but it also comes with a price, namely challenges to dominance and territorial disputes. Everybody wants to reap the benefit of dominance. Being recessive is hard work. Establish the hierarchy, fix your place in it, and maintain it by dominating those under you. Find alternative ways other than strength of arm and sharpness of sword to find your niche. Invent mumbo jumbo, sing, tell stories, invent a deity that speaks only to you, or develop a pro forma that becomes a dynasty, but don't work. Let others do the work for you.

There is an expression, 'the elephant in the room', the idea being that in the course of your discussions or thinking, you mustn't mention the humungous great pachyderm that is the elephant. I've asked repeatedly, how do you kill an elephant? Well, the elephant in the room is *Homo sapiens*, and we are already engaged in our own slow-motion Armageddon, so the real question is, how do you not kill an elephant?

So we reduce our carbon footprint—brilliant. Now contain that figure for the next billion people, and the next and the next and the next, and then tell me that you are not treating the cause, only the symptom.

I employed a Romanian engineer to work as a technical scheduler involving architects' plans. She asked me, 'What is this? I don't know what these are.' What she was referring to was a nursing home. In trying to explain what they were there for, I realised that the sum total of your life's possessions are broken down to a three-foot wardrobe and a bedside cabinet. While you may consider this to be the conclusion to a journey and the fun was in the travelling, the old adage of 'You can't take it with you' currently applies years before the event of your passing. Equally I could suggest that we have returned full circle to the Welsh coal mine where the workers could not be tempted to work all week. They had no need for wealth; their needs were met by a three-foot wardrobe and a bedside cabinet. They don't have nursing homes in Romania; they look after their own.

As a race, what we do matters. What matters, we don't do.

Aneurin Bevan launched the NHS on 5 July 1948 and laid out three basic principles which were set out at its core:

- that it meet the need of everyone
- that it be free at the point of delivery
- that it be based on clinical need, not on ability to pay.

These three core values have been added to with a supplementary seven, which seek to keep the values current. Found on the NHS website, they make for interesting reading. Search for the NHS constitution on the Internet and you are bound to find it. I will not reproduce these seven supplementary values for fear of breaching copyright, but if you have a mind to, feel free to seek them out and study them. Consider the gap between the stated objectives and the reality that you experience on the ground. In my own case I found it necessary to wear a tight-fitting hat. This seems a frivolous comment, and indeed it is, but the hat stopped my eyebrows from slipping to the back of my head. You can only raise them so far before gravity takes over.

I have reproduced my own comments that were evinced from my own reading. They are in order. Your challenge, should you care to accept it, is to align my thoughts to the NHS doctrine. I know what you are thinking: *Lazy bastard wants me to write his book for him.* But there's the rub. If you complete the reading of *How to Kill an Elephant* and cannot write your own, then I have failed in the process of enlightening you.

1) This could best be summed up as the slow-motion suicides who have and will continue to ignore all of the help and assistance offered. Well, you've offered particular attention to these groups and sections. What have you got to show for it?

2) What this could also say is, 'We have a limited budget, and every day we have to make decisions on affordability and outcome. Therefore, we select who does and does not receive treatment.'

3) What this could also say is, 'We have ticked all the boxes on the quality assessment form and know that every patient has received the highest standard of care because the paper confirms this.' (If you cannot look after yourself in hospital, the NHS won't do it for you.)

4) Because with 1,200,000 workers, you have to train them to the pro forma and learn how to tick boxes.

5) The leaders and managers of the NHS organisations have one primary intent, the satisfaction of their own pay scale, maintaining their own job security, and staying within the pro forma.

6) This can also read, 'We will continually research and develop new ways of treating a new range of conditions which we can add to the pot of ever increasing expectations to be aligned with an ever increasing demand, to be met with a finite budget.'

7) But at what stage, if ever, do you accept that the advice, help, or training, call it what you will, is falling on deaf ears and being ignored?

8) Of course the 1,200,000 employees do not partake of these public funds and are not actually served in any shape or form by the NHS, are they? There are no NHS managers earning over £400,000 per annum (and I scored a hat-trick for the winning side of the FA Cup Final).

9) Given that the NHS is corrupt and lies to Parliament, its own patients, and its own staff, and given that its own staff lie to the NHS and the government accepts these lies and promulgates them for its own purpose, just how responsible, transparent, and honest can you claim to be?

10) Somewhere along the lines of 'I've ticked the boxes', it introduces a multidimensional integrated structure to support all elements of the 'health programme'.

11) Sounds good, doesn't it? How do you rank breast cancer over MS, or a child over a geriatric? And how do you rank the wage bill for the management team over the needs of your patients? I know: 'You're worth it.'

12) And the last thing we will tolerate is any dissent, no ripples in the pond of tranquillity, no one actually speaking their mind. And you certainly know how to value staff—well, some of them anyway. Have I mentioned artificial empires before? Yes, I think I have.

What the NHS Constitution offers is the written pro forma, the recipe for satisfying the needs of the pro forma writers.

Classic herd mentality. The whole herd follows its neighbour just a few feet away with no vision or idea of where they are going or what for. But we have a plan, a mission statement, and we know it works because we have the corrupted data to prove it. We ask questions, collate/corrupt the resulting information, and deliver it within our framework of responsibility and pay grade. Herd members, when they reach the edge, turn back into the grouping, but not before others have been encouraged to step out, before they in turn fold back in, but the drift is created and the whole mass moves under its own momentum without any control, any Great Brain, any purpose. The larger the herd, the greater the 'purpose or effect' and the less the control. Ask any member of the NHS how they incorporate these central tenets into their daily work lives and the reply will translate into, 'I'm just doing my job.' Please look, observe, study these 'Nightingales' of the health world. Look for the compassion, the care, the 'going the extra mile', and contrast it with the staff at McDonalds or John Lewis or your local pub, and then see how this organisation shines out as a beacon of compassion and caring, not. You don't have to go to a poorly (officially) performing hospital (they ought to take some 'magic' medicine; it would make them better); any one will do. Watch the staff go by.

Collecting and collating all the data to support the lie is a significant part of what the NHS does. I don't want a glossy holiday brochure telling me how nice the pool is when I can't see the bottom for filth, or how good the hotel food is when it's covered in flies. The brochure doesn't make it right, but management believes in its own lies. And if managements pushes its values hard enough, some will believe. Let me put a different argument to you: Become a Great Brain, take over control of the NHS, and accept all the power and responsibility you need to change it for the better. Where do you start? How do you persuade the 1,200,000 staff to work differently? Do you start like Jamie Oliver with a small collection of individuals, later launching them out into the wider community like a virus to infect and enthuse the workforce? Ah, what you really need to do is change the culture! The game is called Chinese whispers; the message is corrupted on the way down and corrupted on the way back, and the activity is absolutely fucking pointless. It is all things to all people and means nothing. The NHS exists to serve the NHS. People receive help as a by-product. The product is the NHS.

So where is our great herd going, towards the fabled land or the precipice in the clouds?

And what about those members of the herd on the edge, the leaders, the ones that can see the horizon and know where they are going? They are only on the edge because they've strayed there. Their reaction is to fold back in to the protection that the herd offers, because striking out on your own is dangerous at best. What I'm trying to say with this herd example is that as individuals we always look for reassurance and support from our 'people environment', our friends, colleagues, and family and our interaction with everything else human, be it music, radio, TV, or other media, even the cityscape of housing. We all need a sense of belonging, of conforming, of being a herd member; anything less and psychologists would start to take an interest. (It has struck me that psychologists may well be taking a very real interest in me.)

Better To Burn Out than to Fade Away

Does humankind have a future? Sources suggest our population will reach twelve billion by the end of this century. (Sources will be quoted later.)

But after China come India, Brazil, the rest of South America, Africa, and the rest of Asia, all wanting their slice of the action, all wanting their place at the table. But what then? What happens when the whole world is up to standard? I'm not talking about competitive edge here; I'm talking survival, the fight for resources, for energy, water, space, and food to feed our twelve billion mouths. At what stage do we say enough's enough? How many resources can we use before we run out? And when we run out, as run out we must, what's left, the husk of a world full of people and precious little else.

I look at the human project like the life of a typical individual. Starting with nothing, you grow, learn, develop, earn, accumulate, breed, pass on your wisdom, retire, give away, shrink, wonder what it was all for, and die. Your legacy is your DNA and your tombstone. You start with nothing, grab with both hands, burn out, shrink, and die, all as unthinkingly as a mountain gorilla or wildebeest.

Our history could be summed up in a simple sentence: we came, we conquered, we raped, we destroyed, we failed.

What's the idea? At some stage in the future humankind becomes enlightened, learns to accept that enough is enough, stops the endless, mindless growth, the relentless push for survival at all costs? What triggers the epiphany, the realisation that we are fucked?

To lose weight, eat less food.

To stay healthy, avoid unhealthy activities.
To save the planet, use less energy.
Shut shops on Sunday.
Buy less shit.
Form a new herd. Avoid the precipice in the cloud.

The Willingness to Create Momentum

Many years ago before the Internet became such an integral part of our lives, I would occasionally receive letters across my desk from hard-headed company buyers. These letters would tell me about some dying kid's wish to enter the *Guinness Book of World Records* by having the greatest number of business cards. 'Please send your card to the kid's address, and at the same time forward this letter to ten of your own contacts so they can do the same.' Every now and then you'd see a feature in a paper of some poor kid's dad trying to halt the thousands of letters coming through the door to his healthy kid. It was a scam. These letters were called 'chain letters', and others followed. 'Send £10 to the writer of this letter and forward it on to ten of your contacts asking them to do the same. Every time you receive money, post some of it to your original contact and keep the rest.' These became known as pyramid schemes. If you were the originator and enough idiots contributed, then you became very rich, along with other early participants. The problem was that later contributors were left out of pocket. The serious world of financial investors has not been immune from these schemes. You use secondary investors' money to pay high returns to primary investors, then tertiary investors' money to pay high returns to secondary investors, and so on. The schemes grow and grow and are self-funding until confidence is shaken, fresh investment stops arriving, and the deception is exposed. Named after Charles Ponzi, these schemes have continued into modern times. One of the last to be exposed was Bernie Madoff's scheme which collapsed with debts of millions of dollars. Even the cricket team of England lost out to a Texas billionaire's tournament scam.

Now I've often wondered if these Ponzi schemes are a bit more widespread than we care to believe.

I suppose I need to start at the beginning, with the simple statement that I am hungry; it's not a problem because I can gather some wild strawberries. They are not very big, but there are plenty of them. After an hour, I am replete. The problem is that these berries grow under thorn bushes and are difficult to get at, so I invest some time in removing the thorns with a stone tool on a wooden handle. Now it still takes me an hour to collect the strawberries, because they are small, but at least I don't scratch myself picking them. But I spend three hours making the tool, so I am down on my labour. And after every few hours of use, my tool breaks and I have to make a new one. I keep eating strawberries, but on average my labour goes up. This is when I am offered an upgrade, a bronze tool that shouldn't break. The problem is, this new tool is rare and expensive, so it costs me significant labour elsewhere on my part to obtain the tool. Now I can still eat strawberries in an hour, but I am down 60 hours in labour. Sure enough the bronze tool breaks, but no matter; I'm offered an iron tool that is even rarer and more expensive. It costs me 120 hours in labour.

Let's cut out a few stages.

Now I have a strawberry farm of my own. I have just bought a new tractor for soil preparation and for gathering the harvest. But now I have to pay the labour of everybody in the chain of operation who made the tractor so I could save time. This includes the miners who extracted the ore and carted it to the processing plant that was done in lorries on roads that I've had to pay my share of, using coal to heat the ore to convert it to steel to cast the shapes to build the tractor. I also have to pay my share of the costs of the lorries and the drivers and the fuel and lubricants and light bulbs and rubber tappers and the stone quarries. The list is nearly endless. But many things have happened. I no longer eat strawberries, because I can't stand the sight of them. I have a mortgage and a bank loan, my country is in debt, the earth is warming up, I have to compete against cheap labour in other countries, and I wonder if I am in the midst of the biggest Ponzi scheme the world has ever failed to recognise. My labour is now devoted to earning from others so that I can pay my dues to others. But it isn't just me in isolation; it is our entire nation. The thing that I don't understand is why most of the 'rich nations' are below the surface drowning in debt and why inflation is seen as such a 'good thing', in moderation of course.

Ultimately we have to run out of new investors, and I don't just mean people. I mean fossil fuel reserves, iron ore reserves, rubber trees, space, water ...

All I needed to do was spend an hour picking strawberries and I was no longer hungry. But I have devoted my life to extracurricular activities to make my life easier, and in the process I have helped to change this world forever. We cannot put the genie of 'progress' back in the bottle, but we could maybe change its course. But should we call it progress or Ponzi? Our average life expectancy has grown by thirty-five years or so, but at some stage our labour has gone from subsistence to extracurricular for betterment. We might live twice as long, but do we get twice the pleasure? You can spend your life hammering stone to build a cathedral, when you could have spent your life living.

When Half Could Mean All

I called cities humankind's climax habitat to illustrate a point earlier, but in reality they are not. We will not be happy until we have attained a climax world, because we have no off button. We want the Third World to develop so we can sell them products—cars, planes, trains, computers, and the like—using money they have earned from us by selling us clothes and shoes and raw materials and cheap labour. In the process everybody's happy because they have made a living. But again this is wrong, because once we have attained the climax world, we still won't be happy. Far from it. This is because the struggle for survival will continue with nothing in reserve. Humankind is the ultimate Japanese knotweed, cane toad, or zebra mussel; we will not stop our destruction and exploitation of this world that we notionally share until there's nothing left to share. By then we won't need to kill an elephant because all elephants will have long since become extinct. All we will be left with is the pachyderm in the room.

Our lives were ruined as soon as we established the division of labour, concentrated our skills in 'trades', and created a market for resources. 'One of my hours is worth two of yours, so pay me more' becomes 'One of my hours is worth thousands of yours, so make me a billionaire' (a corruption of our dominance gene perhaps. 'I want to be better. Because I get the pick of the spoils, I am more likely to survive').

Why squander this beautiful planet of ours for mindless tat? Ultimately what will we have to show for it? I have tried to illustrate the futility of conservationists (gardeners), charities (people gardeners), governments (self-serving corruption machines), sustainabilitists (let's make a living by making the right noises), and technologists (let's speed up the human project) who only serve to continue more of the same while claiming to make a difference. Give me one example of our modern world actually producing a benefit that I cannot reinterpret as a negative. And there is the rub, for my writing to date is dismissive of our species and its activities without offering any solutions. Equally I have only touched the surface of cause and have concentrated on effect. Is my aim to depress you, to denigrate the human species, or to illustrate another way, a future that we can all engage in? To start to understand the scale of the problem, you have to illustrate the scale of the problem. I will continue to illustrate the futility of our lifestyle and employment choices with a stream of examples, examples which may well touch on your personal orbit and lead to a fuller understanding from a personal perspective. This is no search for a grand conspiracy theory; this is a search for the biology behind our actions.

Politicians talk about leaving a legacy, a headstone that reads 'I was here,' to supersede the castles and cathedrals of a former age, which themselves served to supersede the pyramids and great walls and Stonehenge and other, lesser stone circles. Successful businessmen chase legacies of football clubs and charity giving and stone mausoleums to mark their time on earth. Is there something in the human psyche that deep down understands the utter futility of what we do and seeks to justify our former crimes? We reach our twilight years and ask ourselves what it was all for. Well, what is it all for?

So what to do?

You cannot turn to governments. They are corrupt.

You cannot turn to the UN or the WWF or the RSPB because they are corrupt.

You cannot steer the herd because you cannot be a leader.

Face it: we are fucked.

One of the things I have tried to illustrate is the human condition to essentially piss about. Spend billions on traffic calming without calming traffic; spend billions on diets without losing weight; spend money on charities without solving the issue because the agenda continually expands; spend billions on crime without stopping crime; spend billions on healthcare without providing healthcare ...

We are in the middle of an election campaign where the main topic of conversation appears to be who can pledge to spend the most. Who will end austerity?

Try this for a solution: reduce the tax take by half, end all the non-jobs, provide only the essentials, work fewer hours, reduce the corruption by half, and buy less crap. Save the human species.

1) Reduce the tax take by half. This should concentrate the mind and, at a stroke, end half the corruption. There would be no money for traffic calming or renewable energy or overseas aid or mega salaries for management (back to the *Coronation Street* actors again, £20,000 an episode or £50,000 a year) or council vans sitting on drives with the engines running or paper-chasing management systems. And we would have to release all the able-bodied pets into the wild.

2) End all the non-jobs. This would lead to massive unemployment to rival that of Thatcher's tenure, but so what? The future of humankind is at stake. Why not pay people for what they do, not what they appear to do?

3) Provide only the essentials. Introduce means testing. Every spend needs to be corruption-proof. And remember that you cannot trust the system because it is corrupt. Spend every penny as though it were your last one.

4) Work fewer hours. If you are paying less in taxes, then you can reduce the hours you work. Get a hobby, dig your garden, exercise, cook real food, and above all enjoy life. You've only got one to look back on with regret.

5) Reduce the corruption by half. Reduce the entire government apparatus by half. This is all of welfare, the NHS, education, civil servants, and all the other self-perpetuating bullshit—because they really are not worth it.

6) Buy less crap. Save money. Spend money on what is essential. Forget the convenience food and designer goods and the go-faster stripes on the car. It's all meaningless crap and leaves you with nothing to show for it in the end (a three-foot wardrobe and a bedside cabinet). If you get the hang of it, you could work even fewer hours.

Corruption

We started with the idea of lies to children and have expanded this idea to look at lies within large organisations. The larger the organisation, the greater the corruption. In its simplest form, this can be put down to vagaries in the English language. A simple example would be the expression 'ASAP'. For many this means urgent, quite literally as soon as possible; for others it simply means, 'I'll do it when I can, after I've finished my other jobs.' We can see and agree with the logic in both instances. Now when you have a large multilayered operation where the same expressions mean different things to different people, then it's obvious that some drifting of intent will creep in. But when you introduce targets, then the drifting can become exponential, because the 'target' quite literally becomes the target and information is biased to achieve it. That is to say that the system works to the target rather than the results work to the target.

How often do you sit eating a meal at a restaurant and remark about the meat being a bit tough or the veg a bit cold, only then to be approached by your waitress who asks, 'Everything all right?' Most people reply along the lines of, 'Fine, very nice,' because sitting and having an argument with the staff about the food spoils the occasion. Many years ago I used to drink lemonade by the pint and would order a glass well into the pub's/restaurant's opening hours, only to be served with carbonated water. 'Sorry, we forgot to reattach the syrup line when we cleaned it' or 'It must have just run out' were the common replies. On a large number of occasions I've returned pints of unflavoured water and then been on the receiving end of shitty service because I'd complained. I once visited the same pub for a quick hotplate meal, where you could be in and out in fifteen minutes, on three consecutive days to receive three consecutive pints of fizzy water even though they had been serving lemonade mixers for hours. I once took an area manager of a food pub chain on a short journey and I asked him what the scam was. He hadn't heard of it but did comment that it was known in the trade as 'liquid gold', because lemonade costs only 1p a pint and is sold for £2. I thought at the time, and still do, *If it's worth that much money to you, why don't you make sure you serve the best lemonade in the world, because then you'd increase your sales?* The point I'm trying to make here is that either we are preprogrammed to give a polite reply to the question or we've learnt from experience that complaining spoils the occasion. Most of us only actually give the truth when it's really bad and the event is completely spoilt anyway. From the food management's perspective, they've asked the question and gotten a satisfactory reply. From the customer's perspective, you just don't go there again. You vote with your feet.

Throw in a dose of gratitude for treating your ills or stitching your wounds back together and you've automatically overcome the tendency to give an honest opinion about your treatment at the hospital. My problem is that I have an overactive mind and I cannot resist analysing the staff and their activity while in hospital, whether visiting patients or being a patient myself. You wait for hours and hours and hours for something to happen, and are given false hope of progress and activity, until suddenly, and for no apparent reason, you are fit to go. After you've waited four hours for your pills to come up from pharmacy and a further two hours for your discharge letter to be signed, you can go. They talk about geriatrics bed-blocking and creating shortages in hospital when their normal practice blocks beds for others without the need for geriatrics. Having sat and waited with 'geriatric' relatives, I know that the delays are horrendous. Once you are deemed fit to go, you then have to be assessed by occupational health, who tie in with social services, who have to ensure that your support equipment and support care package is in place before you can finally go, after the four-hour wait for pills and two-hour wait for the doctor's letter. This process can take several day. Through it all you wait and wait and wait and wait. Imagine booking in to a hotel for a weekend and taking five days to check out. For how long would such a hotel stay open? But this is a hospital and there are management systems in place, and nothing can move without paperwork. Of course the curious thing is, the day before you entered hospital, you had no support equipment and no support package in place and you'd managed quite well for many years, but this happy state of affairs ended upon admission, because they don't want you to return unnecessarily; you might block a bed.

Management would describe this as a multiagency commitment to providing healthcare in the community. We are very successful at preventing unnecessary readmissions. But if this meant two two-day visits as opposed to one seven-day visit, would it be a bad thing?

As luck would have it, I've attended a five-year routine health check at my local GP surgery. Bloods and urine samples were taken, and an interview about lifestyle followed. 'Don't you get fed up with patients coming in and ignoring all the healthcare advice?' I asked. The answer came in two parts: 'I have a duty of care to all of my patients' and 'I document the fact that I have given advice so we can't get sued later.' So the government edict of providing healthcare advice to raise health standards in the UK has been corrupted to 'produce a system to protect us from litigation'.

But if you stop and think, the litigation is corruption as well. Let a solicitor handle your no-win, no-fee case for compensation and get your dues, but only after the solicitor has taken his or her cut at £250-plus per hour, which is likely to exceed your monetary redress.

This is also corrupt in the sense that this parasitic process is attacking the very system set up to help you. As a consequence it corrupts the system in cost and objective. 'We could have employed extra staff to prevent this unfortunate incident from occurring to you, but our budget is taken up fighting lawyers.' 'Our paper-based systems record the course of your treatment within the NHS to protect us against litigation, but at the same time this system corrupts the patient care because staff are all too busy "ticking boxes". But you can't sue us, so it's a success.'

One of the seven core values of the NHS is to offer value for money and to spend its budget taken from our taxes on healthcare. But this in itself is a lie. If the systems are in place to prevent litigation, then there should be no litigation because the system prevents it. Because there is litigation, by definition then the system to prevent it must be inadequate. So which is it: do you have a system, or do you not have a system? How many resources are diverted to prevent something that you don't prevent? We are back to 30 mph signs again, I'm afraid.

In the run-up to the 2015 election campaign, one of the Labour Party's proposals was for the introduction of the mansion tax, designed specifically to raise a sum of money estimated at £1,200,000,000 to pay for 'thousands more doctors, nurses, midwives, and home-care workers, and to guarantee that patients in England will wait no longer than one week for cancer tests and results by 2020'.

Let me return to the subject of Ponzi schemes. (Although, if you've been paying attention, you might consider that I never left this subject.) The idea that the whole world is engaged in one may be too great for you to grasp, so let's look at something a little smaller. Remember that a Ponzi scheme works by using new member contributions to pay earlier contributors' returns which are not justifiably earned. As a simple example, investor A is promised a 10 per cent return, but only 5 per cent can legitimately be achieved. He receives his 10 per cent by taking money from the capital of investor B, who in turn takes her 10 per cent from investor C. The whole arrangement works very well until you run out of new investors or until the confidence is lost and later investors want to remove their capital.

With the advent of the state pension, we have created a Ponzi scheme whereby the original investor A's qualified by reaching the requisite age without having made any deposits. Those workers who continued to work paid these original 'returns' until they themselves retired and became investor B's, who looked for their returns from younger workers who would become investor C's. There is no cash reserve ring-fenced and allocated for the payment of state pensions; it all comes out of existing workers' payments. This arrangement is fine only as long as you have a sufficiently high number of new investors. If you return to the population crash demographic scenario predicted from 2050 to 2100, then you realise that the scheme has to collapse. Now while the baby boomer and enhanced lifespan overhang has begun to be addressed by increasing the retirement age, this doesn't alter the fact that we have a pension Ponzi scheme. In countries that have suffered economic collapse, Bulgaria as an example, private pension provision was threatened with 'privatisation' by the state to maintain pensions that would otherwise not have been met, so you cannot guarantee your own protection by making your own provision either. (It didn't happen, but a juicy pot of cash reserve will always be extremely tempting to a government in crisis, and they can always put it to the vote.)

An alternative scenario would be one akin to a reverse Poland situation. I'm not talking about the very first pocket calculators with their reverse Polish notation; I'm talking about the hundreds of thousands of young Polish nationals living and working in the UK. All these economic migrants are earning British pounds and paying British taxes but are not paying into the Polish state pension fund. What would happen if hundreds of thousands of young British workers were to depart our shores and travel with returning Poles to Poland other than the collapse of our pension Ponzi scheme?

Get Your Hair Cut

To save the planet we all need to become hippies with alternative lifestyles. But this doesn't work either, because these people live from benefit cheque to benefit cheque but avoid the trappings of a materialistic life.

Let us consider an alternative lifestyle that might actually save the planet.

Everybody in the UK is intent on making a living. This excludes the young, the old, the ill/disabled, and the rich who have retired or never started work in the first place. Of those who work, 40 per cent of their effort goes into paying taxes to support the UK public sector. The remaining 60 per cent goes to support the UK infrastructure of the private sector. But of this infrastructure, what percentage goes towards non-jobs, these being advertisers, journalists, political commentators, MPs, civil servants, highway engineers, health advisers, dieticians, planning consultants, planning officers, council highways staff, sustainabilitists, ecologists, charity workers, judges, pollsters, and the managers (that's an oxymoron) who controls and instructs all the foregoing, to name a few. In essence we are supporting an entire chunk of the UK economy which does nothing apart from occupy space and demand resources in buildings and energy and travel infrastructure. It is all fraud. Imagine a scenario where all of these non-jobs were made redundant. Let's take a massive slice out of UK plc's overhead and return the savings to the workers. Reducing taxes by 50 per cent would shorten the workweek from five days to four, and the workers would suffer no ill effects. By removing the non-tax-funded non-jobs, such as advertisers, journalists, overpaid actors, and planning consultants, the cost of all goods within the UK's advanced economy would be reduced and we could all survive by earning less. Therefore, we could reasonably begin to think about working less than four days a week—let's say three and a half days. This means that parents of young children only need reduced childcare because they are available for half the week. How many times have you heard parents say, 'The costs of childcare mean it's hardly worth me working at all'? Shops could shut on Sundays, and people would have the time to walk, exercise, cook meals from scratch, and improve their health by living better. A shorter workweek would produce less commuting. Cars would last longer, roads wouldn't wear out so fast and would require fewer repairs, and we'd use less energy.

Refuse to treat the slow-motion suicides. Once the early attempts to change their lifestyle have failed, embrace them back into the fold only if change is made. Obesity, drug addiction, and the like is their problem. Why should it be mine?

Gingerbread Houses

Construction of buildings in the UK is governed by the Building Regulations Advisory Committee, who ensure compliance with structural and design criteria. For many years now we have had a succession of revisions, the aim of which is to ensure that the carbon footprint of every new dwelling is reduced in steps to zero. This is achieved by insisting on low-energy fittings, high-efficiency boilers, high insulation values, and renewable energy designed into the new-build structure. This process has stalled of late because of the economic climate, but it will continue to its goal in the future. A similar process has occurred with step changes in tax rates for company cars and car tax for higher CO_2 emissions. The higher the CO_2, the more you pay. This has been reflected in new car designs and the technology they employ, such as stop–start, eight-speed gearboxes, and lean-burn engines.

So why, if it's good for the UK and good for the world, don't we introduce similar laws to reduce, in step changes over several years, the refined sugar, saturated fat, and salt from our processed foods and increase the amount of wholegrain foods, again in step changes? What vested interest in the food industry would we be upsetting? Every food manufacturer would be working to a common standard, none would be disadvantaged, and every consumer would benefit. Could you actually legislate out common forms of diabetes, obesity, and heart disease, and increased stroke risk? If the principle is good enough for houses, why isn't it good enough for people? Don't we matter? Why is the entire food industry governed by 'people have the right to choose, and we are ensuring that information labels are there on all packaging so that people can make an educated choice'?

Strange how we legislate against battery farming of chickens, veal crates, and live animals in transit, yet we can't legislate against unhealthy foods. We must be able to choose. You can't choose heroin or cocaine because it's dangerous, but you can choose alcohol or fat or sugar.

Every now and again some superbrain statesman declares, 'We have to protect generations unborn. They will never forgive us if we get it wrong.'

How many of us have had a proper genuine banquet? Go to one of the great houses, Chatsworth House for example, and search for the banquet house. Don't look for any grand room, because what you are searching for is a small, discreet, even cosy feature, which at Chatsworth is on the roof. If you were in the most favoured select inner circle of your hosts, you may have been invited to the banquet, where you would be served with the most expensive foods available, so expensive that the foods would often be served wrapped in gold foil. Have you ever wondered why your Christmas cake is served on a foil-covered board? It's a throwback to this prized banquet ritual.

So what was this rare and special treat to be served in gold and only to the most special of friends? Sugar. I live in Leicestershire. Leicester is sometimes described as 'the diabetes capital of Europe' because it has such a high concentration of sufferers. This is put firmly down to the high South East Asian population. The suggestion is that this ethnic population is less well-adapted to the 'Western' diet with its high sugar levels that their bodies' regulatory systems become overloaded and cannot cope.

I am trying to suggest that we Caucasians have no great history of high sugar consumption either and will end up only slightly behind the curve.

Let Us Stack Some Bricks

Isn't it strange that every climate change conference held around the world involves hundreds of air flights burning carbon fuels in pursuit of carbon savings? Not forgetting the new forest to be planted to offset the carbon. Every ecologist/conservationist burns fossil fuels to produce their work in the name of saving the environment.

We seem to have an unbelievable ability to chase our own tail. We laugh at our pet dogs for indulging in this most basic of stupid actions, but we are the masters by far. Let's build a house—an honourable pursuit—to provide shelter and safety from the elements. I am reminded of a tale about a quarry worker in Derbyshire who, every day at the end of his shift, took a rock home with him. After many months of this activity, he had built his own house, at which time he was approached by the local lord, who asked him, 'Whose rock is that and whose land is that?' 'Yours, milord,' was the reply to both queries. 'Good,' said the lord. 'You can live in it rent-free, but when you die, it's mine.'

To build a house requires the land, the materials, and the labour, nothing more, nothing less. Build your own house on your own land with materials to hand and it can be done relatively cheap, if you end up with a little bit of a hovel. But for the purposes of my example, let's go back to those 'kick about' specialists in place and time on Salisbury Plain. Ownership wouldn't have been a problem. Clear an unused space and stake a claim; enter the neighbouring woodland and source your timbers; cut some riverside reeds for your thatch; and mix some mud, dung, and plant stems with a framework of sticks to build your walls. If it falls down, start again. If it's damaged by the weather or age, repair it. Use the local tried-and-true skills to construct something that you know works. And as an added bonus, it was all zero-carbon.

Fast-forward to today and build your house. Begin by working several years to earn enough spare cash to save and provide a deposit, then apply for a mortgage and commit a further twenty-five to thirty years of your labour to complete the purchase.

Then you pay for architects, engineers, estimators, interior designers, noise consultants, traffic consultants, environmental consultants, ecological consultants, planning consultants, planning officers, building regulations officers, checking engineers, environmental health officers, ecology officers, highway engineers, planning committee councillors, council solicitors, council elections, polling officers, vote counters, election commission supervisors, election leaflets, posters, structure plan designers, government planning inspectors, appeals systems both for and against, Section 106 Agreements (off-site traffic improvements: roundabouts, zebra crossings, traffic light improvements, refuge islands, playground equipment, bus subsidies, social housing subsidies, and school classrooms, to name a few), support staff, receptionist, telephonists, IT support technicians, cleaners, computer designers, computer manufacturers, computer program writers and designers, postal services, telephone engineers, furniture makers, carpet makers and fitters, cleaning equipment suppliers and manufacturers, cleaning chemicals and suppliers, calendar makers and suppliers, diary makers and suppliers, 'human resources', and health and safety officers.

All of these people need access to food, water, waste disposal, offices, transport vehicles, electricity, gas, roads, railways, paper, money, pensions, schooling and education, and management supervision.

I am absolutely certain that I have missed many individual tasks, but I think you can begin to build the picture. Now before we go any further, let me make two statements:

You want a house to live in.

You haven't even laid one single brick yet. (While I've been building the picture, no one's been building the house.)

The good news is, when your house is eventually built, in the near future, it will be 'zero-carbon' and environmentally friendly. The bad news is, none of the above identified tasks is, so why all the bollocks in trying to achieve such a thing?

Now if we start the actual construction process and think about materials manufacture, supply, labour, and all the support structure for the physical process, we are going to magnify the task manyfold but still claim it's to produce a zero-carbon house. In the final analysis, 80 per cent of us live in a caravan, so what's so fucking special to warrant all of these bollocks? Ask yourself why houses are so expensive to buy in the UK today. If you didn't pay for all these tasks, there would be massive unemployment—and you would be one of the unemployed, so you couldn't buy a house then either. We really are back to a chicken-and-egg situation here, aren't we? Damned if you do and damned if you don't. But my earlier answer to which came first, the chicken or the egg, involved a dinosaur somewhere. Can you spot the dinosaur here? The whole process of providing a roof over your head dwarfs even an elephant in scale. What I need is something akin to a diplodocus.

But in this process of magnifying the creative needs of the support industry of house building and many, many other tasks, without any question, Planet Earth is damned to suffer the slow motion death inflicted by a trillion worthless actions. And any amount of 'working to save it' is counterproductive, because it just means more and more of the same. In the process, without any question, Planet Earth is damned. What we need to do is stop working and start living. You can start by leaving the Third World as the Third World and decide how we can join them. The human race is looking straight down both barrels of a Planet Krypton situation where we will mine the resources of the planet to such an extent that it will explode, although in the non-Hollywood version, our planet will implode with the collapse of resources. We've headed off the hole in the ozone layer. (Although reports are surfacing in 2018 to suggest that the Chinese had got their fingers crossed when signing up to the banning of CFCs.) We preach about the greenhouse gas and climate change controls, but we do nothing about the conservation of water, food, and habitat or population growth. Instead we busy ourselves, as only humans seem able. If *Homo sapiens* are really the 'wise men', then why aren't we wise? End the Ponzi scheme of endless growth (for growth, read destruction), ignore the gardeners, and let it be.

France is described as 'the sick man of Europe' because of its poor productivity, yet its population has high life expectancy, in spite of a high-fat diet. I won't hold France up as paradigm for how we should begin to proceed, because the French lie; they lie because their lifestyle is propped up by the rest of the EU through the Common Agricultural Policy that unfairly subsidises their economy.

So what to do?

You cannot turn to governments; they are corrupt.

You cannot turn to the UN or the WWF or the RSPB because they are corrupt.

You cannot steer the herd because you cannot be a leader.

Face it: we are fucked.

In the 1980s our company made prestressed concrete products. Building Control would often ask us for calculations to prove our structural claims. Our standard beam sections had resistance moments of strength, and these were recorded on two sheets of A4 drawing paper covered with finest copperplate mathematical writing for each beam. I took a lunchtime telephone call from the senior engineer of Manchester Metropolitan Building Control who had found a mistake and wanted to talk to our senior engineer. He explained to me that, having studied prestressing while at university, he'd wanted to see how the engineering had moved on since his day when he found a mistake. The beginning and end was correct, he said, but the error was in the middle of the calculation. It transpired that when the somewhat dog-eared calculation sheets were renewed, an error had been made in the copying; hence the problem. This error had been present for nearly two years before it was picked up. The point I am trying to make is, hardly anyone looks at, reads, or checks the calculations in support of building design. Everyone is content to 'stick something on file', as it covers their backs. So thinking back to my 300-millimetre-high pile of £300,000 planning consultants' supporting data for my two-hundred-house planning application, who, if anyone, read it? At best they will have skipped straight to the summary/conclusion page, and at worst they wouldn't have bothered at all.

My son who is a landscape architect regularly produces 100,000-word submissions in support of planning applications for this one aspect only. To put this into context, this is around 160 pages of A4 paper typed at a point size of 12, but it still only represents some 40 millimetres of my 300-millimetre-high stack. The one saving grace is that most local authorities encourage the submission of applications electronically, so at least some trees are spared from the papermaking process.

Now one other thing that the vast majority of these planning consultants, both poacher and gamekeeper, share is that they all have a university education. Have you ever wondered where all these degree-qualified students end up working after graduation? It appears that we are actively creating jobs for these students that never existed before. Given that the number of students graduating is significantly higher than in my day, they must find employment somewhere.

Another common feature is the drift of these consultants from one side of the fence to the other, gamekeepers-turned-poachers if you will. A lot of later-year planning officers take their council pension early and then immediately set themselves up as self-employed planning consultants with a direct in on first-name terms with their often former junior colleagues who remain in post at the council.

In the 1980s and 1990s we could successfully build of the order of two hundred thousand houses a year without all this expensive bollocks, so what gives? And before you make the argument that planning committees made up of councillors are getting more educated and enlightened, in my experience nothing could be further from the truth. I have sat in the public gallery and heard the application officer explain to the council committee that planning permission for a business park was granted seven years ago, only then to hear a councillor state, 'If we are not careful, we will end up with a business park here.' I genuinely feel that I know more about the application, having heard the introduction, than the people responsible for making the decision, who had also sat through the same introduction. This is only one of many examples I could give.

What we have entered into is an arms race between the planning consultant and the planning process, where the whole process is getting more and more complicated and more and more expensive to achieve the same result of granting consent for building houses. At what stage do we arrive at a state where only PhDs need apply to work in this field? When only PhDs can earn enough money to buy houses?

The counterargument to this is that we need to protect the resources we have, and that means justifying all of our construction activities by conserving what little we have left. But when we say 'what little we have left', we mean arresting the decline of ancient woodland, meadow, and other 'poster boy' habitats, and mitigating the impacts of all the permitted developments. I say again that this is simply a form of gardening, favouring flowers over rose bushes and lawns over trees, and not a serious attempt to halt anything, because the presumption is for development and the blind pursuit of the global Ponzi scheme. The way that we attempt to mitigate by using more and more resources to protect more and more resources means that we will inevitably fall foul of our ambitions. All these university-educated experts demand more and more resources to qualify and operate in the pursuit of protecting resources. If it weren't so stupid, it would almost be funny.

We seem to be back to the suggestion that education is a device for keeping the kids off the streets. Why else do you need to invent highly educated non-jobs for highly educated graduates? What purpose does it serve?

Not forgetting the highly elevated salaries that these highly educated graduates demand in order to pay off their highly elevated student loans.

It's a curious thing, the Internet and emails in particular. My account is regularly bombarded with spam, unsolicited crap that fills my junk file and slips through into my inbox, waiting to trip me up and seduce me. I say seduce because the object of these 'spam' contacts is either to sell me something I don't want and probably isn't genuine or to allow them to infect my computer with a virus of some description that will command my computer to yield secrets or relinquish my control over it. These messages are a bit like the so-called tart cards that appeared in telephone boxes in cities thirty years ago advertising the services of a prostitute/masseuse with glamorous images that probably bore no resemblance to reality (I never checked). But the really curious thing for me is the fact that my account provider has designed their system to include both a spam selection button for ease of disposal and a spam filter to catch and automatically sideline some of these bogus emails. These spam mails are unsolicited, illegitimate, and potentially harmful to my computer's health and my bank balance. I need to be protected from them, and my account holder has very thoughtfully sought to protect me. Now I say that they are unsolicited (I haven't asked for them), illegitimate (they may not be the genuine article), harmful to my computer (because they can prevent me from using my computer as I intend), and harmful to my bank balance (because I can lose money to these scams). These would be the same account providers that fill my computer with unsolicited legitimate bandwidth-filling/computer-blocking adverts designed to part me from the money I have in the bank. I am so glad that my account provider is so keen to protect me; I can rest easy in my bed at night.

It's similar to the corruption argument between the Mexican drug cartels and the official Mexican government. One is seen to be corrupt and illegal, while the other is seen to be corrupt but legal.

But as I have written earlier, we all have our own inbuilt spam filters to separate the important from the ever so expensive bullshit, so why do my Internet account holders persist in wasting the earth's resources by harvesting advertising revenue?

Revolution! Well, Not Really

Do you remember the petrol blockades of September 2000? The grandly named Farmers for Action and truck drivers blockaded a number of refinery and fuel depot sites around the UK and virtually brought the country to a complete standstill in just a few days, with food shortages and hospitals poised to cancel all but emergency treatment. I got the distinct impression that the majority of participants were shocked and embarrassed by the impact that their campaign had wrought and pulled back from the brink with a sixty-day ultimatum to the government that was promptly ignored and produced only a half-hearted response for fear of doing it again. I thought this was a wonderful example of a crowd response to an irritation leading to a 'revolution' that petered out because no one actually wanted a revolution, just to have a good moan. I know legislation was rushed through Parliament to prevent a reoccurrence of the blockade action, but no repeat of this action has been attempted for fear of an unwanted revolution.

So what is the true figure of advertising? How many billions worldwide is devoted to this task of separating us from our money by separating money from our suppliers? Where did it all begin, with the inn signs and guild signs of yesteryear, perhaps? We started with the informative: 'This building will sell you ale and provide a bed for you and your horse for the night (separate rooms, I assume) or have your boots mended or sold from here.' The old shops in Chester give an idea of the old sheltered stalls where your goods were displayed and you only had to walk around a local market to hear the 'Six bananas for a pound!' cries from the stallholders.

Your customers walked past your door and either bought your goods or didn't, and word of mouth would bring customers from farther afield to sample your quality/cheap wares. With consumerism came demand and more importantly 'disposable income', the difference between your incomings and your outgoings and the ability to indulge your petty needs and whims. Yet the strange thing about 'disposable income' is its inbuilt ability to find a home outside your bank balance, to 'burn a hole in your pocket', whereupon life's little luxuries become life's necessities and the circus begins. The more you earn, the more you spend, constantly in near equilibrium but always edging towards spending more than you earn. We have a tendency to expect our future prospects to be greater than the reality and often end up in debt as a consequence. And this is all driven by desire implanted in our brains by advertising (if it didn't work, they wouldn't do it) and marketing, creating demand for things we didn't know existed, let alone needed, until we edge ever nearer to departing this life, when suddenly the reality check 'You can't take it with you' kicks in. So we are brainwashed/hazed by constant unrelenting exposure to these clever people in suits who know our every desire, if only because they created the desire in the first place. We seem to have replaced the cult of religion (if it ever existed) with the cult of spending money. Now the reason I say the cult of religion, if it ever existed, is because the people living through those historically significant religious times probably didn't feel particularly religious. They built cathedrals because they were paid to, not through the love of God, but maybe out of the fear of him. We are steered by the labels attached to earlier eras by historians, looking for princesses who try to understand the Premier League, without understanding that kicking a bladder around a patch of ground gave people something to do. But if that is true, who today understands that we are in the 'cult of spending money' rather than simply living for today?

So today's high priests are those clever Great Brains who preside at 10 and 11 Downing Street and pore over export–import data looking to 'balance' the Ponzi scheme of the global economy while they earn a living from it. What would happen if we just stopped, neglected to buy the latest fashions and the latest technological gizmo, and failed to replace our cars until they croaked out at the roadside? Individually we would all become richer, our pockets filling with unspent money, our banks wallowing in excess cash with no one to lend it to, our need to work forty hours a week evaporating into factories empty of workers but filled with product that nobody wants—sorry, needs—while the treasury's coffers shrivel and the corruption that money feeds, starves. One of the arguments about controlling the corruption within Mexico is to choke off the supply of money to the cartels and thus reduce the power over the people.

Figures reported in the media suggest that Google has, in the fourteen years up to 2014, produced over $300 billion of advertising revenue, but in 2014 alone it produced $59 billion. Facebook has been slower to market but has produced nearly $32 billion of revenue from mainly digital advertising in six years, with over $12 billion in 2014. If we take the figures for 2014 alone and assume that both parties failed to grow, then over the next fourteen years one trillion US dollars of revenue would be on the cards. Now the point about advertising is that it's a non-job, an overhead, a tax if you will, over and above the cost of manufacture and distribution that is all paid for out of our wages. Don't forget that these figures exclude the actual cost of advert design and generation, the advertising agencies and the studio skills and actors that so glossily portray the simple mantra of 'buy me'. When I was responsible for advertising my company's products, I used to evaluate the cost against the effect. Our product in tonnage terms was cheap and my budget small, but in essence I could afford to spend only in return for result. I had to sell an additional 200 tonnes of concrete for every £1,000 I spent just to break even. Now I know that my product was cheap, but if we were to think in terms of motor cars, which cost, say, £30,000 and weigh 1.5 tonnes, then cars would be valued at £20,000 per tonne or roughly $30,000 per tonne. Yes, I know that cars can be cheaper or much more expensive, but this average will suffice. If you take 2014's revenue from Google and Facebook at $71 billion and divide that by cars at $30,000 per tonne, then that represents 2.3 million tonnes of cars worth of value. In simple terms, we have to build 1,500,000 cars for free just to pay for Google and Facebook alone. But obviously we cannot build them for free, so if we took a budget of 10 per cent for advertising, then we'd be comforted to know that we have to produce 23,000,000 tonnes of vehicles just to break even. But we still haven't paid the production costs for the adverts. If you prefer to think in terms of biscuits at £1,000 per tonne or $1,500 per tonne, then we can begin to talk about 47 million tonnes of biscuits for free and 470 million tonnes at 10 per cent of revenue to break even. You may still think that this an inappropriate approach because a lot of advertising is pitched at very expensive items, let's say a Rolex watch which might retail for £25,000,000 a tonne, and therefore the cost of advertising is miniscule. But you still have to sell the biscuits to make the money to buy the watch; otherwise, you can't afford to buy it. So no change there really, is there? Take the maths one step further with the biscuits and say that the sole ingredient is flour made from wheat which grows at the expected yield of 4 tonnes per acre or 10 tonnes per hectare: 47,000,000 tonnes of free biscuits is the yield from 4,700,000 hectares, which translates to 47,000 square kilometres of intensively farmed cereal crop (3,760 STUs). (What the hell is an STU? It's not sexually transmitted, but read on.)

If we were talking about a biological system here, a superorganism, then we would think in terms of a parasite or a cancer spreading beyond control and weakening the whole by diverting resources away from what matters. But we are talking about a biological system here. It's called Mother Earth, and we are accelerating towards its demise though our blind stupidity. If you want to find a non-biological model for Earth, then check out Mars or Venus and see what you can sell there. Something that would be a great seller at either of these venues would be a life support system. How long before we need them on Earth?

Don't forget that this $71 billion doesn't end up in the bank; it ends up in the hands of the investors and employees who share in the dividends and salaries and use it to spend on the planet's resource of their choice, not just cars and biscuits but fuel and clothing and housing and education and holidays and whatever else takes their fancy.

Now maybe I'm missing an obvious 'trick' here; as we get cleverer and cleverer, we will suddenly and spontaneously realise the error of our ways and will have invented a solution to our problems. We will learn to control and limit the earth's population, we will learn to use our resources only for what matters, we will learn to respect nature for its own sake without the need to concentrate on the 'poster boys', and we will learn that there is only one earth and it's the only one we will ever have. The problem is, if you agree with my arguments, that we have learned nothing. So what are you going to do about it?

If you look at our recorded history on earth, you can view it as a series of progressive steps that have moved our species to our present position. Each step is normally in response to a crisis, let's call it a bottleneck, an impasse that has caused a reaction. You could even start with the climate change in East Africa that forced us out of the trees, followed by population pressures that forced the migration out of Africa and led us to donning clothes to protect from the cold, cooking to aid digestion and maximise the calorific content of our meagre food supplies, building shelter to protect from weather and predators, creating tools to extract resources with reduced effort, creating language to coordinate and organise, practising agriculture to cope with increased population densities, waging war to cope with increased population densities (plan B), coming up with religion to control and threaten the populace, instituting education to maintain control over the ignorant, building cities to cope with population growth, implementing sanitation (eventually) to cope with disease, leveraging science to maximise our toolmaking ability, and instituting government and the rule of law to maintain control over the ignorant, all powered by our insatiable appetite for life.

Nope, sorry, I can't see an obvious 'trick' here, only a constant conveyor belt of self-congratulatory destruction, but only after we've finished with everything else first. Let me ask these questions. When will we stop; what will be the catalyst; when will we realise that enough is enough?

An elderly friend once told me what he had heard about the timber merchants in Claybrooke and of another firm in Leicester. Way back at the turn of the last century they had agreed to a deal whereby Claybrooke would handle all orders south of the city and the other firm would do the city of Leicester and the north of the county. This would all be illegal in this day and age. I came across an article in a local history book that talked about the parish council being upset because they had just paid to resurface the road only to discover very shortly afterwards that the road had been rutted by a steam engine. Claybrooke woodyard was accused of doing the damage but refused to accept responsibility.

My company made structural concrete products. I would sit at my desk designing and pricing against customer drawings. An integral part of my task was to calculate the haulage costs for every job. I used to think that if we did all the concrete in Leicestershire, we would be much cheaper with the reduced haulage costs, and this was in the days before we worried about CO_2. We supplied concrete nationwide from Hastings to Cornwall, and Pembroke to Thurso, and often drove past our competitors' factories to get there. Of course, back in the day, before the advent of trucks and trains and canals, this would never have been a possibility and all materials would need to be sourced and supplied locally. You can easily see the different brick colours in the regions, as all bricks would have been made locally with local clay, for example. Today, however, we have embarked on an international globe-spanning supply network that sees Australian ore being shipped for processing and manufacturing to China before arriving as finished products in ports like Felixstowe. We even used to buy specialist steel reinforcement from Italy because it was the cheapest source, even though it had been made in South Wales before being shipped out to Italy. The answer to the question why is straightforward: haulage costs are cheap enough in this day of container shipping to make it economical to span the globe with trade links. Of course not so very long ago all the ocean-going vessels were almost zero-carbon, made mainly of wood with sails that suffered from the vagaries of the propelling winds and were labour-intensive. Today, even with the largest, most efficient high-tech diesel vessels that achieve around 50 per cent efficiency, all are heavily dependent on fuel and carbon-producing manufacturing methods. So the question should be, is it good for the planet?

While we are busy protecting the environment with our planning consultants and ecologists rushing around the countryside in their expenses-funded motor cars, we fail to realise that in the process they are damaging the same environment. We are aware of the damage caused by pollutants to the natural stone buildings within our towns and cities, and we are aware of the damage caused by 'acid rain' produced as a consequence of our coal-burning power stations.

There are also reports about the damage our pollution does to the hearts and lungs by fuel burning, and in the past to our brains because of the lead content. (Reports are beginning to emerge of particles of pollution being found in the brain. Have we found any solutions?) I wrote earlier that my favourite orchids are rarer in the UK because we are on the edge of their climate range. The nearer you get to the Mediterranean, the more in number and type you will find. Climate change knows no boundaries. We are aware that our domesticated farm animals are producing voluminous quantities of methane, a particularly strong 'greenhouse' gas. We try to determine the impact of climate change on the fauna and flora within our own pet projects and generally find what we are looking for, more funding. We seek out more efficient power, heat, and light producers as a sop to our carbon footprint while continuously making more of them. Let me define 'sop' for readers not familiar with the term. A sop is a thing of no value or concession given in appeasement while the main concerns are being ignored.

But all of these measures essentially allow for one basic premise: business as usual. Continue to extract and utilise the available resources in a more 'efficient' way, but, and this is important, continue to extract them because we cannot sustain our lifestyles without them.

So even though we protect our ancient woodland and meadows from the chainsaw and the excavator, we continually degrade and ultimately destroy them. It's like putting a tiger in a cage. It's still a tiger in tissue and fluids, but it ceases to be a tiger. The predator in mind and spirit, just an empty husk. We now talk of 'natural corridors' linking sites of biological heritage to create a more natural environment that will allow for the transfer of genes from one population to the next, a mosaic of interconnecting biological reserves as the next 'solution' to our degraded habitat. But this cannot protect against temperature or pollution or simply an unwillingness or inability to move along these corridors. Once again we are back to gardening, because we will measure the poster boys but be unable to address the concerns of the entire ecosystem. In spite of this, conservation groups, charities, planning committees, and government will champion their actions as a success, a way forward, and take both the plaudits and the fees for achieving this.

In reality all they are doing is managing the destruction. You can put a fence around your chicken run to keep the foxes out, but you cannot stop the pollution or temperature or viruses or fungal spores from entering the run and harming your chickens. So why do we pretend that we can protect nature within its own fence?

Our Future Could Be
Going Up in Smoke

All Ponzi schemes work on one premise, confidence. Remove the confidence and the scheme collapses. Our governments and charities and agencies all combine to maintain confidence, as they must, for fear of the collapse. Is it better to collapse the scheme now or in the future? Is it better to wait until there is nothing left in the tank, no new investor in labour, money, or resources? Ask Superman about his home planet and think along those lines. And isn't it ironic that it took a scientist to point out to the Krypton government that their world was finished, when it was a product of all the scientific improvements to extract resources that brought it to the point of collapse. Three cheers for science …

I remember reading about a new meat source that could produce reasonable yields of meat on marginal land; red deer are 'wild' animals that were pushed towards the margins and thrive there. I continued to read the article until I got to the bit about the breeding stags that were offered for sale at £20,000–£30,000 because you needed real quality to boost your deer stock. I thought at the time that this would be self-defeating. By improving the quality of the herd by selective breeding, you would begin to engineer 'super' deer that would no longer thrive on the margins but require prime grassland instead. This would reverse the logic for investing in them in the first place.

I have also given a lot of thought to zero-carbon housing and have come across articles relating to straw-bale-built homes. This struck me as a good source of a highly insulated natural material, which may also be regarded as a waste product, that can be locally sourced and erected on-site using basic labour. Sure enough I came across straw-built homes for sale in the UK under the banner of 'negative carbon', because they act as a carbon store and can be combined with solar power for further enhancements. The issues relating to straw bales are those of load-bearing capacity, protection from damp and the weather, and worries about vermin and mortgages and insurance. The solution offered by this 'eco' firm was to incorporate the straw into engineered timber frames complete with glazed walls and full compliance with the building codes, and then to sell them for around £200,000 each.

Take a quick look at Wikipedia under 'straw bale homes' and you will discover that they were used in nineteenth-century America to get around local difficulties on the plains when there was a complete shortage of wood and the soil was unsuitable for the construction of sod houses. Walls were built and rendered to the inside and outside with what is essentially mud. Very many of these structures, including a church, stand to this day.

Further research yields data on super-high compressed straw bales with increased load-bearing capabilities and the provision of specially shaped bales for decorative window features and the like. I get the distinct impression that if these engineers had been alive several thousand years ago, what we now see as the wonder of Stonehenge would have simply been the leftover scaffolding for the real project. (In reality Stonehenge is the leftover scaffolding for the site's real purpose, a purpose we struggle to identify without injecting princesses.)

Provided you keep the straw dry and sealed within render, the issues about vermin, damp, and fire all disappear, but of course our current-day clever chaps have to engineer a cheap everyday simple waste product into a highly engineered expensive non-solution to zero-carbon housing. If they had been alive on the plains of America 150 years ago, they'd have died from exposure. In truth what they are attempting to do is offer a profitable alternative to traditional housing under the mantra of eco-friendly zero-carbon which will have no cost benefits to the buyers when compared to the cost of a real straw house (built of mud and straw.) But it will give the occupiers something to dine out on for years, as they've done their bit to save the planet. Why don't they engineer a modern-day solution to igloos?

I'm sure these would benefit from super-compressed ice with specially shaped blocks around the triple-glazed feature walls, giving residents the perfect view of six months of continuous winter night, if only the engineers could work out how to fireproof the igloos and arrange the deliveries from the factory.

I watched a TV documentary many years ago which was examining the archaeology of slave cabins in the Deep South of the United States. The archaeologists had come across a peculiarity with lines of old rusty nails left in the soil. After some research, they determined that these wooden cabins/shacks had been fitted out with wooden chimneys, as the cost of stone chimneys was prohibitive, and that the people who built them had landed on a novel solution for this obvious inbuilt fire hazard. They built the chimneys to lean out away from the stone hearth in the cabin, and the chimneys were propped to stop them from falling over. In the event of a chimney fire, you simply removed the props and the chimneys collapsed away from the cabin, saving the expensive structure. Several buildings had evidence of multiple fires. This suggests that the solution worked for many years.

Simple problems require simple solutions. We seem to have developed an overly clever approach to what should on the face of it be simple solutions. Part of the problem, I believe, is value. Never sell anything that undervalues the going rate. If the going rate for a new house is £200,000, then sell your new improved design for that amount or thereabouts. We are back to artificial subsidies maintaining empires, only in this instance it isn't government grants but market prices that maintain the empire. Think what would happen if all new £200,000 houses were suddenly reduced in price to £100,000 and were zero-carbon to boot. The entire industry would collapse, banks would fold due to negative equity in all existing housing stock, and a massive recession would take hold. We would then moan about the 'greedy bankers' putting their own profit before the nation's needs and expect our government to step in and save the day. In short we would trigger a collapse in the Ponzi scheme we call the housing market. Whenever we have a 'correction' in the market, that is to say when prices have outstripped their sustainable position and the market is 'adjusted' by prices falling to the 'correct' sustainable price, then we inevitably trigger a recession in the wider economy (or vice versa, when the wider economy triggers the drop in house prices). In one sense, then, we actively expect and demand that our Great Brains, for the benefit of us all, maintain the Ponzi scheme. We therefore not only have a Ponzi scheme set up by the con artists, but also it's actively protected and preserved by the participants, you and me. We even had Chancellor Osbourne threatening a reduction in house values within the Brexit debate as a device to buttress support for remaining within the EU (because a reduction in house prices and an increase in affordability for the houseless masses is a 'bad thing').

But why stop at houses? Why not look further afield at cars? If you built a Model T Ford now, today, with all the advances in engineering and robotics and material improvements, how much could you produce one for? But of course our modern-day engineers wouldn't be satisfied with this means of travelling from point A to point B. They'd want to add the bare essentials of air con, airbags, traction control, lane position indicator, cruise control, satnav, automatic clutch, heated seats, headlamp washers, central locking, heated windows, electric windows and mirrors, alloy wheels, reversing sensor, automatic parallel parking system, built-in telephone, stop–start, automatic braking system, power steering, and power brakes, to name a few. So I suppose the simple answer would be circa £20,000, because that's what the market demands, but this hasn't answered my question: how much could you produce a Model T Ford for in this day and age? In the 1970s our company vehicle fleet consisted of Ford Cortinas that came in as basic 1.6-litre engines for £5,000, but for the 2.0-litre Ghia the price was doubled to £10,000. Did one get twice the car for twice the price?

Our technical manager lived next door to a Leicester car dealer who sold Peugeot and Mercedes cars from two sites. He stated that his dealership sold twice as many Peugeots as Mercs but only made half the profit.

Another anomaly that these cars throw up is that of weight. The old Cortina weighed three-quarters of a tonne, whereas most modern cars weigh between 1 tonne and 1.5 tonnes, in spite of the fact that they have thinner steel panels, lightweight engines, and extensive plastic. I recently bought a magnetic 'GB' plate for the back of my car for a trip abroad and had to resort to Sellotape to stick it to my plastic boot. A couple of years ago the town of Hinckley suffered a freak hailstone storm which shattered roof tiles, conservatory roofs, and greenhouses and battered all exposed motor vehicles, leaving multiple 'divots' in the bodywork and smashing windows.

A friend of mine had a father with three cars on his drive ranging in age from five to ten to thirty years. The two younger cars sustained dents, whereas the older car emerged unscathed. Nowadays car designers have to incorporate all of today's safety features and super-efficient engines to reduce CO_2 emissions. This raises two questions really: were the old cars really that dangerous, and how much more efficient would they be if they were half a tonne lighter?

We have such a propensity to cause harm, and I don't mean wars and other acts of aggression, although these are bad enough. No, what I refer to is the casual manner in which we are prepared to inflict damage on our fellow humans without, it appears, a second thought. I have already written about the food industry and its pursuit of profit over health, but there are other, simpler examples. We all know that drugs like heroin and cocaine are injurious to health and life, and yet people will make, distribute, and sell these products along with all the other illegal substances. So-called 'legal highs' were readily available over the counter in shops throughout Britain and were sold 'responsibly' under the label of 'not for human consumption' until they were banned in 2016. The shopkeepers knew what the product would be used for and also knew that it was dangerous, and yet they were essentially prepared to play Russian roulette with their customers' lives in the interest of profit. A similar argument relates to the manufacture and distribution of alcohol, although in this case the product can be used in small quantities without ill effects and it's not the purveyors' fault if people cannot control their appetites. For every thousand who can, dozens can't. What defence is possible for the tobacco industry, given that the line between safe and dangerous exposure is vague or non-existent? Explosives manufacturers can point to all the benefits of their products in the mining, quarrying, and construction industries, and it really isn't their fault if their products are utilised by governments and terrorists alike to kill. Yet strangely we've gone all conscientious over the supply of drugs to America for use in lethal injection executions, because it's 'morally' wrong. Car manufacturers go to great lengths to build passive and active safety features into their vehicles and cannot be held responsible for the actions of idiot or careless drivers. Yet the same car manufacturers can read reports of thousands of annually estimated respiratory deaths associated with car fuel pollution and basically shrug their shoulders, because it's outside their control and they don't force anyone to use their product. We introduce strict controls on hazardous materials and enforce the health and safety of our workforce, but we don't hesitate to 'export' these jobs and risks without the controls to cheaper alternative factories in the developing world, where not only the labour is cheap but also is life. We do the same for working hours and working environment where our factories, designed and built to strict codes, cannot compete with structurally inadequate 'sweatshops'. We import cheap goods by exporting misery, all in the name of making a living. What is it within our consciousness that allows us to compartmentalise these issues into things that matter and things that don't? Is it simpler than that? The flock of starlings has birds on the outside that are vulnerable to predation. The birds on the inside are protected by being on the inside, but only because their 'fellow' flock makers are at risk. The trick is, make sure you stay on the inside and ignore those on the outside, but this only works as long as the inside is far enough away from the outside. If the flock loses its critical mass or density, then all become vulnerable and the protection generated by the effort required for the flock assemblage is lost. In a peculiar way, is this the definition of charity: offer support to the birds on the edges, but don't change places with them?

All of this is done without any real thought. Imagine what would happen if every real or virtual shopkeeper in the land decided that they would no longer support the slow-motion suicides of smokers and put principle before profit. Is there a difference between a corner store or supermarket selling cigarettes and a 'specialist legal high' outlet in anything except the speed at which some people die as a consequence of their actions? Imagine what would happen if we really tried to hurt people. But we don't need to imagine; our history is peppered with mass killings and actions likely to result in mass death. We all know about the Nazi final solution, but most of us don't know about the millions of Russian prisoners of war who were mainly starved or frozen to death, or the mass starvation inflicted on the Ukrainians by the Soviets during the post-revolution collectivisation process. The conclusion one can come to is that our caring, sharing society isn't; it simply maintains the facade while doing all it can to obtain a living off the sweat of others, those at the edge of the flock. So what happens to the great human project when the reserves (space, water, food, and other resources) run out? Drought, starvation, disease? The world is currently split between the rich and the poor, the clever and the uneducated, the takers and the givers, the strong and the weak. One can guess where the shortage will be felt most. So as we accelerate to the inevitable conclusion of ever increasing growth, we head towards the day of reckoning where we wage war on the weak because we can, we take from the poor because we can, we treat our clan with medicine because we can, we take water from the thirsty because we can, and we take food from the hungry because we can. If you consider this to be too dark a conclusion, think about what we are doing today around the world.

We employ poor ignorant people to work in our cheap-goods-producing factories around the developing world and fill our shops with choice and profit while denying the poor the same.

We whisk our Ebola-suffering volunteers away from West Africa to specialist isolation units and treat them with experimental drugs which cannot be used in Africa in case we need them for our own outbreak.

We import salad crops from warmer climes, effectively importing all the water that these crops took to grow, while leaving a shortage of this resource for the local poor who struggle for water and food.

We fill their fields with birdseed crops that boost their balance of payments figures that are so important to their poor and hungry people in their attempts to join us in the advanced world.

So in fact we don't have to accelerate to the inevitable conclusion of ever increasing growth, because we are already there in the war for resources and the division of life.

What solutions are there to this happy state of affairs that I have described? You cannot turn to governments; they are corrupt. You cannot turn to our agencies or charities; they are corrupt. You cannot appeal to the people because they don't exist, because they have no voice. Only the greatest brains have a voice, and they are corrupt.

I said before that if you want to lose weight, eat less food; you don't need dieticians. if you want to use less energy, use less energy; you don't need energy-saving devices. If you want to live healthily, live healthily; you don't need ignored health initiative after health initiative.

How do you use less energy? Our existing premise is to be more efficient. We improve the miles per gallon of our motor vehicles while building more and more vehicles; we lower the wattage of our light bulbs while making more and more for a rising and advancing population; we use fossil fuels to make thicker and thicker sheets of insulation to protect our homes while we build more and more homes for our increasing population. Think back to the medieval market charters; six and two-thirds miles was the criterion: a third of your day to get there, a third of your day to do your business, and a third of your day to walk home. Jump in your motor car and drive fifty miles to work, driving past thousands of houses belonging to other workers who all live nearer to your workplace, and ask yourself how many similarly qualified workers and how many similar jobs that you can do you have driven past. If you had to go to work by bike or on foot, how far could you possibly travel to make it worthwhile? I used to read the James Herriot books, *It Shouldn't Happen to a Vet* and the like, and one of the things that struck me was the description of the great social event of the year, the annual arrival of the veterinary supplies representative.

James was eager to see the progress of medicines and equipment over the year. The special occasion was always marked by a dinner. One of our regional managers always referred to our sales representatives as 'travellers'; he was hearkening back to a not too distant past where sales reps kissed their wives goodbye on Monday morning, jumped on a train to a town, slept in the railway hotel, and walked around their clientele before returning home to their wives on Friday night. This time was also known as 'the golden age of bigamy' because you could have one wife for the week and another for the weekend.

We condemn our own workers to unemployment because they are not poor enough. They won't work for a dollar a day, so we export the jobs abroad and pay the transport costs instead, and then we pay compensation in redundancy and unemployment benefits using taxes that the workers themselves provided. If you work in the 'wrong' industry in Britain, then your very success in earning money causes your demise and transfers the burden to your fellow taxpayers. We moan about bankers putting profit above any moral responsibility, and yet everyone does it given the chance. If you watch two labourers fill a wheelbarrow, you often see that one pushes and one shovels. You can bet your life that it's easier for one than the other. Charity bosses demand your £3 a day to save/change somebody's life while demanding their £200,000 salaries. Government ministers and local government councillors approve budgets that effectively 'steal' your money without any real sense of accountability (measurement, blessed measurement, meaningless blessed measurement). In the meantime the EU is busy exporting the poor from their own member states to other member states so that they can undercut the local population to make them poorer.

The very first house my partner and I looked at was in Barwell, Leicestershire. It was up for sale and was within our price range at £10,250. (I'm only talking 1978.) It was an end-of-terrace. The biggest drawback was the absence of a kitchen. It did have a cooker outside in what could best be described as a lean-to greenhouse, and it had the benefit of planning for an extension. What put us off in the main was the fact that it was right next door to a hosiery factory. We worried about the noise. We were told that noise wasn't an issue because the factory was only open during the day, when I and my partner were both at work. Nowadays the same house has a kitchen, but what it doesn't have is a factory next door, because the factory has long since shut down and has been replaced with housing. But unlike before, noise is now a problem, because when the occupiers are at home, so is everybody else. What we have seen in the last forty years is the formation of planning zones whereby employment and housing needs have been isolated into their own areas. This has seen large industrial estates develop, some like Magna Park or DIRFT (Daventry International Rail Freight Terminal) isolated from all local housing or on the edge of town with good transport links via road or rail. Perversely, all the railway goods yards have disappeared into car parks or builders merchants or something similar and are no longer available to service freight. Many years ago I quoted a contract for the supply of precast concrete to the Channel Tunnel. Preference would be given to goods supplied by train. I searched in vain for a goods yard where I could load my product onto a train for delivery. When eventually I was offered the use of another company's sidings, it was on the basis of 'only if we are not using it ourselves'. Road freight or road haulage as we think of them only really came about as a result of the end of the Second World War and the vast number of army surplus trucks and drivers that returned from abroad. These were followed by Beeching and Geddes Axes, which demolished the rail network into the trunk route rump that we know today, which is hopelessly inadequate to cope with national distribution of goods even if we wanted to use it for that purpose.

By accident or design we have embarked on policies that have destroyed the relationship between work and home. Gone are the days when the workforce walked or biked or nipped in by bus to the local factory. The relationship was understood perfectly by the Victorians, who often supplied the housing for their own workforce. Workplace and worker residences often sat side by side. The World Health Organisation no less has produced noise guidelines for residential areas and maximum night-time noise levels which provide for a good night's sleep, so important for the maintenance of health. These levels have been adopted by planning authorities up and down the country. This has further driven industry away from residential areas, while at the same time advances in building design and insulation, along with health and safety regulations, have all conspired to make factories quieter and quieter, even though they are now further away. Before local residents complain about noisy manufacturing operations, Health and Safety has already stepped in because they worry about the workforce standing next to the machine rather than the residents dozens of yards away. Yet in my experience, often the loudest noises emanating from an area are from the noise of traffic travelling (from residential areas to industrial ones). During the 1980s we removed all the silencers from our machines to exaggerate the noise our factory produced to try to prevent a housing development from appearing near our boundary. (At the request of the planning officers.) At the planning inspector's hearing, the sound consultant advised that the main background noise was actually coming from the M69 motorway two miles away.

We have just spent the last forty years designing motor car commuting into the work–home equation and now want to apply the reduced CO_2/ improved mpg sticking plaster to the problem of energy reduction. You don't need to be a mathematician to work out that twice as many cars producing half the emissions equals no change. And while we may not see a redoubling of car numbers in the UK, you won't see a doubling in China either. It will be tenfold, twentyfold, or two hundredfold, and then there's India and Africa ...

One solution would be to make travelling prohibitively expensive, which would force individuals to seek either work nearer to their homes or homes nearer to their workplaces. This card has been played to an extent with things like the fuel duty accumulator. This was designed to increase tax above the rate of inflation and to make travelling progressively more expensive (while simultaneously pursuing more-efficient cars?). My first car was a 1.6-litre Ford Cortina that used leaded petrol to average twenty-six miles to the gallon. My last petrol car, a 3.0-litre Saloon with an average of thirty miles to the gallon using unleaded petrol, weighed 50 per cent heavier than the earlier Ford. The problem with this non-solution is that under the present rules, if you can't afford to go to work, then you don't have to; benefits or tax subsidies beckon. Or the impact will be felt elsewhere with higher house prices and fewer house owners, because something has to give. So in this instance we can see how tax is taken away with one hand and given back with the other. This same principle applies in all sorts of other areas as well because the system is corrupt between its ambition and effect.

Another planning initiative I have already described is the reduction in car parking spaces at work. This has been further tempered with proposed council-imposed car park workspace fees in cities like Nottingham and the planned introduction of congestion charging in Leicester, Derby, and Nottingham. These would be the same planning authorities that have separated the housing from the factories by their planning policies, thus forcing you to commute. Create the problem and then design the solution; have I mentioned suntans and skin cancer/sunblock?

During this same period, we have witnessed the demise of the high street and the creation of out-of-town retail parks with huge car parks to draw in the CO_2. Shopping is now seen as a leisure activity and not a necessary chore, and is wholly dependent on cars. True, the retail parks all have bus services, but the retail parks only work by drawing in customers from a huge catchment area way beyond the remit of bus routes. (I suspect servicing these sites by public transport alone could be done, but the end result of travelling time would be akin to the medieval market model.) Planners have awakened to this problem and 'resistance' to out of town parks has risen, but the damage has already been done. Here again the corruption of the local authorities can be seen. They need the tax revenue from shop rates, and they need the revenue from council car parks. They support this search for parking revenue with on-street parking restrictions and enforcement and street-side meters. They impose restrictions on free travel by reducing traffic flow with bus-only-lanes (and, strangely, taxis).

I say 'strangely, taxis' because why should a taxi take preference over any other road user, just because they are paid a fare? Taxis produce as much CO_2 as any other car and contribute just as much to congestion but don't pay car parking fees and provide what could be described as an elite bus service for people who don't want to use a bus. Given that the object of all the park-and-ride schemes is to get motorists out of cars and into buses, what gives?

Ten years ago I would stop at my son's university flat in Leeds overnight before commuting the shorter distance to Lancaster over the Pennines. I was amazed to see the continuous queues of traffic heading into the city, whereas my road out was virtually empty. This was reversed at night as I drove back into the city past the lines of cars. So let's move the process along a bit, shall we? Let's ban cars from cities by imposing total shutdown and see what we get. Will everyone jump on a bus, or will the city die? No, better to tax the motorist by congestion charging and parking fees and then moan about the car problem and offer up fresh initiatives to solve a problem that the councils and planning officers don't actually want to solve, a problem that they themselves helped to create.

When my kids were young and the inevitable sibling rivalry turned into argument or fights, I would step in and ask them what their ambition was. 'Is your plan to kill?' I would ask, because, 'You should stop and think what you are trying to achieve.' So, Mr Local Authority, what is your plan? Is it your plan to kill the motor car or not, kill the city or not, drive everyone to the buses or not, and provide a suitable alternative solution, which you haven't actually got, or not? Or is your plan to stand on the sidelines, whip in hand, taking an enhanced salary for your obvious intellect while you preside over an endless string of failed initiatives? Goodness, you'll be trying to control the 30 mph speed limit next.

During the September 2000 petrol blockade that lasted for all of eight days, the estimated cost of the disruption was put at £1 billion. Imagine the permanent impact that shutting cities down would have. So the options available to the planning authorities are limited and at best can be regarded as inadequate and at worst, worse than useless as they attempt to tinker with the problem. The problem that we have is that the planners are charged with sorting it, but this is outside the scope of planning control and the guidance from central government. You might as well charge the planners with ending poverty or getting beggars off the streets for all the good they can do. But that is the very essence of *How to Kill an Elephant*. We have created elephants that we can't control, and yet we continue to try to behave as though we can.

The way to solve the car problem is to remove the need for cars. For years we have had the carrot of the Internet dangled in front of us. Work smart, work from home, all of us linked in to our virtual offices. Doesn't really help to pick a load in a warehouse or build a car or tailor a suit though, does it?

My wife remembers when her family's household was only one of two in a street of thirty-six who had a car. Nowadays there are sixty, and the days of playing tennis in the street are long gone. But it doesn't have to be like this. How do we break the love affair with the car? We have created a world of a new scale. Gone are the local manufacturers serving the local communities, to be replaced with global empires spanning continents, feeding ocean trade routes and mega ports, disgorging the contents in articulated-sized boxes to cut grooves in our motorways on the way to gargantuan-sized warehouses where the overriding consideration is 'big is better'. At Magna Park, warehouses stand 27 metres high and cover a million square feet of storage area under one roof, drawing in employees from far and wide in pursuit of a penny here and there off the costs. When the Primark warehouse was destroyed by fire several years ago, the local fire officer described it as trying to put out a fire in an enclosed football stadium, something the firefighters had no comprehension of or ability to do. The problem is that the mathematics don't add up, but nobody is doing the calculation because they are too busy making a living to bother. Years ago, attempts were made to keep manufacturing in the UK under the banner of 'Buy British'. Needless to say it failed, because the smart money was realising the potential of globalisation. Shut factories down at home and transport the jobs abroad to developing countries where the grinding poverty is barely aided with the dollar-a-day wages. Cut everything to the bone in terms of welfare and health and safety and building codes and workers' rights, but rigorously promote long working hours and cheap labour. In short, wash away all of the advances of the last 150 years of the Factories Acts and Welfare Acts that were seen as essential for a fairer society. Most importantly, profit from it and make hay while the sun shines. This happy state of affairs will only last so long before the worm turns and costs begin to rise as standards improve in these impoverished countries. Take your profits and pay your taxes to subsidise the unemployed, the very ones your board and shareholders created in pursuit of savings. Pay your taxes so that our government can give billions away in foreign aid to support the poor in these impoverished countries. Pay your taxes to rebuild the ever increasing road network we need to distribute these same goods from ports to warehouse to shops and complain about the noise. Pay your taxes to support the thousands of people who suffer from respiratory diseases caused by traffic pollution. Pay your taxes to employ the planners and environmental consultants charged with solving the congestion and pollution problems. Pay your taxes to force people to use their vehicles less, in the interest of the environment. Pay your taxes to pay for bus lanes and park-and-ride car parks and traffic wardens because your local jobs have disappeared. Pay your taxes to subsidise the UK workers on low pay with housing benefits and income support. Pay your taxes to educate your children for fourteen years instead of eleven so that we can compete on the world stage with high-tech jobs commanding high salaries, seeing as we can't compete with the global market for labour. Ignore the fact that these developing countries are doing precisely that, developing, and the day will come when they have a country with a highly educated, technically minded workforce, at which time we may be their bitches.

And throughout all of this process, remember that we must reduce our dependence on fossil fuels and stop the rise in global warming and protect the environment for the sake of generations unborn.

The summer of 2015 has seen various scientific studies publish reports about what percentage of earth's species are at risk of extinction if the earth warms by just x degrees—real wet-finger-in-the-air prediction at work. But it doesn't matter if the earth warms up or not, because irrespective of this factor our blind insatiable appetite for space, water, energy, and resources will leave precious little room for anything other than humankind. Can no one focus on the real issues? Humankind needs to stop doing what it's doing and find another way, because what we are doing serves nobody any good and ultimately ends with our demise. Now I am not talking of our extinction. I am talking about a 'correction' in the market. The price of humankind's activity is too high and is unsustainable. At some stage the bubble will burst and our market will collapse, as all unsustainable markets do. The questions are, how big will the bubble be, how many billions will die, and what else will be left to share our earth with? Locust swarms take several years to develop and need the right circumstances to reach the biblical proportions that devastate vast tracks of land. But these swarms don't last, because they eventually run out of resources and the population collapses through starvation. What makes human beings such golden bollocks that we think we are immune to the same process? Put simply, the human model is fundamentally flawed and is unsustainable and needs to change radically. Forget the bollocks spouted by environmentalists and sustainabilitists and global warming conferences and treaties and ecologists harping on about grams of CO_2 per kilometre and the rest of the self-deluding and self-serving Great Brains who pretend to offer a solution while advocating what is essentially business as usual.

Consider this: we talk already about the arrival of the tenth billionth person on earth by the year 2056. Predictions extend to twelve billion, but why should it stop there? Travel to Afghanistan and the talk is of average life expectancies in the mid thirties, which would have been the norm for centuries around the world. Compare that with the world's richer economies with figures of mid eighties. On that historical basis we shouldn't consider that our population stands at seven billion; rather, in terms of resources used, it is already nearer to fourteen billion. (I do accept that not all the world's population can currently expect to live to 85 on average, but that is when the developing world will consider that it has reached its potential of equality with the developed world.)

Do nothing and what will the end result be, involuntary euthanasia at the age of 40? Developed nations engaged in genocide for the creation of Lebensraum or water or food. We have a choice to make. Time has run out. The model we have adopted is wrong. All we are 'attempting' to do is pissing in the wind. End globalisation, work locally, use less energy, eat less food, smoke less tobacco, pay less tax, work shorter hours, build eco houses out of straw, sack the Great Brains, become localised, end consumerism, rediscover nature, turn our backs on the false prophets of science, engineering, and education, the engines that drive our destruction, expect less, eat real food, live more.

What is the human ambition?

With sales reps, I soon learnt that one needs a plan. They have to start each day at work with a clear idea of the day's goals. Better still if they have one day's plan for the weeks and even the months in front of them. If they don't have a plan, then they can drift aimlessly and wastefully through the working day that you are paying for, and you would have little to show for it. So what is humanity's plan for the day, the week, the year, the decade? Where do we expect to end up in the future? I suspect we're going to end up with the herd situation again, staying in touch with our neighbours. Is our intention to fill our bellies because we are hungry, build a nest because we are tired, hunt for meat because we have a taste for it, overcome the dominant male so that we can have sex, fight to defend our territory, and fight to increase our territory? But we are so much more complicated than our nearest living ancestors, aren't we? We know that we can pop into a shop or fast-food outlet and satisfy our hunger. We have beds in our houses, play games to simulate the hunt and chase with a token 'carcass' of adulation or congratulations for winning or even taking part, look to impress the opposite sex to enjoy some quality companionship, and stake our claim to territory with fences and boundaries, both physical and metaphysical. We really are so very different, aren't we?

Yes, if only because of our power and ability to damage the earth. When our fruit tree stops producing, where do we go to find another one, over the horizon? We are supposedly 'wise'; this defines us and separates us from all of our previous ancestors. So why aren't we wise? Why are we sleepwalking towards the destruction of our only garden of paradise?

I remember at school how my ambitions were laid out in front of me from the age of 11: do well at the high school and get into course one at the grammar school; do well at my O levels and get into the sixth form; do well in my A levels and go to university. And then the wheels fell off because I had to make a choice. I had reached the end of the conveyor belt and didn't know what I wanted to do. My seven-year plan had lost its focus and had been replaced with the desire for a house, a partner, a proper job to pay the mortgage, children and seeing them grow up and leave the nest and repeat the pattern expected of them—no hippies in my family, thank you. I was even committed to paying the mortgage (it was cheaper than the rent) on my son's flat in Leeds and had the River Orinoco clause built in so that I could renege on my commitments if he buggered off up the Orinoco, leaving me with the financial obligations.

Looking back, I wish I'd done school differently. I'd have taken biology, woodworking, and geography at A level, but I know that this wouldn't have fitted any timetable or potential university degree. It wasn't an option for the school system, but it would have been perfect for me. Because already at this tender age I had fallen foul of the pro forma, the straitjacket of conformity, the definition of 'normal'. But now of course I understand that the normal means support the Ponzi scheme from the cradle to the grave, support the mindless, blind stupidity that is unrelenting growth and desire that will lead to the slow-motion suicide of us all. But it isn't only suicide; it's fratricide with our nearest relatives, genocide with species throughout the world, and ultimately murder for those unfortunates who end up at the wrong end of the war for resources. How strange that the very things that are supposed to save and protect us are all batting for the wrong side. Science, technology, and education all foster the desire for more and provide the tools for achieving it. Imagine turning up for a conference on weight loss to be greeted by a speaker who opens the meeting with 'eat' and closes the conference with 'food', having inserted 'less' into the sentence.

Imagine attending a conference on sustainable energy only to be told, 'Use less energy,' whereas what we get is the advice on CO_2 and thermal dynamics that enable us to do exactly what we have always done, just more efficiently. But the point is, we haven't always done it. We have survived for hundreds of thousands of years without the need to draw oil from the depths, to travel thousands of miles in a day, to journey to the moon and back, to split the atom, or to find the God particle. These are all expressions of having extracurricular time on our hands, time that we don't need for survival but spare time. What would the world be like if we behaved like lions and slept twenty hours a day to conserve energy, if we hibernated like bears to get through five months of winter, if we didn't devote our ingenuity and energy to destructive actions in pursuit of whatever plan we are working towards? (Perhaps it's time to revisit the lifestyles of teenagers with fresh eyes. Their circadian rhythms may hold the solution to our problems.) When do we wake up and realise that we are being busy fools, so wrapped up in working for a living that we don't realise that we aren't living but in fact existing to satisfy the modern-day Ponzi scheme to end all Ponzi schemes? Roll the clock forward a hundred years and imagine the scene: no elephants, no tigers, and no lions in Africa or India outside of zoos and game ranches. How many people will inhabit the earth? What space will be left for anything else. What resource won't be siphoned off and reserved for humankind?

Make it two hundred years and ask what the climate will be, how many people will inhabit the earth, what space won't have been reserved for the stronger tribes, what resource won't have been siphoned off and reserved for the powerful. Make it five hundred years and ask yourself how many billions of people have had to die to make the world population sustainable. Make it now and ask, what is the fucking point? After all, dying ain't much of a living.

The biologists among us will tell you that the overriding imperative to all of our species, and every other species on earth, is to pass our genes down a generation, to achieve immortality through our DNA.

Maybe at some time in the future some species will try to reengineer *Homo sapiens*. Perhaps they'll try to discover what it was among our genes that produced the suicidal, fratricidal, genocidal, murderous streak that made us the most dangerous species this world has ever seen. Then again, I think the biosecurity requirements for this dangerous enterprise would prevent the experiments from occurring. Having said that, if the descendants of our unemployed were ever able to return from Mars in a millennium or ten, my guess is that they'd probably die trying.

Our government, bless them, have introduced 'family-friendly' tests into law making to try to ensure that new laws favour the nuclear family so beloved of politicians. How about introducing 'world-friendly' tests into all of our activities to try to ensure that new actions favour life on earth, including our own?

I cannot believe that I am the first to work this out, or indeed to write about it. Isn't Jimmy Bond in *Moonraker* attempting to halt the systematic destruction of the human race prior to its recolonisation by 'perfect specimens'? Doesn't the Bible talk about a great flood to remove the unworthy so as to provide a fresh, clean start? (While I don't believe in any divine aspect of the Bible, someone, somewhere had to come up with the idea to write it.)

Marks & Spencer announced their ambition to save the world, although they've gone a bit quiet on the subject lately, with their declaration that plan A has to work because there is no plan B. We have no plan A either, so where will that leave your sales figures in the future?

Take all the superfluous effort, the abject bollocks if you will, that people use as an excuse to generate a living at someone else's expense and eliminate it. Save all the resources and energy used by these utterly meaningless activities and cut our fossil fuel use instantly. Stop paying for all of this extracurricular bollocks and discover that you can live for less money in your pocket, buy food without paying for the adverts, slash the tax take from corrupt governments which fund corrupt state elephants, stop funding the Ponzi scheme, and discover that if you work a couple of days a week, you have enough to live on for the rest of the week and don't need to work. And if you only work two days a week, you use less energy because you are not commuting, your car lasts longer, the roads don't get worn out as fast, we don't need to build new roads because there are fewer vehicles, the price of oil drops because there is less demand, and you need to work less to pay for it. We won't need bus lanes because there won't be any congestion. We won't need to educate all these clever sods to do the jobs that didn't exist twenty years ago, before we had so many clever sods in need of employment. We will have the time to care for ourselves and our relatives without the need for childcare and OAP care packages, so won't need to pay for those things either. We could build houses out of straw, secure in the knowledge that there are no large-lung-capacity wolves wild in Britain anymore and we don't need to be sandwiched inside highly engineered wooden panels to comply with non-job building regulations and insurance brokers. How our ancestors managed without building material conglomerates is beyond me.

The Price of a JCB?

I asked before about the price of our ingenuity and the flawed mathematics of our activities. Let's take a JCB excavator, the powerhouse of the building site, and see if we can do some number crunching analysis. Various models are available, but for the purpose of this exercise, we'll make it easy and assume that we're working with one that costs $100,000, roughly £65,000, give or take a currency trader or two. How much work can this JCB do in terms of competing against manpower alone? The work of ten men, a hundred? Which is it? Of course the machine is useless without an operator, the fuel to power it, and the engineer and his vehicle to maintain it and repair it, which translates into another $180 a day in round figures, or $42,000 a year, allowing for holidays. And let's give it a life of ten years. By my reckoning that's a total cost of $520,000 for the life of the vehicle. I accept that the machines can last longer, but the older ones will probably need significant repairs beyond simple oil changes and service.

The question was, how many men can the JCB outcompete in terms of efficiency? That is, how many men can the machine replace? The question of course is biased, because we need to know which men, labourers on a UK building site on $120 a day, or Asian labourers on $1 a day? In the UK the JCB needs to replace 4,333 men days, and in Asia it needs to replace 520,000 men days; that's over 2,000 human years of labour. On the one hand, I think even Jesus would be knackered by now; on the other hand, if the purpose of the labour is to put food on the table, then what have we achieved?

There is a somewhat less than quaint expression used in the military, 'force multiplier', which simply means, 'How can we get our one soldier to kill more than one enemy soldier?' The First World War illustrated the idea perfectly by equipping soldiers with rapid-fire machine guns that would rain down disproportionate rates of death on advancing troops for little 'investment'. Nowadays the military talk about laser-guided missiles despatched from unmanned drones piloted from a continent away via satellite links.

So here in a nutshell we have a JCB, a force multiplier for the digging of holes and the clearance of habitat, an unblocker of ditches, and a lowerer of water tables, in short an earth engine for the restructuring of our planet. Each JCB equates to two hundred years of Asian man labour for each year of its ten-year life. Go back to the Middle Ages and consider the force multipliers then; how strong is your whip arm? Have a browse on the JCB website profile and read the declaration about JCB's sustainable innovation philosophy that says the company seeks to minimise their environmental impact; 'responsibilities to a changing world … reduce our environmental impact' features in the mix.

I think they mean that digging a hole in the ground to fill it with concrete foundations has no environmental impact whatsoever. Stripping the topsoil from a meadow to build houses on it doesn't either, provided of course that you try to shorten your supply chain and improve fossil fuel usage by boosting the JCB's mpg. Not that I have anything against JCB; they illustrate perfectly the corruption between 'We must save the planet' and 'Provided it's business as usual', as long as you throw in some logistic chain improvements (that save them money) and improve the engine performance (that maintains their competitive edge). It's a rather large microcosm of what our governments do: 'Relax. We are on the case. The planet is being saved. When would you like it delivered?'

The website does announce that in 2013 the millionth JCB machine was built—a 22-tonne JS220 tracked excavator in shimmering silver. This machine far exceeds the work capacity and cost of the archetypal backhoe excavator I referred to earlier, and those exceed the capacity of the earlier models from 1949, so we could suggest that on average JCB has in its history generated some two billion years of Asian (or medieval) manual labour. We could almost do with the Bamford prize for peace, or science or literature, to rival that arising from Nobel's alleged attempt to assuage his guilt for providing the means to kill so many people.

It's a curious feature of all of these labour-saving devices, whether you are the operator or designer, or the buyer or the owner of the manufacturing company, that they are designed to let you keep your hands clean, making it so you don't need to sweat to earn a living. How strange then that the same individuals who use JCBs will invest in gym memberships and equipment to make them sweat and maintain their health.

We are at a bit of a medical crossroads at this moment in time, the great saviours of the later twentieth century. Antibiotics are reaching the end of their effectiveness, and no new products are in the pipeline. The pharmaceutical companies are stuck with a dilemma: design a new antibiotic and see it locked away, to be used only as a last resort by the World Health Organisation, or turn their backs on the process. Actually, they are not really stuck with the dilemma, because with an estimated cost of bringing a new drug to market of $1 billion, drug companies are not even looking. The hope is that some clever chappies at university research facilities will stumble across the solution to antibiotic resistant beasties, and then governments will step in to provide the funding. This is important to us all, as predictions are already out there heralding the end of simple routine surgical procedures due to the imminent inability to treat hospital-borne infections arising from antibiotic-resistant beasties.

While we can all see the value of a new range of antibiotic equivalent to one billion dollars a pop, I would ask this question : what do you have to do to generate $1 billion of resources in order to fund the research in the first place? How much of the earth's resources do we have to plunder to release this investment and to see its subsequent return with profit? If all money is based on the value of labour, the promise to meet the commitment to honour the debt by exerting your muscles, then we are looking at one billion days of labour (based on $1/day, the standard Third World unit). What are the true costs in labour and of the subsequent development and extraction of our finite resource?

Every now and again I will come across a brain-teaser requiring some simple maths that illustrates the question I am trying to pose.

You are stuck by an oasis in the desert. Being one hundred miles from the next water source (no mobile phone signal either, so no cheating), you must walk to safety or else perish. The only water-carrying receptacles to hand are some ostrich eggs, each of which carries one litre of water. Your problem is that you can only carry a maximum of four at a time but you must drink one litre of water for every ten miles that you walk. To escape the desert, you must cache filled eggs along the route before returning for more so that you can utilise this resource, as you need them. How many trips and how many eggs do you need to complete the journey safely?

I never did bother to work it out, but we are faced with decisions along these lines on a regular basis. The mantra for all 'rescue operations' is not to put you, the rescuers, at risk while attempting to rescue those in need. In military terms, this is akin to risking the lives of fifty soldiers to secure the life of one downed pilot, say. Macho bullshit tends to get in the way here, and the attempt will be made under the contract of 'they'd do the same for me if I were the one stuck behind enemy lines'. Our fire service take a much more sensible approach, and prosecutions await for any rash actions sanctioned by supervisors in the event of injury.

So, calculators at the ready, what is the cost to this planet of saving a human life? How about twenty lives, or two hundred? Let's get serious: how about two million lives worldwide? Let's find a cure for multiple sclerosis or dengue fever, or hay fever, or male pattern baldness, or ageing skin. After all, we have the resources and are researching the possibilities. Where do we find the resources for these extracurricular activities? Is it in the sweat of human beings or in the plunder of the soil and the air that we breathe? How do we balance the books between our short-term needs and our long-term commitments? At the moment the answer is easy: can I/we/ my organisation make money by doing it? The Chinese believe that they can cure baldness, flatulence, impotence, hard skin, or some other equally bullshit problem by swallowing the ground-up bones of tigers (or is it rhino horn?) or licking a toad at midnight during a full moon to simultaneously find one's true love. Some of us complain about the testing of cosmetics on rabbits or cigarette smoke on beagles, while others complain about the testing of any and all 'cures' on any or all animals. What people don't complain about is the multimillion-pound budgets to be spent on researchers, equipment, and facilities that magically appear out of thin air and that never have any impact whatsoever on the environment. How strange that they worry about the little 'bunnykins' with shampoo in their eyes and no tear glands to wash it away with but never stop to consider the rabbit warren bulldozed to create the research facility, or the facility to make the equipment for the facility, or the fundraising activities to raise the money to fund the research, or the roads to bring the researchers from their homes. The list goes on. Years ago the environmental debate went something along the lines of, 'We're not harvesting fish from the sea; we are mining them. When the veins have run out, there won't be any more fish.' Nowadays the debate is focussed on global warming, climate change, reducing CO_2 levels, and 'saving the planet' while making minute alterations to our activities to silence the crowd, which isn't even asking the right question. We all have to make a living, but things have to die for us to achieve it. Oftentimes these 'things' are people. We are back to 'dying isn't much of a living'.

We can search for 'scandal', evidence of industrial skulduggery actioning actual bodily harm on people in the pursuit of money. The Julia Roberts film *Erin Brockovich* is based on such a true premise, and Nigeria is accusing various oil companies of dumping hazardous pollutants in the Delta oilfields near to the local fishermen's villages. We can even catch a hint of *Equus* in our Swedish meatballs and lasagnes, but we ignore the threat of added water drawn in by nitrates and nitrites or the lethally dangerous hydrogenated vegetable oils—trans fats—because we neither know nor care. A poster in my local Tesco supermarket declares that one in seven of the UK population is at risk of developing diabetes, which will be ten million people if it comes to pass. BBC teletext has run news of a report that declares that we can cut our risk of diabetes by 25 per cent by cutting out just one fizzy drink a day. Like I said, we all have to make a living, and 'things' will have to die in the process.

There is a myth that has made the rounds for a very long time about the Dutch purchase of Manhattan Island for $24 worth of glass beads. This is the most spectacular example of getting something for nothing, nearly. In 1626 the Dutch exchanged iron kettles, axes, knives, and cloth, things that were of little value to the Dutch but of much greater value to the Indians, for the use of the land. Actually, they were the wrong Indians; the Dutch later had to compensate a different 'owning' tribe. It would be difficult to put a value on the island in this day and age, and for my purpose I don't need to. Given that the dollar didn't even exist in the seventeenth century, it's obvious that the 'trade' was flawed, although the principle of paying very little for a lot could be rightly upheld.

Looking back, we can see how the Indians surrendered their great asset, the land and water with all of its resources, both now and in their future for short-term technological gains, iron kettles, blades, and the like. With hindsight they lost the hunting, housing, and material rights to the land for generations to come for very little in return. Imagine trying to hunt game or cultivate food or fish in Central Park now, let alone among the skyscrapers. The Native Americans bartered all of their futures away in what to them seemed a good deal. We of course would never be so stupid, would we? It would take more than trinkets to prise us away from what matters, wouldn't it? We wouldn't surrender our rights to land, water, and game in exchange for wide-screen TVs, mobile telephones, or motor cars, would we? We wouldn't systematically destroy natural habitat in exchange for mere trinkets, would we? Of course the Indians who were alive in 1626 have long since died, along with all their ancestors who could have occupied this space. Not only have they died, but so has the way of life they enjoyed. We call it progress, a succession of 'waves' of development that began with the Stone Age and has worked its way through to today. What shall we call it the computer age? So bear this in mind when the Great Brains talk about preserving things for future generations, because these future generations will not exist, not as we understand them. They will exist as a product of their age and all that that entails. All we can do is measure the passage of time and look back with the belief that it was somehow 'better' in those days, while at the same time grasping the new technology which heralds the end of 'those days'.

Humankind still has this insatiable appetite to look over the next horizon, 'Just to see what's there, you understand. We'll come back if it's not hospitable.' We never accept that *this* horizon is worth preserving for future generations, because something 'better' is out there waiting to be discovered. But if there is one overriding consideration to the horizons we currently seek, it is the pace at which we change.

There is talk of the youngest generation 'thinking' in different ways because they have been brought up with computers and embrace them as something to live with rather than something to learn. Every couple of years computers double in capacity and speed. Will our children change as fast? What will they do to our horizons in spite of our ambition to preserve what we have for them in the future? It took sixty-six years to go from powered flight to walking on the moon, and much of this was done without the aid of computers. We now have such a powerful engine of change. How will we use it? I could also validly ask, how will it use us? The technology has such a momentum of its own that it could be described as an elephant and is beyond our control.

Church leaders talk about the sanctity of human life, the great moral taboo that leads us to preserve life at all costs. We have national laws to protect us from murder, manslaughter (involuntary or not), and assisted suicide, plus international laws against genocide. I say assisted because there isn't a lot of punishment if you've killed yourself. Some countries ban abortions and rule against selective tinkering with our DNA, while others allow it but only under strict controls. The EU has banned capital punishment from our statute books, and the protection of life is paramount or so you would think. I was accosted by charity collectors endeavouring to recruit my wallet for the Dogs Trust who march under the banner of 'We never put a healthy dog down.' It seems that the sanctity of life can cross the species divide as well.

You can never go home after any sizeable absence; move back to any town or even the parental home to find the place has moved on and the differences sit uncomfortably in the mind. So take a hint, all you 'boomerang kids', and don't come back. Joking. So it is that we leave the nest to find out what's over the horizon, essentially on a one-way trip that we kid ourselves we can return from. It is deeply held within our psychology, is not something we can either control or limit, and is itself a great engine of change within the human species. For this reason alone I would suggest that any attempt to save something for future generations is doomed, because anything we preserve, we preserve only for those future generations to plunder, just as we plunder that which has been left to us. How to reconcile the relationship between humankind and the planet then, to find the equilibrium between need and resources?

Wasteful Enterprises

I have tried to demonstrate how we engage in entire industries whose entire purpose is to generate an income by 'being'. Charities, advertising, and even the 'health' industry and its hangers-on such as diets have been examined. But these ignore the activities that serve no real purpose at all and simply pander to fashion. One of these is of course precisely that, the fashion industry, which ushers in the seasonal must-haves that we can't possibly live without, *darling*. I saw some photographs in the *Leicester Mercury* of workers from the 1920s, a 'gentleman' bricklayer and his labourer. The bricklayer could be distinguished from his labourer by the fact that he wore a tie and a three-piece suit, while the labourer had a two-piece suit and no tie. Back in those days most men had a new suit bought annually for Sunday best and moved the previous year's on to work duty for the next twelve months. I occasionally view the shop windows of the so-called aspirational shops with the displays full of designer shoes, bags, and the like that people bust a gut to acquire, and I despair at the astronomical prices. Why does one leather bag that takes three hours to 'run up' command a price many times over a near identical product without the label? I have referred to 'The Emperor's New Clothes' before as portraying a con trick to part the gullible from their hard-earned cash. While I accept that for many this cash isn't in fact hard-earned, simply available through luck or circumstance, the fact remains that to generate this cash, ultimately it takes the sweat of human beings and the plunder of the earth's resources. So while it would be 'obscene' to parade around with a genuine tiger-skin bag in most of Britain today, the fact remains that spending 'obscene' amounts of money on a label is acceptable. They say that money doesn't grow on trees, but where in fact does it come from, if not the trees and fields and minerals and muscle that is used to 'win' it? True, the purchaser of the item may well have not gotten his or her hands dirty, but somewhere in the line of money production many will.

There was a fashion not so long ago where shoes with extremely long pointed tips were all the rage. I used to think that whoever it was who had come up with the original design had done it for a bet to prove how stupidly gullible people really are and how easy it is to part them from their money. In parting with this money and generating wealth for the designers and purveyors of these luxury 'con trick' goods, we launch this money into the orbit of fast cars and private jets and the 'perfect' lifestyle enjoyed by the rich and famous. We may lust after the glamorous lifestyle of Brad and Angelina, (or Richard Burton and Elizabeth Taylor, or should we settle on the Kardashians) but we should remember that their lifestyle is built upon the paying public parting with their cinema admission price or TV licence fees, money scraped from the pot of wages from laying bricks or cleaning windows or building cars.

In reality, the shoe design was in fact done for a bet. If the design was adopted, then the designer won; if the design was rejected, then the designer lost.

In 1837 Hans Christian Andersen's *Fairy Tales Told for Children* was published; this third instalment contained the story 'The Emperor's New Clothes'.

We have a conscience, we care about the world, and we come up with all sorts of delusions to reinforce this claim. Based on a UK population of seventy million, and a number of notable dates a year, we can begin to think in terms of, say:

- 70 million multiplied by fifteen for birthday cards
- 35 million multiplied by 4 for Mother's Day and Father's Day cards
- 60 million multiplied by 15 for Christmas cards
- plus perhaps another 500 million get-well-soon, Valentine's Day, Easter, retirement, engagement, wedding, birth of a new child, sincere condolences cards.

By this reckoning I can produce in this utterly unscientific way a figure of 2,240,000,000 greeting cards together with their envelopes. If we take an average weight of 50 grams for each item, then we have produced 112,000 tonnes of paper/card product which is grown from trees, mechanically extracted from forests, driven by road to paper mills, shredded, pulped, treated with chemicals, processed into paper/cardstock, delivered by road to printworks, offloaded, stored, printed, cut, folded, glued (with glue attached for later use), delivered by road to distribution warehouses, offloaded, stored, picked, delivered to shops, displayed, sold, signed, and sealed before delivery by hand/postie via the nation's transport network to the lucky recipient's door.

We also have to thank our tree nurseries, tree planters, deer-proof fencing manufacturers and installers, deer shooters, tree fellers, plant manufacturers and suppliers, truck drivers, truck and trailer manufacturers, paper processing plant manufacturers, warehouse builders and operators, shop builders, and shop staff. We shouldn't forget our energy suppliers who power the tree extraction and distribution machinery (but only after the staff have driven to the site, the paper mills, and the printers); or the warehouse lighting and heating, forklift trucks, distribution vehicles, and the shops; or the travelling arrangements for the staff to get to work. We mustn't forget the designers, the buyers, the management, the shelf display manufacturers, the printer ink manufacturers, or the suppliers of their raw materials. So what do we do with these mementos of sentiment, these messages from our friendship network? We display them for a decent length of time appropriate to the occasion (a week in our house, except for Christmas, which extends up to Twelfth Night), and then we dispose of them. The 'lucky' ones go into recycling, so that much of the process of storing, processing, distribution, etc. can all start over again. The rest go to the landfill or up the chimney.

But it's all right, this practice has a limited impact on the environment. We use 'ethically sourced' paper from sustainable forests; it says so on the back of the card.

By my reckoning, the 112,000 tonnes of cards are transported from the forest to the woodyard, from the woodyard to the processing plant, from the processing plant to the warehouse, from the warehouse to the printers, from the printers to the distribution warehouse, from the distribution warehouse to the shop, from the shop to your home, from your home to the recipient or the postbox, from the postbox to the sorting office, from the sorting office to the regional sorting office, from the regional sorting office to the local post office, from the local post office to the letter box, and then via the dustbin truck to landfill or the recycling warehouse, and then back to the paper manufacturer for reprocessing. Assuming that you post your card and that it goes for recycling, the 112,000 tonnes travels no less than fifteen times, so in effect the transport requirement has increased to 1,680,000 tonnes for something that we throw away. But it's all right, because it's sourced from sustainable forests. It's just everything else in the chain that isn't, and nobody else gives a toss about it because they are making a living from it.

And I haven't even mentioned the cellophane wrappers and the paper bag they put your cards in at the shop.

This is in fact a variation on a theme I read many years ago about VAT and the trail of this tax through the whole of the tree felling, processing, papermaking, bag making, distributing, shopkeeping system whereby VAT was charged and claimed back at every stage in the process until the product reached the shopkeeper, who gave the bag away and therefore didn't charge VAT on it. This is a wonderful example of work done for no benefit whatsoever.

There is a major debate about the evils of plastic bags and the need to tax/charge them out of existence because of the environmental damage caused by the rubbish and the cost in hydrocarbons of the manufacture. For many years plastic bags have been used for rubbish collection through the use of plastic bin bags, and they have wrought environmental damage on bird populations, but not in the way that you think.

Meat rotting while sealed in anaerobic plastic bags commonly produces botulism-infected produce that is then eaten by scavenging crows and gulls, with deadly results.

So when did cards become so invasive on the friendship network, and will we see their demise through social media and electronic paperless cards via email or even text? The younger generation already think it's OK to 'send' expressions of goodwill via these channels. Long may this trend continue.

A quick search on Wikipedia yields some interesting, if unverified, results: the first cards were printed in the 1840s, and the sales figures for 2008 suggest a US market of 1.3 billion cards, which shows a slight reduction put down to e-cards.

While we might begin to relax at the wastage of greetings cards, we might consider that the personal letter box handles considerably more volume of junk mail, advertising inserts, and stuffers. It would be perfectly reasonable to propose that a further 200,000 tonnes of printed and posted waste finds its way via our doormats to the recycling bin, and this in turn probably enjoys about fifteen journeys through its life, 3,000,000 tonnes of effort, but all, no doubt, printed on recycled paper or paper sourced from sustainable forests.

When, if ever, will some form of accountability that actually means something be appended to the sustainability claims of these resource wasters? Please don't send answers on a postcard; it would be somewhat self-defeating. While we are at it, what constitutes a sustainable forest? Does it simply mean a plantation felled and then replanted after every crop? Sustainable, then, would simply mean that any virgin forest taken into commercial cropping would meet with this criterion as the original trees would be replaced. Equally, any commercially cropped plantation would also qualify, irrespective of what habitat had been destroyed to form the plantation in the first place. Having seen the way that entire hills in southern Portugal were denuded and terraced for eucalyptus plantations for the turpentine/paint industry, I wonder if these would not also qualify as sustainable sourced trees.

For an activity to be considered sustainable, it should be able to continue forever. In the case of a forest, every time a tree is cut down, it should be replaced by a young tree that can grow to a similar state of maturity before it is cropped and then replaced with another young tree. Return to the idea of the Great Wood that covered Europe. This forest could be considered to be sustainable. It had the potential to continually replace any losses through natural regeneration. In 2000, the Millennium Declaration adopted by the UN General Assembly defined 'respect for nature' as a fundamental value, and committed 'to integrate the principles of sustainable development into country policies'. Now I know where the main heat source of the hot air that is driving global warming is coming from. The basic edict appears to be business as usual with a nod to sustainability.

Corruption: the process by which a word, expression, or meaning is changed from its original state to another one usually regarded as false or debased. We can think in terms of alteration, doctoring, manipulation, adulteration, debasement, degradation, subversion, misrepresentation, misapplication, and falsification.

So let us consider these definitions in the context of our Great Brains, be they in government or other organisations, and ask ourselves if we are being served in anything other than a corrupt manner. Now you may already be swearing at either the bloody Tory bastards or the Labour Socialist arseholes, or both if you are north of the border. Or you may hold all of them equally in contempt simply because they are politicians and you never vote anyway. But take the party politics away and ask yourself where we are served by this process. Who is telling the truth? Indeed, does anyone actually know what the truth is? The longstop to all of this thought process will be something along the lines of, *Yes, I know it's flawed, but without government, what do we end up with other than anarchy?* You only have to look at Libya or post-Saddam Iraq and behold what happens when power vacuums form to see the very real pitfalls.

Twenty-five years ago my company had devised a new in-house computer program and was in the process of rolling it out through the office departments. It worked on the premise that once data had been input into the system, it could be automatically converted to support different formats. So, for example, from the schedule of structural components listed for each house design, we could produce a quote for the customer or a pick-list for the loaders or a delivery ticket or an invoice. These required no further data inputting. We even introduced the system to our customers, who learnt that they could place orders with us using our system with virtually no effort. A colleague of mine introduced the system to our invoice clerk, who agreed that the system was much better but said that we would continue to use the old one because 'I understand how to do it this way.' While this was not an option for the company, the issue was quickly resolved when the clerk resigned and left the company. (Honest, we didn't even push.)

So we have a system that is corrupt, that doesn't understand what the truth is, and that props itself up with multiple layers of lies and deceits while at the same time demanding huge resources (by way of taxes) to support itself in the pursuit of 'our interests'. But we cannot do anything to change the system because 'we understand how to do it this way' and we are frightened of the alternative. The government we have did not come into being on a given date in history; it evolved over centuries into what we have today. At no time in its history has it failed to change from one form to another, so the idea that we dare not change it out of fear is a load of bollocks. In simple terms, if you go back far enough, you find that the administration changed every time the king died, only to be reborn under the new stewardship.

Whenever the word 'corruption' is spoken, most think that bribery is typically involved because the definition is set too narrowly in the context of 'conduct by those in power', the suggestion being that money changes hands covertly in plain brown envelopes. In reality the bribery takes a different form. Money is still involved, but it changes hands in plain view, if only we would open our eyes and see it. The CEO of a charity does a really good job and deserves the £250,000 salary; the headmaster of the local school does a really good job and thoroughly deserves the £150,000 salary; the manager of the hospital trust does a really good job and also deserves the £300,000 salary, not forgetting the pension options, of course. But all of this success is built on lies, the lies of performance targets, the lies of the staff below them who are protecting their own positions and salaries, the lies of the mission statements, the lies of the government employers who need to prove how good a job they are doing. We all engage in the process. We are all connected to the matrix and provide our energy to the support of the machine that is systematically destroying this earth, the only planet we have. We are addicted to delusion and lies and cannot see an alternative, so we engage our greatest brains to find the solution to the problem, and they answer with still more delusion and lies that we do not need to believe.

It is in the power of humankind to turn the clock back to the Stone Age with the press of the nuclear button, mutually assured destruction, a mass suicide, but as I asked before, how about a slow-motion suicide, a slow-motion high-speed crash? If we would open our eyes, then we would see that we are in the process of achieving this aim in the interests of 'making a living'. But as I said before, dying isn't much of a living. Sustainable processes can 'go on forever', and we manage our forests to be sustainable while extracting fossil fuels from the bowels of the earth to power the 'sustainable process' that, by definition, cannot go on forever. The use of these fossil fuels is changing the climate and impacting these sustainable forests that by definition cannot go on forever. Our population continues to grow, increasing our demands on the earth and its resources, which by definition cannot go on forever. The fauna and flora we share this earth with are suffering at our hands and are dying out on a par with mass extinction events of our geological past. By definition, these fauna and flora cannot go on forever.

We need to step back, pause for reflection, think, and start a new order, a new methodology of existing that is truly sustainable and not the product of some marketing ploy, while there is still something left on this planet to enjoy.

Take a trip to Calke Abbey on the Leicestershire-Derbyshire border and one of the most noticeable features you will see is the state of the park's old trees. There are warning signs advising visitors to keep away from the trees, which have been deliberately left in as natural a state as possible. Significant portions of many of the older trees are dead and decaying in situ, still within the tree canopy, and any fallen portions have been left to decay naturally. This is a deliberate policy to encourage the full utilisation of the dead wood as 'food and habitat' for the immense range of fauna that thrive on it. Calculations have indicated that more life exists on dead wood and depends on its breakdown than actually lives on the living wood during its life.

Even allowing for a straight fifty-fifty split between live and dead wood, this has to cause us to ask questions about our 'sustainable' forests, remembering the definition of 'sustainable': 'capable of continuing forever'. By cropping the forest and removing the timber for commercial use, while at the same time replacing or allowing the regeneration of forest regrowth, we achieve a sustainable forest. But by definition, we have removed the vast majority of the wood from the forest, thereby denying half of the ecosystem its food source and habitat. Our so-called sustainable forest has effectively been cut in half and will be poorer because of it. So how can anyone argue that the current definition of forests as sustainable is anything other than a sop to public opinion? We have simply given ourselves permission to continue to destroy more of the same, while at the same time framing the destruction with a positive spin.

Or in English, it's the same old bollocks as usual.

When I first got married, my wife and I would trawl through the classified ads in the local paper. Among other things, we bought a dining room suite. Nowadays these adverts have largely disappeared from the local papers, and instead the bulk of this type of purchase is done through the charity shops. But the curious thing is that these sales outlets that are now so commonplace on the failing high streets are utilising unwanted but saleable items to raise funds for their respective good causes. I say curious because their collection rates are so high and the volume of the shops so great that they compete each other into the ground over price and simultaneously damage their fundraising capacity, at the same time destroying the non-charity market for once relatively valuable prized possessions. Fine china and ornament shops have virtually disappeared from shopping centres, and family 'heirlooms' are routinely boxed up and dropped at charity collection points because the intrinsic value has gone. If the apparent 'good' actions of these shops could be weighed against the apparent damage they have done in other areas and the difference established, then I would suggest that these shops have actually caused more harm than good. The irony is that all of the so desperately needed essentials of the consumer age end up priced for relative pennies in aid of helping others.

Chinese Whispers

So what makes a Great Brain? What are the qualifications? We can talk about establishment figures, the 'big knobs', those with their 'fingers in the pies', but this begs the question of how they got there in the first place. More importantly, why they got there in the second? My experience of problem solving often breaks things down to a very simple solution. The answer to the question is often simple; the problem is really knowing what question to ask. Why be prime minister or president or chairman or secretary or treasurer? What particular skill set do these jobs require? You need one skill set to be selected or elected, and yet you need a wholly different skill set to actually perform the job. The measure of success applied is always retrospective and effectively backdated. Hence the French have recently been saddled with Hollande, whose greatest measure of success to date appears to be achieving the lowest approval rating of any recent leader. Yet there is no mechanism to move him on quickly, to promote him sideways, to ease the plight of the French populace, to bring in a new Great Brain to appease the herd.

While the French are saddled with a failed Great Brain, they are also saddled with the rules of their governance, the pro forma devised over the centuries to ensure that the French get what they deserve. This is, in itself, a self-fulfilling prophecy, because if you are unhappy with the result, then you must also be unhappy with the pro forma—and yet nothing is done to change it. The opposition want Hollande to fail miserably because it will lead to their own future success. They need the population to suffer the Hollande term so as to ensure that they are elected with a large mandate to reverse the recent policies and restore the original cherished policies that got them unelected prior to Hollande's success. They need Hollande's unpopularity to wipe their own failed record from the electorate's consciousness to make them themselves re-electable. Have I mentioned the wipe-clean memory of lizards?

In herd terms, it's like when a fresh 'team' strikes out in search of pastures new and the rest of the herd follows. The original leaders are ignored because they failed to find fresh grass and are forced to follow the new leaders while resenting their own loss of power and hoping that the new leaders will fail so that they can regain their leadership role. The system lurches from one pole to the other, with random events dictating success or otherwise. It's tough luck leading the herd in the event of a drought or fire or flood, and yet these are all outside the control of the herd leaders.

And yet there are no herd leaders!

Think of some great enterprise, some massive undertaking that has stamped its significance on the world, look for its Great Brains—the architects, designers, managers, governors, generals—and then ask, 'Where is the leadership? Where, when it comes down to it, is the control?'

My friend and neighbour who works for the Crown Prosecution Service took me to a tour of a local nursing home where his father resided and where my mother could follow. During our drive, he confided, 'We can't lock everybody up, so we have set out our stall to harass and inconvenience the criminal classes.' This is true, we cannot lock every criminal up. And the criminals know it. But we non-criminal innocents rest secure in our beds at night, safe in the knowledge that crime is being sorted.

This begs a wider debate, because what my neighbour had actually said was, 'We will manage our expectation of the criminals' world, and this in turn will enable us to manage the expectation of the non-criminals' world'— in simple terms, 'We will be seen to be doing something. Not only that, but also we will expend huge amounts of taxpayers' money in the pursuit of this solution, which is not in fact any form of solution to the problem at all.'

The news is full of the migrant crisis and the political efforts being made to solve the problem. Borders have been shut and then reopened, trains stations closed and then reopened. Declarations from the EU and the

UN have urged the need for a comprehensive response to the humanitarian crisis unfolding on the edge of Europe to 'solve' the crisis.

We are told that nine million people have been displaced by the war in Syria. In the next breath, the president of the EU talks about a quota system for 190,000 refugees to be accepted into Europe.

But what about the Afghans, the Libyans, the Sudanese, the Vietnamese, the Eritreans, the Nigerians? What about the Christians, the Sunnis, the Shia, the Jews, the Palestinians? What about the homosexuals, the albinos?

At home we have the Labour Party leadership hopefuls displaying placards with 'Migrants welcome here' embossed upon them as they jostle to establish their humanitarian credentials in the eyes of the voters.

I am not about to attempt to solve the problem or even suggest my views on the subject, but what I am going to say is, nobody else is either. The UN, the EU, national governments, and NGOs are not going to solve the problem; they are simply going to manage our expectation of the problem and concoct a defence, a smokescreen if you will, behind which they will spout their rhetoric to assuage our manipulated concerns, while many of the proponents of said action will make a very nice living from it.

Have you noticed how the rule of law has been overridden by the sheer numbers of migrants surfacing on the borders of the EU? Hungary has attempted to impose the law produced by the EU but cannot handle either the volume of people or the volume of media attention it is getting. Austria has opened its borders and promptly announced this to be a 'temporary measure', with the border about to be resealed and the rule of law reapplied. All of these tactics simply attract more and more migrants desperate to enter the fabled land before the representative governments of the people of Europe finally get their acts together and apply the laws they were elected to uphold.

It is human nature to want to help. Unfortunately, it is not human nature to do it properly. How many more displaced people watching the media input over their smartphones and Internet connections will decide that the time is now right to swim across the Mediterranean, or walk across the Sahara Desert, or stow away in a sealed refrigerated truck, secure in the knowledge that a warm welcome awaits them if only they can stay alive to see it?

The media is full of news of Jeremy Corbyn, newfound holder of the poison chalice called 'leader of the Opposition'. We will have a new style of politics, consensus and debate, to establish the democratic wish of the Labour Party to fully represent the wishes of the people.

What's the term I'm looking for? Oh yes, bollocks.

Let's draw a parallel with another great historical princess, and let's refer to Jeremy as Jeremy Christ, the greatest princess in the history of the Christian world. Both amassed a faithful following of believers who hung onto every word spoken; both spoke of social change and a new way of doing things. Both failed miserably, and both were ignored in great measure by the 'believers', who still carried on their business as usual. And both will leave a great legacy. But ask yourself this question: how much of what either Christ espoused is real or attainable? Ignoring the faith element, consider the mechanics. How many individuals did Christ actually address in person? How many heard his dulcet tones, heard his message from the source, remembered the message in clear, concise, accurate form, and then themselves repeated this in exact format to others? The message was often given as stories or parables or metaphors or similes and was always interpreted by the hearers' brains, who gave their own twist and interpretation, or in modern terms 'spin', before repeating the corrupted message to others with the same mental disadvantages. (This doesn't even allow for the complications of multiple translations across numerous languages.)

In September 2015, a 'news' story of great import found its way into the media, that of an antipodean sheep that had evaded capture for a number of years before finally being caught and sheared to provide 4 stone of wool—well, no, actually it was 40 pounds; well, no, actually it was 40 kilos, all depending on whether you garnered your 'facts' from the radio, newspaper, or Internet. For those of you not familiar with the units of measurement, that is 56, 40, or 88 pounds of wool respectively, or if you would prefer, 25, 18, or 40 kilograms.

Now take the message from the messiah—either one; it doesn't matter (messiah or message)—and then add the variable, which for the purposes of this illustration I will call corruption. 'But Jeremy Christ isn't corrupt!' I hear you cry. He doesn't need to be, because we have society to corrupt the message for us.

315

The message can be wildly corrupted, from 18 kilograms to 40 kilograms, and the message can be wildly corrupted by individuals corrupting the message to satisfy their own requirements (think four-hour waits at A & E, ever increasing exam grade results, satisfaction surveys by your local authority).

Throw into the pot the desire for a princess, and suddenly every supporter interprets every statement as direct support of their own view, and the new messiah is born to solve all our problems with a judicious dose of miracles. But where does Jeremy Corbyn sit in my category as a Great Brain? Has he the mental capacity to qualify? His only apparent qualification is his popularity, the very qualification that raised Hollande to the presidency of France but which rapidly evaporated in the spotlight of public scrutiny. September 2015s BBC teletext suggests that the president of Brazil currently enjoys a public approval rating of 8 per cent. This is the same Great Brain elected on a wave of public support with a majority of votes to secure her position, who once enjoyed approval ratings of ninety-two per cent.

Curiously, this begs the questions, what in fact do the herd find to justify their voting pattern, and why does the reality so quickly sour the promise?

Are the herd capable of cognitive thought processes to determine the suitability of our leadership, or do they react in a tropismic manner to the most basic of stimuli? Is *Homo sapiens* tricked by the radiance of the peacock's tail, the dance and livery of the bird of paradise, the width of the stag's antlers, the loudness of the lion's roar?

In the absence of any other qualification than these tropismic responses, we are condemned to the acceptance of mediocrity, secure in the knowledge that our Great Brains are precisely not either great or indeed brains; they are simply the incumbent who has floated to the top.

So why do we hold such stock in the levels of organisation that we pride ourselves in? The whole edifice is corrupt, rotten to the core, in a self-fulfilling, indulgent, self-reinforcing, and ultimately earth-destroying drive to succeed.

We construct entire industries of reason that exist simply to justify their own existence by endlessly reinforcing their own position of import while simultaneously and systematically stupefying the alternative.

It's now the end of May 2015. Today has been marked by two major events: the state opening of Parliament, and the arrest of several members of FIFA in Switzerland and the start of proceedings to extradite them to face justice in America. Charges of money laundering and other things usually associated with organised crime have been bandied about with the full resources of the FBI and the tax authorities. Many in the British press are quietly celebrating the long-overdue 'day of reckoning' for the FIFA bandits and the end of their alleged corrupt practices. The talk is of the Russian and Qatari World Cup tournaments being withdrawn and new venues selected.

Consider the basic requirements for our life on earth, a temperature and climate conducive to our weak carbon- and water-based structures that sit so snugly in the Goldilocks zone that exists between Venus and Mars. And yet we occupy only a portion of our world; not for us the tonnes of pressure at the sea floor, the flesh-destroying extremes of the Sahara Desert, the polar extremities, or even the extreme conditions deep beneath our feet in rock. This is not to say that these zones are devoid of life; far from it. Extremeophiles are constantly expanding our knowledge of life on earth. But while I talk about the Goldilocks zone of our planet, I should in fact be talking about the Goldilocks zone that humans occupied prior to our terraforming expertise development. Current theories suggest that early human beings left Africa on a coastal tour that took us to Australia before further expeditions were made into mainland Asia and Europe. Taking this suggestion at face value then, it would be reasonable to suggest that our expeditionary force could only survive in a narrow belt supported by shoreline resources. Even in more modern times, taking Australia as an example, occupation belonged to the coastal fringes and coastal communities before the interior began to be explored. By definition, this Goldilocks zone of available exploitable habitat left the rest of the biosphere to develop under its own devices, devoid of the attentions of axes and ploughs and JCBs and rifles. Even in Europe the Great Wood could have presented itself as a nearly impenetrable barrier, riven only by rivers and geological features that allowed access to opportunity. Humankind has exploited the world so rapidly since the advent of agriculture that if we measured our activity in the lifetime of some of the species we notionally share this world with, then we would arrive at a truly astonishing conclusion.

If you contrasted the day-long life of the mayfly with the lifespan of long-lived trees and the pace at which their lives differed, then some interesting ideas surface. I have asked the question of how long it would take superior alien life forms to destroy our civilisation city by city (as depicted in films such as *Independence Day*). The answer ignored the breeding potential of a city of people a day for a 50-year span (based on a third of a million deaths a day and 6 billion people). With our present population estimated to be in excess of 7 billion, the same rate of attrition would take 60 years, which, as luck would have it, is higher than the life expectancy of an average Russian man. If we treat this 'average Russian man' as a lifetime, then we could compare this with the lifespan of a Great Basin bristlecone pine, *Pinus longaeva*, which estimates suggest can exceed 5,000 years. A standard life could be determined as 60 years or 5,000 years, depending on the species. What humankind has achieved as a species in terms of damage to the earth could be quantified as having occurred in 10,000 years, which translates into just 167 standard Russians lives or *two* Great Basin bristlecone pine lives. Given that our activity is accelerating with every new technological invention and live birth, would anyone like to hazard a guess as to what the third Great Basin bristlecone pine's life will witness? The sad fact is that for many of the species on earth, they have already been subjected to the invasion by some superior alien life form, us.

While browsing through Simon Schama's *A History of Britain*, I came across an estimate suggesting that the population of pre-Roman Britain matched that of Elizabethan Britain some seventeen hundred years later. What did these ancients know that contained our species so adequately, or rather what do we know that has slipped us of these constraints? Are we back to 'when was the golden age of farmland birds' question? When was the golden age of humankind? Pre–Industrial Revolution or pre–Agrarian Revolution, or do we suppose that it is now during the technological revolution, the computer or digital age? If the sole aim of the human race is to breed uncontrollably beyond the realms of the earth to support us in a truly sustainable state, then we have succeeded beyond our wildest dreams, except that we have no dreams, wild or otherwise, no control, central or otherwise, and no aim—indeed no purpose other than to be. We advance like an all-devouring host of locusts intent on the mindless destruction of all that we can reach, but with the technology to exploit beyond the reach of hands to plunder the resources of oceans and even time itself through the extraction of fossil fuels and the destruction of our future.

Intelligence is the quality of being intelligent or clever; it is the ability to think, reason, and understand, instead of doing things automatically or by instinct. The herd displays no intelligence, no ability to think or reason, and instead does things automatically or by instinct.

Anyone watching nature programmes featuring shoals of fish under predation must understand that this behaviour of hiding from danger behind your neighbour is without rationale. Many years ago a friend asked me a question in the form of a joke: imagine how you would feel if you were in the nest one day, curled up and secure with your siblings, when you asked your mother what you were, what you were called, only to receive the answer?

'You're an edible dormouse.'

The nature programmes even refer to the shoals of fish as 'bait balls'. Billions of sardines swim up the western side of South Africa in vast shoals stretching for miles in length, but they react to predators by hiding among themselves in tighter and tighter balls, providing perfect feeding opportunities for all and sundry. By the time whales, dolphins, sharks, tuna, seals, seabirds, and so forth have visited themselves upon the shoals, entire 'balls' will have disappeared. The apparent strategy for survival is fundamentally flawed, but it persists because of the sheer pace of reproduction, which simply swamps the available predators with such immense numbers that the fish cannot all be eaten, so the species will survive to be eaten another day. One can speculate that at some time in the fishes' evolutionary history, 'bait balling' was an effective strategy for survival and that this mechanism is 'hardwired' in, to the obvious detriment of the individual in the current predation environment.

My question would have to be, what would an intelligent fish do?

To be honest, this is a rhetorical question, because what I am really interested in is, what would an intelligent person do?

Let us consider our strategy for survival. What tricks do we bring to the table to ensure the future of the human species?

The sardines survive in spite of their apparent survival strategy rather than because of it.

In my youth I would read the odd western-themed book of fiction and occasionally come across accounts of buffalo hunting whereby hunters would attempt to maximise their kill rates by achieving a 'stand'. This was an attempt made to shoot the leader, and then the next in line, and so on, the idea being that if you could eliminate the leader, then the herd would simply mill around on the spot waiting for one among them to show some initiative. As soon as one did, the hunter would dispatch that animal, and the process would be repeated until the entire grouping had been killed. I cannot vouch for the historical accuracy of this process, and I would argue that there was some anecdotal evidence to support the stories, but in reality it doesn't matter whether it was true or not.

Hitler's rise to power in 1930s Germany was characterised by the systematic removal of rivals or opposition leaders from the scene. Communists were outlawed, along with trades unions, and many communists and trades unionists became the earliest occupants of concentration camps for 're-education'. Hitler's personal rivals within the Nazi Party would simply disappear through state-sponsored murder (the Night of the Long Knives, as it became romantically known, for example). The invasion of Poland by the Germans saw the removal of all intellectuals, thus creating the Germans own human version of the buffalo stand. When Russia invaded the east of Poland, Stalin's secret police moved quickly to arrest and remove all potential leaders. Most were interrogated before being shot and consigned to mass graves at Katyn, with Germany and Russia blaming each other after the war.

When Warsaw staged an uprising immediately before the Soviet reinvasion in 1944, the Soviets promptly stopped for rest and resupply while the Germans systematically razed Warsaw, and the Polish will for self-determination, to the ground.

Pol Pot's murderous regime in Cambodia under the Khmer Rouge saw the mass slaughter of millions of intellectual giants whose crime was being a potential threat to the regime, essentially anyone with an education. (At last, a reason for education other than to keep kids off the streets during the day?)

Against this background, then, we can see that humans really do have herd leaders, if only in the sense that we miss them when they are gone.

Conversely, the American aim of bypassing Paris in a dash to the German border was derailed by the French uprising in Paris and the desire to ensure that Paris was liberated by 'official French forces under de Gaulle' for fear of a communist takeover. (One set of leaders supplanting another?)

Many of the Jewish ghetto leaders were complicit in ensuring the smooth operation of the German aims, although to be fair they didn't have much of an alternative. Hollywood has its own take on similar issues whereby governments withhold information from the population for fear of creating panic or a breakdown of society/law and order in the event of an impending apocalyptic asteroid/alien invasion or similar event. Because, of course, the actual apocalyptic event won't have any impact (sorry, I couldn't resist it) on society or the rule of law. Some of the Nazi death camps were very adept at maintaining normality, with arriving trains greeted with flower beds and live music as the occupants were separated out into the 'die todays' and the 'die another days'.

We trust our 'leaders' even though there is but scant evidence to justify this position. This may be because these Great Brains can see the way the wind is blowing and will always revert to what is important to them—and that, of course, is them.

And yet the biggest questions asked about or after the genocide of the Second World War are how and why it could happen. How and why would seemingly normal people be so readily corruptible to indulge in such heinous acts? But really the question should be asked differently. What we should be asking is, why doesn't it happen more often? I nearly used the term 'inhuman' rather than 'heinous', but I realised that in fact the actions were in fact intensely human, perpetrated by one group of humans on another. But even now the lessons of these actions are being largely forgotten. True, the Jews are very active in promoting the Holocaust through museums and exhibitions and through films such as *Schindler's List*. As an afterthought, almost in passing, reference is sometimes made to the extermination of Gypsies and occasionally, when the topic is explored in greater depth, the extermination of homosexuals, the mentally and physically disabled (even 'purebred' Germans), communists, and trades unionists. But this ignores the tens of thousands of slave labourers conscripted in from around occupied Europe, including some from the Channel Islands, forced to work in brutal conditions to construct and support the apparatus of war.

The Second World War saw the end of one conflict and the start of another between the West and the Soviet Bloc. Consequently, the Soviet efforts during the war have largely been airbrushed from our Western histories. We ignore the plight of the millions of Soviet prisoners of war who were effectively left to die of starvation, disease, or exposure at the hands of their captors. Here 'captors' is plural because those lucky enough to survive the Germans then had to survive Stalin, who treated all captured prisoners as suspect citizens in need of interrogation and re-education.

We argue about whether the Americans or British won the war in the West (not to forget the many other nations of the world, Canada, Australia, New Zealand, India, Brazil, and so forth), when in reality examination of the headcount of soldiers facing each other across the battle lines shows that the victory is unequivocally Soviet. Yes, I know, we worked to supply and support the Soviets with the materials of war; in truth nothing is straightforward. We learnt at school that the two atomic bombs brought an end to the conflict in the Pacific, but once again we ignore the simultaneous near destruction of the million-soldier-strong Japanese army by the Soviets in Manchuria in a mere handful of days.

What I am trying to illustrate here is the way our leadership corrupts the reporting of events to suit its own requirements and how our herd instinct kicks in to support these prejudices.

While it's true that the Second World War cannot happen again, I would not be so sure about the Third World War.

In trusting our leaders, are we not hiding among ourselves? Have we found our human bait ball?

Leadership by 'celebrity'. Who the fuck cares who copies or seeks to emulate David Beckham, or Jade Goody or any one of thousands of supposed celebrities? Is this not the ultimate example of how the human mind is captured by baubles of sparkly lights?

What is the meaning of life? This is an age-old question whose answer is buried within the human mass, packed tightly in the centre of the bait ball.

We have the solution for global warming, disease, hospitals, health, security, and economics. Our leaders tell us so, so we can relax and sleepwalk into oblivion. So when our leaders tell us that the Germans are emptying our ghetto to move us to a better place, we believe them, because when it comes down to it, the alternative is beyond our ability to contemplate. Even when we are perched on the edge of the trench, waiting to be shot to join our companions who went before us in the queue, we wait compliantly, shocked into immobility. Is this when belief in a higher place steps in? We can die hopeful in the belief that we will be going to a better place? Is this the ultimate delusion?

Are we no better than the ant colony, unable to think and reason for ourselves, trapped eternally in our responses to the chemical signals from a mindless queen, surviving because of our numbers rather than our survival strategy? The hardest thing to do is let go, to abandon the beliefs and cherished ideals seemingly hardwired into our brains, and consider a different way. Abandon all material wealth and devote your life to God; consign yourself to a more fulfilling life to serve a higher purpose. Have you not simply replaced one chemical queen for another?

Remove yourself from the city rat race and semi-retire to the country, or cash in your wealth and retire fully to the life you'd always promised yourself.

Become a hippy and subsist from benefit cheque to benefit cheque while effectively stealing from your fellow, if less enlightened, neighbours. Piss in the supermarket meat fridges in front of shoppers and then await your prize in the rubbish skips behind the store.

But none of this is sustainable, because you cannot exist without the efforts of others. We cannot all be monks or retired people or hippies without the Ponzi scheme of support from others. If everyone pissed in the meat counter, then the supermarkets would either continue to sell the meat with urine (does urine contain nitrates and nitrites?) or would stop selling meat. Either way, the free meat supply would dry up.

What would happen if medicine were removed from the earth? At what point does a health improvement become a hindrance? A recent TV programme, I think it was called *Big History*, suggested that caffeine drove the Industrial Revolution in Britain, a bold claim but, as ever, via a convoluted route. Tea and later coffee saw the mass consumption of boiled water, freeing the city populations from the scourge of waterborne diseases, allowing for the rapid expansion of cities and the workforce to drive the factory output. The birth of epidemiology and subsequent improvements in waste transfer and processing completed the process. (Beer consumption also reduced the prevalence of waterborne disease, but I guess this message is not for the educational target audience.) Yet the Romans understood the importance of sanitation a couple of thousand years earlier.

Now we had the means to construct the first megacity of the modern era, which brought with it a new set of problems. 'Extreme' is the word that springs to mind: extreme poverty, squalor, working hours, and working conditions, and new diseases that flourish in crowded environments, including consumption, or tuberculosis if you prefer, and even legionnaires' if you move on a century or so. And with the megacity was born new subcultures of pickpockets and prostitutes and drunkards and poppy-based laudanum addicts and addicts to today's drugs. Now that's not to say that these types of people couldn't develop outside of cities. The moors outside Lutterworth, now bisected by the modern A5, were once described as the 'haunt of footpads and highwaymen' and were specifically mentioned in the second Enclosures Act to bring them into civilisation by fencing the land. But what could occur was the 'industrialisation' of these activities, relying on the anonymity of the herd denied to village dwellers, where everyone knew everyone's business. But if we stop and think about the pickpockets and drunkards and drug addicts, we realise they are all effectively 'pissing in the meat counter', parasites on society, demanding their living from others. We have engineered the demise of one pathogen, disease, with another pathogen, crime.

Workers are split into two categories, the first for people who generate wealth by their own physical labours and the second for people who generate their wealth from the physical labour of others. You can place factory floor employees in the first and the factory owners in the second. You can place bus drivers in the first and all the pension operators securing the investment for their retirement in the second.

We are back to the argument about effort multipliers and the generation of wealth. Are you paid an hourly labour rate for product produced, or is your pay independent of hours worked and effort exerted? In this latter category, think bankers who are paid bonuses, stockbrokers, fashion designers, and people paid for ideas.

As our technology has advanced, we have moved from relatively small conflicts arranged and settled in short order at one extreme to the industrial wars of the twentieth century. Fates of nations were once settled in an afternoon between several thousand men hacked and hacking their way through the opposing forces, leading to the removal of the incumbent or challenger. Disputes could fester beyond this event until the defeated sympathisers were either brought to their knees or quietly dispatched of. As the scale of society grew, so did the conflicts, until disputes could range across the countryside and last for several years, the Wars of the Roses or the English Civil War for example. Add a few centuries and wars could traverse the globe, with navies fighting to secure territory and wealth, and vast armies under Napoleon fighting from Egypt to Moscow to Brussels and Portugal.

The Cold War saw the invention of MAD, mutually assured destruction between nuclear superpowers, and perhaps the realisation that war was no longer an option. The Vietnam War, the breakdown of Yugoslavia, the Iran–Iraq War, the invasion of Kuwait, the First Gulf War, and the Second Gulf War, chased on its heels by the Arab Spring, the Syrian Civil War, and the rise of IS, to name a few, proved beyond doubt that war was still very much an option.

Lord Whatshisname

Why do we have this innate ability to skirt around every issue? Are we frightened of taking a decision? Newspaper and other media are perfectly content to provide reams of in-depth analysis without it being either, in depth or analysis. Occasionally an article will be written that attempts to expose the real cause of an issue and begins to hint at a fundamental cause, before shrinking back from the conclusion as though the writer is about to disclose the identity of Lord Whatshisname from the Parry Hotter books. It's as though the truth must never be revealed, as if we are not capable of learning it for fear that our brains will turn to mush and civilisation will end as we know it. Maybe it's because an ant will always be just that, an ant, incapable of deviating from its preordained path, even though an alternative has presented itself and the ant is in theory perfectly capable of grasping it.

So what is so precious about this civilisation as we know it? Our sole function in life appears to be to survive to pass on our genes. The cost of this must be borne by others, other people and other hinterlands. I first learnt the term as a geographical expression to describe an area utilised to support an otherwise local enterprise. How could Birmingham survive without the water piped in from the Welsh mountains? How would London survive without the constant supply of agriculturally sourced foodstuffs from around the city, both home and abroad? The time was when cattle reared in Scotland would be driven by foot over drovers' roads to fatten up in Aylesbury Vale before completing the journey to London markets for slaughter. At the time food was patently visible, whereas now it is shipped seemingly effortlessly within just another anonymous refrigerated truck that we give no second thought to. Another term for this innocent geographical expression would be 'Lebensraum', or 'living room'. Students of Hitler will recognise the sinister implications of this innocent term (by living room, we don't mean lounge), as it is the name given to the Nazi ambition of conquering all land in Europe to the east of Germany. Of course for the land to be available for German exploitation, the small issue of the indigenous populations would first need to be addressed by the 'Germanification' of the territory. In simple terms, non-Germans would either be killed or enslaved to the German state.

The Germans made the mistake of taking the land by force and attacking their weaker neighbours, who ultimately turned out to be their 'stronger' neighbours.

Of course if they had simply invested in the countries by buying their foodstuffs and other raw materials, they would have achieved access to this hinterland without a shot being fired. While this would not have seen the ravaging of the countries by war with widespread destruction and sudden death suffered by people, it would have seen the widespread ravaging and destruction of habitats and resources suffered by nature.

If you think it's unfair or an extreme example to single out Nazi Germany, then think of Americans and their drive west through their weaker indigenous neighbours and the destruction of the vast buffalo herds and the habitat that they had created. Or, if you prefer, consider the colonial ambitions of Europeans through South America, Africa, or Asia. But think about the destruction of rainforest in South East Asia for the cultivation of palm oil to fuel our vehicles in Europe and you realise that the same old process is continuing today. We are just more polite about it.

I said before that walking cattle from Scotland to London was much more visible in its day, but I don't really believe that to be the case. With the majority of people living and dying within a few miles of their birthplace, London and Fife might as well have been Sumatra and Timbuktu.

Occasional news reports highlight the Chinese invasion of Africa whereby this resource-hungry and cash-rich nation is carving its way through former colonial states with trade agreements to build roads and power stations in return for access to resources. We regularly read 'Made in China' labels on our telephones and computer games but give no thought to 'Mined in Angola' or 'Imported from Australia'. Prince Harry will pop up to front a Save the Elephants campaign, and those in the know will point to the new rich in China soaking up the world reserves of poached ivory or rhino horn without giving a single thought to the destruction of square miles of jungle for the extraction of copper ore, or the ploughing up of vast acreages for cereal production so that we can feed our garden birds.

But what happens when we run out of hinterland, when there is no more vacant and derelict resource to plunder, no more tiger reserves or wilderness? What then? What will be the price of our blind, thoughtless pursuit of passing our genes down to the next generation?

Lord Voldemort is alive and well and wreaking destruction of our world (the only one we have), and we ignore it. Don't talk to me about the WWF or the RSPB or the UN; they are all deputies of Voldemort, deflecting our attention from the issue, using smoke and mirrors to create the illusion that they are doing something while at the same time making a good living from it. You cannot save a rainforest by gardening; you save a rainforest by saving it. You cannot take a timber crop from a 'sustainable forest' without destroying the forest. To remind you, the Australian definition of 'sustainable' involves the idea that the thing in question can continue forever. This in spite of the fact that over half of the life in a forest lives on dead wood, the same wood that is being removed for timber use. To lose weight, don't go on a diet. Eat less food.

BBC News is full of the refugee/migrant crisis and the 'humanitarian disaster' washing up on Europe's shores. The BBC's sole consideration seems to be that these people are all legitimate individuals looking for a better life away from conflict and oppression. But doesn't legitimate mean that the individuals are legal, that they are complying with sovereign state laws?

The sheer force of numbers is beyond the capability of border controls, and every individual is essentially outside the due process of the law. Therefore they are illegal. Here we see the human equivalent of the 30 mph sign and the legal framework's inability to enforce it. Laws, then, appear to be voluntary and only apply when enough people choose to observe them. But we already knew this within the context of the 30 mph zone. Today, 23 September 2015, EU ministers have agreed to impose mandatory quotas on member states to accept 120,000 non-EU individuals into their midst.

Britain, along with a small number of states, has an opt-out clause, so this ceases to be our problem, until, that is, these individuals are registered as EU citizens, at which point they have the right to travel, reside, and work in any member state, Britain included. It appears then that the right of all EU citizens to work and reside anywhere within the EU applies equally to anyone from anywhere in the world. This figure of 120,000 fails miserably in the attempt to solve the issue given that 500,000 refugees have already arrived this year and the numbers are accelerating.

What we are witnessing is the breakdown of the pro forma. The standard method of dealing with this problem is failing, and no alternative is available. The same could be said of the Greek debt crisis. So what do we do? We need our Great Brains to construct the solution and resolve the issue. What becomes patently obvious is that our Great Brains are neither great nor brains, but simply the holders of the office, pretending to occupy the space, talking the right talk and filling the ideas vacuum with expenses claims and summit dinners, without the vaguest clue or idea about what to do. Parallels can be drawn with pre-war politics in Britain and France, where the individuals charged with protecting us sleepwalked into the greatest conflagration of the twentieth century because they had no clue of how to deal with Herr Hitler. My mind is drawn once more to the ant colony. The queen produces chemicals which the workers obey. No thought is involved, because every ant knows its place and allotted task and follows its pro forma.

Consider the humble hedgehog that has evolved to deal with predation by curling into a tight ball of spike-protected food at the first sign of danger.

Consider the humble sardine that has evolved to deal with predation by forming into a tight ball of fish-protected food at the first sign of danger.

Consider the humble passenger pigeon that had (note the past tense here) evolved to deal with danger by going to have a look at what the problem was.

Consider the humble *Homo sapiens* that has evolved to deal with danger by going to the ballot box and voting for the most decent popular chap who vaguely represents some of their views so that this man then could attend a special place where he could talk endlessly about solving the world's ills while being solely motivated by his desire to continue to protect his position by proving that he truly 'represents' his constituents.

Consider the humble politician who has evolved to deal with predation by forming into a tight ball of like-minded politicians at the first sign of danger, secure in the knowledge that if they stick together, they can't receive any individual blame—and there is usually a good dinner as well.

Strange that they call these events summits. 'Summit' to me evokes images of the peak, the apex, the highest point, the place with the greatest view. I may be doing the nomenclature a disservice here, because if you think about it, after the summit the only way is downhill.

The problem with the hedgehog is its inability to deal with one-tonne lumps of metal hurtling blindly along the black tarmac thingies.

The problem with the passenger pigeon is going to have a look at shotgun-wielding hunters intent on its destruction.

The problem with *Homo sapiens* is its inability to differentiate between the need to have a government and what the government in point of fact does.

The problem with our politicians is the need to form a consensus that satisfies everybody while satisfying nobody. Consensus, by definition, creates a fudge.

Where is the hope for the human race?

(I should not place all the blame on shotgun-wielding hunters for the demise of the passenger pigeon. After all, they thought nothing of laying down poisoned grain or of dousing nesting forests with fire accelerants before setting fire to them.)

A few years ago I read an article that reported on corporate lying in Japan. I'm afraid I don't have the names, but the principle is correct, so I will relate the story anyway. One of the world's largest manufacturers of aircraft seats had devised a test rig at its factory to test the exhaustive safety requirements of this product before it would be allowed to be incorporated into aeroplanes. These tests were independently observed by one of the aviation authorities that would sanction the use of these seats in planes. After several years of compliance, disgruntled employees of the company alerted the authorities to the fact that the results from the test rig had been collected by an intermediate computer which modified the results to indicate compliance before displaying the corrupted data on a second computer screen that the independent observers had access to. In short, the testers lied. But this airline seat manufacturer didn't only make seats;

it was also one of the largest component manufacturers of car parts in Japan. I often wondered if there was any connection between this fact and the plague of car manufacturer recalls that we have seen in recent years.

Alongside the illegal immigrant story headlining the news programmes has come the revelation that the world's largest car manufacturer has admitted to fitting a 'cheating device' to all of its diesel cars (eleven million of them by their own admission, or should that be emission?) which corrupts data from the engine performance relating to pollution emissions. In short, they lied.

Now Volkswagen has, as an opening gambit, ring-fenced nearly £5 billion to cover the costs of redress for legal representation and compensation to address this problem. Now, of course, we patrons of the earth's atmosphere are rightly shocked and outraged at this company's cynical actions to lie and deceive us into a false sense of security. Fair enough, we should be.

But how many of us are disgusted at the target levels set by governments in the first place? How many of us trust the authorities to set safe levels? As part of the brouhaha surrounding the revelations, various pressure groups and their scientific advisers have come out of the woodwork, announcing such things as, 'It is estimated that thirty thousand premature deaths occur in the UK annually as a result of exposure to elevated levels of nitrogen dioxide, and this figure rises to five hundred thousand throughout Europe.' Or it might have been fifty-eight hundred premature deaths annually in the UK; take your pick.

Now if we accept these figures at face value, then we also accept that Volkswagen has been complicit in a large share of these premature deaths.

Let's put them on trial at Nuremberg for crimes against humanity. But while we are at it, don't we also need to bring to task the food manufacturers who lace our food with sugar, salt, and hydrogenated vegetable oils? And the cigarette manufacturers? And the shopkeepers who peddle these noxious products?

But if we do that, do we not also need to prosecute everybody who drives a diesel vehicle, or rides a bus or train, or does their shopping at shops stocked by diesel trucks?

What we will do is crucify Volkswagen because they have been found out, and it will be business as usual for everybody else. Problem solved.

But while we construct the crucifix among the world's courts and government commissions, ask yourself this question; when governments set targets to measure and control any aspect of our lives, who checks to see whether the test results are corrupted? The first thing we managed mortals do is work out how to lie, to corrupt the figures so that the results bear out the policy requirements. That's why 95 per cent of us are seen within four hours at A & E.

It's also why our children are getting better and better at passing exams.

But you should also ask questions about the policy and who benefits from the lie.

Education is a political issue that can sway the electorate in their voting patterns. Health is a political issue that can sway the electorate in their voting patterns. Who then is the greatest receiver of benefit from the lie, the managed or the management? In the case of health, the benefits are reaped across the full management strata, from every junior manager to the leader of the government basking in the successful policies. The people who do not receive any benefit are the customers, the patients, and the taxpayers who pay the bill, and the entire management structure is designed to hide this fact from these people.

Of course, the major argument espoused by the election pundits for the voting pattern is centred on 'trust': whom does the electorate trust to best serve our needs?

So where are the disgruntled employees? Why does no one alert the authorities to the corruption between the measuring device and the computer printout seen by the independent monitors? The reason Volkswagen is in the Katanga is that one administration, the American one, can generate extreme benefit for itself by exposing a failure in somebody else's administration, be it the EU standards of testing or the company's own corrupt practices. Additional benefits will also surface with multimillion, or will it be multibillion, fines/court settlements, decreasing sales from a foreign car manufacturer and a boost for the United States' own domestic ones. And, of course, look how good our administration is. You can't pull the wool over our eyes, can you? Time, by now, has run its course and the true scale of Volkswagen retribution will have surfaced.

There has even been the suggestion reported by the BBC that the German government knew about this scandal several months ago and did nothing.

Now of course all of our administrations must prove to us that they in turn are fit to govern us. Open season will be declared on Volkswagen. Yet commentators are coming out of the woodwork stating that they have known of discrepancies between published and actual field tests for ages and have been lobbying governments for a response. These discrepancies fell into the 'unofficial truth' category and were thus ignored by all the departments that will suddenly be launching their own exhaustive enquiries into the 'newly discovered' data while they beaver away to prove how competent they are to receive their manager salaries.

You only have to look at the US administration's response to Hurricane Katrina in New Orleans to realise that their agencies are just as useless as ours, but here they have the moral high ground and don't care how much damage they do to Volkswagen.

It's Thursday, 24 September 2015, and the papers are leading on Volkswagen with, 'British lawyers are lining up to launch class actions to sue VW for misrepresentation. Bankruptcy is not out of the question.' I wonder if any of these unsuspecting victims of this monstrous fraud salivating at the prospect of a few thousand quid in compensation will ever stop to think what the value of their vehicles will plummet to in the event of VW going bust.

We live in a big society, which supposedly only functions because we have big organisations looking after our needs and requirements. Without the mass of government and NGO's, and the government departments of health, welfare, education, and so forth, our lives simply couldn't function in this modern world.

What I have tried to illustrate is that our modern world is seriously flawed and in many ways defunct of any rhyme or reason. Our sole ambition, which as a statement is wrong because we simply don't have an ambition … our sole action appears to be to pass our genes down a generation, and we will do precisely what we need to do to achieve this. But we don't intend to do this either, so our sole tropismic response to light and air and food is to procreate, and we will do whatever we need to achieve this. But we don't achieve it; a tropismic response just happens without thought or intelligence. So, we live, we eat, we drink, we seek shelter, we breed, and we compete for these resources, and the successful ones just are. There is no higher purpose, no great plan, no rhyme or reason, just birth and death with a span of time between these events.

We concoct great fantastical illusions of purpose and reason supported by grand enterprises to support and reinforce these notions, but when it really, really comes down to it, what do we do? Strip away the facade of life and living and expose the core of *Homo sapiens*, and what are we left with? Imagine life the day after the apocalypse, when the aliens have returned to their craft to plunder the next planet, when the plague that has killed 99.9 per cent of the population finally abates to leave the genetically immune to inherit the remnants, when the alarm finally sounds the all-clear from within your nuclear bunker. Life begins afresh, day one of the new order. And on this day consider your priorities, food, water, shelter, security, life. Contrast that with day 3,650,000 of the old order (ten millennia)—yes, yesterday.

Consider your priorities: mortgage payment, credit card bills, pension savings, job promotion, designer clothes, car insurance, anti-wrinkle serum, moving to the right school catchment area, starring on *The X Factor*, giving to charities ...

Reconcile these 'ambitions' with saving the planet, halting global warming, preserving the biodiversity of the earth.

Those of you old enough to have investments in property or stocks or pension funds may worry about bubbles. This innocent term is used to describe situations or circumstances where the supporting infrastructure is unsustainable or without foundation. The Wall Street crash of the 1920s springs to mind, where too much money was secured against too little asset. Then there was the dot-com bubble at the turn of the century where investors spent billions on companies that couldn't warrant their value. The Dutch drove their state to bankruptcy over the buying and selling of tulip bulbs in the 1600s. Every housing boom outstrips the supply of money from potential purchasers, leading to the inevitable crash when the inflated bubble bursts.

Ultimately, every bubble is underwritten by the actions of human beings. No, ultimately every bubble is underwritten by earth and its resources, and we are writing cheques that the earth cannot honour. We live in a bubble; our society is a bubble, our civilisation is a bubble, and our pathetic attempt at sustainability is a bubble. Indigenous Amazonians, Papua New Guineans, Kalahari Bushmen, and Australian Aborigines have the nearest claims to living a sustainable existence. For centuries we have striven to indoctrinate these 'poor' people with a proper purpose in life, a higher order of belief in our God, consumerism, and our spiritual one as well. On day one of the new order, who would you wish to have as a neighbour?

Start by looking for things that are real, that have substance and meaning.

Education Is a Force Multiplier

We have looked at the military ambition of multiplying the effects of individual soldiers on the battlefield. Give them a machine gun over a rifle, a breech-loader over a muzzle-loader, and a cartridge over a powder horn and a wadded ball. We have looked at JCB and the impact of their machinery on the labours of human beings.

But the greatest force multiplier of all is education. Our learning establishments seek to explore the developments and histories of humankind and relate them to today's generation. Not for us the delays of trial and error; we can build on all that has gone before. We don't have to invent the wheel or capture fire or discover penicillin; we can exploit the labours of others. But at what point does our exploitation of the idea become the idea exploiting us?

When villages were small and their supporting hinterland similarly sized, our problems were within scale. Firewood could be collected by hand from the local wood and carried to the hearth. Harvests would be gathered and carried to the stores by hand. As villages grew and towns developed, the range of hinterland increased, and we arrived at the six-and-two-thirds-miles rules for market charters: a third of your day to travel to market, a third of your day at the market, and a third of your day to return home. As the ranges increased, so did the logistical problems, and we arrived at the problem of how to cross the dry desert from oasis to oasis with x number of ostrich egg shells and y trips to create the 'water bridge' between fresh water supplies.

Donkeys and other beasts of burden were employed as a force multiplier. One donkey carries the firewood carried by five men; oxen and a cart, the firewood carried by thirty. In this glib statement we have invented the wheel and the cart, have domesticated our beasts of burden, and have learnt the needs of husbandry. We have invented roads and fords and then bridges. We have invented potholes and tolls and taxes and tarmac and highway maintenance and traffic jams and canals and railways and motorways and aeroplanes, because the wood was too far away from the village to collect firewood easily by hand and on foot.

Just as a thought, why didn't we plant a wood nearer to the village?

Now we commute for hours to work and back, travel long distances for meetings and holidays, import our wood from around the world, and employ our time machine to plunder the prehistory of wood in its fossilised form of coal or oil or gas. Now we work for forty hours a day over the span of forty years to pay our mortgages, utility bills, health bills, and taxes and expect the 'wood' to be waiting for us in the fuel tank or copper wire when we return home.

If this is what education has done for us, what have we achieved? It is wrong to blame schools solely for this exercise in chasing our tails, because in essence what we have described is our culture, our civilisation, our pro forma for living en masse. We could sum up our civilisation's achievement to date, as a simple failure to answer the original question, so far in fact that we don't even know what the original question was.

The question was, 'What can I do? By the time I have returned from the wood with firewood, my fire has gone out.'

The answer is, 'Move nearer to the wood, plant a wood nearer to the house, and restrict the size of the village to keep the available fuel supply within range. Balance the population with the labour of human beings and the range of the hinterland.'

Nowadays we fail to tackle the problem from every conceivable direction. Our solutions achieve nothing. We invent a stove that uses less wood (there are charities in Africa supplying stoves to subsistence farmers), we improve the fuel efficiency of our motor cars (CO_2 or NOx? That is the question), and we build straighter, wider, faster roads with flyover intersections (forgetting, with the M25 being a prime example, that if you build a better road, more people will use it).

Is this the answer to the question, 'Given that the population of Elizabethan Britain was similar to the population of pre-Roman Britain, what did they know that we didn't?' What they didn't know was that their lifestyle was sustainable(ish). What they did know was that life was hard and short. How to reconcile the two, a truly sustainable lifestyle that's both long and comfortable? Part of the issue, even then, was the differential between the rich and poor, the lord and his peasants, the tax collector and the workers.

The lords provide the repository and the driver for the knowledge, the history, technology, and arts and crafts. They had the wealth to commission the different and unusual, let's call it the extracurricular activities, above and beyond the needs of survival.

A television programme looking into the history of castles suggested that the Normans constructed a lot of motte-and-bailey castles after the invasion of England. The commentator suggested that on average a new castle was completed once a fortnight for ten years. These tall mounds within a larger moated area represented a major investment in manpower to construct such huge earth structures subsequently topped off with wooden stockades. Yet the English prior to invasion had not felt the need to construct them. The conclusion would have to be that some amount of persuasion was required to achieve such a massive programme of works.

We can readily imagine the forced-labour element and the threats of force attached to the request to 'drop everything else' and bend your back to this imposed imperative. Yet the peasantry were probably quite used to these 'requests' for tithe of food, labour, and artisan skills, and the only real difference would have been the directness of the imperative and the surnames of their commanders.

These castles were not intended to provide any benefit to the indigenous English; rather, they were to enable their subjugators to rest easy in their beds and dining halls beyond the range of militant uprisings. But if the castles served to prevent the 'militant uprising' by imposing an insurmountable threshold to the insurrection, maybe the lives of the indigenous English were improved in that they didn't die needlessly mounting unorganised spur-of-the-moment attacks against a superior force.

But why stop at the castle? Why not build a pyramidal tomb to ensure your afterlife? A cathedral to atone your sins and guarantee your ascension?

An empire to salve your monstrous ego and create your legacy? How about a business empire so that you can buy that private Caribbean island? But before you accuse me of a lurch to the left, consider that the 'benefits' filter down through the classes. Every lord needs his lieutenants; every artist needs her patrons; every worker needs his workplace. Given the choice, even slaves would welcome the benefits of working in the master's house rather than the field. But it would be fair to say that the distribution of the benefits is uneven in the best traditions of Ponzi schemes.

The problem we seem to be suffering from appears to be one of hierarchy, and an unswerving ambition to avoid being at the bottom of the pile. Studies of hyenas indicate that status is hereditary. A hyena is either born with social status or not. Other animals acquire status by growing big and strong and fighting for the right to breed, take the pick of the food, and enjoy shelter from predators. But this ability to grow big and strong is once again primed by hereditary forces at work. A dwarf buffalo is hardly likely to compete with normally sized bulls, and the whole breeding selection process is based on might and physical presence on the stomp. But other factors have to be in place; otherwise buffalo would continually grow and grow in size and the prairie would be covered in elephant-sized buffalo. I will suggest that buffalo are constrained in size by their environment—let's call it their hinterland. Their physical bulk is controlled by their ability to source food throughout the four seasons and their ability to outrun predators. Yet there is a place for dwarf buffalo, well, strictly speaking, dwarf elephants, as there is fossil evidence of miniature elephants stranded on newly formed and shrinking Mediterranean islands where their bulk was constrained by the environment.

Island habitats are a rich source of miniaturised versions of mainland species and are prime examples of the constraints of a hinterland on a species. This is not a message that we seem to have any grasp of on our island earth.

It seems that we as a species, in spite of our supposed intelligence and free will, are controlled and condemned by our prehistory. True, we don't succumb to an annual rut (although the idea could prove quite interesting). Instead we spread our imperatives out throughout the entire year, and indeed our entire lives. Is this the drive to own a bigger house, a better car, and the smartest clothes and to have the prettiest partner, the cleverest children, the biggest salary, and the most exclusive holiday destinations? Was life so much simpler when all we males had to do was beat our chests to make the loudest noise, grow grey hairs, and fight the odd opponent over the years? But if (a big if) our lives were akin to that of gorillas, how to reconcile the needs of the small family unit with multiple family units congregated together? How did we adapt from families to villages to towns to cities to nations to federations (if the EU has its way)? How to reconcile the need to survive in ever greater densities with our DNA that betrays our sentient ideas? We credit the ancient Greeks in Athens with the creation of that most cherished principle of democracy. We champion the British model of government via our 'mother of all Parliaments' (don't raise this fact in the Isle of Man; they might scoff in your face) that has been adopted around the world. (Don't mention this in North Korea either, as it may be injurious to your health, wealth, and happiness.) Yet when it comes down to it, how do these cherished ideals impact any of us as individuals? Vote for the losing party and you are disenfranchised—no democracy for you. Better still, refuse to engage and don't bother to vote; you still live in a democracy.

Rest easy in the knowledge that our lower house is monitored and overseen by our upper house of failed ex-MPs, party donors, retired ministers, and fairy-tale promoters (bishops), but heavens above, at least we had the sense to get rid of the hereditary peers!

How fast could passenger pigeons evolve to fly away from danger? They only had fifty years or so and failed miserably. How fast can hedgehogs evolve to abandon their spines and learn to run from danger? They have only had one hundred years or so and numbers have slumped dramatically. (The jury will still be out on these beasties. Is the culprit the motor cars, toxic slug pellets, or the changing face of British housing and shrinking gardens?) So how fast can *Homo sapiens* evolve to live in cities? While cities have existed for several millennia, the modern dependence and corresponding abandonment of the countryside has been a recent invention tied in with the Industrial Revolution. And it is still accelerating. At some time in the last century, I've forgotten exactly when, the deeply philosophical question was asked of London children, 'Where does milk come from?' The answer, as we all know, was 'From bottles.'

One of the recurrent themes of time travel is the need to tread lightly on the visited past so as to avoid altering the present and future. In the simplest dramatic form of time travel portrayed in films, it is essential that you don't kill your younger father; otherwise you wouldn't exist in later times. How easily we grasp this principle and swallow the reality of the 'fact' while simultaneously ignoring it. If we truly grasped the notion, then we would fight tooth and claw to prevent the extinction of the flora and fauna that we notionally share this earth with. Our problem is that we do fight tooth and claw, but we just don't know what we are doing within our gardening operations. Can I suggest the need to tread lightly on the present that we are only visiting?

Changing the subject slightly, a number of 'children raised by animals' cases have been documented in recent years, and even some cases of caged children kept isolated and hidden from society by their own parents. These children have been 'rescued', placed in care, and subjected to medical supervision and scrutiny. Language specialists have studied these unfortunate 'experiments' and have been amazed at how quickly these individuals can learn vocabulary. But while they can learn the meaning of words, it appears that beyond the age of 7 they are unable to learn grammar. It's as though the 'bones of the brain' have fused and are incapable of reprogramming past this age.

Now this set me to thinking. It occurred to me that we are just seven years away from losing one of the great skills that separate us from non-human animals, namely language. If all children were removed from their parents and raised in machine-operated crèches, along the lines of *The Matrix*, then all of our cherished language skills would disappear.

In the context of the film, Neo would have no language skills and would be unable to communicate, let alone save the human race. And if we had no language skills, our centuries of acquired knowledge would be placed outside our reach and civilisation as we know it would cease to exist. Add this to the list of impending apocalypses. But what am I saying? We have been here before, unable to interpret Egypt's written text until the chance discovery of the Rosetta stone. South American cultures used knots to record … what, we don't know; the skill has been lost. How many other languages and cultures are buried in ignorance, set to tease scholars of the past into the future? Bored in church one day while my wife arranged flowers, I read the introduction to the revised Bible of James the First and the justification for revising it. Words change their meaning over time. The quoted example was 'alleged', which then meant 'proven' and now means 'unproven'. Or if you prefer, the King's description of the new St Paul's Cathedral was 'awful' in that it filled the King with awe.

How many other cultures and languages and skill sets have been lost to the triple ravages of the Conquistadors, the Jesuits, and European diseases? (I will let you arrange them in order of deadliness.) So where does this leave us in relation to the 'milk bottle' and the development of the urban over the rural questions? How fast can humankind evolve to handle this changing situation? This is, I think, a trick question, because in reality I believe that there is a major distinction between adapting and evolving. Humankind adapts readily to changing circumstances, and it is this skill, perhaps above all others, that has enabled our species to thrive. But within the context of Volkswagen cars, we are yet to evolve to tolerate high levels of NOx gases and their subsequent impact on our health. Evolution is something to be measured in centuries or millennia. Yes, I know, African elephants are 'losing their tusks', but this is down to selective breeding (all elephants with big tusks are removed from the gene pool as fast as poachers can find them), whereas I would be really impressed if elephants learnt how to fly.

So there we have it: humankind adapts but hasn't evolved to suit our present lifestyle. Our genetic history is omnipresent within us and controls our apparent free will, our instincts, our need for herd leaders, and our fascination with shiny baubles. Is it my turn to be 'the One', Neo?

Can I give you the option of taking a pill which will see you rejected from the Matrix to be flushed away as detritus, never to return? Our minds, our free will, and our decision-making processes are controlled by the machine program we call our DNA, which packs us blindly into the depths of the herd, complying with the edicts of the pro forma. There is another way.

Step one would be to set a goal, an ambition for the human race. Let us consider the first imperative to be one of population control. With an ever increasing population, we outstrip our hinterland ever increasingly. Ignore the argument raised by the sustainability lobby; to all intents and purposes, think of them as dieticians. If you want to reduce demand on the earth's resources, then reduce the demand, pure and simple. If you want to lose weight, eat less food. The simplest way mathematically to reduce demand is to reduce the population. A more difficult way to reduce demand is to remove the demand. This requires the dismantling of the consumer society. In short, we need to adopt an entirely different way of approaching our lives and lifestyles. There is a short description of the life of a person looking at the centres of sensual pleasure. It goes something along the lines of anal, oral, sexual, oral, anal. Babies start by enjoying a good dump, before moving on to everything tested in the mouth. Teenagers seek out sexual pleasure before old age, where pleasure is taking in food and drink. Then, in very old age, the only pleasure left to us is enjoying a good dump.

But this can be observed differently in terms of aspirations and demand. Babies need food and comfort, toddlers need toys (often the box they come in is good enough), children need education and toys and birthday parties, teenagers need social media, sex, drugs, and rock 'n' roll, adults need jobs, partners, children, houses, cars, holidays, pension schemes, and all the other paraphernalia that substitute for life. And then we die, leaving nothing but ashes or worm food. And the trail of destruction that marked our passage through the world, we leave to our inheritors.

Following on from our inability to brush our teeth or our preference to take a diet pill over eating less food, why do we have car insurance websites? We will pursue savings over our purchase price, whereas if we were really interested in saving money for ourselves and indeed other road users, we would be better off by learning to drive properly. Everyone moans about the price of car fuel, and yet a significant proportion of the driving population treats the appearance of the single green light as though Lewis Hamilton has just seen five red ones go out. Are we not treating the symptom rather than the cause?

A neighbour of mine left his job with one of the Big Four banks and went freelance as a consultant in America. He's been there a number of years now and returns frequently to his family home. His consultancy firm works by providing tailored computer program scripts to his employers which are designed to 'iron out bumps in the road'. If I've understood properly what he was saying in a crowded bar one evening, they work to preserve the business of disgruntled customers. The sales/complaints department of his employer receives calls from unhappy people and are sanctioned through the scripted programme to respond in such a manner to maintain the business.

Values vary from business to business, but for the purpose of the explanation we will ascribe a cost of $100 to generating a new customer. This would be spread across advertising, selling, the administration of opening an account, and the like. So it follows that every time you lose a customer, you have also lost your $100 investment, and it will cost you an addition $100 to generate the replacement customer needed to maintain your company turnover. In principle then, allow your response team to offer freebies or discounts of up to, say, $90 to keep the customer from cancelling their contract with you. The skill is to know when to offer $10 and when to offer $90. It would also follow using the Pareto rule that 20 per cent of your effort secures 80 per cent of the result, and it will get more and more expensive to find new customers.

(There is a message here for the NHS, who spend millions of pounds defending against malpractice claims, when a significant proportion of complainers simply want an acknowledgement and an apology.)

We are taught that competition keeps our prices down. Without competition, monopolies establish themselves and drive our prices up. What would happen if there were only one supplier for a product nationally and no choice existed? Let us look at something like fixed landline telephones, which used to be the case under the General Post Office (GPO). Under this happy state of affairs, the GPO could generate whatever profits they could get away with because there was nothing to hold them in check. (For the purposes of this exercise, let us ignore any regulatory bodies that may have existed.)

The GPO had no need to advertise or market. It's not like they were selling wheels; we all knew what telephones were for. They had no need to maintain vast telesales operations, or concoct a multiplicity of deals and discount structures to entice new customers or bamboozle existing ones. I remember reading a telephone directory introduction in 1985 and being amazed at the claim that BT (formed in 1980 from the hiving off of telecommunications from the GPO) employed 250,000 workers in the UK. Nowadays the figure is nearer 100,000. Anyone unlucky enough to suffer a breakdown of service will know that calls are routinely routed to India, where the heavily accented English is a real barrier to customer satisfaction.

One can argue then that the introduction of competition has increased the costs on the BT business through the need for sales and marketing. It has increased the cost to the UK economy by the wholesale closure of home-based service centres, with customer service being exported thousands of miles offshore. Have we as customers seen any improvement in prices or choice since the end of the monopoly? The range of offers is too confusing to enable a straight comparison to be made, and most of us take a semi-blind stab at the answer. Now I understand the argument about monopolies, but what I don't understand is the answer that we have adopted. Yes, we now have a choice of provider; yes, we have a choice of prices; yes, we have a range of service levels. Each provider wants to provide a near equivalent choice of price and service, and each is locked in perpetual 'combat' with each other. If any one company were to establish a dominant position in the market by being the best for price and service, the other companies would seek to claw their way back into the market by upping their game. If the dominant company then used its size and position to undercut its competitors to secure even more of the business, then this would be deemed unfair and the industry regulators would step in to ensure that competition was maintained. But if the one company really was the best in every respect, why must some customers effectively be forced to use the poorer alternatives? We seem to have arrived at a position whereby the industry regulators will step in to punish excellence and reward incompetence. We are actually doomed to mediocrity by punishing success. But we have also forced the new combatants to devote resources to marketing, which has either suppressed profits or increased costs, the latter of which have been pushed through onto the customer. My question would be, have we customers benefitted in any way from the end of monopolies, or have we simply exchanged one set of price constraints for another?

Considering the telecommunications market, the industry saw virtually every city and town street dug up for another service provider to install cables as an alternative to BT's existing network. The cost of all this duplication was borne by us, the end user of these services. I accept that monopolies can abuse their position with price and service levels, but have we simply replaced one bloated provider with several others who have to

345

indulge in sales, marketing, and infrastructure construction to establish and maintain their market share?

Curiously, the very saviour of the poor customer from the abusive attentions of monopolies is government and government-appointed commissions. I say 'curiously' because these bodies are themselves monopolies. Who will protect us from them? So it seems we have arrived at a peculiar situation where monopolies are bad for us because they have the potential to abuse us, while at the same time they are good for us because they are looking after our interests.

Because monopolies can charge what they like, they can become bloated and indulgent, as I illustrated before with excessive secretarial employees in the water and electricity industries. So who polices the state monopolies and their bloated indulgences?

Straight Talking: Honest Politics

At the annual Labour Party conference, 1 October 2015, Jeremy Christ has given his first speech to the party faithful. These are presumably a different faithful from the ones who attended previous Labour Party conferences under Tony Blair. If a leopard changes its spots, is it still a leopard?

What struck me as immensely cynical was the lectern slogan behind which JC gave his speech.

Two lines and four words:

- Straight talking
- Honest politics

By definition these terms can best be described as mutually exclusive. You are either honest or a politician; you cannot be both. You either talk straight or you 'spin'. If you don't 'spin', you are not a politician. The speech was delivered both to the party conference and the media and was constructed to provide sound bites for the TV and quotes for the newspapers. What was presented was a good old socialist message to a good old socialist assembly.

Buried in the BBC's text page under Technology was a simple report about Samsung TVs, which appear to have been found using a 'cheat device' in a similar vein to VW cars. The TV recognised when it was under test and reduced its energy consumption to give an artificially low figure. An unnamed manufacturer's TV also recognised that it was under test and also automatically reduced its energy consumption. It would appear that cheating is perhaps more widespread than thought.

The reason that these manufacturers are soon to be in the Katanga is that they are foreign-owned companies importing goods into the EU and they have fallen foul of our strict compliance rules. Isn't it strange that a manufacturer of goods whose energy output can be physically measured can be taken to task by an organisation whose output cannot be physically measured, and that has strict compliance rules that are continually cheated by the members and employees who have a vested interest in ensuring that their own performance is enhanced and rewarded? When has an EU account ever been signed off as true and complete? Maybe the problem that VW, Samsung, and others have is that they have allowed standard methods of operating in the public realm to enter their corporate world. Let us have the investigations and inquests into these shady goings-on on both sides of the spectrum, public and private.

But of course this will never happen, because ultimately it would be like asking the judges to judge the judges, the invigilators to invigilate the invigilators, the government to govern the government. Whom would it serve to expose the shortfalls of the system, and how would you do it when every tool, scam, lie, cheat, spin, measure, and mechanism is part of the same system?

Very occasionally a report is made, perhaps once a decade, where an employee tasked with saving money in a business or public authority recommends that they should lose their own job as part of the money-saving exercise.

Ignoring the obvious ignorance of said employee's personal redundancy pot or contractual severance package or other salient circumstances, the fact remains that this sudden rush of apparent honesty merits comment in the media.

If you turn the equation around, the question would then be, 'How do you justify your continued employment within this organisation?' This in essence is the issue, because every employee every day proves to their peers that their employment is justified, whether it is or not. So every day, every employee sets out to 'cheat' to a greater or lesser extent to justify their continual employment. But does it stop there? Do women not cheat by masking their visage behind 'beauty products' (sorry, I forgot, men are using these products too) and disguising their scents to smell like Kylie, while men seek to smell like 'Homme' or something similar?

Pile on the latest fashions, drive the right cars, read the right newspapers, own the right smartphone, use the right apps, drink the right lager out of the correct bottle, turn the right colour orange, live in the right street, send your kids to the right school. … Wouldn't life be easier if we just grew our own peacock's tail? Some of you will be thinking, *What a miserable git you are. Life would be so dull if we all wore the same clothes and drove the same black cars and lived in identical tenement blocks.* (Whereas we all settle for this year's colours and live in caravans.)

Thumb Up or Thumb Down

England lost to Australia and, because of their earlier loss to Wales, have failed to qualify for the quarter-final stages of the Rugby World Cup. Brendan Rodgers has been sacked as manager of Liverpool FC. The rugby performance inquest will begin after the tournament has finished, but calls for the coach to go are already loud within the media.

This process is constantly repeated for all aspects of sport in the UK and beyond.

The basic tenet of sport is that it is a competition. There is a winner and there is a loser—fact. Even the draws add up to win or loss at the end of the season. Unlike in the gladiatorial contests of ancient Rome, nobody is supposed to die or even get seriously injured. Indeed, umpires are instructed to protect the 'combatants' from injury, whether it's scrums collapsing or boxers unable to protect themselves.

Why not bring death back to the playing fields? Let rugby teams play each other to the death. Football managers should be euthanised for failing to win the league or cup. Long-distance runners should run until the last person left alive. Sprinters could be pitched against mechanical greyhounds that kill the slowest runners. Javelins could be thrown backwards and forwards between opposing competitors, the longest throwers able to impale the weaker ones.

Perhaps then we would have something worthy of all the bollocks and bullshit that the media and the audience subscribe to. But, in fact, we would have no bollocks or bullshit, only winners and dead people. You either perform or die trying.

Have we replaced the spilling of blood and loss of life with virtual death? Assassination by written word, death by cutting remark, capital punishment by public opinion? What else could possibly drive the column inches, the televised sports punditry, the pub arguments, other than the desire to resolve the national obsession? England lost, and with these two little words, the debate continues. I cannot say 'start' because the debate began with every match and tournament leading up to the World Cup.

Every journalist will spout their learned opinion, scrutinise the failures, expose the unrecognised truths, and resolve what exactly? How to sell more newspapers: establish the journalist's own reputation outside the bloody pit of combat and regurgitate the same old bollocks over and over and over again. For 'rugby team', insert 'football', 'hockey', 'cricket', 'netball', 'synchronised swimming', 'conkers'. For 'manager', insert 'selectors', 'specialised coaches', 'facilities', 'governing body', 'government funding' …

The herd buys into it, like the nineteen-times divorced man walking down the aisle knowing that *this time* his bride-to-be is perfect, again. Like the destitute gambler placing the next bet knowing that *this time* he'll win big time, again. Like the drug addict injecting heroin for the last ever time, again. The media builds excitement and hope into a huge wave of possibility that crashes onto the shore before receding, to disappear back from whence it came, but only after obliterating all trace of its passing so that we, the consumer, will be ready and eager to engage in the next wave. What is it about the human mind that accepts this trivia as essential? Are these the programs that trap the human body within the matrix and make our anaesthetised minds compliant and resistant to reality? Is this the reason for Stonehenge, the minds and bodies hijacked by some unattainable dream to be moulded into someone else's 'column inches'?

The debate has already began as to Brendan's replacement at Liverpool, a scene oft-repeated throughout the high-pressure world of the Premier League. Names are already being touted for the next saviour, names of 'respected' figures within the industry who are often 'resting' between contracts, or should I say sackings. Because, as we all see, if we bother to look, each new 'bride' for the oft-divorced club is the one for the perfect marriage made in heaven.

What is the reason for the human mind's ability to indulge in these extracurricular activities that trap us in a loop of effort, mental or physical?

Is it simply herd instinct? Are we once again condemned by our genes to build pyramids on the rugby pitch? Is football the new religion?

Was religion the new Stonehenge? Was Stonehenge the new reason to avoid wandering over the horizon in search of something new? Was this because there was no new horizon—the land was full, occupied to capacity by neighbours, hemmed in by the Atlantic Ocean?

When our ancestors look back on the twentieth and twenty-first centuries, how will they record our presence? It's easy to think of our great libraries, engineering achievements, computer databases, the Google archive perhaps. We can look back into the nineteenth, eighteenth, seventeenth, sixteenth, and fifteenth centuries with a lessening degree of certainty, but the fifth century? Consider our hindsight from the ninetieth century, what will our contemporary knowledge be then?

Will the football league and Stonehenge be seen as being on a par with each other? I know the fixation seems a bit over the top, but bear with me. Each is a physical manifestation of our extracurricular activities, and each represents a considerable investment in labour and expense of resource. If you could generate an assessment of 'effort' for both, divided by the population at the time, which would represent the greater workload, which would command the greater slice of the pie? In both, the real benefit is concentrated in a relatively few hands. These would be the ones who don't get dirty, the architects, the lash experts, the prima donnas on their £250,000-a-week 'wages'. Yet for all our current fixation (OK, mine if you prefer), for centuries Stonehenge has been largely ignored and only resurfaced on the nation's radar in recent times. It's a bit like the blue plaques appearing on nondescript buildings declaring, 'I woz ere.' Learned individuals or history anoraks, take your pick, suddenly decide that our mundane lives will be enlightened with the knowledge that some dead person (usually) lived, died, suckled a tit, went to the loo, ate dinner, and otherwise got on with their own largely mundane lives within said pile of organised bricks. 'Er, they were real people living in a house similar to mine. Well, I never.'

This celebration of a life well lived calls into question the validity of all the other occupants, dead and living, who have engaged in similar activities in the history of the building, and indeed all others in the street and town. 'Your efforts were not worthy; no need to celebrate your contribution to our heritage.' Yet success is a lottery, a chance for recognition spurned or taken, or even ignored. We could just as easily have a plaque that reads, 'Here lived a princess.'

Are these blue plaques not all some form of distraction designed to make something out of nothing? History buffs justifying their existence by spending money, somebody else's, on such trivia?

The celebration of a long-dead celebrity.

How Much?

We have looked at the Ponzi scheme as a financial issue, the way that today's pensioners are supported by today's workers, and the system only works because enough new investors pay into the scheme to satisfy the current needs, with no provision made for the new investors' future requirements.

Let us look at the Ponzi scheme of effort.

Previously we considered my anorak approach to the cost of my lunch by the tonne, including the fact that potato crisps were £16,000 per tonne thirty-five years ago. Now if the object of the exercise was to eat a nutritious potato as a snack, I was paying over two hundred times more than the price that the farmer gets, the ex-farm price for the privilege. I had determined that the price of a microwave oven would quickly be amortised over the first bag of spuds if I had gone down that route. You will remember that my reasoning at the time for the extreme markup of price was that crisp buyers were paying for the marketing campaigns, which meant we were paying for ITV and its adverts.

I also looked at the income streams of the big Internet providers which translated into the value of millions of motor cars and indeed millions of hours of labour and millions of tonnes of materials.

Manufacturers have an expression, 'adding value', which they use to describe the processes that they perform and the ambition of making things more valuable and, in turn, profitable. Indeed, taking the potato as an example and by starting at the factory gate with a delivered value of, say, £90 per tonne, every process, including cleaning, peeling, slicing, cooking, packaging, selling, and delivering, has a cost, and these costs and more—a profit—have to be recovered from us, the customers. Without this simple equation holding true, all companies would effectively cease to function, as cost would outweigh the effort and you cannot expect to buy crisps at £90 per tonne. As you toil away at your workstation, you expect to receive a financial reward for your efforts, wages, which in turn enable you to pay for your lunch and other needs.

If you ate only potatoes, your 30-gram snack would have cost you a quarter of one pence in round figures before you cooked it. By extrapolating this argument with other foodstuffs, then the 100 grams of oats boiled and mashed in water would cost you 0.9p a serving (porridge). So who wants to eat potatoes and porridge every day? Yet many will happily snack on crisps or chips and the new trendy food offering of pots of porridge and pay hundreds of times its value for it. Is it an accident that 'convenience' is spelt with 'con' at the beginning?

What we have here is best described as a horse-before-the-cart scenario. Which came first, the chicken or the egg? If your food costs pennies, you wouldn't need to work for as many hours as you currently do. If you didn't pay the high prices for the food that you currently pay, then the manufacturers would cease to exist. They would go out of business, and their workers would be out of a job, not earning wages and being unable to pay high prices for food as well. Because when you pay two hundred times the value of something, you are also paying all the wages of all the people in the supply chain. One of the TV adverts of late has featured some celebrity chef quizzing the origins of a lump of meat that has been 'locally sourced within seventy miles' as a virtue. Let's do a little bit of maths here. Pi multiplied by the radius squared yields a figure of over 15,000 square miles.

Gives a whole new meaning to 'I'm just nipping out to the shops', doesn't it? And this is advertised as a virtue. While we are on the subject, there is a 'wee catcher' pad for those 'unexpected moments' which has a capacity of 'over twice as much as you may need'. I have one of those metric/imperial conversion charts for use when cooking, and I can't find a figure for 'as much as you may need', let alone twice that. Perhaps it's a measure used by alcoholics.

I remember a reply to 'Please, boss, can I have a pay rise?'

The boss's reply was to delete his employee's holidays, weekends, non-working hours, etc. Through judicious double counting, the boss proved that the employee didn't in fact do any work for him and therefore didn't deserve a pay rise.

So the question is, what do you work for, other than to pay the wages of everybody else, who in turn work to pay your wages? In one sense the whole purpose of education is to learn one of two things:

- how to be compliant and work for others
- how to control and exploit other workers.

We appear to have forgotten the question again (a seemingly common occurrence in human beings): why do we work? When someone was very ill and still came into work, I used to hear the comment levelled at them, 'You work to live, not live to work.' It seems we work to earn money to make our lives easier by paying other people who are working to make our lives easier, while they themselves are making their own lives easier by taking a cut of our earnings.

My mind strays back to the ant colony and the mindless activities of its population. Yes, it works. It works very well, but no thought is involved.

Homo non-sapiens?

What is it MeEnroe used to Say?

Wednesday 7 October 2015 saw the 'crowning' of this year's *Bake Off* competition winner Nadiya. Today saw a double-page advert in at least one tabloid newspaper congratulating Nadiya and suggesting that she have a day off from baking and buy her cakes from their supermarket instead.

Now forgive me if I'm wrong, but aren't the basic premises of the programme to encourage people to be DIM (do-it-meselfers), to take delight in personal achievement, to produce non-mass-produced individual food products that taste infinitely better than the dross served up by supermarkets and their mass producers of food? Imagine presenting unboxed supermarket food to the judges and the reactions that you would receive. 'WTF? Are you taking the piss?' (Notwithstanding the fact that the entire premise of the programme is taking the piss.) Whoever the brainstorming marketer behind this multi-thousand-pound advertising idea was deserves to be taken out at dawn, tied to a post, and shop-bought-sconed to death.

Let's have a Slimmer of the Year competition where the winner is congratulated in an advert by a fast-food restaurant chain that offers a year's free dining at an all-you-can-eat fast-food restaurant.

Let's have a gun manufacturer award this year's Nobel Prize for Peace with an AK-47 and twenty thousand rounds of ammunition.

Or we could just take the view that the supermarket has skipped out the unboxing and judge presentation stages and has simply gone to the 'taking the piss' stage.

Nihilism

We work hard to pay our taxes to support all the non-jobs and all the non-workers. We work hard to pay the wages of all other workers, who work hard to pay our wages in turn. Where is the rhyme or reason to this circumstance?

Nihilism: a rejection of government, religions, organisations, and belief in the power of the individual instead.

One individual has no power, no power to steer the herd, no power to resist the flow, no power to make a difference. This is an oxymoron, since power resides in the masses. To have influence and make a difference, an individual needs to recruit supporters, who become an organisation.

I reject the power of religion and government and organisation because it is corrupt and is corrupting. Human nature is both corrupt and corrupting. We cannot see beyond the peacock's tail, so how can we make cognitive decisions based on whims? How can I change human nature? Without changing human nature, Jeremy Christ is doomed to failure. However pure the message and intent, it cannot survive exposure to the light. I am the new messiah, and my message will serve others long before I am resurrected and long after. I asked before how we ended up with Sepp Blatter running world football. As events have transpired, it seems that he won't be for much longer.

I could just as easily ask how we ended up with the pope, or any other 'leaders' of the Christian world, ending up running the Catholic Church. But who is actually leading whom? If the pope behaved in a non-papal way, how long would he survive before being assisted in his ambitions to join his Master? Pass the fruits of corruption down the line so that all within the organisation feel the benefit and accept the ruling hand, while simultaneously controlling the ruling hand by that same acceptance. It is a closed loop that can be applied just as easily as the governance of the NHS, or education, or the Labour Party, or Save the Children, or cancer research ...

The one thing that all these organisations have in common is the ability to do good. There is a scene in *Braveheart* where the king issues the order to fire arrows into the Scottish rebels even though they are in hand-to-hand combat with his own mercenaries. When warned that it will kill his own troops, he replies with the comment that it will kill theirs as well. We could line all Basques up against a wall and shoot them as a cure to the Basque Separatists problem. We could ban the passage of all Muslims into Europe or America. We could exterminate all badgers in a given area to reduce the incidence of bovine tuberculosis, whether they carry the disease or not. At last I have touched on a subject that has sparked a debate with impassioned arguments raging on both sides. We could feed all the birds in Britain with supplementary feed grown in fields around the world because it is of benefit to our birds. We could collect all the charitable donations in the country and allocate them to cancer research because it will accelerate the development of cures for cancer. Or we could ban the collection of all donations for all cancer research and only allow research in non-cancer fields. We could force every child in the UK to study at school for fourteen years, or we could force them to study a minimum of ten. Shock and horror, but there was a time when that was all that the pro forma demanded. Were they so wrong then? And if they were wrong then, does that mean they had the monopoly on being wrong? Are we wrong now? How then do we do the calculation of benefit over cost?

The chief executive of a hospital trust/charity can earn some obscene amount of money while pleading for more resources so that her organisation can save more lives. Of course the same chief executive could save lives by forgoing a chunk of her payment package; after all, 'just £3 a month can transform lives'. But the chief executive has already done the calculation and has concluded, *I'm so worth it*. And that, by and large, has ended the debate.

How do we measure success? By the size of our pay packet, the number of our children, the number of certificates of higher education we have, the age at which we die, the number of years we live in retirement, the number of gold medals we've won at the Olympics, the amount of money we give to charity, the number of Earth-like planets found in other galaxies?

But straightaway this question is flawed, because of the term 'measure'. The instant we measure something, we corrupt it and lie about the answer.

Look around you. We talk about this being the technological age, the computer age, the age of science. Should it be the laser age, a reflection of the tools we use for cutting, or the plasma age? Did the ancient Brits think they were in the Stone Age, or did they content themselves with being alive and fed and warm and safe?

We should perhaps call the current period the age of measurement, of quality control, of questionnaires, of league tables, of effectiveness of governance, of mpg, of equality, of freedom. We have gone from 'I've earned enough to live on for this week' to 'I can afford to go on holiday this year', or 'I can afford to change my car in three years' time', or 'I can pay my mortgage off in twenty-five or thirty years', or 'I can afford to retire in fifty years' time'—all this in a few centuries.

But while we call it the age of measurement, we are by my definition living in the age of lies. If you accept that a chimpanzee is not an imperfect human but is in fact a perfect chimpanzee, then how would you measure the chimpanzee? Its tool use is primitive, it can't hold a conversation with us, it survives unaided in the wild, it needs a shave, and its table manners at teatime are atrocious. We place it in the 'league table' as our closest living relative. We study it to try to understand our own evolutionary origins. We train/educate chimpanzees in research facilities, we observe them in their natural environment, we kill them for bush meat, and we even try to establish their rights in court. But the only measure required for the chimpanzee is that it is a 'perfect chimpanzee', nothing more, nothing less.

Appreciate them for what they are.

How to Kill an Elephant, if it ever sees the light of day, may well end up with 'The Number-One Bestseller' emblazoned on its cover as the publisher attempts to turn a profit from their investment by appending a 'peacock's tail' to the 'wodge' of tightly packed printed paper. (There goes my publishing deal.)

(I self-published.)

We are measured from before the moment we are born to beyond our moment of death, some of the time to ensure our compliance and some of the time to ensure our non-compliance. We need to comply to be educated, but we need to non-comply to be brilliant and original. Where would Albert Einstein or Stephen Hawking have been if they had complied throughout their lives? How perverse that their non-compliance becomes the new standard for compliance for future generations of students. Compliance is a block to development, a handbrake on advancement.

Think back to the earlier point about the pope having to comply in a papal manner to remain as pope in the eyes and hearts of the faithful.

Pablo Picasso, Salvador Dalí, John Constable, Henry Moore, Auguste Rodin, they all had to break the mould, fail to comply, stretch the boundaries, refuse to be measured, refuse to accept the lie. So where does this leave any of the popes, great liars? The papacy is the greatest lie in the history of Catholicism. Yet Catholicism has the longest, continuous ruling history of any of the controlling regimes.

What does this tell us about the human condition, that belief in fairy tales overtakes all other considerations? In fish terms, Catholics are sardines condemned to the bait ball.

The very essence of measurement is to document and control, but at the very instant of measure, the essence is to lie and regiment order into that which cannot be ordered. How can you take a thousand kids and turn them into a league table, into a series of grades, of opinions, of labels, of prejudices, of needs to perform to satisfy the pay grades of their captors?

Captors; are they not trapped within an orbit, spinning against the restraint of measurement, moulded into the shape of that which is desired? But the desire is not to create better children; the desire is to create a satisfactory compliance with the measurement dictated by government and corrupted by the entire apparatus we call education. How can you take a thousand patients and turn them into a league table, a series of grades, of opinions, of labels, of prejudices, of needs to perform to satisfy the pay grades of their captors? Captors; are they not trapped within an orbit, spinning against the restraint of measurement, moulded into the shape of that which is desired? But the desire is not to effect cures; the desire is to create a satisfactory compliance with the measurement dictated by government and corrupted by the entire apparatus we call the health service.

Measurement allows us to balance and manage resources to solve the problem, to achieve the greatest return for the effort, to effect the greatest benefit for the customer. What's the word I'm looking for? Oh yes, bollocks.

Measurement allows us to balance resources to manage the problem, to allude to the greatest return for the effort while rewarding the process rather than the result. However poor the result for the pupil or the patient, the process will be satisfied, and all who sail within the good ship of measurement will be rewarded, because it is designed precisely to corrupt the truth with self-congratulatory, self-serving reinforcement. All teachers bask in the glow of another record-breaking year of results, along with another record-breaking year of resources spent. All pupils and parents bask in the glow of their record-breaking achievements, along with yet another record-breaking year of failing to address the needs of employers. Blue tits synchronise their breeding activities to coincide with the emergence of caterpillars, which in turn react to the emergence of spring leaves on deciduous trees. Sparrowhawks in turn synchronise their breeding activity to coincide with the emergence of the single-brooded blue tit fledglings as an easily obtained source of food for their own young. Our schools 'raise our young' to fledgling stage before abandoning them to the fate of 'sparrowhawk' employers. We learn that we have among the longest 'childhoods' of all mammals and use the stability of the family unit to provide a safe learning environment to equip them with the tools and skills necessary to survive when independent, an investment that is currently corrupted into fourteen years of 'education' before abandonment into the wider world.

Is measurement the root cause of our modern-day malaise? We take a series of answers to questionnaires, convert these answers into a score, and then attribute these scores to a table of 'predictive' responses that fit our pro forma. Give a scale of answers, 'strongly for', 'mildly for', 'neutral', 'mildly against', 'strongly against'. Score each in order with a numerical value from 1 to 5. Add the cumulative score for the questionnaire and rank it against another scale. The lower the score, the more supportive; the higher the score, the more against. Medium scores place you in the middle ground.

Collate this data from a population and extrapolate the score to determine policy or decision making or suitability. But what does this information mean, if anything?

Let us take a simple example: ask twenty people in an indoor environment to indicate whether they are too hot, just right, or too cold.

Use this information to set the room thermostat.

Ask the same twenty people if they are wearing heavy, medium, or light clothing.

Ask the same people if they are engaged in strenuous physical activity, are engaged in moderate physical activity, or are resting.

Ask the same twenty people if they have entered the room from a cold store, a room similar to the one they're currently in, or a sauna.

Ask the same twenty people if they are male or female, young or old, healthy or unhealthy.

All of these supplementary questions have an impact on the original question, and all alter the requirements for the room thermostat.

In the real world, some of the people will be sitting next to open windows, others will be wearing coats, and some might even be suffering from hot flushes.

Ask why the room thermostat is being set. Is the purpose of the questionnaire to set an ambient temperature to provide the maximum comfort for the majority of the people? Is the purpose of the questionnaire to determine if you can reduce the room temperature to save energy costs?

What I am trying to illustrate is the utter futility of asking the question in the first place, as the answer is to all intents and purposes meaningless.

What was the purpose of the questionnaire in the first place? All participants would be reasonable in their assumption that the room thermostat had been set at the correct level. The management would also be reasonable in their assumption that the room thermostat had been set at the right level. The questionnaire has created a false 'truth' that becomes very difficult to address.

Wrong Way Round

If you buy a car with a high mpg rating, then you are doing your bit to reduce greenhouse gas emissions and are helping to save the planet. In reality you are simply slowing the rate at which greenhouse gases are being generated. If you want to reverse the process, stop travelling by car and use less energy. Invest in carbon-sequestration technologies and remove carbon dioxide from the atmosphere.

But that doesn't mean you should travel by bus. You should stop travelling by motorised transport. 'But if I don't travel, how do I go to work and earn a living? How do I obtain the goods and produce I need to survive?'

This is where the great lie takes over.

Listen to government and the self-appointed, self-serving experts and learn that you should:

- Use low-energy light bulbs.
- Use low-carbon forms of travel.
- Calculate the carbon footprint of your goods and services.
- Buy sustainable wood products sourced from sustainable forests.
- Recycle waste, be it food or material product.
- Buy locally sourced product.
- Turn your TV, computer, and other electrical appliances off of standby.
- Invest in non-carbon-producing sources of energy such as wind turbines, solar panels, and heat pumps.
- Invest in modern heating systems to reduce your energy needs, and thoroughly insulate your dwellings and workplaces.

(I think it was Charles Darwin, but I might be wrong, who landed in Tierra del Fuego at the tip of South America and stumbled across the natives who were wrestling in the snow. They were scantily clad with flesh exposed, without appearing to suffer any ill effects. We can't go back and ask them, because in a handful of years their population disappeared under the onslaught of European diseases that they had no defence against. If humans can adapt to these conditions, why do we have the need for heating anyway? As one of my son's lecturers once said to him, 'There is no such thing as inclement weather, only inappropriate clothing.')

If you engage in all of these initiatives, will you save the planet?

If you engage in all of these activities, will you think that you are saving the planet?

If you fail to save the planet, whom will you blame: the government, other British people, Americans, Australians, the Chinese, the Indians?

So far this year an estimated six hundred thousand migrants/refugees, take your pick, have entered Europe to secure a better life. Can we imagine that the Afghans, Eritreans, Nigerians, Sudanese, Libyans, Vietnamese, and the like are not going to engage in the high-energy European lifestyle and that their newfound carbon footprint will not exceed their old one by many times?

All of these 'planet saving' enterprises are illusions designed to placate the 'needs' of the people, while simultaneously placating the 'needs' of business and the needs of government, and singularly ignoring the needs of the planet. Before you jump to any conclusions about business overriding the 'needs' of the people, consider that what the people need is the ability to earn a crust by working for businesses.

Tick the right boxes of the questionnaire and leave everybody happy with the compromise of higher energy bills or fortnightly dustbin collections or a landscape dotted with wind turbines. Does anyone genuinely believe that our publicised attempts to limit the rise in world temperature are going to succeed?

Will future generations blame this generation for the environmental catastrophe that we will land them with, or will they simply continue on the same path of mindless exploitation of the earth's resources? Did nineteenth-century Americans give a damn about the passenger pigeon? Did the seafarers of the day care about the great auk? What have we learnt, our educated, intelligent generation, that gives us any hope for a brighter future?

Name one adopted initiative that will not merely 'make a difference' but will actually solve the problem.

The scale of the problem makes killing an elephant seem like child's play. We've moved way beyond stopping *Apollo 11* ten seconds after lift-off, by hand, on our own.

The human mind has a wonderful capacity to compartmentalise issues into pigeonholes of 'problems I need to resolve', 'problems being solved by others', and 'problems beyond my scope of influence that I can ignore'. In reality, I have just lied. The human mind has a shit capacity. It cannot cope with or consider the enormity of the problem, let alone conceive of the solution. It reverts to its comfort zone and adopts a pro forma, a recipe, a standard format for dealing with unrelated issues that will probably do the same job. After all, it's not as if anyone's life depends on it, is it?

Do you remember the actor Charley Boorman, best known for teaming up with his mate Ewan McGregor and travelling on motorbikes from London to South Africa (as documented in *Long Way Down*), or across Eurasia and America (in *Long Way Round*)? Boorman's father cast him as an abducted white boy captured by an indigenous South American tribe in a film (*The Emerald Forest*), and transported real-life South American Indians to his jungle film set in Mexico for the purpose. In a magazine article, the elder Boorman commented on his son's dyslexia and mentioned how both his son and his on-screen abductors looked at things. 'No,' he said, 'they really looked at things, as though their inability to read was replaced with an intensity of thought and need to understand beyond the shorthand of the written word.'

Now, I've said it, we use the written word to convert all we do into recipes, the root cause of all pro formas, the end of thought, the cessation of personal endeavour, the shortcut to oblivion, the codification of original thought into a reference library where scholars can learn the lessons while simultaneously ignoring the question. Our human bait ball cannot see beyond the boundaries of our existence, cannot tear itself away from the queen's chemical controls, cannot see beyond the peacock's tail, all while mocking the ostrich's alleged defence mechanism and simultaneously perfecting our own version of it.

But if this truly is the age of measurement, then we have a problem with the nomenclature, because 'stone', 'bronze', and 'iron' are used to describe the cutting-edge technology of the respective eras, whereas I would suggest that the age of measurement describes the antithesis of cutting in that we have settled on a very blunt instrument indeed.

Basic tenets:

- Human activity on this planet is unsustainable.
- Nothing we are presently doing will alter this fact.
- Nothing we are presently planning will alter this fact.
- All of the purported solutions are illusions designed to give plausibility to the masses to enable them to support their leaders.
- Our leaders, our Great Brains, know only how to preserve their positions, not how to solve our problems.
- Our Great Brains use our problems to secure their positions and maintain the status quo.
- None of our Great Brains know how to kill an elephant.
- None of our Great Brains even know why they might need to.
- All of our Great Brains are custodians of their own positions before those of the masses, but to be fair, they know better than we do.

Our only 'hope' lies outside our control, perhaps in the form of a global pandemic that scythes through the population, an alien invasion that 'harvests' humankind as we did the American bison, an asteroid impact, a volcanic event to rival the Deccan Traps, perhaps the supervolcano beneath Yellowstone National Park, or maybe the one under southern Italy.

October 2015 saw the much vaunted launch of the 5p 'tax' on single-use plastic bags designed to reduce the amount of plastic ending up in landfill and spread across the countryside, distributed by the wind. Scotland has just reached its first anniversary of this tax and has declared a reduction of 85 per cent in issued bags. While wandering around my Asdalavista store doing my weekly shop, I came across a notice relating to single-use plastic bags in the grocery section:

Customer Information – Bags for unwrapped food

These bags are free when used for unpacked or loose food.

The single-use carrier bag charges (England) order 2015 applies to plastic bags used for packaged or wrapped products.

—The Department for Environment,
Food and Rural Affairs

No charge is made for the bagging of 'loose products'; charges will only be made for bags if bagged items are double bagged. So if you buy a single courgette, you can still have a bag, whereas if you buy a three-pack of courgettes already sealed in a plastic bag, you will be charged for bagging said item. My eyes strayed along the shelving and couldn't fail to notice that around 80 per cent of the items are pre-packed in sealed plastic bags. Not only that, but also the plastic is of a different grade and the bags need to be ripped or cut open, rendering them useless for secondary use.

Reading the report from Scotland, one would think it reasonable to assume that the blight of the single-use plastic bag has been solved and that our Great Brains have earned their keep and saved the planet.

Prior to the trip to the supermarket, my wife had purchased a birthday card and a sheet of wrapping paper, which were duly despatched to her in a heavy-duty plastic bag around two foot square with the words, 'I'll give you one of these because we don't have to charge for them.' This is because the heavy-duty bags are designated as multiple use and are therefore free of the 'tax'. And also because the shop assistant was a moron.

But we have also learnt something from the Scottish example: the impact of a 5p charge on people's behaviour. With this simple act, the vast majority of the herd has changed its behaviour and 'supports' the government's ambitions to 'save the planet'. But the herd could have altered its behaviour without the 5p charge, so the only logical conclusion is that the herd isn't prepared to shell out 5p when it doesn't have to. In our household, single-use plastic bags are reused as refuse bags to collect waste in between trips to the dustbin. We may well get to the stage where we end up buying plastic bags to use for this purpose, but these won't be counted in any government statistics.

At the same time, Jamie Oliver is attempting to persuade a Great Brains committee to introduce a tax on sugar in fizzy drinks. Somehow

I can't see a 5p tax imposition on sugary drinks having any influence on the herd. Yet curiously, while the plastic bags have no direct impact on the quality of human life (ignoring possible suggestions of chemicals adversely affecting human health), sugar is rapidly being singled out as a major inhibitor of a long life.

We arrive at the perverse situation where we won't waste 5p on something that we think should be free while at the same time we will ignore a 5p or similar charge on something that is positively harming us (should that be 'negatively')?

I am old enough to remember the introduction of free plastic bags at the supermarket.

Prior to that, we would place a large plastic basket, normally used for laundry, in the boot of the car and empty the smaller items from the shopping trolley into it, with the larger items left loose. This basket was then carried into the house and the job was done. Free bags were withdrawn after a while and then reintroduced on account of customer demand. My point here is that what began as a perk, an advantage, rapidly became a deep-seated right that could not be removed for fear of the customer's reaction. Tales of supermarkets running out of trolleys and wire baskets have cropped up in the media as shoppers have failed to deal with this imposed 'impossible' situation they have been subjected to.

And yet this 'deep-seated right' to free plastic bags can be overcome with a 5p cost threshold. Would it have worked at 3p or 1p, I wonder? This example speaks absolute volumes about the human condition, if you would but give it a minute's thought.

This then leads back to Jamie Oliver and his 'sugar tax'. We need to ask the question about the necessary threshold to achieve an 85 per cent reduction in sugary drink consumption.

Our Great Brains have determined that this would be unworkable, saying that instead the industry should be given time to address the issues on behalf of the consumers. That's a bit like asking publicans to organise a solution to drunkenness, or gun manufacturers to find a solution to gun crime.

ASAP

As soon as possible. For many the interpretation of this expression is that what ever the matter is should be treated urgently. It can also mean, 'Do it as soon as you've done everything else first.'

The Food and Drink Association's slogan, 'Sustained growth', appeared on a TV ad during October 2015. I believe the expected interpretation of this phrase is that the industry will act in the interests of the planet in providing responsible policies for growth. It could just as easily be read as 'sustaining growth', which the Australian government defines as something that 'can go on forever'. I would suggest that this slogan means that the FDA intend to grow forever, whatever the consequences.

I took delivery of my new computer today.

In fact it was scheduled for delivery yesterday, and I received various emails and texts to advise me and to enable me to track its transit to my door. After a no-show, I was not surprised to see, 'You had requested the delivery to be rescheduled until today', because, and I quote, 'The driver was unable to find your delivery address.' To clarify: the driver couldn't find me, which was then translated into my instruction to reschedule for the next day, all without a single reference to me. I must be telepathic, while the courier company in fact is telepathetic. Statistically, I have now gone from a disappointed waited-all-day-in-vain client to another perfectly satisfied customer at the push of a computer key in the customer satisfaction computer menu (but not my computer, because it arrived a day late). Whereas in truth I never was the former; statistically I have only ever been the latter. This sort of begs the question: if you are simply going to lie about your delivery performance, why bother to measure it in the first place? Think of the time and personnel wages you could save if you shut your quality control section and removed drop-down menus from your processes.

To be fair, the courier's website does say that if I miss the scheduled delivery, they will attempt delivery again the next day. I just assumed that this meant I had to miss the delivery, not that they had to miss me.

I can also, through the use of the technology they have provided for me, track their inability to deliver parcels on time, secure in the knowledge that they will still provide perfect customer service and attract more customers.

I wonder where they learn their skills from.

If measurement is the primary reinforcement of our lies, how do we measure without corruption? What can we trust? How can we pin the tail on the donkey? At the very beginning I stated that I was prompted to write *How to Kill an Elephant* (have a monumental moan) by an apparently innocent questionnaire from a neighbour researching for a master's degree. I couldn't answer the questions because I knew that the answers would steer her in a particular direction, towards a conclusion that was neither merited nor justified. I 'knew' that this was not a reasonable result, and therefore I wouldn't partake in the 'game' asked of me. OK, I'm an awkward git, I admit it.

But we are all subjected, every minute of every day of every year, to the 'game'.

Our lives are measured by the clock, 86,400 seconds a day, 2.2 billion of them in our 70-year lifespan. Wake at 7.20, catch the 8.15 bus, clock on at 8.30, have coffee at 11.00, take lunch at 1.00 …

Drive at thirty miles per hour (you never know), sit at your 21° desk, tuck in to your thousand-calorie lunch, run five kilometres on the treadmill, buy a birthday card …

Mark a cross in a ballot box, listen to music (feel the beat), mow the lawn, bury a loved one, buy a new car …

There was a time when all you had to do was stick a rock in the ground to mark the longest and shortest day.

What happens if you don't measure things?

What happens if you don't lie?

How can you tell if you are lying or not?

Is this where we need some form of ancient Greek riddle to solve the problem and save our lives?

If you measure, you lie; if you don't measure, then you can still lie, but there is no way to prove it. To prove it, you need to measure; therefore, you lie.

Measure = Lie

Lie = Measure

The bigger the calculation, the bigger the lie.

The bigger the organisation, the bigger the lie.

Big organisations need big lies.

The bigger the lie, the truer it becomes.

The truer it becomes, the bigger the lie.

Each feeds on the other, reinforcing, self-serving, creating its own momentum, taking command of all that it touches.

And then at some point, some moment in time, the lie fails and the truth collapses.

(The apocalypse.)

England can win the 2015 Rugby World Cup.

(The final whistle.)

Tune in for the next one. We'll be peddling our living again.

A newspaper report claims that over half of teachers have thought about quitting in the last six months.

(You've still got double maths on Monday morning.)

The lie becomes the product, the reason, the justification, the purpose, and it becomes the industry that traps and envelops us. The adverts become the truth, the reason, the justification.

Compare your lives with those of others, buy the same clothes, frequent the same locations, drive the same cars, smell the same ...

You too can live the life of your celebrity heroes.

What is wrong with Goldilocks?

She likes her porridge just so ...

She likes her bed just so ...

She used her organoleptic skills, not for her a thermometer or a bed ride meter.

Once upon a time (but not that long ago), Goldilocks was lost in the forest (a really wild forest outside the range of her phone coverage area and hence no satnav) when she stumbled, hungry and tired, upon a cabin. Having knocked and received no answer, she opened the door and instantly smelt food. Before her on a table were three bowls of what appeared to be a mixture of crushed oats, barley, and rye with a liberal sprinkling of previously dried strawberries and raspberries. If she had but thought to read the label, she'd know that these were also fortified with seven vitamins and iron.

She tried the first bowl, which to her keen senses had been heated for two and a half minutes in a 750-watt microwave and was just the right temperature, but then she realised that the second bowl contained honey and almonds rather than red fruit. This too was just the right temperature. She was about to take a second spoonful when she noticed that the third bowl seemed to have some blackcurrants mixed in with the otherwise red fruit. Well, you'd be a fool not to taste this one as well, wouldn't you?

With her hunger sated, she felt an urge to lie down and rest. Her eyes strayed to the three beds. She alighted on the first and was instantly grabbed by the luxury of the memory foam mattress that seemed to support and mould her perfectly. Not surprisingly, she drifted off to sleep.

Somehow, I think my modern tale has failed to attract the attention of preschool kids and will fail to be a bestseller. True, there is some repetition, but my story seems to be lacking in many child educational requirements. But I have to confess that I hadn't really intended this to grab the attention of children.

Children are born wide-eyed and vacant, but they outdo a Dyson at sucking up knowledge and experience. But we cannot allow this natural learning to develop; we must channel the effort into the conduits of the pro forma. And we have to prove the success of our channelling by measuring, because if it isn't measured, it isn't real. We reward compliance with gold stars, or smiley faces, or verbal praise, or better schools, or higher grades, or better universities, or degrees. But along the way, we warp and misdirect these blank databases that are young people's minds and corrupt them with irrelevances. This interference is both accidental and deliberate. When I was a child growing up in the 1960s, our class football teams all wanted to be Bobby Moore or Geoff Hurst or Gordon Banks. (At least they won a World Cup.) But none of us ever gave a thought to smelling like them. None of the girls wanted to marry them or meet up for casual sex in the hope that a marriage proposal would be on the cards. None of us even knew that WAGs existed. None of us even gave a thought that these footballers might be married or have partners. None of us wanted their haircuts, although to be fair, the standard of barbering was universally uniform in that the haircut fitted your head and had probably emerged from the National Service Training Schools. (I don't mean pudding basins. You know what I mean.)

What then is to become of our channelled, warped, and misdirected minds when we reach maturity? The damage is done, the 'brain bones' fused and set, hardwired to the program, unable to differentiate between the real and unreal, the relevant and the irrelevant, what matters and the dross. We have all matured to achieve our designated position in the ant colony, to obey the chemical signals, to protect the queen, the Great Brain. We are all engaged in building our own Stonehenge, our own pyramid, our own Great Wall of China, and none of us realise it or know why we are doing it. It's no wonder we can't understand the motivation and dedication in building the Great Pyramid of Cheops (although I think I've done a pretty good job with the Neolithic World Cup in Wiltshire) and get bogged down with the princess.

A friend of mine has a drink problem. Well, no, she's an alcoholic; that's the problem. And when I asked her why she drank too much, she blamed it on her inability to ignore the peer pressure and marketing pressure from adverts while growing up. My first thought was, *Bollocks*. My second thought was, *As long as you blame others, you will remain out of control*. With my third thought, I began to accept that the blame could be lodged elsewhere.

It's very difficult to avoid the tsunami of marketing and peer pressure for fast-food restaurants and convenience foods and wrinkle cream—sorry, serum. Just as a thought, how much a tonne is a wrinkle? (The wrinkles weight, or the cost of treating one? I'll let you decide.) It takes a stubborn and awkward frame of mind to question the 'chemical controls' pervading the herd and to swim in a different direction. (I've just mixed the ant colony, the herd, and the bait ball in one sentence. Am I brilliant or what?)

A wrinkle has no weight, so if the answer was expressed in light years, would the wrinkle reach beyond all of space?

Back to measurement.

My parcel was delayed for a day at my request (apparently), and I became another 100 per cent satisfied customer. If you are going to lie about the measurement, why bother to measure in the first place?

Our education system is based on measurement, measurement that is a lie, so why bother to measure education in the first place? Why not just lie?

Our National Health Service is based on measurement, measurement that is a lie, so why bother to measure in the first place? Why not just lie?

Our Great Brain is based on measurement, measurement that is a lie, so why bother to measure in the first place? Why not just lie?

Volkswagen's NOx emissions are based on measurement, measurement that is a lie, so why bother to measure in the first place? Why not just lie?

Global warming is based on measurement.

Sustainable forests are based on measurement.

The wholesomeness of processed foods is based on measurement.

The sporting prowess of our national teams is based on measurement.

The comfort level of your office environment is based on measurement.

Your attractiveness to the opposite sex is based on measurement, so paint your face, dress like Greta Garbo, polyfill your wrinkles, smell like a French cutting, surgically enhance your tits and lips, and decorate your skin with artwork—and that's just the men.

I am reminded of a fairly recent report of a man who filed for divorce soon after his marriage because he'd never seen his spouse without her slap on and couldn't stand the sight of the real her. (It was on the Internet, so it must be true.)

But in the process of appealing to the opposite sex, or more correctly your hoped for partner, take comfort in the knowledge that the pursuit of your lie has funded the lifestyles of a multitude of others, from the cosmetic manufacturers, to the dress designers, to the plastic surgeons, to the Polyfilla makers, and to the tattoo artists, to name a few. But why stop there? You've also funded animal research testing laboratories, high street chemists, clothes design university courses, the Student Loans Company, student landlords, ink manufacturers, and tattoo gun makers, and milked billions of bacteria for Botox.

And you've funded local government testing and licencing authorities, trading standards officers, and customs officers searching for contraband perfumes or counterfeit products, council tax bills on commercial and industrial premises, and criminals seeking to sell counterfeit and non-duty-paid goods.

On top of that, you've aided corrective plastic surgeons, tattoo laser removal specialists, laser manufacturers, office, factory, and warehouse builders, local authority planning officers and architects, noise consultants, and traffic consultants, and provided fodder for lifestyle features in magazines and television programmes.

And you've paid for all the energy consumed in all these processes and the extraction, refining, and distribution activities to get the product to where it's needed, and all the motor vehicles and their manufacturers, and all their workers' travelling requirements, not to mention the costs of raw material producers and ocean-going vessels to distribute these base components.

In addition, you've paid for all the research scientists, international summits, and government initiatives to control the generation of greenhouse gases and ultimately make our world *sustainable*.

And you've done all this for a lie.

We cannot all be princesses.

All for the ability to lie to the opposite sex. How much simpler life would be if you wore a paper bag over your head (not a plastic bag, as this costs 5p) and a shapeless smock? At least you don't have to grow a peacock's tail.

I am reminded of a recent report, this one from France, whereby a 'tall, handsome, physically attractive man' enticed willing women to his apartment for consensual sex while they wore a blindfold. One woman removed her blindfold and was so revolted by the bald, fat, ugly 'Adonis' that she took him to court for raping her. It emerged in court that the man had had many sexual encounters with satisfied women who enjoyed sex with their imaginary partner, although by the time the case had reached court, the power of the Internet had turned the case into a 'class action' of 'raped' women.

1) Q: What is the temperature?

   A: The thermometer gives a reading of 21°, and within the
   tolerance and calibration of the device is the truth.' This
   is a valid response to the question.

2) Q: Are you warm enough?
   A: Yes I am; or No I'm not.
   These are valid responses to the question.

3) Q: Are you all warm enough?
   A: Yes we are; and No we're not.
   These are valid responses to the question, but now we have
   a problem because we have recorded yes and no.

4) Q: What temperature do I need to set the thermostat to?
   Now the problem is unsolvable because of the answer to
   the previous question.
   Whatever the answer is, it will not be valid for the entire
   population of the room.

5) Q: What is the temperature?
   A: It's too hot, or It's too cold, or It's just right.
   These are all valid answers, but this is the same question as
   question 1, which no longer has a truthful answer because
   we cannot measure the movement of mercury in a glass
   tube to arrive at an answer. We cannot calibrate people.
   Yet that is precisely what we do, from before birth to after
   death.

If we cannot agree on what temperature the room is, what can we
agree on?

How do we preserve the bird populations of the UK and at the same
time preserve the bird populations of the rest of the world?

How do we preserve the woodland bird species of the UK?

How do we preserve the arable farmland birds of the UK?

How do we preserve the urban bird populations of the UK?

How do we preserve the dense-hedgerow-nesting birds of the UK?

How do we preserve the production yields of arable crops in the UK?

How do we satisfy the ever increasing demand for food of the growing
population of the UK?

How do we preserve the ever increasing demand for housing, recreation, education, healthcare, transport networks, welfare, and employment opportunities for the growing population of the UK?

Show me one person, committee, institution, or Great Brain who can answer any of these questions and I will show you a liar.

If all you are going to do with your measurements is lie, why bother to measure in the first place? Why lay a guilt trip at the foot of every caring individual when whatever they or anybody else does is a lie?

Why stand in line next to the open grave awaiting your turn to be shot? Why accept the status quo?

What is the alternative?

1. Accept that there is no alternative and that we cannot change our ways or have any meaningful beneficial impact on the earth's future.
2. Accept that all the flaunted alternatives cannot change our ways and will not have any meaningful beneficial impact on the earth's future.
3. Accept that all our flaunted alternatives probably make things worse and have a detrimental impact on the future of the earth (think palm oil, imported birdseed).
4. Accept that any alternative that allows the continual exploitation of the earth's biosphere and mineral resources is not an alternative.
5. Accept that controlling the human population of the earth is the single most important consideration for the future of the earth.
6. Accept that reversing consumerism is the second most important consideration for the future of the earth.
7. Eliminate force multipliers from the human toolbox and limit the capacity for exploitation.
8. Return to the Stone Age.
9. Build a house out of straw.

For all our intellect and applied force multiplier of education, what have we learnt? The American beaver, and with it the thousands of beaver-constructed dams and water habitats, was nearly driven from the face of the planet because of the fashion for beaver-skin hats, that is for the vanity of humankind. Studies in Yellowstone National Park have recorded the boost to the ecosystem provided by a healthy beaver population. The ponds and marshes created by the beaver attract a wealth of species dependent on the pacified water. These same studies have also confirmed the damage done by the removal and the restoration achieved by the presence of wolves. The elk population boosted by the absence of wolves has removed the growing tips (the apical buds) from the new tree population, turning the landscape into a bush scape with no new trees to replace the old dying trees.

For the vanity of human beings, elephants and rhinos have to die so that ornaments and trinkets can be made of their body parts. It seems I asked the wrong question with 'How to kill an elephant?' We already know. Perhaps the question should have been 'Why kill an elephant?' But we don't need an answer; we just do it anyway.

In the last decade, news reports emerged of murder cases where the defence argued that the defendants were not responsible for their actions because they believed that they were held within an alternate reality. This became known as 'the matrix defence'. These people could not be held responsible for their actions because they had no independent control and were manipulated by the program. If in the argument you supplant electrical impulses with chemical impulses, then the matrix becomes an ant colony. Change chemical impulses to tropismic responses and you have a herd. This obviously ignores the fact that in nature chemical and electrical systems work in conjunction with each other for the command and control of cellular processes of single- through to multiple-celled organisms and are also responsible for tropismic and 'intelligent' responses. You cannot separate nature from the human position, so why do we work so diligently to ignore and override it? We don't need a higher machine intelligence to trap us in an elaborate matrix of control; we already have our own system of DNA and evolutionary core to do that for us. We parade our supposed free will like a badge of honour (a blue plaque?) from the cradle to the grave as we systematically obey our programming while anaesthetising our 'conscious' processes with fairy tales of religion, democracy, empire/pyramid building, and 30 mph signs. What is it about our conscious mind—you know, the one that separates us from animals—that makes us so special? We comprehend our position as a 'higher' organism superior to the beast of field and water while simultaneously complying with their command-and-control systems. We have learnt nothing beyond the peacock's tail.

Apocalypse

'Apocalypse' is a noun used to describe the complete final destruction of the world, or an event involving destruction or damage on a catastrophic scale.

While you wait for the apocalypse, and there are plenty of survivalists in the land of the free preparing for it, consider whether it will happen within your lifetime or the lifespan of those nearest and dearest to us. Within the human span it is possible to love and care for around five generations of direct bloodline from parent to great-great-grandchild. For most, great-grandchild tends to be the limit. So how far into the future do we care? 'If we fail to solve the problem of global warming, will our grandchildren ever forgive us?' How about our great-great-great-great-great-great-grandchildren, say, in around two hundred years' time? Do we comprehend this idea? Does it register? Is it summed up within 'Future generations will never forgive us'?

You needn't worry; these are rhetorical questions, because of one simple truth.

We are the apocalypse. No need to wait for one; we are it.

Ask a dodo, or a great auk, or a passenger pigeon, or a Tasmanian tiger, or the Great Wood, and ultimately ask your fellow man, 'What does the future hold?'; Crystal balls to the fore.

The answer is, systematic destruction of all extracurricular species on earth that cannot adapt or survive the accidental or deliberate predation from our species. The systematic exploitation of every resource and environment to be terraformed to support our ever expanding population.

The systematic collapse of our island earth hinterland and the miniaturisation (dwarfing) of our species as we adapt to the decline of resources, or the wholesale genocide of 'foreigners' as we strive to secure our share of the dwindling stock of water and habitat conducive to the health of humankind. (That would be human beings from *our* tribe, not *theirs*.)

States downriver of the Nile are already in dispute about available water flow from countries upstream. Israel is no doubt ready to fight to ensure the continuing flow of rivers across international boundaries into Israel. If the Colorado River flowed north into the US from Mexico, you can rest assured that America would have moved to secure access to this resource by now.

'Apocalypse' is a word taken from the ancient Greek to mean, quite literally, 'uncovering', 'a disclosure of knowledge', a 'lifting of the veil', or 'revelation'.

How to Kill an Elephant is apocalyptic in that it seeks to unveil, to uncover, and to disclose knowledge.

What will you do with this knowledge?

Just a few years ago, the thread of the actions of humankind could be summed up as extraction as opposed to harvest. The argument went that we were no longer harvesting fish from the sea or cropping the great whales; rather, we were extracting or mining them, because 'once they are gone, they are gone', and no mechanism exists to replenish them.

Now, just a few years later, all the talk is of sustainability, but whereas extraction is real and absolute, sustainability is a lie, a PR trick designed to seduce and anaesthetise the herd, those of them who give a notional shit anyway. Because sustainability simply means its business as usual, with notional controls and self-rewarding reinforcement loops of hubris (in Greek tragedies, excessive pride towards or defiance of the gods, leading to nemesis. Nemesis is a Greek goddess, a source of harm or ruin, also known as Adrasteia, meaning 'the inescapable'.)

It seems that I have not had a single original thought in this entire treatise.

As you understand the need for pro formas, you also understand the need for lies to protect them. Rochdale has seen the prolonged and sustained (but at least not forever) sexual abuse of predominantly white underage girls by predominately British Muslim men of Pakistani origin.

Rochdale is not alone in this. The pro forma doesn't allow for this variation from the 'norm', and therefore steps have been taken to eradicate the issue by ignoring the claims of the girls and ignoring the action of the men. The norm states that we live in a multiracial society with multifaiths and that we live in peace and harmony free of racism. The Great Brains have determined that the needs of the pro forma outweigh the needs of the victims and that society will gain a greater advantage as a consequence. It is not 'politically' safe to indicate that some predominantly British Muslim men of Pakistani origin see predominantly white girls as fair game without any standing or position of respect. These predominantly British Muslim men of Pakistani origin men saw the predominantly white girls as worthless trash to be subjected to whatever sexual activity the former desired. Whereas the 'team' responsible for the safety of these same girls saw them as worthless trash that could be subjected to whatever sexual activity the predominantly Muslim men desired. Even now, the media cannot refer to the perpetrators as Muslim, they refer to them as Asian. They are not Chinese. Or Hindu's or Japanese, they are predominantly British Muslim men of Pakistani origin. The pro forma allows for a calculation, a balance, between the needs of the pro forma and the needs of the girls, and in this equation the girls were indeed determined to be 'worthless trash' and abandoned for years by the very people we have charged with protecting them.

The laws are very clear about sexual abuse and underage sex. They have been developed over the years to protect and safeguard vulnerable individuals, so in theory there is no issue, no argument about how the protective powers of the state should have been actioned.

Fortunately Dr Harold Shipman was not protected by the pro forma and was jailed for life for the murder of many of his elderly patients. He is believed to be Britain's most prolific mass murderer, with his victims suspected to run into the hundreds. When my mother-in-law died suddenly at her home, my wife and I were first on the scene. I rang it in for the doctor to attend. Never having had to do this before, I was shocked to discover that the first people to attend were not the medical profession but in fact the police. 'Since Shipman, we attend all unexpected deaths,' we were told before they bagged all medication and removed it for analysis.

Unfortunately Dr Harold Shipman had in fact been protected by the pro forma. How else can you conduct a career-long campaign of killing vulnerable people without being caught? Doctors save lives, remember; they don't take them.

When my father-in-law died at my home eighteen months after his wife (two days after being discharged from hospital), I was no longer surprised to greet the police as the first visitors. I was ready for them and knew what they would do and what questions they would ask. Because we now have a post-Shipman pro forma for unexpected death, does this mean that we can sleep safely in our beds secure in the knowledge that any successful attempts on our lives will be thoroughly investigated and foul play will be discovered? I think the police will tick the right boxes.

It would appear that the ancient Greeks had already worked it all out twenty-five hundred years ago and wrapped the whole theorem up in a mythological fairy tale that would serve to guide the memory by providing a story. It appears to have worked partially, as we remember the story but ignore the message.

Chimera

I have written about mutually assured destruction of the nuclear age and the fact that the Cold War has slipped below the everyday consciousness. Recently released files show that British pilots were ready to take their nuclear payload aloft at two minutes' notice. (They slept in caravans at the runway's edge.) The tensions between East and West have abated, although some may argue that the boundary has simply slipped a hundred miles or so east to the Ukrainian-Russian border (if we only knew where that is). Media commentators hearken back to these tension-filled days with remarks along the lines of, 'The modern generation has no sense of what we lived through.' And it's true, they don't.

But it would also be true to say that this generation, and indeed preceding ones, has no idea of the slow-motion apocalypse that we have engaged upon. Who is mad now?

Yes, we talk about global warming and have concocted elaborate smokescreens of plausibility to assuage the masses and protect our Great Brains, but we have done nothing to halt the process.

But what about the UN initiatives, the legally binding agreements?

When a partially state-run car manufacturer (part owned by the state of Lower Saxony) can lie about its pollution levels, and the sovereign state of Greece can lie about its financial levels, and the British health service can lie about its four-hour wait at A & E, from the lowest nurse to the highest minister in the land; when state-employed highway engineers can lie about their usefulness, and scientists can secure funding to support their lifestyles by swimming with the tide; and lasagne beef is laced with horsemeat, our justice system doesn't supply justice. Our schools chuck out ever more qualified kids who are then condemned by the universities for a lack of educational skills. When we need to accept immigrants as a 'good' thing to provide for our old age (I wonder, but not for long, who will provide for the immigrants' own old age), and our police force fails to serve and protect, and our taxes fail to end need; when our food is laced with hydrogenated vegetable oils, and our corner shops make a living by selling carcinogenic tobacco, and our schools cannot discuss obesity for fear of upsetting fat people or triggering anorexia.

So you trust the UN fix, then, do you?

The chimera in Greek mythology has the head of a lion, the body of a goat, and a serpent's tail, and breathes fire. Our modern-day equivalent has the head of a donkey, the body of an ant, and the tail of a peacock, and breathes fire from fossil fuels.

Viva la Revolución

If the opinion polls and surveys are to be believed, then a significant proportion of doctors and teachers have thought about altering their career paths. I don't believe these polls because the measurement is corrupt and used to steer opinion. I have thought about becoming a prima ballerina, but without years of training and many operations, it ain't gonna happen. What is irrefutable is that a very large proportion of the electorate fails to vote in local or national elections. Could it be that these are the enlightened individuals who have worked out that the pro forma isn't for them? Or could it be that they have resigned themselves to the notion that others are more capable of determining their future? Or could it be that they have decided that they have no influence?

George Orwell's *Animal Farm* suggests that all attempts at change are frustrated as the old order is quickly replaced with a new order, which entails more of the same. History is full of revolution after revolution where elements of the herd rise up to change the pro forma only to see the surnames of their rulers change and precious little else. New paradigms are created and then frustrated by a reversal to the old hardwired instincts and practices that caused the upheaval in the first place.

It's a bit like 'The king is dead; long live the king.' Once an ant, always an ant. True, subjugates can become subjugators, but by definition this leaves others to be subjugated.

How is the change motivated? We can look at religious and political doctrines, democracies over dictatorships, militarism, education and hazing or brainwashing, threats of violence causing fear, the force of law, peer pressure, and force majeure, but each leads to a pro forma. Each is corrupted by the recipients to arrive at an 'amicable settlement' between subjugators and subjugates to enable the 'project' to work. There have been exceptions. Pol Pot's murderous regime removed the 'amicable settlement' from the project by destroying the relationship between ruled and rulers by killing any objectors.

Stalin's collectivisation of peasant land saw an estimated four million to eight million die or be disenfranchised in the pursuit of Bolshevism. Remember the kulaks? Many will have never heard of them. These were land- and machinery-owning peasant farmers who naturally resisted the urge to gift their hard-won spoils to the communists so soon after the abolition of serfdom in tsarist Russia. Soviet authorities would answer, 'OK, if you insist,' to the reply of, 'Over my dead body.' Those who survived this process were either starved to death on their own farms or shipped out to Siberia without any provisions and dumped to meet a similar fate.

Strange to think that the middle-range estimate puts the kulaks' treatment on a par with the victims of the German Holocaust, and yet this man-made twentieth-century mass murder has completely slipped by the consciousness of our enlightened generation. I have remarked that it's a wonder murderous treatment of one class of individuals by another group of individuals doesn't happen more often.

Today we have the threat of global catastrophe caused by global warming designed to instil fear and to be used as a tool to steer our herd away from the precipice. It brokers no argument as nations throughout the world sign up to adopt new protocols and enforce change on their populations. Our leaders are busily filling the empty wheelbarrow, which, as ever, is facing the wrong way. They have the unenviable task of squaring the circle, simultaneously protecting growth while restricting energy use. We employ our smartest engineers (not highway ones) and technologists to find more and more efficiencies to reduce our energy consumption.

China has just announced the end of its one-child-per-family ruling over fears about its ageing population and its population stalling. China will fill itself up with Chinese, whereas Germany will fill itself up with refugees/migrants from outside the EU. Ignoring the economic arguments about maintaining growth and propping up the Ponzi scheme, another consideration is Germany's maintaining the strength to defend its borders from foreign invasion. Failing countries will be left to the mercy of stronger nations and their growing populations, who may take by force what Germany is volunteering. So the net benefit to Germany is?

The Coal Question

Many vehicles are being designed to burn less fuel per mile, whether that is cars, trains, ships, or aircraft. Ignoring the fact that the number of these vehicles keeps increasing, we also arrive at another paradox in the human mind. It's akin to the use of a slimming pill over eating less food. Because our vehicles burn less fuel, we can afford to drive further and faster and more often than we would have done. At no stage in the process do we have to modify our behaviour, because the 'fix' has been achieved by others, our Great Brains. True, the amount of fossil fuels consumed will have been reduced in the short term until the numbers catch up, whether at home or abroad (just how many Indians will begin to drive in the next fifty years?), but how does that solve the problem?

Say we had a standardised population over a fifty-year period; consider how the use of the motor car has grown out of all proportion in the UK alone. During this time period we have gone from single-car families to multiple-car families, and the number of zero-car families has been reduced.

Figures taken from the Society of Motor Manufacturers and Traders and the Royal Automobile Club Foundation suggest that the number of motor vehicles will have doubled in around fifty years since 1971. I know, we haven't reached fifty years since 1971, but if this proves to be optimistic, simply throw a few more years into the pot and you will arrive at the same number of vehicles.

The political appeal, and therefore the attraction for our Great Brains, is that we the herd do not have to modify our behaviour. We simply have to buy the newer, legislated for, efficient vehicles when our old ones wear out. Our Great Brains have engineered a solution that translates into business as usual—sorry, sustainable business as usual. The maths is appealing. Take a 30 per cent reduction in fuel consumption and plot that on your nation's stated-CO_2-reduction graph. Ignore the potential for increased driving due to its increased affordability; ignore the need to build more roads or repair the existing ones that wear out quicker; ignore the damage done by the release of saved money into consumers' wallets—'Cor, I've saved £1,000 on fuel this year. Let's fly to Disneyland in America'; and ignore the reduction of the affordability threshold for owning and running a car and the potential for increasing car ownership.

William Stanley Jevons published a book *The Coal Question* in 1865 in which he stated, 'It is wholly a confusion of ideas to suppose that the economical use of fuel is equivalent to a diminished consumption. The very contrary is the truth.' His argument, made 150 years ago, and which is sometimes referred to as the Jevons paradox, was that as the fuel efficiency of coal engines increased, the amount of coal consumed also increased. You needed less coal, so the engines became more affordable to run and you could use more of them to do more tasks that were thitherto too expensive to consider. While I have hinted at the complexity of the issues revolving around the increasing use of more-efficient energy sources, it might be reasonable to consider the full remit of questions on a par with the understanding of a rainforest habitat. So our Great Brains latch on to the presence or absence of tigers as the gauge of the ecosystem's health and focus equally intently on the 30 per cent (projected) reduction in fuel consumption. The empirical case is made (30 per cent), the bonuses are paid, the speaking tour is booked, and the memoirs are sold to the highest bidder. In the meantime our CO_2 production continues to rise. We don't need a diet; we need to eat less food.

When my son was at Leeds Met University in the early noughties, I stumbled across a car reduction innovation whereby there were camera-supervised roads where single-occupant cars were banned and only two or more occupants were allowed the privilege of driving on the road in question. This may have eased the congestion on the particular road locally, but at the same time it likely drove (forgive the pun) single-occupant vehicles onto alternative routes, which increased the distance travelled.

If I took my other son out with me for a drive, then I could use these roads with impunity, but any potential benefit was far outweighed by the fact that I had increased my payload and fuel consumption for the full length of my 220-mile journey.

We continue to rely on the magic wand of technology to extricate us from our perceived difficulties. We want hydrogen fuel cell or electric motor cars to reduce our travelling emissions to zero, while ignoring the emissions produced in the production of 'off-site and out of mind' fuels which magically appear at our point of use. We ignore the emissions generated in the manufacture of the vehicles and all the processes involved in raw material sourcing and conversion, and then we spend billions on advertising to create the unsustainable dream, billions that have to be generated from somewhere.

There are two options for providing education in remote places. In Montana, 15-year-olds may drive to school or church under supervisory licence controls. In Australia's outback, where ranches can be larger than sovereign states, education is provided at home over the radio. The scheme began in 1948 and has spread to all but two states, with modern satellite broadband offering real-time interaction between teachers and fellow pupils, who may be five hundred miles away from each other. What I am trying to illustrate is the fact that there is more than one way of skinning a cat in providing an education. Our system seems to have the sole aim of keeping kids off the streets during the day. No, contradict that, our system seems to have the sole aim of keeping them on the streets at the start and end of the school day. If the system can work in Australia over vast distances yet still allow for interaction between staff and pupils, why can't the same system be used closer to home? Supermarkets can monitor the activity of checkout staff by recording the keyboard activity at the tills and can control the work ethic, so why can't teachers observe the work ethic of pupils remotely? Unlike the classroom, the keyboard has no hiding places and teaching could be accurately focussed on where input would be most usefully deployed.

All Roads Lead to...

The problem in Rochdale was that the pro forma was incapable of accepting an 'inconvenient truth'. Now where have I heard that before?

I have tried to expound numerous examples of how we are condemned to stupidity by our genes.

We are a chimera, an animal pretending to be something we are not, intelligent, sapient, elevated above the beasts, when we are blinded by the dazzle of the peacock's tail, stupefied by the princess in the room, anaesthetised by the queen's chemicals, and bound by tropismic responses. Imagine the carnage a trained chimpanzee armed with a fully loaded AK-47 would do to its family group given that it has no conscience and no consideration for the consequences of its actions.

But you don't need to imagine the carnage wrought by Americans armed with shotguns pointing at passenger pigeons when we know the former had no conscience and no consideration for the consequences of their actions.

And you can see the carnage wrought by us, armed with JCBs and jumbo jets and education and having no conscience and no consideration for the consequences of our actions. Because the overriding consideration of all of the fixes appended to our stewardship of the earth can be summed up in three little words: business as usual.

And all the effort and enterprise expended in our efforts to 'save the planet', cure cancer, end poverty, limit temperature rise, save plastic bags, save the children, feed the birds, cut NOx, or grow green fuel—take your pick—simply exacerbates the problem. If you want to lose weight, eat less food.

I have a solution to our traffic woes. We could increase the speed limit on all UK roads by ten miles per hour, double the width of all major roads with increased lanes, employ more highway engineers (they are really clever after all), and construct new flyover interchanges at busy junctions.

Actually, I have more than one solution. We could reduce the speed limit to fifty miles per hour on motorways and forty miles per hour on non-motorway roads, drop to twenty miles per hour in residential areas, and fit all vehicles with speed limiters controlled by satellite technology. Further, we could dismantle all flyover junctions, reduce the size and number of lanes on all major roads, and then sack all the highway engineers, the clever ones anyway (so that would be all of them).

Plan A will boost NOx and CO_2 production, swallow up our green and pleasant land, raise our taxes, boost our employment of university graduates and the civil engineering sector, and enable us to commute a bit further every day. We would be able to source our meat locally from within an eighty-mile radius (19,000 square miles) and distribute imported goods from the other side of the world at greater speed.

Plan B will cut NOx and CO_2 production, reduce the pressure on our green and pleasant land, reduce our taxes, boost our employment of non-university-educated workers living locally to their place of work and markets, and localise all of our activities in a vein similar to the six-and-two-thirds-miles market charters of past ages. It will reduce commuting and improve the quality of life.

I should confess that this idea, and other suggestions dotted throughout *How to Kill an Elephant*, is designed to be a question answered with a single letter. Consider each question along the lines of the children's game of hangman, where the object is to guess the word. Every failure leads to the construction of the gallows and then the body, and your task is to guess the word before you die. The word has eight letters, and the first and third letter is *e*: e – e – – – – –.

Unlike M & S, I have a plan B.

'Plan B is preposterous!' I hear you cry. So plan A works then, does it?

A man once turned up at the patent office with a new style of mousetrap. It consisted of a flat wooden block with a razor blade set vertically across its middle, with a piece of cheese at one end. When asked, the inventor explained that you placed the trap near a mouse hole and when the mouse leaned over the razor blade to reach the cheese, it would slit its own throat and die.

The patent engineer, being a patient sort, kindly explained that he knew that mice happened to have tough jugular veins and that the trap would not work.

The next day the inventor returned with a modified trap and was ushered in to the patent engineer's office.

Again the engineer explained that the trap was the same and that it wouldn't work. 'Ah, but you see I've modified it,' explained the inventor.

'But it's exactly the same,' said the engineer.

'No, no, no,' replied the inventor, 'I've taken the cheese off. Now when the mouse comes out and leans over the razor blade,' he explained, 'the mouse will exclaim, "Now where oh where has that piece of cheese gone I saw yesterday?" It will scan its head from side to side looking for it.'

At what point, and you have to concede that at some stage it has to happen, do we stop building bigger roads with faster speed limits? Is it when the real challenge and skill is to drive on the green stuff because most of the ground is covered in tarmac?

You may well be asking if I consider plan B to be a serious solution.

So, you think plan A is then? Because that is more or less what we have been doing for the last ninety years.

Every now and again some Great Brain or other will declare, 'We cannot continue to build new roads; we have to find a different solution.' Our problem is that the question is not how to ease the traffic problem; rather, it is how to eliminate the need for the traffic in the first place. Set this against a backdrop of globalisation and cheap imports from across the world and I would suggest that we seem to be barking up the wrong tree. The trick is not to know the answer; the trick is to ask the right question.

The Romans (and they were not the first to build roads), we are told, built roads for the rapid transit of troops used in the subjugation of indigenous tribes. But as we all know, all roads lead to Rome, because once said tribes had been subjugated, the real reason that would truly justify the expense of time and effort came in to being, namely the transit of goods from around the empire. If the sole purpose of road building was to enable you to fight, then where was the profit? Roads enabled commerce, and commerce generated employment and wealth, whereas war cost money, disrupted commerce, and led to idle hands that could be filled with swords and spears to be pitted against you, while being temporarily un-idle before being permanently so (or so you hoped if you were Roman). While the idea of being able to rush your troops from garrison to conflict has a certain momentum, has no one considered that your enemy could utilise these roads just as effectively?

Napoleon has been similarly credited with the creation of the national network of roads around France for precisely the same reason. Yet it was Wellington sitting astride the road to Brussels who thwarted Napoleon's last great ambition. So for all the vaunted success of Napoleon's foresight, roads could be seen as instrumental in his downfall.

A major disadvantage of the US strategy in Vietnam was their failure to control the roads from the north to the south of Vietnam. The French were driven out of Indochina because the natives refused to be subjugated thanks to a lack of control of the roads. They simply walked through the jungle/countryside and attacked the French as and when. For all their machinery of war, the Americans were bypassed by the Ho Chi Minh Trail, which behaved like a multichannel river delta meandering across national borders and enabling the North to take the war to the Americans.

My point is, roads are a two-way street. For every advantage they provide to the road builders, they can also provide a disadvantage. We built roads and railways to enable British goods to be distributed to market at home and abroad. These same roads and railways are now being used to enable goods from around the world to be distributed throughout Britain.

Nowadays fresh flowers and vegetables are flown daily from Kenya to Frankfurt for road distribution throughout Europe, including Britain. Baby sweet corn, French beans, and roses are cut and prepared in Kenya today to appear in our supermarkets tomorrow. Our market gardeners have seen their markets collapse in the face of this competition. Some would shrug their shoulders and blame it on a failure to be competitive.

Kenyan labour rates are cheaper, and their growing conditions are often more conducive. With warmer temperatures and irrigation, several crops are possible each year. Yet, whereas local market gardeners could walk to market with a horse and cart, Kenyan market gardeners need a fleet of airbuses and articulated lorries, all climate-controlled to maintain freshness, and voluminous quantities of fossil fuels. Kenyan workers will live in straw houses and walk to work and demand low wages. British workers live in masonry houses that cost thirty years' of hard labour and pay high taxes to support our distribution network and our education, health, and welfare systems. Yet we are considered the ones to be blessed.

NHS

Limitless need and limited resources. Do the maths.

Fifty thousand years of being a bush, ten thousand years of being a tree. Five million years of being a family, ten thousand years of being a herd.

Do our family instincts betray our herd instincts? Can I now argue that humankind is in fact not a herd species and suggest that maybe this is the root cause of our inability to function cohesively? Whole countries go to war, and yet eyewitness after eyewitness will relate to the fact that they fought to protect their own life and the lives of their immediate unit. They cared about their comrade next to them, not king and country as we are led to believe. We team up to identify small sections that we can cope with. We don't support the Premier League; we support individual teams and we isolate individual players. We can support the national team, a squad of 20 or so players, but not the 360 that make up the Premier League or the 1,500-plus that form the four divisions. We support our fellow supporters on the terraces, and they in turn share in our groans and triumphs. Football becomes an abstract concept beyond the scope of individuals; it becomes 90 minutes of theirs versus ours. Sepp Blatter may as well be on a different planet (you think), and for him football has an entirely different meaning. He doesn't support football. Who would truly support the game and subject it to the summer heat of Qatar? Football is an abstract concept to him. He will define his role as organiser, administrator, and fixer. He doesn't care who wins or loses; he cares about whether he survives or not. He cares about being a subjugator and preserving his position and securing the spoils, while trying to secure the tacit approval of the subjugated. Do we arrive at the position whereby he needs to be seen to be fulfilling the task without actually achieving it? All the paying public supporters want is a venue and a tournament. How, why, and where is beyond their orbit. Just give them a time and a place and they will turn up. But cock up at your peril. Qatar is one insult too many, one imposition too great for the 'lovers' of the game.

Even the arbiters of opinion, the media, have failed to dislodge Blatter from his position as they witter impotently at the sidelines: 'Hey, ref, did you forget your glasses?'

'Penalty. How could you miss that?'

'Foul.'

Can we use this model to expand out to government and administration elsewhere? The greatest players in the land can make crap managers and fail miserably at other attempted tasks, because kicking a bag of air around the park is no qualification for other skills. Winning the beauty parade and becoming MP or PM doesn't qualify you for the job in hand; it qualifies you for winning the contest.

If we talk about orbits, then we can begin to illustrate the problem from a clearer perspective.

Our human organisation is separated out into strata or orbits, each with its own purpose and reason for being. We could divide into rich and poor, governed and governing, learned and ignorant, and old and young. The options are immense, and each stratum can be further divided, because there is obviously a huge gulf between Bill Gates and a penniless beggar, between Manchester United and Carnforth Rangers, between David Cameron and a parish councillor.

There are links between adjacent orbits, call them permeable membranes, that separate but at the same time allow for transit across the barrier. The model works until the equilibrium across the permeable membrane is disturbed. Think of thousands of migrants pushing through the membrane of Europe's borders. What once were sealed are now exposed as permeable. And there are conduits that purport to cross the orbits. These are the opinion peddlers, the media, and even celebrities who escape their orbit to occupy a 'higher plane'.

Is this a model of a multicelled animal, groups of cells organised to perform allotted functions, receiving resources, excreting waste and essential product across the cell wall, each acting independently while at the same time cooperating with no visibly obvious command-and-control centre?

In a paper published in 2013, an Italian, Dr Eva Bianconi, of the University of Bologna, and her fellow workers produced an estimate of the number of cells in a standardised human being. The figure, which is still an estimate—and other estimates exist across a significant range which could include another seven noughts—postulates that this standard human contains nearly forty trillion cells. It's enough to make you believe in God and intelligent design. Well, maybe not.

Back to orbits.

At the very end of my A-level chemistry course, I was enlightened with a 'lies to children' version of the periodic table and its structural origins. Each line of the table represented an orbit containing electrons, so the top line had only one orbit. The third line down contained three orbits, and the sixth line had six orbits.

The first vertical line grouped all the elements that contained the first electron in the orbit, and the last vertical line grouped all the elements in which the last electron that filled the orbit was present. To explain things in the simplest terms, elements in the first line are very active because they are unstable, and they try to shed the spare electron by forming more-stable compounds. Conversely, the last column contains a 'full' orbit of electrons with elements that are very non-reactive because they are stable. By the same token, elements in the penultimate line are 'desperately keen' to achieve stability by 'stealing' an electron from another element to form a compound with a complete orbit. In its simplest form, the top line contains two elements: hydrogen, which has one electron and an incomplete orbit, and helium, which has two electrons and a complete orbit. You only have to have seen the historical film of the Hindenburg disaster to begin to understand the difference between the reactive properties of hydrogen and the much safer helium alternative. No one is attempting to put a helium fuel cell in your car either. Back in the day, the United States controlled the manufacture and supply of helium but wouldn't supply the German airship industry with it.

With this basic child-level understanding of the process, I suddenly understood why metals behaved like metals and were keen to shed electrons, and why non-metals like fluorine and chlorine were so keen to gain electrons and behave in the way that they do. The final column of the periodic table contains the noble or inert gases, basically the stable content elements which have no incentive to change.

Why the chemistry lesson?

It strikes me that it should be possible to construct a periodic table of human orbits. Actually, this would need to be a series of tables to reflect the different needs and aspirations of the human condition (from birth to near death, for example).

I have mentioned the activity of the elements to either end of the rows of the periodic table and need to explain that those in the middle can react in either direction in that they can achieve stability by either losing or gaining electrons, while at the same time 'leaning' towards their nearer row end. These orbits of electrons are described as 'shells', with elements seeking to achieve stability by completing their nearest full shell by addition or subtraction.

Nuclear physicists will no doubt be scoffing at this model illustration, but I have described it as a lie to a child.

If we take a sample of people, how can we categorise them? Do we have takers and givers, with some who can go either way, while the smug buggers, the contents, have no desire to participate? Do we have leaders and the ones who want to become leaders and the ones who are content to be led? If we combine the first two, do we have leaders siding with the givers and those wanting to be leaders siding with the takers?

The contents, are these the stateless oligarchs, the billionaires who transcend borders and exist outside the active table?

Can we find a pattern for individuals akin to the that of elements? Some are only happy when they take; others when they give. And having taken or given, do they then achieve a more stable state? Do they need to form a compound to achieve equilibrium? Does alcohol or heroin complete the orbit and induce stability?

Do the givers seek to participate, the parish councillors, the PTAs, the WI committee members? The takers seek to receive.

Another thought strikes me with the near-40-trillion-celled standardised human. This beastie grew from a single cell, a fertilised egg. That's 1 to 40,000,000,000,000 in 30 years—phenomenal. Wow. On the one hand, it all happened by accident (Shakespeare and monkeys), but on the other, it is all controlled by our DNA. Never underestimate the power of DNA.

Celebrity is a pro forma for living, the mindless standardised equivalent of/alternative to having a personality of your own. A clone, a genetic replica.

It's 10 November 2015. The papers are full of two earth-shattering stories. One, international athletics has condoned widespread drug taking and cheating in return for cash incentives. Two, 'the rise of the public service fat cats'.

Both cases are indicative of the widespread corruption of the human race and the desire to gain an advantage by exploiting other individuals. Why is anybody surprised by it? The august bodies who proposed to bring an end to these issues are identical to the very bodies they are trying to reform.

What a miserable existence I am proposing. No birthday or Christmas cards, no computer games, no cheap clothing from the Far East, everyone dressing the same with no latest fashions or variety, no celebrity lifestyles to emulate. Yet all these things, these everyday essentials that we just couldn't live without, are all relatively new inventions. There were no cards before the 1840s and no computer games before the 1980s, although fashion setters and celebrity trend icons have existed for centuries. Who would not want to aspire to be Jesus, or conversely, who would want to copy the drunken laggard who failed to provide for his wife and nine kids? Yet life went on. The modern generation has no monopoly on entertainment or enjoyment. We can travel thousands of miles in a day but see only the beginning and the end, with nothing in between but clear skies above an ocean of cloud or water. The time was when the journey was the adventure and not the arrival. We can now immerse ourselves in realistic computer games, whereas we used to immerse ourselves in our imagination.

Have we lost the ability to think, to imagine, to entertain ourselves without the constant attention of animated pixels on handheld devices? Are we richer for it, or have we surrendered our souls to the computer chip, the new messiah to guide our life principles and secure order and compliance into the future? Are we embedded still further within the matrix? When out and about, look for yourself at how many people cannot walk down the street or park or sit in cafés or trains without the constant need to interact with disembodied voices or written text. I remember when our house telephone was first fitted. I had a well-to-do great-aunt whose telephone number was Matlock 9. This new generation of ours would behave as though they had their oxygen supply cut off if you suggested the suspension of these toys. You don't have to imagine. Just take the removal of free plastic bags from shops and multiply it by a factor of 10, 50, and 1,000.

Is this the source of the human failure, an inability to surrender an advantage without taking up a suitable replacement? Who would surrender their candles before the advent of the gas lamp? Who would surrender the gas lamp before the advent of electricity? (Have we gone full circle? Trans fats were developed as a cheap alternative to the more expensive tallow wax but ended up in our food instead as a cheap preservative to save lives.) (I think there's a bit of a clue here. When a food additive works by killing all bacteria and removing the need for refrigeration or other preservatives, it might be reasonable to consider that it may be injurious to our health.)

How can we possibly endure a life without computers or mobile telephones? This question never troubled the 1930s generation. Did the world stop spinning on its axis back then? Does this mean that I am advocating the removal of this technology?

I will leave it to you to decide what is important or not. Do you want to preserve Christmas cards, cheap personal commuting, designer clothes to die for, the right to breed without limit, the welfare state, pensions, the fifty-year work ethic, our Great Brains, our unfocussed education system, our highway engineers, the Ponzi scheme of all schemes, our time machine, our apocalypse?

Should we focus instead on what is important; food, shelter, health, conserving resources, unplugging the time machine, disconnecting from the matrix, living a life worth living, trying to live a life independent of the profit of harming others.

Karl Marx talked of the need for the workers to take over the means of production. His aim was to prevent the exploitation of the masses by the industrial factory owners. We've seen what happens when individuals produce according to their abilities and take out according to their needs. Human nature kicks in and suddenly the needs-to-abilities ratio hits an imbalance.

I remember reading about a house sparrow (*Passer domesticus*) nest discovered in a house eaves very late in the year, well past the end of the breeding season. Within the nest was a fully grown chick that was trapped by the leg and had been unable to leave. The parents were so driven by the need to feed that they had fed the captive youngster for months and kept it alive. While one could talk about the altruism of the parents, one could just as easily talk about the instinctive drive to feed and an inability to switch off from the predetermined course of action. There are other examples. The hedge sparrow (dunnock; *Prunella modularis*) and long-tailed tit (*Aegithalos caudatus*) are known to accept the assistance of close relatives in the rearing of their broods, as are other celebrated species around the world.

A recent documentary featured a disabled orca or killer whale that was supported by other members of the pod, which enabled the youngster to survive.

The queen ant lives a life of comparative luxury, with all her needs of food, cleanliness, and security provided by her asexual daughters, but with four provisos: she had to build the first nest and raise the first brood, and if she fails to provide the eggs and the pheromones, she will be dispatched and replaced.

Are there any species out there that tolerate entirely non-productive group members?

Commerce revolves around the generation of wealth. Take a raw material, convert it into a product of more value, and pocket the increase in value. *How to Kill an Elephant* (you're reading it, so it must have seen the light of day) started life as a seed in the ground, which developed through time, and with water, minerals, and sunlight, to form a tree. This tree was processed to make paper, was bound into a printed document, and landed on a bookshop shelf near you. What value in monetary terms did the seed have, given that it was one of perhaps five thousand produced by the parent tree that year alone? What value in monetary terms did the paper have before my thoughts were drizzled over its surface? I know, you are already thinking what value my thoughts have, but we are a long way into the book, so I must have piqued your interest.

Personally, I believe *How to Kill an Elephant* to be the most important book since the Holy Bible, so not very then. (Because, what I write demands fundamental change of the human species, beyond our scope to wrestle control from our DNA.)

But if the seed had not come from a tree but a grass, the same process would have produced a sprinkling of cereal grains and, after processing, maybe ended up on your spoon at breakfast this morning.

This is where the great irony creeps in. If you accept that the earth's resources are being plundered to fund the advertisers and highway engineers and all the other previously identified non-jobs (a sample anyway; the list is greater than my imagination), then the cure to all of our problems is money. Let me explain. If the tiger population is nearing critical level, then we can utilise resources to protect them. In English, give money to the Tiger Project or WWF, or subscribe to the local zoo's breeding programme. If people are dying from heart disease, give money to the British Heart Foundation, pay your taxes to the NHS, or take out private medical insurance. The problem with this is the source of the money. If you are a food manufacturer, an employee thereof, or part of the supply chain, and at some point in the manufacturing process the food is debased or laced with fat, sugar, salt, or trans fats, then you are creating a large proportion of the heart disease, so you can afford to treat its symptoms.

Or for heart disease, read diabetes.

Or for diabetes, read cancer.

Remember, there are two ways of adding value, either improving the product to an acceptable standard or debasing the product to maintain an acceptable price. Given that we are into a price war between the Big Four and the discount supermarkets ...

Alternatively, drive to work in your motor cars so that you can treat lung diseases caused by exposure to pollutants.

The post-war generation saw the creation of cheap prefabricated houses built of iron frames and asbestos-cladded walls.

Why not a straw house?

Stand on a motorway bridge, observe the traffic flowing beneath you, and try to count the number of vehicle.

Daily flow rates of 150,000 vehicles are not unheard of, so let's do a bit of maths; 150,000 ÷ 24 = 6,250 per hour, or 100 per minute. Of course flow rates will vary between peak flow and the early hours, but this figure will do.

The Highways Agency assumes 15 per cent of the vehicles will be large goods vehicles. We are now furnished with eighty-five cars and fifteen large vehicles per minute. Ascribe a value of £12,000 per car and £40,000 per truck and you can calculate a figure of £1,620,000 worth of vehicles per minute, or £97 million per hour. As we have been quite free in our assumptions as to the average value of each vehicle, and assuming that you are not burning the small hours while watching the motorway, I think it reasonable to suggest that at least £100,000,000 in vehicle value is passing beneath your feet every hour. This allows for some depreciation of the values from new. Of course we could view this from a different perspective and measure the motorway in distance rather than time. Allowing for all the HGVs to be doing the speed limit of sixty miles per hour and the cars to be doing seventy miles per hour (you never know, it could happen), and given the difference in monetary and speed values between the two vehicle types, it is possible to approximate the value of £100,000,000 to an average distance approaching sixty-five miles.

This equates to the total value of all vehicles on the M1 between the M25 and the M6 at a reasonably given time.

Let us consider the monetary needs of tigers. And you didn't even know they had pockets, let alone purses. Money is needed to create reserves, to provide guards against poaching, and to build roads for ecotourists and hotels for them to stay in. It's needed to provide research facilities with radio collars and field scientists; to provide education facilities to enlighten the local population; to move or translocate locals from one side of the boundary to the other; to compensate farmers for lost livestock; to pay hunters to take out rogue man-eaters; and to provide yet more education facilities to enlighten rich Chinese patrons seeking to find a cure for male pattern baldness or impotence or whatever. Money is needed to fund international breeding programmes and studies. It's needed to fund charity campaigns and charity administrations. Indirectly it's needed to provide the aeroplanes, runways, terminals, airport buses, duty-free shops, safari gear outfitters, camera makers, printers, travel agents, magazine adverts, brochures, and all the facilities and their supporting facilities for all the workers to earn the money to be able to spend it in search of that elusive tiger experience in the jungle.

Local economies benefit from the presence of ecotourism and learn to love the tiger for the extra money earned from tourists for their food and service requirements.

Can you spot the silent witness here?

It's camouflaged to blend in perfectly with its surroundings. You need to see it in just the right light and from just the right angle; otherwise you'll miss it.

Tiger preservation is a good cause. There, I've said it. But that's not the witness.

Let me pull a random figure from my imagination to account for the monetary needs of tigers. But before I do that, consider where all this money comes from. After all money doesn't grow on trees, does it?

Do one of the following for example: (*a*) work as a coal miner, (*b*) work as a book printer, (*c*) work as a dining room suite manufacturer, (*d*) work as a scaffolder on a construction site, or work as a forklift truck driver in a warehouse.

This will do for now, five jobs all generating money so that you can contribute to a tiger charity or indulge in a bit of ecotourism. But what am I saying? The person in example (*a*) works with fossilised trees; (*b*) works with wood pulp; (*c*) works with timber; (*d*) walks around on wooden planks; and (*e*) moves a multiplicity of products on wooden pallets on and off of wooden truck beds.

So in fact money does grow on trees. Would these be the same trees that the tigers rely on to survive? Maybe not, but teak patio furniture could well qualify.

What I am trying to establish is the link between wealth creation and the destruction of habitat to achieve it. Your timber product may come from sustainable ethical forests from around the world that have never seen a tiger in the history of the beast. But while we search for the tiger or absence thereof, do we not ignore the silent witnesses? These are the unknown animals and plants that have yielded their place in nature but remain ignored by us as we strive to earn a living and chip in to tiger preservation.

So what figure shall we set aside for tiger preservation? How much good would £100,000,000 do?

When my two boys were young, I told them that the reason we had roadworks was because there wasn't enough tarmac in the country to surface all the roads at the same time. Tarmac from every hole was lifted and recycled before filling another hole somewhere else, I said. To my delight, they queried this idea, because it didn't seem right to them.

So the reason we have a problem with tigers is because we have to dig holes in forests around the world to make forests in wild places because we haven't got enough forests to fill all forests up at the same time. But if my 7- and 9-year-old boys could see the folly of this action, why can't adult intellectuals, charity bosses, governments, and charity contributors?

I have commented on palm oil cultivation for biofuel and the clearance of large forest areas to accommodate it. After all orangutans have no need for forests. But people have complained about this, haven't they? Why not have a break from reading, a cup of tea perhaps, coffee maybe? But this ignores all of the previous forest clearances to provide for tea plantations, coffee, banana, chocolate, rubber, and that old favourite, the ethical, sustainable timber plantation. Macaws and gibbons had no need for them, let alone the thousands of nondescript 'lesser' species that make up the biosphere.

Let's stay with the tarmac argument, because if we haven't got enough tarmac to fill all the holes simultaneously, what do we have enough of? Have we enough space on earth to accommodate seven billion people and at the same time accommodate all the other species that we share this earth with? Have we enough money to accommodate all the health needs of our UK population and at the same time accommodate all the other needs for housing, welfare, defence, and education without digging holes in them?

The simple answer to both these questions is no, so why do we try?

I am a life member of Warwickshire Nature Conservation Trust and have been for nearly thirty years (although this state of affairs might be altered if they read this). I receive their newsletter via email. A major campaign they are currently pushing is entitled Every Child Wild, and this has been adopted nationally by all the wildlife trusts. The concern is the identified disconnect between the current crop of youngsters and their failure to engage with wildlife and wild places. Very few children play outside of homes and gardens, and the figures have been dropping over the years. When questioned, a significant number of parents of young children, along with other adults, felt it was important that children be exposed to wildlife and wild places. Yet the statistical data produced indicated that in spite of the apparent concern, the adults in the main had done nothing about it.

One swiftly-jumped-to conclusion is that the majority of 'concerned adults' simply don't give a shit.

Are these the ones who fastidiously brush their children's teeth? But don't worry, there is a diet pill that binds with fat to solve this problem. Give your time or money to the trusts, and they will provided educationally orientated supervised visits to wild areas. Coventry Public Health, BBC Children in Need, Tesco Charity Trust, and the Four Winds Trust all contributed funds which enabled thousands of schoolchildren to be structurally exposed to wildlife. You cannot fail to notice the money-collecting antics of BBC Children in Need with its extensive TV coverage and advertising for weeks ahead of the telethon. People and organisations the length and breadth of the country contribute so that, among other things, children can be exposed to wildlife. Of course you could keep your money in your pocket and use your fundraising energies to take your kids for a walk in the countryside.

The tiger is a predator, with short bursts of speed and activity, high-energy food, a short digestive system, and body and senses attuned to the hunt.

The antelope is an herbivore, with sustained activity at low density, low-energy food, a long digestive system, and body and senses attuned to flight from predators.

Each has developed its own strategy for survival. Both live on a knife edge between life and death. At some point in their evolutionary origins, a 'decision' was taken to take one route or the other. On the face of it, the tiger has it easy; it 'pops out to the shops for fast food', and within its lifespan it achieves success in the hunt. (If it doesn't, that tends to be the end of its lifespan.) By comparison, the antelope spends many hours eating, chewing the cud, and digesting to meet its calorific needs, and within its lifespan it achieves this aim. Two methods of extracting energy from the sun, each in balance with each other, each dependent on each other. Both evolved to be in equilibrium with resources.

There is of course an equilibrium between predators and prey, between carnivores and herbivores. We cannot all be carnivores, ultimately because the entire system revolves around solar energy via photosynthesis and the generation of digestible plant matter that primes the system. The ratio of prey to predator is firmly established by natural correction in populations. A natural glut in prey and the subsequent glut in predators is soon followed by a collapse in predator numbers in response to a collapse in prey.

So where in this natural balance of things do we place human processes?

A New Paradigm

Why do the media refer to terrorists and their claims of responsibility? Wouldn't it be preferable to refer to their actions and their claims to irresponsibility? Why not give terrorists' attempts to gain media attention an immediate negative reaction and reportage?

The real issue throughout *How to Kill an Elephant* is the lack of a solution. It's one thing to identify and denigrate the existing methodology of the human condition and another to proffer the solution.

Can we identify the issues, the barriers to generating a new paradigm?

1. We are preprogramed in our behaviour by our DNA.
2. We are unable to change course and alter our behaviour.
3. We have an inbuilt resistance to loss, embracing betterment, and shunning deterioration.
4. We rely on our leaders to steer us in the correct directions.
5. We ignore our leaders by corrupting the message they give us.
6. Our leaders corrupt our needs in favour of their own.
7. We cannot understand our behaviour and are tropismic.
8. We miss our leaders when they are gone and ignore them when they are present.
9. We have accepted the mantle of being a tree when we are only qualified to be a bush.
10. We breed uncontrollably.
11. We have developed force multipliers.
12. We have an inbuilt ability to exploit our fellow man.
13. We have an inbuilt ability to exploit everything else.

14. We are *Homo non-sapiens.*
15. We rely on others to make the changes that we all need to make.
16. We adopt pro formas and recipes as model solutions.
17. We adapt to everything but evolve to nothing.
18. We are absolutely relentless.
19. We have invented time machines.
20. The systems and methods we design as solutions exacerbate the problems.
21. We are romantic and are blinded by princesses.
22. We are blind to our actions.
23. We are the apocalypse.
24. We are blind to our problems.
25. We have no ambition, no higher purpose; we just are.
26. We do not know how to kill an elephant.
27. We perceive only one threat at a time.

Have you noticed that whenever a human disinterest story is featured in a newspaper, it normally includes a reference to the state of the victim's finances by stating, 'In their £ – – –,000 house in a nice/exclusive area'. It's as though the value of the upset/trouble/shock/scandal/crime increases with the wealth of the individual(s) concerned. 'Ooh, it's such a shame. They'd risen out of the gutter and deserved better than that.'

It's a form of measurement.

Sitting idly in the local out-of-town shopping centre car park today, I was suddenly struck by the idea that this state of affairs occurs in other areas as well. In October of 2015, the government phased out the requirement for paper tax discs to be displayed in all car windscreens. The reason given was the use of computer technology and the link between databases and ANPR (automatic number plate recognition) cameras. While some of us choose to adorn ourselves with the latest designer gear and accessories and associated bling, nobody patrols the public realm with an 'I earn [this much money] as a salary' badge on the basis that this would be a step too far for 'decent people'. True, some will interpret the level of bling as a measure of an individual's wealth, but others who can't and don't want to know the difference between real and knock-off will simply pity the poor creatures and wonder why they don't spend their money on something important instead.

A proportion of this desire for status symbols and the associated misguided spending pattern is down quite simply to peer pressure and the need to prove one's worth. We have a government intent on dissuading us from using our cars with various schemes in place or due to be enacted. We have congestion charging in London and Durham and its threatened introduction into Nottingham, Derby, and Leicester, park-and-ride schemes appearing on the edges of many cities, and threats of workplace parking charges, to name a few. Throw in the CO_2- and NOx-reduction initiatives, along with previous government inflation-beating fuel duty rises, and you could be forgiven for thinking that cars are a bad thing.

So why at the height of the post-2007 recession did the last government pay out hundreds of millions in car scrappage schemes to promote the purchase of new cars? The official answer was to remove the old technology 'gas guzzlers' from the streets and replace them with more-efficient vehicles. The other reason was that the Society of Motor Manufacturers and Traders (SMMT) lobbied the government to artificially stimulate the sale of cars in the depressed market. So, are you serious about reducing the number of cars on Britain's roads or not. Note the absence of a question mark, because the last question isn't really a question.

So what triggered this thought process while I was sitting at Fosse Park? It's the 'I earn [this amount of] salary' badge that is displayed on the car number plate, namely its age. If you removed the age indicator from the number plate, the argument for the need for a MOT when the car reached three years of age would be catered for by the same mechanisms as the car tax. By removing the age badge, the peer pressure to change would diminish overnight and, heaven forbid, we might keep our cars for longer periods. The smart bling merchants already personalise their plates to keep the non-car fanatic from knowing the truth, so it is not essential for the age to be displayed anyway. No doubt someone will crawl out of the woodwork (should this be forest?) to 'prove' that the advantage of burning less fuel will outweigh the disadvantages of the entire manufacturing process from extracting the ore to shipping components halfway around the world.

It has occurred to me that my solution for obesity, namely eat less food, could be further simplified by 'don't get fat in the first place', a variation in the process that cuts out all the health implications, so its merits should be considered.

STUs. It Had to Happen Eventually

Annual tea production worldwide is given as an estimated 5,000,000 metric tonnes. Iran produces 84,000 tonnes from 32,000 hectares, which suggests a yield of 2.625 tonnes per hectare. The amount of 5,000,000,000 tonnes would utilise an estimated area of 1.9 million hectares, or 19,000 square kilometres.

Coffee production for the world's largest producers is estimated at over 7,000,000 tonnes from an estimated 7,000 square kilometres. Adding the two areas under cultivation together, we arrive at an estimated figure of 26,000 square kilometres of land that was formerly 'virgin habitat' before its clearance for cultivation, which in turn enables us to get our several fixes a day.

Fortunately the most popular form of beverage is water, with tea second and coffee third. Collectively, tea and coffee production is utilising a minimum of 26,000 square kilometres of what would once have been natural habitats.

Now to put that in some type of perspective, our taste for tainted water accounts for habitat that could provide a home for an estimated 2,080 tigers. Finally, a definition for an STU, a standard tiger unit, and not, as you previously thought, a sexually transmitted disease. This is based on a figure of one tiger per 12.5 square kilometres. (This figure has been trawled from the Internet. I haven't quoted the source because it is simply one estimate for one habitat and I have to put my peg in the sand somewhere. Siberian tigers would need a bigger area, but the people in those areas don't grow tea or coffee.) Now we all know that tigers are not distributed throughout the beverage production range and tigers are not the only animals to utilise this potential area. If we considered another poster boy/cat, the leopard, Kruger National Park estimates yield a figure of 30 per 100 square kilometres. In leopard terms, some 7,800 cats could survive in these areas deprived to them. Of course this also ignores the vast array of other species extracted from the cultivation areas. Instead of giving money to WWF and the like, I suggest that you would have more impact if you stopped drinking tea or coffee and drank water instead. It brings a new meaning to the expression 'I could kill for a cup of tea,' doesn't it?

My guess is that these figures could be doubled, at least, if we added orange juice, whisky, beer, wine, and the many other beverages that we survived without quite happily for millennia.

What a pleasureless world I describe, no tea, coffee, beer, wine, and the like.

What a pleasureless world I describe, no tigers, leopards, Tasmanian tigers, giant pandas, and the like.

So am I seriously suggesting we turn our backs on all of life's little pleasures? The question is, which one, tea or tigers? But I already know your answer. Fancy a brew of apocalypse?

The Dangers of Frig Spaces

During the 1970s my family would take an annual holiday caravanning around Western Europe. I remember staying on a campsite in Belgium when the proprietor enquired if we wanted any milk. We should have known really; the bucket in the crook of her arm was a bit of a clue. Sure enough, a short time later, she had corralled a cow in the adjacent field and we soon received a supply of body-temperature milk. I have to say that warm milk on cornflakes is one of the acquired tastes that I never acquired. Back in those days, caravans didn't come with fridges, and milk was purchased daily and stored in buckets of tap water to try to keep it cool.

As a child in the 1960s, my wife was despatched to the local store to buy some frozen peas. It was only a short walk of 100 yards. Her young mind occupied itself by twirling the paper bag around. Sure enough, one time the dampening paper bag failed and disgorged its contents along the footpath. (The peas still ended up in the saucepan though.)

Who would have thought that the advent of cheap refrigeration in households could have had such a profound impact on our everyday lives? We are talking Japanese knotweed here. The introduction of these machines triggered a reaction beyond the inventor's imagination.

I will try to produce a list of variations that could be laid at the foot of the humble fridge.

- the demise of the local shop
- the birth of the weekly shop
- the birth of the supermarket
- the birth of the supermarket oligopolies

- the demise of the small mixed farm
- the birth of agribusiness
- the demise of the hay meadow
- the birth of the monoculture silage ley
- the demise of traditional management of small family farms
- the demise of thousands of miles of farm hedgerow
- the demise of the varied hedgerow height rotation from tall to short and layered
- the reduction/readjustment (take your pick) of farmland bird populations
- the creation of holes in the ground from which the iron and copper ore, and energy and oil for plastics, come from
- the creation of the holes in the ozone layer
- the filling of redundant holes and alternate habitat opportunities with waste refrigerators before we became so enlightened as to recycle them
- the birth of massive warehouse sites to service the supermarkets
- the creation of an air bridge between Kenya and your local supermarket (for example)
- the demise of local dairies and creameries
- the advent of the disconnect between farmers and customers
- the demise of the family butcher, local abattoirs, and local livestock markets
- the debasement and poisoning of our food
- the obesity epidemic
- the reduction in our life expectancy
- the rise of the motor car, essential for the weekly shop.

I am sure I will have missed some, but this list will do for now. In justification of my list, I would suggest that refrigeration provided the catalyst, the opportunity, for these changes to take place. Now I am not suggesting that cheap refrigeration was the direct cause of all these actions, but it would be difficult to see how these things could have played out without this catalyst/progenitor.

My wife's parents lived in a 1930s-built house with a stone thrall, a large mass of stone in a cool cupboard protected from the sun to keep food cool.

My grandparents had a cool room in their limestone-built house with a meat safe, a perforated zinc cupboard where meat was stored to keep the flies off it.

Our engineers have waved their magic wands of force multiplier over the once humble fridge. I remember my parents buying their first freezer for the storage of frozen foods in the early 1970s. We had just moved house and now had the space for one of these contraptions, which soon became two and then three.

I used to drive a lot on business and was often accompanied by area representatives on visits to clients. Talk would be varied, and turned one time to my companion's friend who owned a Porsche sports car. I cannot name the model or date, but I can remember the conversation. Every time my companion's friend went to overtake another car, he had to turn the air-conditioning off; the unit took so much power from the engine that he lost most of his acceleration. My own experience with Volvos (yes, I admit it) is that I noticed a significant drop-off in performance if the car was fully laden with passengers and luggage and the air-con was operating. Later models had an automatic switch that noticed the extra demand from the engine during acceleration and temporarily turned the air-con off until the load was reduced. I wonder, but not for long, if air-conditioning is considered when publishing CO_2/km figures. An overt cheat device.

And I haven't mentioned air-conditioning in homes, offices, and shops ...

Having studied architects house drawings for decades for a career, one of the little jokes incorporated into the drawings is the insistence on calling fridge spaces marked on drawings 'frig space under'. Irrespective of the drawing scale or space available to annotate the drawing. If you are too young to understand, ask your mother what it means.

The Truth Is Out There

The beauty of the Internet is the amount of information nearly instantly available at your fingertips. Various bodies provide information on wine production and the estimated figures of yields. The different bodies give different figures that can produce a range of results. It is because of this that I do not quote the sources. The information is out there for you to research should you wish. In one sense the figures are not critical; rather, it is the idea that is important. Wine yields vary greatly between the growing regions, and it is very difficult to extract anything but an average. At best I arrive at an estimated range of worldwide wine production of around 43,000 square kilometres or 3,440 STUs (tigers).

Italy and France, being the world's largest wine producers, have never in their European history been trodden on by wild tigers, but how about by bears and beavers and white admirals and stag beetles?

But for all my talk of STUs, we are still left with the fundamental problem. Take all the tea in China and coffee in Brazil and end its production tomorrow, and two things will happen:

1. the land will be reutilised by the locals to grow something else, which won't be tigers, and
2. whatever has adapted to life in tea and coffee plantations will likely be disadvantaged by the movement to the new crop.

We are back to the question about the golden age of birds.

In the UK we have no concerns for tigers, but we could consider a conversion table for STUs, which could work something along these lines:

1 STU = 12.5 square kilometres = 4 standard leopard units = 25 standard red deer units = 25 standard fox units = 2,000 standard frog units = 10,000 standard tree units = 10,000,000 standard moth units = 10,000,000,000 aphid units, or × standard units of something else that you care about.

We could just as easily work in ancient meadow or ancient forest or ancient heath units. The point is, the level of destruction encountered by our desire to taint our water requirements is quite simply extraordinary.

Now double this for another seven billion people, or double it for extended life expectancies, or double it for increasing wealth, or keep the population the same and increase obesity levels. Whichever way you choose, the STUs get eaten up, never to be seen again.

We have had public information campaigns extolling the virtues of saving water and energy. Turn the tap off while brushing your teeth, boil only as much water in the kettle as you need, turn your thermostat down one degree—you know the type of thing. Now consider the impact of obesity on the earth's resources. Let us take a figure of two thousand calories a day for an adult male (ignoring the differences between sedentary office workers and building labourers) and then consider the energy wasted in adding several stone of excess body weight. You have the energy not only in the food itself but also in the growing and cooking of it, and don't forget the impact of extra weight on vehicle performance and CO_2 and NOx. Obese people are effectively going through life with the tap left turned on and the kettle permanently filled. Our Great Brains launch schemes whereby large energy consumers have to buy carbon credits to reduce their impact on the world stage, but we ignore the millions, or should it be billions, of individual users. And all of this waste (waist) is effectively dug out of the earth's resources and underwritten by tigers. Imagine you and a group of friends have gone potholing, and while you are underground the ceiling collapses, hermetically sealing the tunnel. Think about the conservation strategies you would adopt in terms of conserving oxygen, water, food, and energy while you wait for your hoped for rescue. Now make the pothole a bit bigger and call it Planet Earth, which is hermetically sealed (barring any asteroids and the like) with a finite resource of oxygen, water, food, and energy, and consider your options. Go back to the pothole and fill it with your friends and family, and then consider how desperate you'd become. Worse still, be a total stranger in the pothole filled with someone else's friends and family and consider whether they'd bother to choose straws to see who should be the first volunteer to die to preserve what's left. Instead of a pothole, consider that you're in Israel watching the rivers dry up before they reach your borders, or in Egypt watching the empty channel south of Sudan. Human nature wouldn't allow this? Consider that the one hundred thousand German prisoners of war captured at Stalingrad who survived the march to a POW camp in Russia learnt to eat their fellow prisoners. Any cannibalism was punished by the Soviet guards, who would beat the offenders to death. The really clever Germans (trainee highway engineers? Ouch! I go too far) alerted the guards to any low-scale cannibalism so that the body count would be increased before they themselves tucked in. Prisoners were fed on whole grain, which passes through the human gut. One enterprising hut learnt to sieve the whole grain from the shit and generated enough to offer a black market of food to other huts. History does record that only around five thousand lived to see Germany again.

While on the subject of wasted energy, it would appear that I have strayed into the man-made global warming acceptance camp. My belief is that global warming, if it exists, is simply a symptom of the greater problem that we are mindlessly terraforming and destroying our precious island resource in the pursuit of pursuit. We have no common aim or ambition, no control or purpose in life, in spite of our supposed higher order of being.

GDP

The World Bank provides data for the per capita GDP for countries around the world in standardised US dollars. The data for 2014 assume a midterm population. In last place, if that's the way to describe it, is Burundi, with a figure of less than $300 per person, but twenty-six other countries score less than $1,000 per person.

The two richest countries on the list have relatively small populations. The reason for their success is essentially down to oil. It is easy to establish that the wealth is generated by 'digging a hole' into the earth's history while simultaneously digging a hole into the earth's future. China, rapidly developing to become the world's largest economy, has seen phenomenal, thirty-seven-fold growth over the last thirty-two years. How can we reconcile the difference between a few hundred dollars in the 1980s with, say, $50,000 per person in today's values in 2040? We are not making any new wealth, simply extracting it from one source, be it past or present, and stealing it from our future. But the impact is not contained within China's borders. The growth is fed by an insatiable hunger for raw materials and energy won from holes all over Australia, Africa, Asia, and Europe, not to mention America's demand for cheap goods, to fill our empty shop shelves and empty souls. In Krypton terms, we are rushing headlong into a honeycombed planet on the brink of implosion. Where is our Jor-El to construct a spaceship to send our sons to a fresh hope and future?

Does the solution lay with the bottom of the league table, the native of Burundi who has a negligible impact on the earth and its resources? The population of Burundi consists of people who are generally poor and uneducated with poor healthcare and a reduced life expectancy, and yet the population rate is increasing at around three times the global average. Burundi is short of the force multipliers of education and technology, but its impact on the natural world is still locally profound—not much room for tigers there. I say locally because the average Burundian will have negligible impact on the lives of Americans, whereas Americans will have a definite impact on the lives of Burundians by providing copious amounts of aid. In English, this translates into digging a hole elsewhere in the world to fill the one in Burundi. Let's destroy our wild places to feed Burundians as they in turn destroy their wild places.

Am I suggesting that we should let Burundians starve to death? Would it be fairer to condemn all unconceived children to a non-existence? Burundi has suffered from major political upheaval and has an ongoing civil war, both of which have only contributed to the country's problems. A large proportion of the population has been displaced and is unable to grow their own food. The country is in a mess and needs to find a solution. It hasn't got one yet and is unlikely to find one that operates within the confines of the world solution pro forma. The world, for all its supposed intellect and ability, has no solution to offer other than spreading the load.

Spreading the Load

How does the human herd work? Wildebeest and similar groupings work on the principle that there is safety within the herd and danger at the fringes. Baboons will protect the weaker individuals and take responsibility for the safety of the group, while at the same time preserving hierarchy over subordinates. They demand the best food and shelter but will 'saddle up' when the call is made to drive off defeatable foes (leopards). Or it's every monkey for itself when faced with a lion pride or tiger. Where would we place humans in this scale? Superior humans demand the best food and shelter and will offer 'security' to subordinates.

How does medicine work? Are we tricked into accepting an illusion? All the advances are outweighed by evolution, by the arms race between bacteria and viruses, and indeed by our own DNA.

Remember the lesson about the passenger pigeon and the question about films like *Independence Day*? How many cities would the aliens need to destroy to put a dent in the human population? In my previous answers, I deliberately ignored the breeding potential of humans. Check out https://www.worldometers.com and load their running commentary on births and deaths daily and annually to discover the answer. Today, it doesn't really matter what specific day today is, the breeding potential of our species will be around the figure of 250,000 births in excess of deaths. This will hold true for the next ten years or so, not because the population will level out but because the figure of 250,000 will need revising up.

Many years ago I attended an illustrated talk on one of the local nature reserves, Narborough Bog. At the end of the slide show, the speaker asked for volunteers to help remove an overhanging willow protruding onto the neighbouring recreation ground. Much to the warden's obvious surprise, I volunteered, and spent one Saturday labouring away at the cut lumber to stack it within the reserve's boundary. As part of my reward, the warden took me to the location of the two flower spikes of early purple orchid, *Orchis mascula*, which had not quite come in to flower. I returned a week later to photograph the orchids but discovered that a slug had beaten me to it and had munched through both flower stalks. I continued around the reserve as my guide had taken me the week before until I came across the assistant warden, who barred me from my circuit and basically accused me of disturbing the wildlife by entering a restricted area, the area that he was in. I was young in those days. Unlike now, I didn't metaphorically rip his head off and tell him not to be such an arse. Suffice it to say that was the end of my volunteering days at the bog.

Looking back at this event has set me to thinking (like I need any excuse) about the beneficiaries of my interest, which led to the beneficiaries of my labour, which led to my dissatisfaction with the process, while at the same time providing the benefit for the assistant warden and his ego trip. Where in this process was the benefit for the wildlife? He could walk down the path without disturbing anything, whereas I couldn't. But there is a wider issue. You have to know that the wildlife is there if you are to be able to take the necessary steps to protect it (rose bed or lawn), but at the same time you are not allowed to look at it for fear of disturbing it. Here we have a basic dichotomy: we have to know something is there to want to 'protect it', while the very action of protecting it disturbs it—and human curiosity cannot leave well alone.

As part of the earlier talk, the warden's great feature was the fact that a sparrowhawk nest was on the site. What made it so special was that two females were cohabiting a single nest and being fed by one male. This type of event had only ever been recorded once before, he claimed. And in spite of the fact that you need a licence, and in spite of the potential damage he could have done through disturbance, the warden had climbed an adjacent tree and photographed the two females incubating side by side on the nest.

While orchid hunting in Spain and France, I often find plants that I cannot identify even though I sit with guidebooks in hand. Most can be split into one of two groups, *Orchis* or *Dactylorhiza*, by the way that their roots grow. *Orchis* has typically two bulbs ('orchid' comes from the Greek word for 'testicle') in the ground, while the other group has fingerlike rhizomes. I can halve the problem by splitting my find into one group or the other, but I can only do this by exposing the root and harming or potentially killing the plant. In spite of the fact that I have travelled a thousand miles or more in pursuit of these plants, I refuse to place them in harm's way to satisfy my curiosity. I also take great care in foot placement and always scan the ground for barren approaches to the plants before approaching. Would this qualify me as an assistant warden?

Let us consider another measure of success, not GDP but GDPC, the latter being gross damage per capita. Who in the world causes the least damage to the world per person?

What Price an Olympic Medal?

It is a curious thing that we celebrate originality and protect it through copyright. Recent years have seen a variety of court cases between musicians claiming ownership of sampled or copied music. Even J. K. Rowling was accused of copying. Sometimes the claimants win; sometimes they don't. Other intellectual property is put out there for apparent public consumption, and the process is accepted for the publicity it generates. We have shades of a reverse pyramid scheme here whereby the base of the pyramid earns a living by recycling the intellectual property of the idea generators at the top. Does this explain religion? Those not clever enough to devise the message secure a living by 'faithfully' (sorry) reproducing, replicating, distorting, corrupting, and ultimately lying, as they 'belong' to the thread of the original concept. Is this the basis of building a pyramid? The need is planted in the mind of the ruler, actioned by the architects and engineers, steered through the recruitment/slave market, encouraged by the whip swingers, sponsored by the quarry owners, and ultimately beloved by the tourist guides and camel ride suppliers. Is it conceivable that all the effort, and the toil, the sweat, blood, and tears was expended to satisfy the delusions of one man, even though a god? Where in the grand scheme of things does the pyramid sit? Who takes ownership for it, the pharaoh, the quarry owners, the high priests, the stonemasons, the designers? After all, they all made a living from it, or an after-living in one instance. For pyramid, read cathedral; castle, earthwork; Stonehenge, the football league.

Let's change the topic to life-saving innovations for cars or to combat pollution. Nils Bohlin invented the car seat belt and is credited with saving hundreds of thousands of lives. Strange that we celebrate the invention of this life-saving device while simultaneously ignoring the life-taking device, the internal combustion engine and resultant car pollution. We compartmentalise the options open to us: travel as safely as possible while ignoring the fact that the act of travel will have a negative impact on the health and life expectancy of the general population. The benefits outweigh the disadvantages, but we never give any thought to the disadvantages. 'Yes, I know smoking is dangerous, but I'll face the consequences if and when I have to.'

Obesity is a force multiplier. When considering the Western waistline, think of it as the Western wasteline instead. Our lifestyle is obese to profligacy.

Consider the condemnation of the Russian state-sponsored drug cheating of athletes and compare that with the millions of pounds spent by Team GB on legitimate training techniques, facilities, science, medicine, and psychology. Then ask, considering the scale of things, which is the cheat. Every Olympic medal won by Team GB in London in 2012 cost an average of £4.5 million. We have removed ourselves from state-sponsored cheating with drugs and replaced it with state-sponsored cheating with technology, science, and facilities. To prove what exactly? That British athletes are the best in the world, when supplied with world-class training, coaching, facilities, and equipment and the taxes to pay for it all? I understand the argument for being the best in the world when it comes to our armed forces. We expect our fighting men and women to set forth onto the battlefield with the best technology, equipment, and training that our taxes can provide. What does an Olympic medal prove?

Bill Gates is celebrated as a great charitable giver, a paragon of virtue, among the richest men in the world, and he is simultaneously responsible for one of the biggest holes in resources. How fucked up can our sense of purpose be? Philanthropy.

How can we afford it? Where does the cost ultimately lie, other than in the resources of the earth? The media is full of announcements about one rich person or another divesting themselves of vast sections of their wealth in pursuit of a better world. My first thought is that they have appointed themselves as head gardener. 'We'll have a flower bed here, lawn sweeping down to the orchard housing my new collection of apple varieties, a box knot garden on the south terrace ...'

My second thought is that they have no gardening qualifications, no intrinsic skill or ability to achieve the desired aim, and indeed no idea of the consequences of their actions.

My third thought is that being rich gives you no rights to command or dictate or steer the good ship Earth on its course through history.

My fourth thought is that their wealth is generated from the sweat of human beings and underwritten by the resources of the world. How dare they announce their philanthropic gifts as a benefit to the planet?

My fifth thought is that their extreme wealth is obscene. They pretend to provide benefits to selected sections of society or habitats when they have been responsible for destruction to natural habitats on a global scale. We considered the labour days generated by JCB machinery based on one-dollar-a-day labour rates. Consider the labour days required to provide Bill Capability Gates and his brethren with $100 billion, so that they can feel good about themselves by returning it to select pet projects.

The Human Excavator

BBC teletext Science page: US bee population collapses by 25 per cent over a five-year period as biofuel surges. This is a perfect example of digging a hole in one thing to fill a hole elsewhere. But we aren't digging a hole in one thing, bees; we are measuring the population of bees, their importance to agriculture, and the cost of their absence to the cost to human food production, while simultaneously ignoring the detrimental impact of these processes on all other similarly challenged species which have no apparent impact on the cost of our food production. We are back to gardening again.

The researchers say that the conversion of land to grow corn for biofuels is a key element in the decline of the bee population.

I have demonstrated how we should revile the seller of legal highs who hides behind the legal framework of 'not to be used for human consumption' while perfectly aware of his peddled products' intended use. I have asked the difference between this shopkeeper and the corner shop selling tobacco products essentially under the same legal framework. I have illustrated the harmful practices of our food manufacturers in the pursuit of profit over product. I have illustrated the lack of control of government.

Why be an apical bud? One could view life as a struggle for resources: minerals, water, energy, space, and health. And it helps if you are at the top rather than at the bottom of the pile (assuming you don't live in a compost heap). We learn that rainforest plants have multiple strategies to exploit any opportunity to reach for the light, to gain access to space and the energy from the sun. The apical bud could be seen to be the leader—indeed this term is used to describe the principle shoot of a tree—and enjoys all of the advantages of light and space that the plant's vigour can produce. In the race to the sun, the secondary buds, it can be argued, are simply there to provide support to the leader to enable the whole to achieve its maximum potential. The leader's success or failure determines the entire future of the tree, whether it becomes a forest giant or an also-ran bush condemned to a subsidiary existence. Ultimately with trees, the earlier side shoots are disposable and simply serve to provide resources during the growing process, until they become redundant and fade away once their job is done. Are people so different from trees? Where would we place the pope or a king or a president in the hierarchy of buds?

Yet the strength of a tree is not measured by the apical bud; it is measured by the girth of the trunk, the diameter of the canopy, and the tree's fecundity. There is not a lot of advantage being the king among five people, unless of course you are the last five people left on earth. Yet it may have been the king who sanctioned the pushing of the nuclear button, the shortcut to Armageddon for those too bored or impatient to wait for our present course of action to reach the same conclusion. At the first sign of danger, El Presidente of the US is whisked away by the secret service security apparatus to a secure location where the button can be actioned. One can imagine that El Presidente will be securely tucked away in an underground hermetically sealed bunker to ride out the ensuing maelstrom of destruction. So that's all right then.

But pull back from this extreme example and think about the club captain with his reserved car parking space at your local golf club, the executive boxes with an extra 40 millimetres of arse space at the football ground, the chairman of the committee with the casting vote ...

Think about the purveyor of legal highs, stripping resources from the expendable side-shoot customers with only one care, to ensure that he himself doesn't end up as a side shoot. No, forget that, I need a different term for 'care', because in reality such people have none. They are bankrupt of this expression. But I cannot differentiate between the purveyor of legal highs and the purveyors of legal tobacco or legal processed meat. Or how about the CEO shutting her local factory down and shipping the work to the Far East to save a shilling or two? It seems that humankind has one overriding intent, and that is it, overriding. Look to the 'care merchants', those who proclaim their intent to provide succour and support to the damaged and deprived. We are talking religions and charities here. Recent reports proclaim that there are over one thousand charity bosses in the UK earning in excess of £100,000 per annum. What would you do with £100 million to salve the needs of your customers? (Tigers would need handbags, not just purses.) This begs the question for me: when you give to a charity, who are you supporting, the poor unfortunates or the management? Notwithstanding the arguments about gardening and digging holes in one place to fill a hole somewhere else, I cannot in all honesty see any purpose for the present charitable organisations that we have. They are corrupt and corrupting and are an industry in themselves. It's nice to know that they all enjoy tax breaks as 'registered charities'; it's enough for me to want to start my own religion. Let me attempt to answer my earlier question: what would you do with £100 million to salve the needs of your customers? The simple answer is, you would spend it. You have an obscene pile of money that you need to shift, and shift it you will. Imagine you were in charge; how would you account for the spend? With £10 in your pocket, it is quite easy to understand what you have achieved. Now try £1 million, let alone £100 million.

A recent report on Britain's Davis Cup tennis triumph focussed on the inadequacy of the cash-rich management's ability to purchase success for Britain's tennis professionals. David Lloyd builds tennis centres for a living and queries where the £30 million spent on the national tennis facility went when he builds them for £10 million a time. Imagine, that extra £20 million could have won us an extra 4.4 medals at the London Olympics.

But all of these questions and suggestions pale in insignificance when you consider the source of the resources. I had a visit from a Jehovah's Witness today, and it reminded me of a former sales representative of mine who gave 10 per cent of his earnings to his church, presumably to save his soul. What he could have done was work four hours less a week and enjoy his present life more before the energy in his body was recycled back into the universe from whence it came. What I could have done is made less concrete product and remained just as profitable because of my reduced wages bill. That would have meant that the holes in my local sand and stone quarries, and the more distant limestone quarry next to the cement factory and iron ore/reinforcement processing plant, could have been minutely smaller than they were. Not forgetting the hole in the oil reservoir. But it's not all negative. After all, we did produce more vehicle pollutant and CO_2 to get the raw material and finished product from its source to market. There really isn't enough tarmac to fill all of the holes in our roads simultaneously. Until we realise that we are chasing a lie and chasing the end of the world in the process, we really don't have a future.

Now, fill all the holes in the National Health Service.

Fill all the holes in the welfare state.

Fill all the holes in the education budget.

Fill all the holes in the overseas aid budget.

How about filling all the holes between the ears of our Great Brains?

You, if you are clever, will be asking why I don't respect my former sales representative's right to spend his earnings where he wanted. 'What about free will?' I hear you ask.

Would this be the free will of the starling flock? The free will of the bait ball? The free will of the wildebeest herd? The free will of the buyer of the latest must-have gadget, fad, perfume, or anti-wrinkle cream? The free will of secondary buds? The free will of football fans to support FIFA?

How about the free will of tigers to roam the earth for the next few millennia?

It's enough to make you put the kettle on. Fancy a cuppa, anyone?

My quick chat with the Jehovah's Witness did yield me another insight. If you accept the concept that humankind has been on this earth for a couple of million years as a bush and perhaps ten thousand years as a tree, then his need for his church seemed to me to be satisfying his need for an apical bud. This leads to another thread of ideas. Have we spent millions of years looking for an apical bud because DNA seeks out dominance, 'the fittest' as in 'survival of'? (Not dissimilar to the plot of the *Minions* movie.) Yet because we have not evolved to our position, simply adapted to it around ten thousand years ago, we are incomplete. Modern humankind is considered to be one hundred thousand years or so old, with some form of developmental epiphany around sixty thousand years ago, so that still leaves a gap of fifty thousand years of bush existence. Our 'apical buds' that we cherish so religiously (sorry, I had to) are as real as a peacock's tail. Our DNA condemns us to a life of false prophets whom we pursue with such fervour and endeavour as we wait in line to surrender our existence.

Are we not, metaphorically speaking, waiting patiently for our turn in the gas chamber, with no apparent alternative? I accept that we cannot cheat death. The question I ask is, why do we so earnestly cheat life? Why, in our so cherished activity of cheating life, do we so readily destroy and deny life in its many forms?

Do the lucky ones among us believe that they can indeed cheat death? All it needs is faith or the suspension of logical thought as they surrender their souls to a chemical signal from an imaginary 'apical bud'. Prehistory is blissfully silent as to the massaging of souls prior to our more enlightened age. And before you attempt to demonstrate the presence of princesses in cave art and 'Venus' figurines, remember that donning rose-tinted glasses to travel back in time simply transfers modern prejudices to our forebears and has no more relevance to them than intergalactic space travel has to us. Remember the Neolithic World Cup and ask yourself how we ended up with FIFA. Alternatively, think about FIFA and ask how we began with Stonehenge.

I watched *Jupiter Ascending* the other day. The basic tenet of the film plot is that the greatest resource of all is time. I won't elaborate beyond this 'spoiler', but I will attempt to elaborate on the principle. If we are lucky enough to enjoy our eighty or so years of average life expectancy, how would our lives change if we could look forward to eight hundred years instead? Terry Pratchett wrote humorously about the juxtapositions of the mayfly and the ancient pine tree, the mayfly speaking of the fact that 'you don't get suns like we used too' as dusk approached at the end of its single day of adult life, and the pines having short conversations that take years to pass.

Time really is our greatest resource. We are all born to die. It's the bit in the middle of birth and death that matters. Why squander what life we have in the pursuit of absolute irrelevances? Why squander the life of this precious jewel of a planet on absolute irrelevances?

Yet when threatened by a terminal diagnosis, we implore our doctors and scientists to dig a hole in their own lives to fill the hole in ours. We expect them to dedicate their every waking moment to winkle a cure out of the fabric of their time. We expect others to dedicate a section of their working lives to pay their taxes, to dig a hole in their lives to fill the hole in ours. 'But we are altruistic!' I hear you cry. Or do we seize the opportunity to earn a crust by capitalising on the shortfall of others? Remember, your tobacco supply is only as far away as the nearest corner store.

Let's turn the question around. Instead of eighty years, let's make it twenty. What would your attitude be now? Would you jealously guard your resource and commit only to things that really mattered, or would you squander it in pursuit of perfumes and a university education? How long would you tolerate adverts in your TV programmes and nuisance calls on your telephone, and how long would you save up for your mortgage deposit?

As Einstein would say no doubt, 'Time is relative.' As I walk around churchyards (not a hobby), I read the dates and feel sorrow for the young lives lost and take some comfort in the longer lifespans. Yet we seem hell-bent on filling the gap between birth and death with extracurricular nonsense fomented to provide others with a 'living' by digging holes in our own lives. We work 40 per cent of our time to pay our taxes. We fill our shopping baskets with consumer goods to fill our empty lives as fast as they, in turn, empty our wallets. We delude ourselves with the 'value' of our purchases and lifestyle in a birth-to-death helix of stupidity until, near death, those of us with some notice look back over our lives and realise the utter futility of our actions. 'You cannot take it with you' is often quoted with barely a thought to the actual notion being uttered. As you lie on your deathbed, life suddenly comes into sharp focus and the mind concentrates on the important things in life, friends and family, while 'lifestyle' disappears into the insignificance that it deserves. When faced with death, many who emerge with a reprieve suddenly claim to have a fresh perspective on life and vow to live their lives afresh and concentrate on the things that really matter. Well, here's a news flash: every soul on earth is faced with death. When are you going to achieve a new perspective on life and concentrate on the things that really matter? Start by being a selfish bastard. Refuse to be swayed by the beggar on the high street. Refuse to be swayed by the charity box shoved under your nose. Accept responsibility for your life and lifestyle and guard it jealously. Ignore adverts and shun the agents who carry them into your field of perception. How many people slam the telephone down on nuisance callers for interrupting their day-to-day activity while accepting broadcast adverts and printed media that bombard them, constantly infringing on their time?

Understand that when I say the beggar in the high street, I mean the shops as well as the scruffy con artist on his pitch. Understand that when I say you should ignore the charity box shoved under your nose, I mean the shops as well as the collection box. Consider that the clothing designer is a con man charity collector whose purpose in life is to part you with your labour to fill his own life with goods and opportunity. The term is 'adding value': take a raw or partially processed product and process it another stage while increasing the value and cutting a further slice out of your labour when the price is paid. (Are you happy to pay £16,000 a tonne for a sliced and flavoured potato?)

Understand that when I say your labour, I mean your time, the most precious commodity in the universe. You cannot make any more of it, but you can certainly shorten it through lifestyle choices.

When we received enquiries for quotation, our rather archaic response was headed with, 'Thank you for your invitation to treat.' I don't pretend to understand all the legal niceties of this expression, but I think about the price attached to the product for sale. The designer handbag displays a ticket with the implicit intent of, 'We offer to supply this bag to you for the consideration of £ [insert stupid figure here]'. The buyer in taking the product to the tills has the implicit intent of parting with £ [insert stupid figure here] to complete the contract.

If no one buys the designer handbag, it will appear at a later date with A ticket of £ [insert less stupid figure here] in the hope of finding a buyer. But remember, it isn't £ [insert stupid figure here]; it is time, in the amount of [insert stupid figure here]. The goods are offered in exactly the same way as the charity collection box on the counter. You have the control to decide whether to contribute or not. It is voluntary.

Ah yes, but what about the rich among us? They can afford it. They can only afford it because they strip a disproportionate amount of resources from others. ('What, you want £16,000 per tonne for a potato?') Or resources from Planet Earth. Rich people are thieves of time, yours and the earth's. They have the ability to squander not only their own labour but also the labour of all others who provided them with their wealth.

Work on the principle that every action is digging a hole somewhere else to fill your hole. And while you are busy filling your hole, others are industriously emptying yours for their own benefit (taxes, education, interest payments, crime, fast food …).

A World Run by Highway Engineers

The recorded history of humankind can be summed up as a continual series of roadworks as we move a hole from one place to another.

The recorded history of humankind can be summed up as a continual series of roadworks controlled by highway engineers.

The recorded history of humankind can be summed up as a continual series of roadworks that fail to solve the problem and simply treat the symptom.

The recorded history of humankind can be summed up as a continuous closed loop of development to solve a problem while at the same time expanding the scale and complexity of the problem and failing to solve the problem. This can be described as struggling to survive, but whereas our ancestors struggled to feed, shelter, and protect themselves and breed, our current crop largely take these basic requirements for granted. (Burundians and others will take exception to this statement, no doubt, while at the same time enjoying a population growth rate three times the average rate.)

In *Superman*, the home planet has been mined to such an extent that it has become unstable and implodes. Could mining be described as 'digging a hole'?

The Prime Directive

Trekkies will be able to quote the prime directive by heart.

The basic premise is that superior life forms should not interfere with inferior developing life forms on the many planets that they encounter on their travels.

Conveniently, this doesn't apply to Planet Earth and its many 'alien civilisations' of organisms that are not *Homo sapiens*. Generations of *Star Trek* fans have accepted the morality of this course of action without questioning the impact of our actions on our only Planet Earth. Given that we are all hell-bent on extracting an advantage from our fellow bipeds, what chance does everything else stand? The definition of 'alien' appears to be 'anything non-human from another world', or more technically something that doesn't share our DNA strands and comes from another planet. But *Homo sapiens* cannot separate itself from the DNA strands that form every other life form known to humankind on this earth without denying our origins and having a negative impacting on our own future.

Chickens have teeth, remember? It's in their DNA, a throwback to the dinosaurs and beyond. It is no accident that we have two eyes, a backbone, a nervous system, four limbs, and all the other characteristics that define us as us. Our pre-bipedal gait was defined by the earliest fish that left the water to crawl on land millions of years ago. We still carry fragments of that DNA within our own. Without it we simply wouldn't be able to walk. We think of evolution with an analogy, the 'tree of life', with many branches developing from a common core, ultimately sourced from single-celled organisms. We can see the birds, mammals, fish, flowering plants, and so forth as distinct branches of the tree and can further divide the apes from other mammals and humans from the apes. Charles Darwin was savagely ridiculed in the press for his assumptions that humankind evolved from the apes. His head was superimposed onto an ape's hairy body with a plethora of scorn. The problem with the 'tree' analogy is that it suggests that each species is essentially at the growing tip of each twig or stem and can only grow in one direction, divergently. We are beginning to understand that our DNA is corrupted by fragments of DNA from viruses and other pathogens and that this DNA is replicated and passed on through our generations. Chicken DNA is modelled by the dinosaurs and their teeth, but this only illustrates one tiny facet of chickens' historical biology. For all the divergence and apparent gulf between us and daffodils, we have a common ancestor and a common basis for life. In our tree of evolution, we think we are separated by diversity, whereas it would be fairer to say we are more connected by our origins.

We have not progressed beyond the Stone Age in our reasoning abilities. We have settled on an apparent tried-and-true order of doing things that served us during the Stone Age but has no chance of serving us now or in the future. We do not have the intellectual tools to survive in the twenty-first century and or to solve the problems that our activities produce. With hindsight we quickly skip over the earlier epochs of human development as primitive, ignoring the fact that they lasted for hundreds of thousands of years. What about the pre–Stone Age? Our ability to seize on the new, superior advance belies the fact that it takes millennia to achieve what we would now consider to be baby steps. Our thought processes are still trapped in the Stone Age.

The Stone Age lasted for millions of years and predates *sapiens*. Indeed it is normally split into early, middle, and late eras to reflect the earlier *Hominins*, Neanderthals, and modern humankind's influence.

Consensus on the Anthropocene is hardening in geological circles with a start date proposed of 1950, whereby the activities of humankind can be identified in the geological record for the future. That is to say that a physically detectable layer of geological deposit is forming in the strata of the earth as a direct consequence of our actions. This ignores the absence of fossils of recently and soon-to-be extinct species that we are denuding this planet of. We do not stop to consider our past with the due diligence that it requires. We have designed stone tools that have been fashioned in a similar pattern for millennia. Archaeologists will date styles of fashioned rock to periods, which information indicates precisely how our minds have worked for generation after generation. We are trapped into repetitious thought processes that bind us to our place for centuries.

Ars Warts (i.e. *Star Wars*) strikes me as a perfect example of our inability to think beyond the pro forma. Imagine the scenario: space travel enables movement between galaxies in limitless space with infinite opportunity to exploit an infinite number of habitable planets, and yet we have the same old conquest-and-control bullshit with the evil empire against the freethinking liberal republic. If you are a billionaire, how much more do you need? How about a trillionaire; how many noughts will make you happy? How would your wealth be measured? Precious rarities would be extractable from a multiplicity of sources and lose their scarcity value. Ignoring the fact that the liberal republic only exists as a counterbalance to the evil empire, its other purpose is to provide an alternative means of employment for an alternate grouping of Great Brains so that they can forge a very good living without getting their hands dirty.

For all the fantastic leaps forward in technology, large sections of the action revolve around storm troopers leaping out of landing craft to face down the machine guns on the beaches. I know the original films were described as the first modern westerns, but why bed it in cavalry-charging tanks territory?

Except, of course, this tried-and-true formula is safe and risk-free as it becomes the most successful film in cinematic history. And that says it all really. We have no imagination beyond our latest stone arrowhead design that has served us for millennia. But the human species, through sheer weight of numbers, has developed into a planetary scourge worthy of the evil empire that will destroy entire planets without as much as a backward glance or cognisant plan.

A commonly held belief is that the Clovis point was the only man-made weapon point used by North American Indians from fourteen thousand to twelve thousand years ago. This gives this technological tool a lifespan of at least two thousand years before alternate examples surfaced. What I am trying to illustrate here is the fact that this simple (to our eyes) technology survived unchallenged for one hundred generations of human beings.

Imagine the situation today. Take a trip down to the local supermarket and peruse the vast range of point sizes and styles, and then compare these with other ranges on the high street and the Internet.

But consider this thought: our intelligent Stone Age predecessors enjoyed a thought process that survived unchallenged for one hundred generations. This would be anathema to our present culture, which swaps its TV sets for the latest advances. You'd have to consider Christianity to find a similar two-thousand-year thought process that has maintained its near original form. The Egyptians built pyramids for an estimated one thousand years.

Homo sapiens is established in fossil evidence around 160,000 years ago, and it took around 100,000 years for them to leave Africa to expand the human range. The suggested timeline is a move east across the Red Sea 60,000 years ago, leading to the colonisation of Australia by 50,000 years ago. Subsequent waves of migration led into the Middle East and then on into Europe and Asia, with a final occupation of America around 15,000 years ago. It took only around 1,000 years to reach the bottom of South America. The last 60,000 years of human history could be summed up as a period of exploration, but this suggests a deeper purpose, so I will refine this summation to a process of simply 'seeing what's over the horizon'. Even today our scientists probe ever deeper into space with manned and unmanned probes and ever more sophisticated telescopes with ever more wavelengths. Or we plumb the depths of the deepest oceans or seek out the minutest of particles with microscopes or the hadron collider because, at its simplest, as a species, we are never content with what we have. We have this unwavering ability to break the toy to see how it works while ensuring that it will never work again.

It has been said that some of the scientists working on the Manhattan Project developing the first atomic bombs believed that the resultant chain reaction generated by the first test explosion would never stop and would see the destruction of our planet. And yet we detonated it anyway.

We can build pyramids for fifty generations, maintain a stone tool pattern or Christian belief for one hundred generations, or explore our horizons for twenty-five hundred generations. We have terraformed through agriculture for five hundred generations. It would be easy to conclude that we have the rapier-quick mind of a stalagmite.

But we can go further: the singular drive of all living organisms is to survive and replicate, a process that has occurred for billions of years. At no point in the entire history of life has a conscious thought questioned or surfaced to provide a conclusion to this process. Life just is, a self-fulfilling autonomous process of energy capture that replicates itself until our local star ceases to exist in its current benign form, or until a similar 'endgame' occurrence.

Why do we seek out new horizons? What is the imperative that drives this action? Is it simply a faulty gene that makes our species incapable of contentment? I hate to say it, but it's as though our innocence was stripped from our souls, we became embarrassed by our lack of clothes, and we were thrown out of the Garden of Eden. Was this the great event that occurred sixty thousand years ago that converted *Homo sapiens* from an unassuming animal into the 'Great Ape' with its Great Brains, the destroyer of worlds? Why is it our species, and not the chimpanzees or gorillas, that has dominated and exploited the earth's resources? Would it help us to understand the process if we took our clothes off and threw them away? (You go first. I'll follow.)

I do understand that our position as dominator is up for discussion. Our numbers are pitiful when compared with the number of insects or nematode worms or bacteria. And one could argue that any or all of these alternate life forms would destroy all before them if they were given the opportunity. The difference, I would suggest, is that humankind is exploiting the opportunity already while being (sub?)consciously aware of it. We are all waiting for somebody else to do the heavy lifting for us to make it all better, be it national government or charity or world agencies, but we know that cheese wells will never exist, so why delude ourselves?

Is it simply a faulty gene, or have we learnt it? Is it hereditary or nurture? Ask the advertising industry, and they will hold their hands up and simply say that they never forced anybody to do anything they didn't want to. And they would be right, but they have created the environment and the peer pressure which has had the same effect. Was the hypothetical forbidden fruit actually an advertising campaign with the tag line 'Go forth and find your fortune'? (Or designer rags?) The final irony is that the emperor was right to don his new clothes, although he did it for the wrong reasons, and by a quirk of fate was disabused of his ambitions by an innocent child.

But children are born innocent and have to be programmed to accept the present reality. Every child is born into the Garden of Eden and has to be taught the shame and desire that motivates the human race. And it is our Western culture that is accelerating the process, with ever shortening childhoods and ever increasing thresholds of desire and need. Every child is born into a world where their desires are met by food, warmth, shelter, and security, and every soul is satisfied at life's end by meeting the same requirements. It's the bit in the middle that does all the damage. I know; it's the bit in the middle that provides the food, warmth, shelter, and security for babies and the elderly to enjoy.

But we have considered food and its modern-day version of expensive, processed, poisonous, terraforming, habitat-destroying, and extinction-driving impact ...

But we have considered warmth and the earlier climate-changing destruction of forests together with the later climate changing-destruction of ancient forests via our time machine ...

But we have considered shelter and its modern-day version of thirty years of bonded serfdom as we strive to pay our mortgages or someone else's via our rent ...

But we have considered the lack of security, where we pay for police and cameras and the (lack of a) justice system and security alarms as we tolerate crime at every level ...

Tell me again what the bit in the middle is designed to do? How many horizons must we cross to find the solution that we desperately seek while desperately ignoring it?

Evolutionary Cul-de-Sac

The giant panda is described as a species that has entered an evolutionary cul-de-sac in that it has selected a diet that barely keeps it alive and has a reduced fecundity that renders it on the verge of extinction. Pandas are doubtless unaware of their predicament. Left to their own devices, they will probably outlive the human species. Part of the problem will be one of measurement, something akin to the 'fact' that bumblebees are too heavy to fly or kangaroos have a diet too poor to enable them to travel long distances. Is *Homo non-sapiens* aware of its own evolutionary predicament? By comparison, humankind has entered an evolutionary cul-de-sac in that it has selected such a preposterous diet and achieved such high levels of fecundity that it will not rest until all resources are severely compromised.

Ars Warts illustrates the pro forma of thought that saddles our species. Let us call it the command-and-control chain, because when command is exercised and control required, it is this chain around our brains that is jerked. In the days before the Sky continuity announcer hijacked the credits by talking over them and shrinking them to half size or less so that the next film could be promoted, one could read some of the small print that used to talk about the 'original motion picture' and its right to copyright protection. Of course, what it should actually have stated is that the owners of the copyright reserve the right to regurgitate the same old storyline with fresh players for a fresh audience's consumption, taking the line of simplest risk and grasping the maximum reward. Disney used to release their 'children's classics' on a seven-year cycle so that each new generation could view them as 'new releases' and Disney could keep earning from the original investment.

The message appears to be one of 'follow the matrix, stultify original thought, hijack and take ownership of any successes, and above all protect your own position'. At the same time, the finished new product is championed for its fresh new original risk-taking approach. Why, as we reach out for the new experience horizon, do we settle for a version of the one we've already got? Is there a default gene that kicks in and declares, 'I was only joking'?

I used to criticise one of my sons along the lines of, 'You want to know what's for lunch while eating your breakfast.' I should have gotten him a job at Sky, because when sitting down to watch this week's premiere (Sky's pick from among the Western world's weekly offerings that they think are so earnestly correct for us to view now), the viewer is assaulted with alternative titles on alternative channels or alternative times for films that she really ought to see *now* and that Sky will repeat at the film's end before the viewer has had chance to pause, reflect, and swallow the present bolus of entertainment that she has just spent ninety minutes chewing. Imagine if you will a tome of books written by a prolific author whereby all the works are contained within one cover and simply continue without chapter or page break from one novel to another. Alternatively, don't imagine, but enter the world of the box set TV fest where entire series can be enjoyed (endured) in one sitting. I have never been able to determine why art galleries devote vast wall space to small paintings in glorious solitude. Why not fill the structure from floor to ceiling with every available artwork for the masses to enjoy? Surely there is no advantage to considering an artwork in splendid isolation, framed within its own space and given time to contemplate its merits. Smokers out there will know the 'pleasure' of the first of the day, the first drag of each new cigarette, the little reward that you have saved up for yourself since the last one. How would this experience compare with that of a chain smoker, who ignites each cigarette from the stub of the earlier one? How does a drunk reconcile the taste sensation of the thirtieth drink of the night with the first?

I heard a comment yesterday from an overweight and overexposed individual. She suggested that her 'condition' just proved that she had 'lived', as in 'lived life to the full'.

So having lived life to the full, why does everyone bleat about the consequences and expect everyone else to provide a solution to their self-inflicted condition?

Betrayed by Our Appetite

Why do our journalists examine in great depth various aspects of what should be areas of great concern for the general population and then fail and withdraw from their obvious conclusion before arriving at it? Journos will bleat about individual charities or executives and identify flaws and failures without ever questioning the principle of charity and the effectiveness of it in its entirety. All are blinded by the 'good causes' and the obvious benefit of charitable activities. Yet they will espouse arguments about individuals and off-piste activities such as court cases against fox hunts or actions to gag whistle-blowers or other dissenters which lead ultimately to the bigger question of why have a charity in the first place without ever breathing the question. It's like they set out the ideas that 3 multiplied by 3 equals something without ever finding the answer. Is it because the readership is not prepared to accept the conclusion and therefore the guided missile of argument has to be diverted to miss the target for fear of losing the readership?

I was canvassed by a Jehovah's Witness recently, and I have reminded you of a former employee of mine who gifted 10 per cent of his earnings to his church. The latter had formulated his voluntary contribution to his chosen aim and further backed this up with time spent doorknocking and canvassing others to join his club. He arguably devoted 10 per cent of his waking life to his religion in pursuit of redemption in the next sleeping life/death. He had the potential to dig a five-year-wide hole in his life and in the lives of those he loved in exchange for what, a pro forma, a roadmap for idiots, searching for fairies at the bottom of the garden.

The calculation has been made that every cigarette smoked reduces your life expectancy by eleven minutes (240,000 cigarettes at 25p each, for a total of £60,000, is equal to five years of life). Consider the £60,000 cost of the cigarettes and compare this with the notional pre-tax average £30,000 salary per year. Each average smoker will have devoted an additional two years of his or her working life to the generation of this resource, so the average reduction could be considered to be seven pre-tax years and not five.

In these simple terms it is possible to see the hole dug in a smoker's life. Many will take this information on board to either quit or reduce their smoking, or indeed never start in the first place.

While we legislate and tax and treat smokers, educating them on the evils of their activities, we singularly ignore all the other evils that strip time from our lives as we 'live our lives to the full'.

In a final irony, as we legislate, tax, educate, and treat smokers, we allow them to strip time from our lives as well, whether we smoke or not.

Let us consider a mathematical exercise without actually doing any maths. Let us look at thieves of time and consider them within the context of our own individual lifetimes.

Start with the premise that we all have an allotted time on earth, specific to every live birth and unknown to each individual. We can talk about average life expectancy, but this doesn't help you as an individual other than in the calculation of your life insurance premium.

At the same time, consider that your own reasonable life expectancy could fall into any of eleven decades starting from ages 0 to 10 and ending at ages 100 to 110. If you knew that you wouldn't reach your tenth birthday, would you want to go to school for six of those years? If you knew that you wouldn't reach your thirtieth birthday, would you stay on at university until you got your doctorate? If you knew that you wouldn't reach your fortieth birthday, would you commit to a thirty-year mortgage spend? If you knew that you wouldn't reach your seventieth birthday, would you hive off a percentage of your earnings for a pension?

Occasional news items feature teenagers who are dying from an incurable disease and are desperate to continue their schoolwork and sit their exams. Why? A plus in English and maths doesn't end up carved on the headstone. Excelling at school will not make the disease go away.

I know that the counterargument to my train of thought is that school is where their lives are, including their friends, their support network, and a focus is needed to relieve the mind of the impending surrender of energy back to the cosmos. Quite often these individuals and grieving parents set up a charitable foundation with a simple end to provide a TV set for the hospice or holidays to Disneyland in an effort to alleviate the pain and pointlessness of it all by making a contribution to help others.

This examples I have described could be summed up as a compliance with a pro forma, a standard format of life and living that removes the need for conscious thought or effort. 'Let us follow the conveyor belt to its inevitable conclusion and be comforted by the normality of it.' If you knew at birth that your child had been chosen as the blood sacrifice to appease the gods on his or her sixteenth birthday, would your behaviour change? Would your desires gravitate to the other extreme of the bucket list? (Note that I already slipped Disneyland into the discussion earlier.) Let us add a nod to *The Bucket List* to our otherwise pro forma–filled life and pretend that time with our dying child is precious within the context of the other pro formas within our lives. These would include the need to pay our mortgage and our taxes because, after all, life has to go on.

Seeing Past the Avatars

There is talk of a billion-dollar-a-year 'love scam' whereby vulnerable people are targeted via dating sites on the Internet. The situation is that victims seeking love are seduced by criminals into forming relationships with non-existent avatars that offer the perfect relationship partner. Often these criminals pose as members of the armed forces or workers in remote places where enforced absence and a lack of personal control over their circumstances prevents them from ever meeting the object of their affection. The victims themselves have often been on the receiving end of recent emotionally upsetting incidents in their lives, covering disease diagnoses, spousal deaths, divorces, or difficult break-ups with long-term partners. They are emotionally vulnerable and easily manipulated by what are essentially hardened con artists whose only intent is to gain a monetary advantage over their victims. The crimes are particularly nasty in that they play on human emotion to effect a result. Often the technique employed is to start small with urgent requests for small gifts of money for a child's birthday or school trip, before ramping up to significant cash demands to finish university studies or complete a cash-strapped work project. Individuals can lose life-changing sums of money to these scammers and can also lose houses and jobs through their desire to financially support the objects of their love. Even after the scams and the avatars are exposed, many victims are still trapped 'in love' with the non-existent objects of their desires. Emotionally, they have been drawn in to an unassailable cul-de-sac of love where the object of their desire does not exist. Avatars are carefully constructed to reinforce the needs of the victim and to blind said victim to the reality of his or her situation. The Internet lends itself to this mental manipulation with cloned photographs and only text to study and relies, quite simply, on the victim's willingness to remove reason from his or her often irrational beliefs.

There is a contradiction in intent—isn't there always?—whereby money is volunteered towards legitimate demands and the 'crime' is only discovered after the exposure that the money has not gone to its desired end result. It's not unlike the prostitute who appears at the police station to complain of rape. 'When did this happen?' asks the desk sergeant. 'Tuesday,' comes the reply. 'But it's Monday. Why did you take so long to report it?' The reply comes back, 'I didn't find out the cheque had bounced until today.' The crime only exists because of the knowledge of the deception, whereas without the knowledge of the deception there is no crime.

The billion-dollar figure is an estimate, with many victims too embarrassed to admit their gullibility. Therefore the true figure will never be known. Victims are even persuaded by their avatars to provide compromising videos of sexual actions which yield avenues for blackmail. We are, many of us at least, prepared to surrender sentient, rational thought processes to artificial worlds created on the Web.

What strikes me are the parallels between these 'love scams'—no, that's not the term to use; these scams targeting the emotional frailties of the human mindset and the other emotional scams that dwarf these relationship cons by both their financial scale and orbit.

These other scams often begin with urgent requests for immediate financial assistance to address a pressing concern. They start small with requests for one-off payments which are often converted into regular 'subscriptions' of support. Emotional reinforcement of the process is achieved by the regular provision of 'tokens' of reward that remind the victim of the 'correctness' of his or her actions and to ensure the continuation of the financial support. Victims are targeted with emotionally charged voice-overs against a backdrop of emotional, crisis-demanding, urgent-attention-seeking scenarios. Slick campaigns constructed by professional agencies are designed to tug at your heartstrings to elicit the desired response. The earliest stage is always the formation of an emotional response. Thereafter rationale is often dispensed with as the 'client' is manipulated into continuing the financial commitment. At what stage, if any, does the realisation that the monies have gone astray and you have been the victim of a scam occur?

I could start with my overworked example of killing foreign birds by saving our own. I could look at donations to save a dying child in famine-torn Africa ending up in the hands of the same agency whisking economic migrants from the territorial water's edge off Libya to landfall in mainland Italy, and through these actions ensuring that more will attempt to follow suit by risking life and limb crossing the Sahara Desert or failing to achieve the twelve-mile boundary. These would be the hands of predominantly fit and able twenty-something men in search of a better economic prospect who have paid people traffickers for the short boat ride to salvation. I have already referred to the excessive pay packets enjoyed by the top one thousand charity bosses of a minimum of £100,000 a year. (That's at least £100 million 'gone astray' for starters.)

How easily the human mind steps across the non-existent chasm between the thought processes of scam or scam, legitimate or illegitimate, right or wrong, doing good or being done bad to. We have official truths, ably supported by governments that aid and abet charities with tax breaks and even financial rewards, and official lies of scam artists levering financial reward from the bank accounts of duped and misled individuals whose only mistake was to have become emotionally involved and blinded by the skilful playing of heartstrings. Every day we are presented with forks in the road and decisions to turn left or right, wander over the horizon or stay put, fund the actions of others as they fund themselves, or selfishly ignore the demand of extracurricular demands on our planet. Often these charities/criminals pose as members of the armed forces or charity workers in remote places where enforced absence and a lack of personal control over their circumstances prevent them from ever meeting the object of their affection. 'Caveat emptor' is the phrase you are looking for, 'let the buyer beware'; and you hadn't even considered that you were buying something. As you plug in to the matrix of emotional need, are you not surrendering your time and life to the 'machine' that seeks to destroy our planet? Caveat emptor relates to the disparity between the information known by the seller and the information known by the buyer. One can imagine the discontent if you had bought an earth-saving VW diseasel from your local garage only to later discover that the seller had hidden salient facts from you, salient facts that would have coloured your purchasing decision and led to your subsequent dissatisfaction when a different reality was made known.

So it is, within the confines of the 'love scam', that money is prised from the wallets of smitten adorers who are unaware of the true nature of the product, the 'affection', that they are buying. Is this not also the case with charities that headline the plight of selected individuals so that the entire apparatus of the business model behind the charity can be maintained? It is a business model that seeks to hide the salient facts from its buyers as resolutely as any con or scam artist. Have we progressed at all beyond Edward Longshanks saying, 'But we'll hit theirs as well,' in reply to the raised concern that archers would rain down death on one's own combatants engaged in hand-to-hand fighting with the enemy? I am reminded of the scandal that reached the media several years ago about a 'rehousing' scheme set up by a zoo for unwanted pet rabbits, guinea pigs, gerbils, rats, mice, and derivatives. The pets were 'rehoused' within the stomachs of the zoo's large snake collection. Some staff were distraught at the sound of small mammals being beaten against a board to stun them before being fed to the snakes after closing time and had complained to the media. (No whistle-blowers here; it can be a dangerous occupation.) Yet various 'pet charities' routinely euthanise healthy pets that they are unable to find adoptive homes for. Research charities will fund experiments with animals while others seek to end the practice. They can't both succeed. Yet the same charity-funded researchers will hide the use of animal experimentation from its donors lest the disconnect between stated intention and actual intent be discovered.

It's nearly time to introduce you to Hannibal's brother—no, not the one with the elephants; the other one, Lecter, you know, Caveat Lector, or in English, 'let the reader beware'. (Hannibal's surname was corrupted, with the *o* becoming an *e*, when he entered US Immigration.)

Where else can we find these non-existent avatars, I wonder—but not for long (wonder, that is)? Maybe it's the avatar that says, smoke this. Or inject this. Or paint this on your skin to become beautiful. Or eat this. Or drive this. The list is maybe only limited by your imagination, except, and this could be key, it's your imagination that is suckered by the avatars.

Killing Strangers

We kill strangers so that we don't kill the people we love. We are familiar with Bangladeshi sweatshop factories burning down with the workers locked inside, or multistorey buildings collapsing due to illegal building practices. Our retail outlet chains are quick to wash their hands of any responsibility and talk about 'ethical' buying practices. Yet the whole business case of sourcing clothes from Bangladesh is based on cheap overheads and costs, which converts into cheaper products in UK warehouses. For cheap costs, you can read low wages, long hours, minimal health and safety, poor health, and overloaded and overcrowded factories. In China, the economy is underwritten by the provision of cheap energy sourced from among the most dangerous coal mines in the world reinforced with horrendous levels of life-shortening pollution that doesn't discriminate between rich and poor. All of these are circumstances that would not be tolerated in Britain. We are killing strangers so that we don't kill the people we love, but it's all right because we can buy cheaper goods. But it's all right, because we can kill strangers as long as we don't kill the people we love. We can kill strangers, out of sight, at arm's length, and without compunction. It is the natural order of things. It becomes their side against ours, and deep down we all know how that one ends. But we don't, because we accept this established position at home as well without a thought, also accepting that our doctors have to make choices based on age or availability of resources or, quite simply, by the lottery of waiting lists or the cost weighed against the benefit.

I stumbled across a Russian website called Stop a Douchebag, or Stopham as it's known in anglicised English. This organisation has taken upon itself a direct action campaign against the traffic abuse inflicted on city residents by motorists who engage in unlawful activities. There are some five million cars in Moscow alone, and the multilane roads often move at a crawl. Muscovites get around this problem by driving at speed along segregated footpaths, even though they are occupied by pedestrians. Shortcuts through city parks are commonplace. I have learnt from the subtitled videos that the concrete barriers used to prevent vehicle access are removed in the winter to allow the passage of snowploughs to clear the footpaths. Stopham, which translates into an anti-rudeness campaign, will block the passage of these vehicles and politely request the drivers to reverse and go back to their entry points. They do this by blocking the passage of vehicles while on foot. Reactions from the drivers range from apologies and reversing to ramming the Stopham pedestrians and carrying them on the bonnet at speed for hundreds of metres until they are stopped in the congestion. All the events are captured on video and posted on YouTube. The sanction that they use is to stick large adhesive posters onto the car windscreens. The more belligerent motorists will have their screens covered with three or more posters. Now, you can either applaud Stopham's actions or condemn them as idealistic activists who should find something better to do, it doesn't matter, but it does give a wonderful insight into your average Russian.

Stopham also organise action against illegally parked cars that block footpaths or block pedestrian crossings or simply double park in live traffic lanes. (And the people who do this wonder why the roads are so congested.) They also target the illegal car parks that spring up across the city. Offices and businesses will pay for 'parking attendants' to control public roads and car parks to ensure that only their employees or customers have access to the 'private car parks'. Others simply barrier off public space and then charge for the use of 'their' car park. Stopham fund their activities by raising revenue via YouTube which pays for the stickers and replacement cameras. They have even received verbal support from Vladimir Putin himself, but they receive no official support. And they target everyone from rich to poor to government officials, and even diplomats who attempt to abuse their positions.

As an aside, I was driving through my local town past the senior school when I came across a small group of sixteen-plus students who were enjoying a packed lunch on a bench just down from the school. As I passed, a girl stood up and threw her rubbish to the ground next to the rubbish bin. (Our education system certainly engenders a sense of respect for the environment.) As ever, this set off a train of thought in my mind and I formulated a strategy for approaching this girl, purely a hypothetical one of course. I would ask the girl why she had thrown her rubbish down by the bin instead of placing it in the bin. I would then ask the girl what she would do if I slapped her face. The expected response to the second question would no doubt be that she would call the police and expect them to charge me with assault. This produces an interesting dichotomy, because she wants to ignore the law when it comes to littering but then wants the protection of the law when it comes to assault. I know that I have already touched on this idea with 'outlaws' being outside the protection of the law, except that this point for discussion is somewhat overboard for littering when compared to assault.

Back to Russia: the irate motorists will emerge from their cars, swearing, pushing, punching, threatening, and often producing weapons ranging from baseball bats to pistols and knives, and even fragmentation grenades in one instance. They are usually overpowered by the Stopham activists, who release the individuals as soon as they calm down. It is at this point that the 'injured motorist' then calls the police for assistance because he has failed in his attempts to achieve his goals. This is in spite of the fact that his entire behaviour has been caught on multiple cameras. The offenders are always the physical protagonists, and the activists are models of politeness, given that their campaign is against rudeness. Many of the videos finish with the drivers having their licence details being taken by the police and fines being issued for driving on pavements or parking illegally. As a final twist, the women are often the worst for resorting to violence, even though the activists refuse to fight back, unless the woman has attacked women within the group.

This whole scenario exposes much wider issues, highlighting the obvious corruptness of the police, who will walk past all the illegally parked cars and footpath-driving vehicles seemingly reluctant to take action until goaded by the presence of cameras to do what they are employed to do.

I will call it a 'sister organisation' that goes under the name of Lion Versus; their aim is to highlight the ignorance of the laws about smoking and drinking in public places that were introduced some years ago. Activists will politely request that smokers extinguish their cigarettes and move to the designated smoking areas in future and also protect the women and children who walk by from their smoke. Once again reactions range from apologies to swearing, threats, pushing, and punching for daring to impose restrictions on their illegal activity. The activists, who do not hold back, will tackle police cadets smoking on train platforms as well as full policemen and security guards. Once again, videos frequently end with smokers being removed to the police station for processing.

I am reminded of the Sting song lyric whereby he expresses the hope that the Russians love their children too, in the belief that this could prevent a third world war. On the evidence presented by the obstinate, ignorant Russian smokers, drinkers, and motorists, I think they would press the button to launch the nuclear missiles and worry about it later. These characteristic belligerent responses are seemingly endemic to Russians and can be seen in the posturing and rhetoric of Putin.

But when I cast my mind back to earlier sections of *How to Kill an Elephant*, I realise that I have identified similar characteristics here as well. The motorists caught speeding will bleat about, 'Haven't you got some proper criminals to catch?' While our motorists typically stay off the footpaths, they will use the hard shoulder on blocked motorways and will ignore speed limits and weave between lanes, undertaking to gain an extra car length. Not to forget the chaos that ensues at the school gate when single-minded parents (should that be single-thoughted?) fight their way through the process of dropping/collecting little Johnny. It appears that all one has to do is impose a time condition on a restricted area for our underlying human nature to surface. My dad used to take me to school late, big deal, but if I went by official bus, that was always late anyway, so what was the difference?

Trawling through other websites on Russia I have stumbled across other demographic reports that make for interesting reading, especially within the context of earlier chapters. Two out of every three Russian men are drunk when they die. As a nation, Russians' alcohol consumption is among the highest in the world. Moves to increase the duty on imported beers have driven the populace to cheaper spirits, and imposition of controls on spirits has led to the drinking of perfume and illegally produced alcohol as cheaper alternatives. A significant proportion of children are born with defects caused by alcohol or tobacco, and only 30 per cent of births are recorded as healthy. Life expectancy is low; even a 15-year-old boy in Haiti (with a per capita income of less than $1,800) can expect to live three years longer than a 15-year-old Russian boy. Drug addiction to Afghan heroin is rife, as is HIV infection from shared needles. Half the army conscripts are unfit for full service due to ill health. Russia also has the second-highest number of multidrug-resistant TB sufferers, coming in behind India, whose population is around nine times greater. If only 30 per cent of babies are considered healthy, then within the context of the 'runt of the litter', 70 per cent of the population have already been condemned to an early grave. Even ignoring further impacts from alcohol, tobacco, and heroin, their lives cannot be redeemed whatever action is taken. The Organisation for Economic Co-operation and Development (OECD) and World Health Organisation (WHO) estimate that over 30 per cent of all deaths in the Russian Federation are a direct consequence of alcohol. This compares with 3.2 per cent in the UK. *Russia Today* suggests that there are two million registered alcoholics in Russia. An estimated fourteen thousand people are killed in drunk driver accidents annually, which should be compared with the provisional figure of 290 deaths in the UK in 2012—with roughly half the Russian Federation's population of 140 million.

Custom and Practice

At what stage does an idea or invention become custom and practice? The Clovis point existed for two thousand years, pyramid building for one thousand. At some stage each 'idea' or 'invention' became the norm, with no further mental process required. Each became a pro forma, a standard methodology, a chemical control from the queen ant that blinded the workers to the need for independent thought. It is reported that some sections of society are awaiting the rebirth of the Messiah. The year 2000 was seen as a likely anniversary date, and indeed the Messiah may be walking among us now, biding his time to announce himself to the world, or finish his schooling obligations, anyway. (I think David Icke has largely been discounted now.) Now God might be thinking, *For fuck's sake, I gave you one. How many more do you want?* My own peculiar interpretation is that the section waiting avidly is composed of highway engineers who, having invented double yellow lines, are now dreaming of a rebirth of something omnipotent, something along the lines of a double red line whereby they really, really mean it this time. Something-all powerful that will not be ignored.

It Must Be Important –
Look How Much We've Spent

The BBC has just produced its long-awaited report into the child abuse activities of Jimmy Savile and Stuart Hall to much fanfare and airtime. Dame Janet Smith, a former high court judge, spent three years and £6.5 million to produce a thousand-page report in three volumes. Major 'sound bites' decry the fact that the BBC was in awe of its celebrities who wielded too much power and were untouchable. Ignoring all other aspects of the report, I will simply make some simple comments. If the problem at the BBC is the fact that celebrities wielded too much power and were above reproach, why has the same BBC employed a 'celebrity' judge to produce a report that can then been paraded to the media on the basis that the said 'celebrity report' will be considered to be beyond reproach? If the 'truth' is the 'truth', then does it matter who produces it? And why does it have to be a retired high court judge who can wield her celebrity status to generate sufficient gravitas? In this action, the BBC has confirmed that it is still in thrall to the control of celebrity and has deliberately set out to employ an 'untouchable', irreproachable entity to prove that they have in fact learnt absolutely nothing.

Notwithstanding this fact, we are back to (*a*) a £6.5 million bill and (*b*) a thousand-page report that is then broken down to a fifteen-minute news item, which will be reduced still further by the average viewer to become the sole element carried forward in the minds of the populace. In English, this translates into 'read the executive summary' and file the rest under 'proof that we are serious about it and have the volumes to prove it'. Oh, 'And haven't I justified my paycheque.'

Just one other thought: the belief that the BBC was unable to control its celebrities and that opportunities had existed to halt the accused activities was stated before the enquiry started. Why spend £6.5 million to learn something that was already known?

Unjust Distinction

While I was at a hospital clinic the other day, my eye was taken by a diabetes poster advertising the idea that you shouldn't discriminate against diabetics on the basis that diabetes doesn't discriminate.

Information found at https://www.worldometers.com states that 85 per cent of diabetics are type 2, and of these 90 per cent are obese or overweight. Type 2 was largely a disease of adult later life but is now being seen in obese children prepuberty in the USA.

Of twenty-two industrialised countries, the US has the highest obesity levels. Two-thirds of the people over 20 years of age are overweight, and nearly one-third of people over 20 are obese. Taking this information on board, do you think it would be reasonable to suggest that diabetes does indeed discriminate in the vast majority of cases?

But the medical profession is forced to discriminate every single day on the grounds of insufficient resources, so why not discriminate between those who help themselves and those who plunder their health through their activities? Too harsh? Consultants working in the field have declared that obesity (which often manifests itself as diabetes) is a ticking time bomb that will bankrupt the NHS. The counterargument to this is that we must increase the resources to suit the needs of the patients. Good luck with that one. Why not have a proper adult debate about the problems generated by our lifestyles, diets, and self-inflicted injuries? What am I saying? Such a forum does not exist, nor will it ever for fear of upsetting someone. We will just let them die instead because we lack resources. All those who die become silent witnesses to the hypocrisy of our organisations, victims to the blind stupidity of universal access to healthcare without any obligation to protect your own health. I am reminded of a case recently whereby a farmworker was fatally injured when he fell from a large crop box held off the ground by a telehandler during a transit movement. One would think that the farm owners would rightfully be prosecuted for gross negligence, until, that is, you learn that the accident happened the day after a training course that highlighted the perils of being carried in crop boxes suspended at height by telehandlers. Why is it possible to ignore health issues on a personal basis when prosecutions can apply whenever a corporate body is involved?

Workers have to be protected from loud noises in factories and are regularly assessed for hearing loss by employers so that in the event that hearing deteriorates over the course of a career, blame can be ascribed to the employers who failed to offer adequate protection. Yet the same employees can spend hours in the evenings and at weekends listening to excessively loud music without any control. Now, in this example, help is at hand for the employers, because nightclubs tend to damage the hearing at different frequencies than industrial noise and blame can be attached to the appropriate cause. A further analysis would suggest that if your employer force-fed you food to the point of morbid obesity, you could claim against your employer for damaging your health, and yet if you simply ate off your own back, then the slate is wiped clean and the full resources of the state are mobilised to 'manage' your personal condition. Meanwhile the resources raised by taxes are finite and limited and care is provided in a discriminatory way to the recipients based on the lottery of distribution. Seems fair to me, or rather it doesn't; it seems obscene to the extreme. Our Great Brains' solution is to raise more money through taxes or to ration taxes and resources to universal requirements. One could sum this up as robbing Peter to pay Paul, but a simpler description would be the analogy that we dig a hole in one resource to fill a hole in another. Stir in the endemic corruption found within all large organisations and the problems of supply and demand are exacerbated exponentially.

Guidance Note

Our police cannot discriminate on the grounds of race, creed, sexuality, religion, etc., but they can discriminate by way of an entirely random process on the grounds that it is random. Leicestershire police decided only to investigate burglaries at houses with an odd house number (or was it even? Because I live in Leicestershire, I want to leave an element of confusion). At a stroke they had halved their workload or doubled the misery, depending on your perspective. Given that my house has no number, only a name, I did wonder. Of course as soon as this was made public, the police had to alter it for fear of their random discrimination becoming a guidance note for successful burglary careers. (Or did they?)

Empire-Building

It's the end of February 2016 and little logos are appearing on TV adverts promoting companies' association with the upcoming Summer Olympics in Brazil. It strikes me that here is a perfect example of how to build a pyramid or Stonehenge. Start with an idea, enthuse your supporters by promoting its merits, and begin a construction programme. Start with the high priest, you know, the one with a direct conduit to God. Well, a direct con, anyway. Sell your idea to your earthbound god as a suitable vessel for the eternal afterlife and then stand back and admire your handiwork while cementing your position of power over your earthbound god. Before you know it, all sort of wheels will have been set in motion. Architects and engineers will have been recruited and training programmes expanded to provide a succession of bright new minds for the future. Stonemasonry, transport management, and slave recruitment drives will have similarly been enacted. What I mean to say by mentioning slave recruitment drives is that armies have been despatched to conquer neighbouring territories to ensure a steady boost in manpower. What's a little death, destruction, and subjugation between neighbours after all? And as an added bonus, a bit of plunder wouldn't go amiss, would it? Before long, the scale of the project develops into an entire nation's principal economic drive while simultaneously impacting conquered neighbours. And then of course it would be such a shame to make all that talent redundant. Might as well build another pyramid.

Moving on from the pyramid building Egyptians, today of course, we are much more civilised.

Our armies of marketers have scoured the earth and cornered the revenue streams from multinational companies that wish to be associated with the Olympic brand. True, there has been no blood spilt, but resources have been stripped from all over the world to support the extreme vanity of the project. (Holes to fill holes.) Governments have released resources for training facilities and coaches, essentially to claim bragging rights to the also-rans. An entire circus of pundits, media personalities, engineers, and cameramen will have descended on Rio. These,. together with the national networks left at home, will polish the product and fill our airwaves with superlative expressions of superhuman endeavour. Newspapers and magazines will fill with written accounts of triumph or condolences for the 'it was a wonderful learning experience'. 'Of course, I'm really targeting 2020.'

If you listen very carefully, right on the edge of your hearing you might just hear the sound generated by a high-speed train rattling along its soon-to-be-built track. HS2, as it is known in England. I think *The Simpsons* town of Springfield has its own monorail.

Back to the girl at the bench who threw her rubbish down by the side of the rubbish bin. I remember watching a televised football match at Bradford City's ground, where the wooden grandstand erupted into flames, killing 56 and injuring at least 265. I still remember the emotion in the commentator's voice when he announced, 'That poor chap's hair is on fire,' as the latter climbed the final barrier to escape onto the pitch. King's Cross station suffered a fire in an escalator shaft that killed 31 and injured 100. The sloping channel housing the escalator acted as a chimney, drawing air from the tunnels and accelerating copious quantities of blinding toxic smoke into the station concourse. In both instances the root cause of fire was the accumulation of litter dropped through the grandstand or on the escalator. These occurrences were both identified as management failures, and in neither instance was any blame laid at the feet of the thousands of previous occupants of said station or stadium who had mindlessly discarded their litter. It does occur to me that generations of grandstand cleaners perhaps had simply pushed litter through the open risers and this practice had never been queried.

While driving through France on some motorways in the south, one sees signs in a variety of languages asking motorists to take their rubbish home and not discard it by tossing it out the window onto the verges. Workers have to risk their lives to collect this rubbish; this threat of consequences is used to bring the message home. Will we stop dropping rubbish? Once again we have evidence of our very own personal dispensation card, only this time it is recognised by government. If a motorway worker were to die picking up rubbish, it would be the employer's fault for failing to provide safeguards, while the littering motorists would escape without a care.

Warped Cooperation

Just when you thought I couldn't mention speed limits again. I was driving towards a local village today when I was blinded by the flashing headlights of various approaching vehicles who were earnestly warning me of the presence of a mobile speed camera. This process was repeated during my return trip thirty minutes later. I know that this warning practice is illegal and motorists are occasionally prosecuted and fined, but as ever the incident began a thought process. What the 'courteous' drivers were doing was warning me so that I didn't break the law by speeding. What the 'courteous' drivers were doing was effectively aiding me in my illegal speeding activities by helping me to avoid detection. I wonder where the line would be drawn. Would you bend down to pick up money dropped by a mugger and then hand the money to the mugger in view of the victim? How about handing a dropped item back to a shoplifter with the comment, 'Excuse me, you seem to have dropped this from the inside of your coat'? How about handing a murderer his knife back mid murder, after he dropped it? (Blood is slippery after all.) How about helping the drunk to his feet and placing him in the driving seat of his car? How about witnessing the disembowelled body of a child mown down by a speeding car? We seem to have arrived at a warped sense of public spiritedness where aiding and abetting criminals is seen as socially acceptable. The message seems to be that the restriction on speed is unjust and therefore socially acceptable to ignore. But when we say speed, we don't actually mean speed; what we actually mean is time. Time is what we gain by speeding, an extra five minutes in bed, ten minutes extra up at the pub, a few minutes extra to relax. Strange that we risk life and limb, albeit often someone else's, to gain a few minutes a day without ever considering how we spend our hours. I am not trying to present myself as a 100 per cent perfect law-abiding driver (I have sped in the past), although I have to say that I have never been able to speed in built-up areas, I blame it on my imagination. I am trying to promote a debate about why we bother to half-heartedly control the socially uncontrollable. Either control it or don't; why pretend? Why pay the salaries of the pretenders? Why dig a hole just to fail to fill another one?

Think back to Russia. Hurtle down the footpath dodging pedestrians to gain a few minutes while, as a nation, you drink yourselves and future generations to an early grave. Why pursue a five-minute saving on your journey while squandering twenty years of life? The maths is appealing, so I have to mention it: take the shortcut through the park, dodge pedestrians on the footpath, and double-park in live car lanes to save ten minutes while you buy a pack of twenty cigarettes that take 220 minutes off your life. I think this could be worthy of a time-and-motion study. But before you laugh at the stupidity of it, consider your own position: work eight hours a week to pay for your motor car so that you can commute for three hours a day and earn enough spare cash to enjoy your leisure time. Don't drive? Spend your money on convenience foods to save time while filling your system with excess sugar, salt, and saturated fat, which will shorten your life expectancy.

Once again I have slipped in an untruth; it isn't that we aid and abet the lawbreaking speeders; it is that they cannot imagine the potential consequences of speeding. No, that isn't true either. They can imagine the consequences; they just don't. They have the ability to do it, but they don't do it. The thought processes required to visualise the effects are in a different orbit than the thought processes required to get to work, the school, the pub, the cinema, the hospital, a premature grave, or what have you and therefore can be ignored. I know this idea has been suggested before by others but perhaps not in this context. To bring the dangers of excessive speed home and introduce it into the orbit of functionality, one would need to deploy a tool of some description, let's say an extremely sharp metal spike embedded in the centre of the steering wheel facing the driver. This is entirely counter to our existing philosophy of a steering wheel mounting an explosive airbag within a reinforced cage surrounded by deformation zones and underscored with traction-control systems and anti-lock braking systems. It is possible to make the case that the 'orbit of functionality' has been trained away from the 'orbit of reality'. While this process is infinitely beneficial in the training of front line troops committed to defending the nation (assuming you are not a front line troop), the process can be seen as infinitely detrimental to the health of car accident victims, particularly the dead ones, as you can't get much more infinite than dead. It has been said that modern-day motorists become immune to risk and train themselves to rely on the active safety features so they may drive faster, and nearer to the vehicles in front. If you were really cynical, you could calculate the relationship between 'safer cars' that protect their occupants from serious injury and the need for crashes to act as 'planned obsolescence'. The market for spare parts and replacement vehicles has to be enhanced by a 'healthy level' of crashes, although fatalities need to be avoided because it's hellishly difficult for new and used car salesmen to sell to the dead. (I bet it wouldn't stop them from trying.)

Can we repeat this process of separating the orbit of functionality from the orbit of reality? For clarity, train the individual to accept a level of thought below the level required to acknowledge the threat.

This would be akin to accepting your lot in a bait ball of fish or erupting from the hive to defend the queen bee, when stinging the enemy results in ripping your poison sac from your abdomen and dying.

But the reality is, we do not even have to train; we are capable of spontaneously achieving this state of nirvana unaided. Why else do tobacco, alcohol, and crack cocaine hold such appeal to so many? As a further example of how the two orbits are separated, consider the graphic photographs of smoke-damaged organs on cigarette packaging and ask, why do people continue to smoke? A recent analysis of our Neanderthal hybrid ancestors' DNA suggests that many traits in modern humans were acquired from our interbreeding with them. Among these traits was addiction. In that sense, then, we really do have a 'suicide gene', albeit a 'slow-motion suicide gene', excepting those among us who choose to enter for the Darwin Awards while under the influence.

Hannibal: An Update

I lied; surprise. Hannibal is credited with committing suicide after capture by the Romans as an alternative to the fate they had in mind for him. Not only was he wise enough to know how to kill an elephant, but also he knew how to kill himself. If only our Great Brains had the sense to know when to retire from the field. I am not asking for blood to be spilt, just for some honesty to be exercised. But our Great Brains slip from elected representation to non-elected representation so that they can continue to harvest the fruits of their deception, or depart to careers in public speaking to promulgate the fantasy of their achievements and knowledge, or assume plum jobs in the corporate world, where their knowledge of how to exploit their fellow man is avidly assimilated into the unpublished mission statement. Yet an entire industry of pundits, experts, analysts, and journalists sit enthralled in the wake of these Great Brains, demonstrating through their obvious intellectual prowess the fact that they don't in fact have any. Not for them independent thought processes. Oh no, follow the pro forma, take the lazy option, and conform to the queen ant's pheromones. It is all that is expected of them, and the pay is good as well. Let me issue a challenge to all you intellects: break the pro forma, demonstrate the folly of your non-ideas, and recognise the emperor's new clothes for what they are. Design a new Clovis point. Deconstruct the pyramids. Raze the churches to the ground. Halt the destruction of our planet for no purpose whatsoever. Eat less food. Eat real food. Ignore diets and every peddler of diets, who only have one goal: to plunder your resources to increase their own.

But what happens when our Great Brains fail to leave? The president of Syria must have a monstrous ego to remain at the head of his shattered country. By any measure, any vaguely responsible individual would have sought an exit strategy, rather than preside over the country he or she supposedly loves, plunged into years of civil war and destruction. What possible pretence exists that yours is the right way? And even if you win, you've lost, with millions displaced and with people, homes, and cities shattered beyond repair. But this is a rhetorical question, because it only needs a little tweak to spin it around. What monstrous ego exists in David Cameron and Jeremy Corbyn that they believe that they are the best answer and solution for our needs? Are we not engaged in a bloodless civil war between different factions intent on wresting control of the TV remote from each other? Which channel to watch and at what volume? Will it be *EastEnders* or *Coronation Street*? What time shall we go to bed? How can you condemn Bashar al-Assad without condemning Cameron or François Hollande or Angela Merkel for exercising their egos? When it comes down to it, all Great Brains are fundamentally flawed, and their very existence demonstrates the stupidity of our species. We are back to herd instincts in spite of the fact that our herd has no leaders.

Let's postulate a theory: balance the aims and ambitions of the welfare state against the reality on the ground and consider that every charitable donation represents a measure of the failure of the welfare state to deliver. Put simply, charitable donation = failure of the welfare state. The equals sign, as ever, is reversible, so: failure of the welfare state = charitable donation. This means that we can add up all money collected by those charities active in the field of 'welfare' and use the figure as a measure of the monetary failure of the state. We can also use this figure to calculate the level of corruption within the state by suggesting that every failure replaced (not rectified, for that implies a solution) by charity represents a reinforcement of the state by suggesting that all needs have been met and therefore the state is doing a really good job. Charities through their 'replacement bus service' allow the state to pretend that the trains are running on schedule, and no doubt this justifies a pay rise for state workers for being oh-so good. Is this why charities enjoy the perks attributed to charitable organisations? Government need the attention of these organisations to fill the holes that they themselves have singularly failed to fill. But in our model of measurement, we have created a self-reinforcing loop of satisfaction whereby the success of charities reinforces the success of government, whose representatives in turn are rewarded by government for being successful. It's brilliant; everybody wins, nobody loses, except of course the taxpayer and the charity giver, so that would be virtually everybody then.

I have earlier described charities as 'self-appointed head gardeners' who effectively decide whether flower beds are preferable to orchards or lawns or rose gardens. I can also describe them as head decorators, because their principal activities include the papering over of cracks and the liberal dosing of decorators' caulk prior to painting. You have a decision to make: give all your spare cash to the government for the provision of services and needs, or give all your spare cash to charities for the provision of services and needs. Each is equally deserving; both are expert at establishing and satisfying need; and neither has a hidden agenda that satisfies their member's needs before those of their 'customers'. Nobody ever got rich by working for either body, and no need has ever been turned down. But at the risk of tying you up in knots, when I say spare cash, I really mean spare time. The spare time we need for extracurricular activities such as cathedral building. How easily we surrender this resource. Why is there only one Stonehenge (or Avebury, or Woodhenge, or Offa's Dyke …)?

You may take a different view: government cannot possibly identify all needs and offer funding to support them. Enter the private sector, which can identify alternate needs and provide funding for them given sufficient support from the public or corporate world. Unlike the government, which has finite resources that cannot satisfy all requirements which everyone universally attempts to ignore, charities have finite resources that can be universally expanded if only enough people would donate. Charities are never replete; they always have an expanding brief of need that far exceeds their original aims and ambition. The famous Calendar Girls sought to raise enough money to buy a replacement sofa for a hospice. Help for Heroes wanted to build a swimming pool. Both evolved into multimillion-pound organisations because they connected with a vein of popular giving. There is a cartoon that appears occasionally on birthday cards and the like that shows two mosquitoes on a hairy arm. One mosquito is feeding and grossly inflated, while the other mosquito comments, 'Pull out! Pull out! You've hit an artery.' This allegorical tale perfectly illustrates the apparent success of our blessed Calendar Girls. But our Calendar Girls didn't explode with the pressure of arterial cash, because waiting in the wings was an eager army of recipient charities that would ensure no spillages and, of course, no wastage.

When at school studying for my A-level biology, I resolved to study song thrushes and their anvils. These birds collect snails and carry them to selected stones which are used by the birds to break the snails' shells, making the interior accessible for consumption. I often heard the tap, tap, tap of these birds as they beat their food on the anvil. As a project it had a lot going for it, not least the ability it gave me to bunk off school and wander around the countryside legitimately. You can see, even then, how I showed such promise at making my work commitments bend to my true needs, and all with the consent of my tutor. During general studies I elected to study the provision of playground equipment in the local parks. I seemed to concentrate my attention on Queens Park, because it was the one nearest to my school and I only had limited free periods or general studies lessons to devote to research. And the fact that it adjoined my girlfriend's college and facilitated extended lunches with her may have had a bearing on it.

Unfortunately for me, I struggled to find thrush anvils when I needed them. After a couple of weeks the weather had turned and I was still in need of a project, so I reluctantly settled for a different avenue. I realised today how amateurish my attempts at satisfying two needs at once were. What I should have done is to have looked anew at the thrush anvils and tweaked the perception slightly. With hindsight I see that I probably could have received a grant even at pre-university stage. My plan was to look for patterns in colour, size, location, and frequency of use of these stones. I now realise all I had to do was alter the nomenclature. If I had but realised that the thrushes were in fact sacrificing the souls of the unbelieving molluscs on the altar of the great thrush god Turdus, I would have been on a winner. It's taken nearly forty years, so what has led to this epiphany? The article in the *Daily Mail* replete with photograph of a chimpanzee wielding a rock in an agitated manner before smashing said rock into the shrine of similarly sized stones placed so carefully at the side of a tree. You must forgive me; for some reason I could not muster the will to read the article, so I cannot in all honesty say that this was simply an expression of amplified aggression often documented in chimps, who will beat their feet on tree buttresses or bash the equivalent of dustbin lids to generate noise to exaggerate their physical presence when they feel the need to assert themselves. Perhaps if the journalists had read a book other than the Bible—say, a Terry Pratchett one, *Soul Music*, which describes itself as a 'story about sex and drugs and music with rocks in'—the gist of the article might have been a little different. Instead of the strapline with which I was greeted, 'Do chimpanzees believe in God?' it could have been, 'Elvis has returned, as a chimp.' I despair at the quality of our journalists, but you knew that anyway.

Sorry, I must apologise. I normally give the Latin for native species, *Turdus philomelos*, the latter derived from the Greek for 'loving song'. And *Turdus*, that every schoolkid knows, means 'thrush' (or something similar). (Only a few schoolkids got as far as the American robin. Your search engine awaits.)

Back to my school days. It was obvious I was showing real promise for public life. The only real issue was that nobody liked me. (Surprised?) (It was mutual.) I have illustrated how easy it was to corrupt a common need to suit a personal one. In later years I could easily have used my skills to obtain a top-of-the-line Range Rover or Jaguar, say, to ensure that the charity I represented didn't come across as a cheap, struggling enterprise unworthy of support. After all, you couldn't possibly turn up to some posh money-raising function in a Vauxhall Astra and anything less than a Savile Row suit, could you? It would be wholly inappropriate for your charity to provide you with a clothing allowance. That would smack of corruption. So much cleaner to take a salary commensurate with your responsibilities. And, as an added bonus, your salary gets locked in for future years, so no need to fight that battle again. And you never know whom you might bump into if you dine at the finest restaurants and holiday in exclusive resorts. It's just win, win, win; it's brilliant for everybody.

Of course I needn't have taken such a mercenary view and gone into charity work. I could have spurned such petty tasks and concentrated instead on higher orders. If I'd done the legwork, I could have voiced the needs of *Turdidae* and become high priest, the conduit of inter-species communication, with the added bonus of their pan- Eurasian distribution.

As the original illustrator of their previously ignored worship, I would have commanded the respect of other human believers and ensured my unassailable position as vicar to the god Turdus. With just a little extra work I could have spread the message on behalf of the *Paradae*; after all, these delightful little beasties entertain many householders with their acrobatic antics. With hindsight it becomes obvious that they have tried so desperately to attract the attention of the human species with their bright colours and cheerful disposition, but we had failed to make the connection. How to design the temple of worship? You need to command respect after all. The stone altar would have to be the starting point for thrushes. And because they line their nests with mud, it would be reasonable to accept the use of stone structures reinforced with mortar. And a choir to celebrate their singing abilities would also be required. Architecture for the *Paradae* is a bit more problematical because the tendency to use wooden structures yields a credibility issue. How to design a wooden structure with a single round ingress–egress point without it looking like a tit box? Perhaps the real skill is to embrace it. After all, many human supporters happily supply such boxes for the tits in their garden. Why not accept a community point of worship on a grander scale? A large section of society already makes offerings of peanuts, suet, and half coconuts without recognising the true calling that provokes them to do so. A little education produced on my behalf ought to see the tithes rolling in for the first temples, on my behalf. I can already see the first schism on the horizon, as many believe the garden robin to be a thrush, whereas modern science is leaning towards them being flycatchers. They also favour the use of old kettles and half-open fronted boxes, which would conflict greatly with stone-built temples. Mind, it's only a problem if you don't plan for it. The trick is, retain control and don't let anybody else hijack your position of pre-eminence. Mine is a divine right after all.

A Security Door with No Lock

Let us consider the purpose of an international border. In effect international borders act as permeable membranes between sovereign states, allowing controlled movement of goods and people. There are two important considerations here. First there is selected movement, and second there is control. At borders, we seek to control the movement of goods that could damage our own economy through oversupply (dumping), disease (foot-and-mouth disease in meat, viruses in fruit and veg, insect pests in wood, etc.), crime (money laundering, illicit drugs, non-duty-paid tobacco or alcohol), and illicit goods (not fit for purpose on the grounds of unsafe design, or counterfeit drugs, designer goods, films, and so forth). It is normal, but not essential, to allow the passage of people as tourists or selected workers to boost the economy, and this is often a 'two-way street'. The common mantra that our Great Brains are keen to exhort involves the need to control the movement of terrorists and their equipment for attack. To this end we have established laws, treaties, border guards, visas, passports, and even, in the past, the Iron Curtain predicted by Josef Goebbels. If we were to travel back in time to the post-Norman conquest, then these barriers could exist at the barricade walls around individual motte-and-bailey castles. Having toiled and slaved (quite probably) to develop these safeguards, one would have to be a complete idiot to fail to shut the gates at night or man the gates at times of trouble.

I am reminded of my visit to the citadel at Jaca in northern Spain. This redoubt, developed in the age of cannons and musket balls, incorporated the latest thinking with deep-walled moats overlooked by multiple gun platforms that would rain murderous projectiles down on any attacking force. Strategically placed to guard the pass between France and Spain, it only ever saw action once, when the Spanish tried to wrest control back from the French, who had walked into an essentially unguarded fort. How stupid can you get? All the design and effort afforded by the Spanish people counted for nought. Worse than that, the citadel was used against them, and therefore its impact passed into negativity. I did not study the history of this site enough to discover the fate of the numbskulls who were responsible for this fiasco, so I cannot comment. I can imagine the fate I would have had for them if I had been in overall charge.

Why have the sovereign states of Europe ever bothered to maintain international borders if, at the first sight of a massive transit of illegal immigrants, they were going to abandon them? I will define 'illegal' as 'without legal recourse to their activity', that is to say without passport, visa, or consent to enter another's sovereign territory. At this point the humanitarians among you will be demanding that their rights as refugees and asylum seekers be upheld. Can I ask that my rights as a member of a sovereign state also be upheld? There are already rights and protocols in place to ensure that refugee status be given to genuine refugees fleeing hostility in the first safe neighbouring country bordering the hostile location. We now appear to have 'policing by consent' at our international borders, whereby difficult decisions are simply delayed, deferred, fudged, and swept under the carpet for another time and another administration to try to resolve. But in this instance it isn't the consent of the sovereign citizens; it is the consent of the illegal immigrants. In the meantime, the trickle of economic migrants has turned into a torrent, with no mechanism to resolve the issue. Consider this: if you cannot turn away 5,000 people a month, how do you plan to turn away 150,000, or will it be 300,000? Do I need to point out that failure to arrive at a solution now simply moves the problem downstream to the next batch of Great Brains, who will screw up in turn while blaming their predecessors?

Some years ago a former work colleague of my father-in-law had his car damaged while parked outside his house. The neighbour opposite had sideswiped his vehicle while turning in to his drive. Luckily the offending driver was a manager at a local garage. He reassured his former colleague that he would 'sort everything'. A week passed and the manager's car was repaired, but no contact was made to 'sort everything'. When he was challenged, the true nature of the manager surfaced. He declared that he had not damaged any car and offered, by way of proof, the fact that there was no damage to his own car. He said that he was under no obligation to do anything. I am quite sure that many of you will have experience of similar events at home or work where you have been wronged to a greater or lesser extent. I will call these events examples of natural injustice. It occurs to me that one of the reasons our species is so ready to indulge in a little 'civil war' or rioting and the like is that it enables many of these little injustices to be righted. Resentment of individuals or classes can escalate beyond the original remit of a 'natural recourse of justice'. I can well imagine how the desire to put an RPG through the front door of your local car dealership's manager has an appeal. While you may not pull the trigger yourself, targets can be discussed and sentiments expressed. The problem occurs when the 'slates are wiped' and it becomes apparent that even more slates have been created. It begs the question as to why we have no recourse to natural justice beyond the hopelessly inadequate legal system surrounded by its flocks of lawyers and bureaucracy. Once again we have a system devised over centuries by generations of Great Brains that fails to address the real issues that we face. If we were to expand on this idea that a failure to address small issues can build up a sufficient head of steam which leads to civil unrest and even civil war, then we would ask why we don't in fact have a mechanism to satisfy the needs. Our Great Brains will declare that, in fact, all the mechanisms are present and correct and that 'everything is rosy'. This would be because the correct and present mechanisms are in place to ensure the continuation of the Great Brains and precious little else.

I realise I am bordering on mention of the *r* word, revolution, but before you have me carried off to the 'mental adjustment bureau', consider what is meant by revolution. George Orwell wrote *Animal Farm* to illustrate the folly of revolution, and Marx and others have summed it up as 'a class of individuals recruited by the disaffected would-be rulers to be used to overturn the existing rulers so that more of the same can continue, just with different names on the pay packets'. In essence, the pro forma is wrong, so let's replace it with a new pro forma that shakes out to be remarkably similar to the original one. I could also declare that the Clovis point is wrong and replace it with a sharpened stone point manufactured with tried-and-true technology. How about, 'It's time to stop building stone circles in a clockwise fashion. The only way to do it is anticlockwise'?

What we did was to stop building stone circles and replace them with stone churches and cathedrals. What we did was to take subsistence living, where you worked long enough to eat for the rest of the week, and convert it to full-time working, where you work long enough to eat to the end of your life (or so you hope. But then if you get it wrong, it would be the end of your life).

Our Way

If we consider Syria to be a failure on the basis that it is not a success, then it would be legitimate to ask where the fault lies. Let us suggest that the fault lies with the political party that has provided the Great Brains and the dynastic control of the Assad family. But could we equally source the blame to the French colonial control which created the conditions for such a regime to exist? If we are going to blame the French, then should we also blame the Turks and their occupation during the Ottoman Empire? A more recent event is the much vaunted Arab Spring, which was 'supported' by Western democracies and lit the fuse of change as the populations strove to throw off the yoke of oppression. Equally, we could cite a failure of the 'international community' and all of its Great Brains, who have failed to suggest or implement an equitable solution. The two constants within Syria appears to be a failure to really address the needs of the people and the success of addressing the needs of the Great Brain, be it Turk, French, al-Assad, or any of the alternative groups of 'freedom fighters'.

But the first problem is to identify the 'needs of the people', when the people have no device or mechanism to either determine their needs or explain them. The default mechanism in all instances is to determine the needs of the Great Brains. Of course the Great Brains go to great lengths to explain their appeal to garner support from the people, but as ever this process soon breaks down to the lowest common denominator. So it is that American presidential candidates can win or lose an election for having a five o'clock shadow, having failed to shave three times a day. Our own, ever so intellectual debate about EU membership has resolved itself around FUD—fear, uncertainty, and doubt—with the spectre of being the shunned kid in the playground wrote large. Once again we are entering arguments of perception which will be decided by orbits of functionality rather than orbits of reality.

We surrendered our rights to self-determination millions of years ago when we became a social animal unable to survive without cooperation with our fellow species. The problem with this statement is the overriding caveat which translates loosely into 'every ape for himself' when the chips are down. There is a Horn of Africa and Arabian Peninsula baboon, *Papio hamadryas*, the hamadryas baboon that lives and feeds in communities hundreds strong. They graze on vegetation, insects, and fruit and roost overnight on steep clifftops that offer protection from predators. They exist in 'one male' units called harems, where females and young live together with non–sexually active males under the direct control of a single dominant male. Two or more harems will unite to form clans, while two to four clans will unite to form bands which may contain two hundred or more members and which travel and sleep as a group. Several bands may also come together to form troops of a thousand or so animals. The whole is still divided into small, manageable harems controlled by dominant males. At best we could consider that our DNA has evolved to this level of social development; at worst, that we never got past the family group stage with an anti-inbreeding mechanism and the willingness to fight for territory.

Now it is true that we live in communities numbering into the millions, but I will argue that our DNA has yet to catch up. What we have is a system of control and governance developed over the centuries essentially by committee that seeks to provide for our needs while its actual purpose is simply to provide control and governance without redress to our needs.

I will state it again: we surrendered our right to self-determination millions of years ago when we became a social animal unable to survive without cooperating with our fellow species. This means that we were still swinging in the trees, or maybe we hadn't even left the ground or climbed them at that stage. So why do we pretend? Why do we recruit the illusion of individual control to subjugate the masses?

This has become our way of life.

But just suppose that we run the clock forward and global warming reaches a tipping point beyond which the planet cannot rebalance itself, and temperature runs out of control and makes the earth uninhabitable for our species. What then? You will have noted my earlier comments about global warming. I am personally not convinced that global warming is the biggest threat to our existence. I am convinced that as a species we are mindlessly terra-forming the world at an accelerating rate, which will make us all poorer going into the future. Like the lizards on the wall in Spain that cannot cope with more than one threat at once, our Great Brains have settled on global warming as the singular threat that will receive our attention until something better comes along (to the exclusion of other threats, such as population growth, which seems to have had its day and disappeared over the horizon). Global warming can take more than one pathway to achieve a significant reduction in the human comfort zone. The thermonuclear arsenals are still there, ready to impact on our species' survival, after all.

This has become our way of death.

What will our achievement be then? How will you reconcile our strategy for survival with our imminent extinction? Instead of 'Will the last person to leave switch the low-energy light off?', we will have the last people alive asking, 'Why?' And what will the answer be? Collectively we will have destroyed the world; individually we will have done nothing wrong. Yet individually we will all have thrown our sweet wrapper under the grandstand. We will all have driven tigers to extinction in exchange for a cup of tea. We will all have dug a hole in Africa to fill the holes in our native birds' bellies. We will all have dug a hole in resources underwritten by Planet Earth.

We could start a little nearer to home. But that will not work, because that is the root cause of our problem.

I could start a little nearer to home.

I.

Waiting for ...?

Search on the Internet for Zack Hemsey, listen to his song 'Waiting between Worlds', and consider his sentiments. He describes three scenarios of an abused girl beaten up by her boyfriend who discovers that she is pregnant, an old man who aims to take his own life to join his lost partner, and a mother sitting at her brain-dead son's bed contemplating the life support being switched off. All three are considering life and are on the cusp of death, be it that of foetus, self, or son. Their orbits of functionality have met with their orbits of reality.

What will it take to bring your two orbits together?

The phrase 'waiting between worlds' could sum up our present position, with the world as it was yesterday and the world as it will be tomorrow. Wait for the Great Brains and the process will continue to accelerate. They have no solution, no ideas, just an earnest desire to survive their occupancy of the high table without having to face any issues, let alone resolve them. They would have you go on a diet, a diet of energy, of food, of resources, when in reality what you need to do is use less energy, less food, fewer resources.

The Centre for Economics and Business Research has produced projected league tables for the world's larger economies that extend through to the year 2030 (the Cebr Global World Economic League Table). They list the top thirty nations of the world. While these jockey for position, the same thirty names generally appear throughout the lists. The figures are estimates or estimated projections. I have taken the trouble to add all thirty together for the year 2013 and again for 2030. Bear in mind the idea that we are bound by treaty to reduce the amount of greenhouse gases so as to limit the projected temperature rise to no more than 2°C. (Who can forget the hype around Kyoto, Rio, and Paris? Except that Kyoto has already been forgotten.)

How then does anyone reconcile the near double growth in GDP with a reduced greenhouse gas production? Notwithstanding this figure with regard to greenhouse gas emissions, consider this as a suggestion. Consider the decline in tigers, elephants, house sparrows, song thrushes, and anything else you may care about and multiply the decline by 200 per cent. Or, if you prefer, consider the size of the hole we industriously dug in 2013 and make it nearly twice as big in 2030 alone. We will be able to do as much damage in 2030 in twenty-six weeks that it presently takes fifty-two weeks to do. If you can't wait that long, then we will do in twelve hours what it currently takes twenty-four hours to do.

Waiting between Worlds

Zack Hemsey describes a mother sitting at her brain-dead son's bedside contemplating the switching off of the life support machine.

It is real, it is dreadful, and it cannot be ignored.

I have never been faced with this prospect and do not even know exactly what the life support machinery does. But I would consider this. The World Wildlife Fund and other conservation bodies seek to provide the life support machinery for endangered species and habitat throughout the world. (And if we are going to continue the analogy, who pulled the trigger?) Consider that the life support machinery is designed to offer short-term assistance to the damaged body until such a time as the body has recovered sufficiently to reassume its normal function. In what world will our life support machinery be made redundant due to lack of need? The world where humankind has abandoned its mindless path of destruction, or the world where the destruction is so far gone that there is no point in trying to continue to save the patient?

It is real, it is dreadful, and we are ignoring it.

It can be concluded that our last-gasp attempts to save the patient are already doomed to failure as our isolated pockets of 'reserve' gradually disappear due to a lack of genetic variation. And while we may strive to mix the genes of the poster boys, we simply do not understand the infinite complexity of the habitat that supports them and that they in turn support. But, and it saddens me to state it, many passionate individuals devote their lives on earth to conservation and care and are compassionate to the extreme, yet in the final analysis they are still gardeners selecting one variety of species over another. Think about your own conservation efforts where you select breeding bird species by offering nest boxes with just the right size hole for blue tits and multiply this choice by thousands of gardens. True, a few will offer a variety of boxes for robins and the like, but taken in the context of our town-planner-induced shrinking gardens without hedges or shrubberies or mature trees for fear that the trees will fall and hurt someone (with relatives having access to 'no win, no fee' ambulance chasers), the net result is human-induced selection of species—and all done without us even trying, or giving it a first thought.

If the problem in Syria can be summed up ultra-simplistically as a failure of the Great Brains to provide for the 'needs' of the people, how can allowing millions of refugees/economic migrants from Afghanistan, Pakistan, Somalia, Iraq, Syria, and the like be reconciled with the 'needs' of the people in the receiving countries? If the problem in Syria can be traced back to historical decisions or the failure to make decisions, then in failing to make decisions now we are creating problems for the future. Today is tomorrow's history. Now the Great Brains will hesitate and imply that they will be damned if they do and damned if they don't. The answer then is that they will be damned, so in reality it doesn't matter which decision you take. Be selfish and defend your tribe, and ignore the 'suffering' of the economic migrants as they would ignore yours. By stripping the potential that their lives can offer from their own homelands, they are being selfish in their desire to have a better life without being prepared to work for it. What we arrive at is a battle of selfishness: be selfish and defend your border, or be selfish and ignore the border and cross it? Which has the greater justification? By offering aid to these vocal witnesses to their disadvantages, you strip aid from the silent witnesses at home as they strip aid from their own silent witnesses left at their home. We cannot house and provide for all who currently live in the UK, so why do we need another two hundred thousand a year from elsewhere? Why dig a hole in our resources to fill a hole elsewhere? The Great Brains have no idea beyond the adopted/inherited pro forma, no mandate, no magic wand, and no clue either. They have no place at the head of the herd either, because the herd has no leaders. Journalists are keen to seize on the sound bite 'Europe is facing the greatest movement of refugees since the Second World War,' as though there are lessons to be learnt from this period of our history. This is yet another monumental example of corruption, 'lessons will be learnt', which is shorthand for, 'We forgot what happened last time but should have remembered'; 'We failed to consider the issue correctly'; 'We want to wipe this slate of failure clean and carry on with a clear conscience'; 'We failed to anticipate and manage the problem in advance'; 'We fucked up, but never mind, it's business as usual.' What lessons were learned at the end of the Second World War? That you can't devote all the resources necessary to defeat the enemy in a 'total war of annihilation' and expect to then have the resources to provide for millions of refugees. Given that the war was still to be won in the Pacific against the Japanese, how can you instantly switch from bullet and bomb production to food and transport and housing needs? The refugees who died of exposure, starvation, and disease as they trekked across Europe were as much victims of the war as they ever were victims of the peace. If the provision had been made for the peace, then the resources necessary to end the war would have been diverted and the war would not have ended when it did, resulting in more casualties of war and maybe fewer casualties of the delayed peace. What lesson would you learn from the Second World War?

Let me change the context slightly. When you leave the road on a forty-mile-per-hour bend while doing eighty miles per hour, lessons will be learned, assuming you survive the 'accident'. Given enough evidence or serious injury or death, the law will take an interest and prosecutions will be made. 'Yes, Your Honour, I was driving at a reckless speed. I have learned from my mistake and will not kill my girlfriend again' doesn't cut it, unless of course you are a Great Brain. Is it because we can survive if the occasional boyfriend gets jailed or banned for a few years, whereas we cannot survive without our Great Brains? Too big to fail, or too corrupt to accept the truth?

If you are caught speeding and it's your first offence, you are offered a speed-awareness course whereby you are exposed to the folly of your ways. What the course is attempting to do is educate the errant drivers so that their orbit of functionality expands to meet the orbit of reality. I have met a couple of individuals who have attended these courses, and they have recalled some information, such as the increased braking distance involved to stop when speeding and the time advantages generated by speeding during short journeys relative to the risk. A similar argument can be made for health campaigns and alcohol and tobacco initiatives. How to raise the mental activity orbits from a lower plane to a higher one?

Let us consider these orbits in a slightly different light. It would be reasonable to assume that our blessed highway engineers operate on a different orbit from that of normal people. I should perhaps describe normal people as a baseline, a reference point, not an average but your typical (un)thinking person. In Victorian times, gentlemen would resort to class or gender to describe orbits; 'poor women's brains would explode if you tried to engage in political discussion with them' or 'of course, your working-class person cannot understand the principles of higher thinking' and similar misogynistic/elitist bullshit. Highway engineers believe that what they do makes a difference to the attitudes and driving ambitions of the general population. The general population simply curses at the unevenness of the ride over speed bumps or the inconvenience of waiting at speed chicanes, or bemoans, 'Not another sodding mini-roundabout.' Indeed there is some recognition of highway engineers' impotence, in that many of the devices utilised represent physical impediment to the otherwise speed-unchecked flow of traffic. But these physical 'barriers' are only a small proportion of the attempted methods of control. The orbit that highway engineers operate in is not the orbit that their 'customers' operate in.

Don't worry, we have a fail-safe. We have the police to entrap and snare errant motorists, with cameras, both fixed and mobile, and manned traffic patrols to observe and educate/penalise, and as a long stop, accident analysis to determine probable cause in the event of life-threatening or -ending 'accidents'. (Of course, there are very few accidents; hence the nomenclature has changed to 'incidents', where the ambition is to fix the blame.) The blame is the difference between the orbit of functionality and the orbit of reality.

Yet before the highway engineers and police are ever let loose on motorists, motorists are themselves the subject of the DVLA, the driving and vehicle licencing authority, whose remit is to ensure that only fit, competent drivers are qualified to get behind the wheel of a vehicle. Theirs is a different orbit altogether from that of highway engineers and police and indeed motorists. In their world all drivers have proven themselves to have achieved a standard of competence and have been certified to be operating in the orbit of reality, or it has been determined that the driver's orbit of functionality is high enough to satisfy requirements for safe driving. That is why we have no 'incidents' and only accidents caused by mechanical failures, because our system ensures that only competent people drive vehicles, and the DVLA workers receive the salaries to prove it.

Have I identified a problem here? The DVLA has an orbit of functionality which is not the orbit of reality. The highway engineers have an orbit of functionality which is not the orbit of reality. Motorists have an orbit of functionality which is not the orbit of reality. Yet the DVLA is charged by government to ensure the orbit of functionality is near enough to the orbit of reality to enable drivers to operate in a safe manner. I will identify the gap between functionality and reality as corruption. Corruption is the measurement between the perceived truth and the truth. Corruption means that we are doing a good job, while truth fills the obituaries and the hospital beds. Given that we seem to be immune from the truth in our activities and happily accept the projected reality of functionality, is it any wonder that we have no clue about our purpose and aim in life? In which orbit would you place our Great Brains? The simple answer is in an orbit of their very own, where their orbit of functionality bodes no discussion about failure and truth is but a dream. Theirs is an orbit in which they know the needs and requirements of the people better than the people do. Theirs is an orbit of politicians and lobbyists and political journalists and pundits who speak the same language and fit the pro forma. Theirs is an orbit where they know best what the people want. That is why we will receive countless thousands of economic migrants into our midst, because it 'is good for us'. Has anyone considered whether this situation is 'good' for the donor countries to see a mass exodus of predominantly young, fit, upwardly mobile males with higher levels of drive and ambition, the very individuals that you need to take your own country forward? Is the real problem that the orbit of expectation is not being met by the reality on the ground in their native countries? Is the reality on the ground fixed by the Great Brains and the corruption they support?

During the Industrial Revolution when British machines were being exported around the world, they were often despatched with a British-born engineer who would install and operate the machinery going forward. The engineer's knowledge and abilities would be transferred to the locals, and economies would develop around the nucleus of the new technology. Here we seem to have a reverse situation, where talent is stripping itself from the base and transferring itself to a 'better life' where development has already occurred.

But we don't have to stop at highway engineers and their clan of co-conspirators. We could substitute teachers or doctors or civil servants or charities or priests or football managers or …

Have a frank and candid discussion with any of these groups and remove the politically correct smokescreen of bullshit that they drape their professional lives in and you will realise that many have a realistic understanding of the reality of their situation. But the reality of their situation is suppressed by the pro formas that they serve, the structure of control, the chemical signals that have to be obeyed. Perhaps the greatest control is the desire to continue to earn a living. Independent thought is massaged out of these individuals until they have the diversity of clones, almost programmed to enact their roles as nest builders, nursery staff, soldiers, or queen attendants, shaped to fit the mould required. Model citizens. Big Brother is alive and well, eager to stamp out any transgressions from the party line, eager to offer you up for re-education. But the reality is not as controversial as I make it out to be, because the vast majority accept the confine of their orbits of functionality, because, quite frankly, that is all they are capable of. This is not to say that the 'lower echelons' of society are incapable of independent thought, because the 'upper echelons' suffer in exactly the same way. How else can you have universities teaching 'sustainability' when no such thing exists in a world where humankind is avidly destroying it? How about teaching students the fundamentals of safe car design without addressing the deaths caused by car pollution? How about donating your multibillion-pound fortune to pet gardening projects? How about introducing yet another initiative to cut the speed of cars in built-up areas?

If Bill Gates, among the richest men in the world, cannot find a use for all his wealth, what chance do any of us have? The simple answer is, there is no use for all his wealth that doesn't damage this world we share. We damaged the earth to generate the wealth to line his vaults. He in turn will damage the earth by releasing his resources on whatever charitable enterprise he endorses. If you really care about people, then don't strip the resources from them in the first place, and if you really care about the planet, then don't release the resources back to the people. Strange, isn't it? Gates doesn't want to saddle his children with the responsibility of all that money. The best possible solution to the mountain of resources that he has accumulated would be to destroy it. That way, it will only have dug a $35 billion hole in the earth's resources once. His chosen path digs a $35 billion hole in the earth's resources twice. Deep down, perhaps, the wealth embarrasses him, and his cultural profiling determines the need for philanthropic enterprises, because, make no mistake about it, enterprises they are. They follow business models and pay salaries and build research facilities and relentlessly push the boundaries of our knowledge. They join the ranks of 'force multipliers', very well-funded ones.

While reading *How to Kill an Elephant*, you may perceive me to be a champion of a miserable, boring life devoid of all small luxuries. If you become one of my devout disciples, there will be no tea, coffee, alcohol, tobacco, suntans, holidays to far-flung exotic places …

If you do not become one of my devout disciples, there will be no tigers, leopards, elephants, rhinos, black-footed ferrets, or, devoid of the fauna and flora, exotic places to visit.

If global warming and rising sea level predictions are correct, there will be no Maldives or Fiji or major coastal cities free from the threat of flooding.

Are you waiting between worlds?

Are you, through your consumption of the earth's resources, not in fact waiting, but actively driving the world to its sorry conclusion? And in spite of the best attentions of the UN, national governments, conservation bodies, and intellectuals (highway engineers), no amount of speed bumps, chicanes, legislature, police, or failed 'orbit adjustment' will reduce the speed at which we do it. Your foot is firmly planted on the accelerator, and nothing proposed or enacted to date will alter this fact. Krypton, here we come.

But to be fair, how concerned about speeding or smoking or alcohol consumption am I? What I have endeavoured to do is illustrate using simple examples ideas that we can all relate to and, through these examples, hopefully understand the abject folly of our actions. As a species we parasitise our fellows, exploiting them for our own advantage. Artistic images of hell often depict lost souls scrambling to reach the surface by standing on the backs of an infinite sea of fellow inmates. *What Dreams May Come*, a Vincent Ward film starring Robin Williams; *Constantine*, a Francis Lawrence film starring Keanu Reeves and Rachel Weisz; and even the *Mummy* franchise starring Brendan Fraser and Rachel Weisz, depict Egyptian hell as a sea of souls desperately trying to scramble over the bodies of others to reach their destination.

Each film depicts an artistic vision of hell.

Yet when it comes down to it, that is precisely what we do in life. You don't need to die in order to fight for an advantage over your fellows. In fact, it helps to be alive, because for many, climbing on the backs of others is an essential requisite to staying alive. How strange that our depiction of eternal damnation is actually a model for eternal living. We have practised it for millennia; indeed we are very good at it. Now you may be thinking that I have gone too far with this assessment, but consider your local high street where the legal high shop is next to the tobacconists, which is next to the off-licence, which is next to the cake shop selling hydrogenated-vegetable-oil-containing cakes, which is next to the butchers that sells WHO-recognised carcinogenic processed meats, and in the alley between the shops a trade in heroin is taking place. Of course your high street may not resemble this scene at all, because yours is full of charity shops and betting shops and fast-food outlets because the local supermarket has put all the other shops out of business. It is possible to consider that this is our mechanism for evolution? How else can we describe it other than 'survival of the fittest'? We can identify the giant panda as a species that has entered a cul-de-sac, having selected a diet so poor in nutrients that it struggles to muster the energy reserve to breed. It has been described as a species on the edge, which may or may not be the truth in the way that bumblebees are too heavy to fly. The idea that such species exist is enough to consider, because if it is not the giant panda, it could be the cheetah or the dodo. As a species, humankind has also entered a cul-de-sac in that we have adopted a diet so rich and so rewarding that one earth cannot support it.

So you can drink and smoke and speed to your heart's content, or as it happens for the first two, your heart's discontent; it is not a problem to me. Surrender your life willingly to your Neanderthal genes but, and this is the important bit, don't come running to me for help when the piper arrives to be paid. Live life to the full and accept the consequences to the full. Surrender your time, your effort, your resources to others who profit from your actions as you in turn profit from theirs. Join the human race to the bottom. We have returned to the story of the prodigal son: live fast and loose and then stick your hand out for more. Many live their lives as a disposable asset, exposing themselves to risk at every turn, and then go to the front of the queue for treatment and absolution. In a world where resources are limited, take the responsibility for your actions. If you want to live forever, live healthily and reject the multiple pitfalls along the path. Turn your back on tobacco and alcohol and obesity and 'recreational drugs' and gambling, and eat real food devoid of poisons. Or don't. The choice is yours, and you are free to make it. But why should I pay for your actions? What gives you the right to dig a hole in my time? Why do my taxes get used to treat obesity, tobacco-induced cancer, sclerosis of the liver, heart attacks the result of arterial clogging? Where is my choice? The myth of universal healthcare is precisely that, a myth. As our ever expanding needs continue to outstrip supply, treatment is rationed by resources, waiting times grow, treatment is delayed, and patients are excluded on the grounds of cost benefit or age, and living a long and hitherto healthy life counts against you.

What *How to Kill an Elephant* needs is a happy ending, hope for the future, something to indicate that our path to global destruction can be diverted and arrested. I cannot give you one. Nothing that has been proposed, enacted, or legislated for will have any impact on our apocalyptic tour of the earth and its resources. No leadership, be it political, religious, conservational, charitable, or science based, can control the uncontrollable and yield a happy ending. Only a natural-borne disaster such as an asteroid strike, bubonic plague with bells on, or a resurgence of the Deccan Traps lava field, or an unnatural act by some power-crazed megalomaniac with a germ warfare suite à la the antagonist in Ian Fleming's *Moonraker*, can hope to save the planet from the scourge of our activities, if only by replacing our scourge with another, 'fitter' one.

Why bother to write *How to Kill an Elephant* then? What purpose does it serve? If I can expose the charlatans who feast at the high table issuing useless decree after decree for the abject failures that they are; if I can bring an end to the employment of 'highway engineers' and all their useless speed bumps and endless, pointless initiatives in every task they engage in around the world; if I can bring an end to the cancer of advertising and the massive waste of resources that it squanders; if I can bring an end to universal healthcare that isn't welded inseparably to individual healthcare responsibility; if I can bring an end to the massive resource-squandering activities of charities and philanthropists and their endless quest to be 'head gardener'; if I can make you question your place in this world and make you a jealous guardian of your time; if I can bring an end to all diets and dieticians and their endless pursuit of career-enhancing non-solutions, whether it's food or energy; if I could bring an end to the relentless quest to educate the uneducable and the resources that such an activity wastes; if I could bring an end to corruption in all its forms; then I might have achieved something, a pinprick, a diet in our resource use that will evaporate to nothing amidst the pressure from our ever expanding population and its waistline and its line of waste. I would hope that you consider *How to Kill an Elephant* to be the most important book you have ever read. It will also be the most pointless.

Consider the typing pool of underemployed women back in the day at Anglian Water, or the electricity company and place highway engineers, planning officers, sustainabilitists, conservationists, Great Brains, and the like in the same category. They are all a waste of space, achieving nothing, but they all demand and receive high salaries (perhaps I should describe my beloved highway engineers as highwaymen, for they truly are a money-grabbing obstacle in the road). Each and every one of these individuals has a hand in your pocket, shovelling money from your wallet into their own. Your wallet is not filled with money; it is filled with time, your time, your expended effort, your sweat, your labour. Earlier in *How to Kill an Elephant* I referred to the day in the year when you have paid your taxes and start earning for yourself. The answer equates to 40 per cent of the year, or May 25. That means you work two days a week to fund the equivalent of three typists fighting in an office to type one letter for the day. And then to cap it all, the typist throws the typed letter into the bin.

Here's an idea: fit an incorruptible electronic chip in every vehicle that records the speed on a hundred-minute loop and prosecute every driver who was speeding and was involved in an accident. Tamper with the chip and the vehicle won't start. Sack all the highwaymen. Keep your time in your pocket. I know, it's a non-starter. It's nowhere near clever enough for our Great Brains. Of course the other problem is the right of motorists to speed and the inability of the chattering classes to accept the imposition of control. We just suffer the consequences. Time to add another letter after e – e – – – – –.

We seem to have developed our lifestyles into a perpetual-motion machine that could also be described as a vicious spiral. We work to pay our bills, we pay our bills so that we can work. We pay our taxes so that they can help us when we need them. So it is we pay for the care of our elderly because we are so busy working to pay our taxes that we don't have the time to look after them.

I used to have a representative who would bring individual ideas or germs of ideas to me. He would get frustrated by my lack of action but would continue nevertheless to try to expand his needs and the perceived needs of his customers. One day he confided in me that he used to get quite frustrated at my apparent lack of action but had learnt that one day he would come into my office and I would give all his ideas back to him. 'It's like I give you individual fibres and you hand me back a rope, a complete strategy.' Your task as reader is to assemble the many fibres I have given you into a rope of your own understanding and, through this process, learn how to discover the questions and answer them for yourself.

All life on earth is dependent on the sun. Well, maybe not all life, but the bits that we involve ourselves with on the surface of the earth do. (There is possibly more life contained in the rock beneath our feet than in all the 'nature' we see.) Energy from the sun is the source of the modern-day energy we consume as food or feed into our 'carbon-neutral' power stations. It was also the source of all our energy extracted by our time machine, coal, natural gas, and oil. All life is dependent on a source of energy, and because we humans have yet to learn how to photosynthesise, we are dependent on plants for all our energy, or on second-hand energy from plants recycled through our meat providers. (For the purposes of this illustration I will ignore nuclear fission, as it's rather unpalatable in your breakfast bowl.) If we remove all plants from the planet, present and past (no coal, gas, and oil), then after we have eaten all the animals that we can catch, we die. But before we die, there is the gap between where we digest our own tissues in the pursuit of energy to stay alive. The realisation of the importance of plants may drift across our hungry minds. With this graphic illustration I seek to highlight the fact that we are dependent on plants for our survival. Everything we do is underwritten by Planet Earth and the energy-capture-and-release mechanism developed by life. Go back several thousand years when the only motive force was the labour of human beings. Then, no food equalled no labour. What I am trying to illustrate is that all humankind's activity is underwritten by plants and ultimately the planet we share with them. If all humankind's activity is underwritten by the planet, then how many holes can we dig in the planet?

I lunch most days in a local café that serves freshly prepared food. I spend my time there observing Church Street and browsing through the *Daily Mail*. I wouldn't spend my money buying that paper, but I don't mind reading somebody else's. I mainly concentrate on the crossword puzzle. The newsagent's shop is opposite, and the owner regularly surfaces for a cigarette. Once again we have the interesting display of behaviour whereby he leaves the shop where it is illegal to smoke and then crushes the fag end on the pavement next to the bin.

Now for some number crunching. Use your search engine, find https://www.worldometers.info, and consider the impact of our activities displayed on the live projections. I keep banging on about obesity and personal responsibility for one's health. This website can help us put some 'fat on the bones'. The following figures are already out of date given the 'live' nature of the data provided, so please look for yourself.

There are 1.6 billion overweight people in the world.

There are 540 million obese people in the world.

The money spent in the USA today on obesity-related diseases is $433 million, and climbing higher than you can follow with the eye.

The money spent on weight loss programmes in the USA today is $166 million.

The population density of tigers is one per 12.5 square kilometres—an STU.

The amount of forest lost today is 12,500 hectares, which equals *1 tiger*

The amount of land lost to desertification today is 29,000 hectares, equivalent to *2.3 tigers.*

The number of cigarettes smoked today is 14 billion = *154 trillion minutes = 293 million years of human life lost, today.*

Money is an expression of hours worked or Planet Earth's resources utilised. It grows on trees, is cropped in fields, is dug from the ground, and is transported by our time machine from worlds long gone.

The money spent on illegal drugs today is $1 billion.

The number of cars produced today is 176,000.

The amount of money spent on public education today is $8.5 billion.

The amount of money spent on public healthcare today is $10 billion.

Today is 15 March 2016, and I don't know when 'today' started, so please read the information for yourself, whatever today is.

Today the world population is given as 7,408,745,317, having grown by 201,379 today.

Read it and weep.

And ask yourself, can nothing be done?

Where is the leadership from our Great Brains?

What is the purpose of our existence?

You Have to Lose to Win

We can stray away from the high street and even the supermarkets on their out-of-town sites and look elsewhere for competition. Olympics athletes compete to reach the highest step on the podium, desperate to outperform their fellows as the latter in turn try to reverse the result. There can be only one winner, but the adulation (and the sponsorship and state funding and coaching help and sports psychologists and bragging rights), while showered on the winner, is dependent on the losers. Who will provide the adulation for a single-competitor race?

Leicester City fans are keen to see their team win the Premiership. The players can expect a bonus or offers of higher reward from other, less successful teams with better pedigrees. The fans can bask in the success of their team and lord it over the also-rans. Who will provide the adulation for a single-team league?

The biscuit manufacturer is desperate to win the orders to fill his factory and his coffers with money and will look to gain any advantage over his competitors. His success is based on another's failure. For his business to flourish, another must fail.

The farmer is keen to maximise the yield and earning potential from her land and will use high-quality seeds, fertilisers, and state-of-the-art machinery to achieve these aims. And through the quality and quantity of her product, she will seek to command a better price over her competitors. In that sense it helps her if they fail or underperform.

A headmaster is keen for his pupils to outperform other schools in the area and will vie to obtain the best-quality teachers and monetary grant per pupil to the detriment of other schools.

The pupils themselves will compete to get into the 'higher sets' with the better teachers and better learning environment, as indeed will their parents, desperately keen to get their kids into the 'right' school (or give up and pay to use the finest private schools that they can afford). Would it be fair to say that competition is an integral part of being alive?

The Herding Instincts of Grass

Consider grass, fairly familiar in one form or another to most of us. Think about the lawn growing in your local park which is manicured and maintained by the local parks department. Each plant is fighting for life. It needs space, sunlight, water, and minerals, and it needs to secure these resources in competition with every other grass plant in immediate proximity. To be really successful as a grass plant, it needs to secure enough of these resources to be able to set seed and pass on its genes to further generations. Failing that, grass has the ability to spread by side shoots and rhizomes to create new self-contained clones of itself over time. While this image of a trimmed lawn is artificially created by humankind, it is possible to transfer this image to the African savannah or Asian steppes or South American pampas where the dominance of grass is maintained by climate, soil condition, or external agents such as fire or elephants which prevent the establishment of shrubs and trees.

It is possible to consider that the grasses 'cooperate' to form a monoculture of different grass species. It is also possible to deny any form of cooperation as simply a coincidence whereby the grasses exploit the opportunity presented to them. One could also argue that grasses themselves have evolved to satisfy the opportunities presented and they are as much a product of the environment as the environment is a product of the grasses. To summarise, every plant is striving for life and competing against its neighbours, and collectively the plants are dominating the landscape and preserving their collective existence.

Would you call grass a herd?

The immediate reply would be no, but think on a different timescale, something akin to time-lapse photography. While we can monitor the movement and growth of the surface leaves, it becomes more problematical to follow the activity of the root system, but it is reasonable to assume that similar activity occurs below ground. Consider a traditional herd, with individual wildebeests jockeying for advantage, access to fresh grass and water, and safety within the herd, while grass leaves will grow and jockey for position to gain access to sunlight, and the roots will do the same in search of water and minerals. It is only the scale of movement that differentiates between the herd of wildebeests and the herd of grass. The wildebeest will be driven by need to travel over the horizon to reach new resources, while the grass will be driven by need to grow over the horizon (shadow) of its neighbours to secure access to sunlight and deeper into the ground to secure water and minerals beyond its neighbours' reach.

This concept may be strange to accept, but if we scale the plant size up to trees, then the idea may become more palatable. Forest giants dominate the canopy, while younger saplings strive for space, water, minerals, and sunlight. Trees at the canopy wish to remain there, while younger trees reach for the skies. We could describe the vertical movement from forest floor to canopy as a vertical migration bounded by the physical characteristics of the trees themselves. A tree could be considered to migrate from soil level to 100 metres during its lifetime, whereas a wildebeest could travel for thousands of miles in circuitous migration through its lifetime. But while the wildebeest may travel thousands of miles, it simply moves on a daily basis from food or water source to food or water source. Its migration is simply the sum of all these days added together. The limits are determined by the respective physical strength of wood and the physical attributes of wildebeest. Garden snails, the ones that have found God anyway and have not had their brains bashed out on the altar to Turdus, may complete daily migrations of several yards as they forage for resources. (Tests with marked snails show that snails will return home from some distance if artificially moved by hand.) I have a colony of garden snails that rest during the day under a stone trough before emerging at dusk to forage, returning as day breaks.

Would you call a forest a herd?

What would be the criteria? Trees act and react in relation to their neighbours, and they travel in search of resources. While 100 metres may not seem very far as a migration over two hundred years, say, it is a remarkable achievement given the pace of growth. And it is successful.

So I will call a prairie a herd of grass, and I will call a forest a herd of trees.

None of you will object to my calling a large number of wildebeests a herd. But if we change the species of mammal and consider the daily migrations in search of resources, energy, space, shelter, water, and breeding opportunities that make up the requisite requirements for a successful species, we would still be able to validly describe it as a herd.

Would you consider for one instant that the herd of grass has a leader?

Would you consider for one moment that the herd of trees has a leader?

Should you consider for one moment that the herd of wildebeests has a leader, given that the fundamental difference between grass and wildebeests is one of scale of movement?

Should you consider for one moment that the herd of the mystery mammal has a leader, given that each herd member is simply engaged in securing sufficient resources for itself and its offspring?

Each and every grass plant, tree, wildebeest, and mystery mammal is fighting its neighbours to secure access to the resources it needs to succeed. There is an element of beneficial cooperation between the herd members, albeit accidental. Grass roots interlock to form a turf with increased resistance to erosion and disturbance, forming a 'blanket' to make it difficult for other species to become established. Trees provide shelter to each other to aid resistance to high winds, and roots can interlock to bolster resistance to erosion and overturning. Wildebeests can provide security in numbers with multiple eyes and ears scanning for danger while others graze or sleep, and multiple bodies scattering from ambush can confuse predators. Synchronised breeding can also saturate predators with 'easy prey', whereas random births provide opportunities all year round. Synchronised fruiting of grasses and trees also saturates seed and fruit eaters with such an abundance of food that not all of the resources can be consumed before launching the next generation. The sheer abundance of individuals in close proximity allows for maximum gene exchange and reduction in inbreeding.

There is also an element of non-beneficial cooperation between the herd members. Fire or infectious disease can spread rapidly between members, and the sudden glut of offspring and seedlings increases competition between them and can compromise available resources. High numbers will also deplete resources and reduce the vigour of individual members. (Conversely, the high genetic spread of multiple herd members may include the very genes that protect some against infectious diseases, reducing the possibility of an extinction event. Conservationists should take note, as populations reduce to low isolated levels, the whole viability of the genetic variation needed to withstand significant disease events is compromised. Tigers may already be extinct; they just haven't stopped breathing yet.)

There is still no leader.

It should be said that the herds are not pure, because within each distinct biomass exist other species, some in herds of their own, such as zebra or Thompson's gazelle, and some on their own, such as woodland and meadow herbaceous flowers, each exploiting an opportunity.

What function then does a herd leader perform? A zebra stallion can protect his harem of mares from mainly other stallions intent on supplanting his genes. A matriarchal elephant can use knowledge acquired from her dead elders and her own experience to guide her herd through lean times of drought in search of food and water. But if she gets it wrong, through no fault of her own she could condemn her extended family to death. As financial advisers are fond of stating, 'Past performance is no guarantee of future performance.' Just because the waterhole was there the last time we had a drought …

But while I want to stay with elephants, I want to play a game called Finding Nemo to illustrate an alternative view. Many of us have seen the Disney film where Nemo goes in search of his mother and is in turn searched for by his dad. Nemo, a clownfish, has been selected for its colour and stylised cuteness for the audience's consumption. Clownfish colonise sea anemones and seek shelter from predators by hiding within the poisonous tentacles of the host anemones. The Disney story couldn't possibly be further removed from the truth if it tried. The colony of clownfish consists of a small number of fish where each fish is held in a strict dominance hierarchy with only one breeding male and female. If one of the fish is lost to predation, then one of the fish lower down the hierarchy scale fills the void. The most dominant fish in the colony is female; she lays all the colony's eggs after having mated with the most dominant male fish.

The really clever bit occurs when the female is removed from the colony. One of the largest, most dominant males changes sex to become female and lays eggs fertilised by the new largest male. This whole wonderful arrangement is automatic and 'preordained' by the clownfish's genes. Do the clownfish have a leader? What does this have to do with elephants?

Elephants have a reputation for memory. It is assumed that elephants learn all about their environment and remember where to travel in times of stress to find food and water. The oldest elephants have the greatest experience and therefore adopt the role of herd leader. What happens, though, if they follow the pattern of the clownfish, not the sex change bit, but the 'preordained' bit? From birth, every elephant is 'assisted' by its relatives to give it the best advantages in life. By assisted I mean that it is cajoled, bullied, and 'steered' by its elders in the desired direction. Each elephant could be considered to learn from its peers and could memorise every push, barge, trunk clout, kick, and prod with a tusk, along with the 'verbal' prompts of vocal instructions, and could be taught a hierarchy as robust as that of the clownfish. Each potential matriarch is held in her position by elder aunts and her mother, until they have all died, and then, like the clownfish, succession to leader/matriarch occurs. At that point the group has a new leader. Or does it? Instead of a leader, what you may actually have is simply an elephant that refuses to be led. The difference between leading and refusing to be led is quite profound.

Earlier in *How to Kill an Elephant* I referred to recent research indicating that elephants can hear in the ultrasound range and can hear thunder as far as 150 miles away. Imagine being in London and hearing thunder in Sheffield. Having heard the thunder, elephants in a dry region will head towards the thunder for the rain. They may even be able to differentiate between dry thunder and wet thunder and thereby be able to avoid wasted journeys.

It is worthwhile pointing out that elephants in London probably can't hear thunder in Sheffield because of all the extraneous noise generated by humankind and their machines. And given enough industrialisation within Africa, this option would also be removed.

If we have only just learnt that elephants can hear to a range of 150 miles, from how far away can they smell rain or vegetation? We assume that the matriarch is using her memory and knowledge to lead the herd to food and water, but actually she may be reacting to her organoleptic senses and, by refusing to follow, effectively, to our eyes, leading the group to resources.

The distinction here is that we do not have a leader, simply an elephant that will not be led. This is not the same thing.

We could look for human parallels, for example a bullied child who, realising that the bullies have all left to go to a higher school, now refuses to be bullied. The aristocracy have a similar mechanism whereby the 'rule' is passed down by death and inheritance. The new lord no longer has to follow instruction, but the title confers instruction on 'junior' members, who observe the hierarchical position of the new leader, who simply isn't being led anymore.

Deaf Spiders

The site https//:www.worldometers.com provides an illustration of the impact of obesity in the United States. Having read the small print, I now understand that 'today' is linked to the clock on your computer, so if you want to know the daily figure for any of their statistics, simply view the site just before midnight on your computer. It produces a monetary figure for the cost of treating illnesses directly attributable to obesity. It also supplies a figure for the cost of weight loss programmes in the US. There is an exceedingly non-PC expression for the control of unwanted pregnancies which goes along the lines of, 'You should learn to keep your legs shut.' The two obesity-related monetary figures taken in mid March 2016 are $500 million and $190 million per day respectively, when in reality all that Americans need to do is learn to keep their mouths shut.

Unfortunately, as I have indicated before, there is no appetite (forgive me; I couldn't resist) for such a solution because, when it comes down to it, nobody can earn a living from such a simple procedure. I could use this for another seminar on weight loss. I could advertise this as a comprehensive weight loss solution so as to avoid the demands of disgruntled former customers upset by 'eat, food, less' although not necessarily in that order. 'Keep, mouths, shut, your' is four words after all, although they do not necessarily make sense in that order either.

One fundamental problem that we have is our inability to distinguish between need and need. We need to gain employment to satisfy our needs without ever considering our needs. By gaining employment we secure monetary reward, which becomes the sole reason for the employment.

The logic behind the employment becomes irrelevant, because the logic is replaced by the monetary reward. My electricity board typist would sit bored for hours on end for the pay packet at the end of the month. She would sit bored in a mind-numbing occupation in cramped quarters dominated by petty boundary disputes with two other women. My beloved highway engineers dream up pointless initiative after pointless initiative. Even they must have some inkling that their efforts are pointless. But the point is, they have been paid for the privilege of wasting their intellect, abilities, and work life, and the financial reward is all that they seek or demand. The reservoir dyke has sprung a leak, so spend your time and effort filling buckets and tipping the water back into the same reservoir. Your efforts are absolutely worthless, but no matter, your labour has been rewarded with money. Government ministers shovel endless amounts of money into the abyss of the welfare state and the NHS and expect them to produce a result. This action does produce a result, namely to pay the salaries of the government ministers and nothing else. Education establishments spend resource after resource on educating the uneducable, but it doesn't matter, because the true product of their purpose is their own financial reward. Issue two challenges to your workforce: firstly challenge them to justify their existence on the payroll, and secondly ask them to justify their imminent redundancy from the same employment. In the first instance you can expect solid arguments for their continuing employment; in the second, they simply will not understand the question.

How about talking to the trees? When giraffe graze on acacia trees, the trees react to the disturbance by releasing chemicals into their leaves which are distasteful to the giraffe palate and the animal is forced to move on. But these chemicals, shown to be airborne, stimulate trees downwind to release similar chemicals, thereby protecting themselves from giraffe before the latter have even reached them. Does the herd of acacia have a leader?

When a basking crocodile detects potential prey near or in the water, its movement into the water attracts the attention of other crocodiles who may follow. Do crocodiles have a leader?

When a rabbit detects a threat, stamps its feet in warning, and flashes its scut (white tail) while running away, it alerts other rabbits, who follow. Do rabbits have a leader?

When an airborne vulture effortlessly cruising the thermals detects a dead or dying animal, it begins a descent to the food. Other vultures detect this movement and join the carousel of descent to the food. If a vulture successfully detects food on its own, it relies on other vultures to enable the feeding process. Different species of vulture have different physical properties, and some have the tools and strength to open carcasses through tough hide to allow weaker species to feed. Do vultures have a leader?

We would never consider trees, grass, crocodiles, or vultures to have a leader, and yet they display an amount of 'coordination' which is effectively an automatic response to external stimuli that could be construed as leadership.

We do, however, consider that elephants have leaders, even though we have no proof that they are not simply responding to external stimuli. There was a joke from my junior school days which went along the lines of:

TEACHER Why are you pulling the legs off that spider?

PUPIL Look, sir, when I tell the spider to go away, it does, but if I pull all its legs off and tell it to go away, it doesn't.

TEACHER And why do you think that is, boy?

PUPIL Well, sir, when I pull all its legs off, it goes deaf.

A perfectly logical explanation, but then so is the elephant using its memory and experience to traverse a desert in search of water that happened to be there twenty years ago in a similar drought. On this basis we should be able to dispense with the services of weather forecasters because we have seen these weather patterns before. In a similar vein, it is like asking a cancer patient what they ate or which chemicals they were exposed to over a twenty-five-year period in search of the cause, although in the elephant's case, the absence of Dr Doolittle's help may be a real barrier to progress.

The most recent Disney version of *101 Dalmatians* contains a scene where the parents 'bark a request for help in finding their pups' over a 'jungle telegraph' of other animals who radiate the request out for miles in all directions until the reply is similarly relayed back. There is the suggestion that elephants communicate at ultra-low frequencies and that they may be able to do so over several miles. Do elephants use a 'jungle telegraph' to ask for and receive information relayed by different animals over many miles? If an elephant asked the question 'Any water over your way?' and received a reply from a particular compass point, they could move towards it before asking the question again when several miles nearer and could then modify the direction of travel to reflect the more up-to-date reply. Alternatively, an elephant simply needs to give the ultra-low frequency sound for 'Yippee, I've found water' for this to be relayed by different elephants over large distances for the same effect. This is similar in many respects to the strategy employed by vultures, which can detect carrion from many miles away simply by watching the behaviour of other vultures. A descending vulture attracts the attention of other vultures perhaps three miles away, and these in turn attract the attention of still others that could be another three miles away. In this way, vultures could be drawn to a carcass perhaps forty miles away because they study other vultures as much as they study the ground. If we take these figures at face value, then we may deduce that a single descending vulture could draw in other birds from a forty-mile radius, an area of five thousand square miles. But there is still no leader.

The Folly of Leadership

In recent years, in certain areas of Africa culls of elephant have taken place. Specimens were killed at random and 'orphans' were left behind. These groups of younger elephants grew up in isolation of normal elephant behaviour and control and became delinquents, causing unforeseen problems down the line. In some game reserves these delinquent males took to bullying and killing rhinoceroses. Wardens were at a loss to understand why the rhinos were dying, until the elephant action was observed. To cure this problem, the more enlightened approach to culling elephants is to cull entire family groups, leaving no 'orphans' behind to become delinquent. The idea is that these younger elephants grew up in an environment with little interaction with mature elephant bulls who imposed discipline over teenage bulls that were 'throwing their weight around' by 'beating up' rhinos. The culls were organised to reduce the pressure on the environment caused by a surplus of elephants and ironically to encourage the populations of rarer species such as rhinoceros. Conservationists acting as gardeners again, favouring one species over another without understanding the consequences. Now we have a different potential for harm, because by culling entire extended family groups, we are systematically removing 'bloodlines' of elephant clans from the gene pool, the gene pool that needs to be as vast as possible to provide protection against virulent disease. Fancy a rose garden, anyone?

Back to the mystery mammal. But you already know I was referring to humans.

When you leave your house, drive to the station, board a train, and walk the last few minutes to work, there is no leader.

When you stop for a takeaway coffee laced with unseen sugar and pay thousands of pounds per tonne for tinted water, there is no leader.

When you drive to the supermarket and fill your trolley with processed and convenience foods laced with salt, sugar, and saturated fats (and no taste, except for the salt, sugar, and fats), there is no leader.

When you drop off your kids at school and they avidly hang on their teacher's every word (normally at a young age when teacher knows best), there is no leader.

When you vote in local or national elections for councillors or MPs or MEPs to elect governance, there is no leader.

When you hunt on the high street for the latest fashions and must-have goods, there is no leader.

There is no leader, so why do we do it? Convention? Because everybody else does it? 'I am a programmed robot and I must follow my programming'? Because we are a social animal and this is what we do? 'I am controlled by chemical signals emitted by an unseen queen and I am compelled to obey'? 'It's not my fault that I can't see past the Clovis point'? Because it's easier to fit in? Because it's what we need to do to survive? Because it's how we survive?

Let us consider a perhaps halcyon age when food grew in abundance on trees and bushes, when prey animals were widespread and in such numbers that our activities hardly made a dent. Our technologies of stone tools and fire and clothing made our lives comfortable and left time for extracurricular events.

If I were to propose a return to these standards of living, then the immediate problem is one of space and resources, or more simply, the sheer population of our species outweighs 'nature's larder'. Our species has gotten around this issue by refining and developing through artificial means the quantity and quality of nature's larder through agriculture reinforced with terraforming. Not for us the confines of forest clearings when we can clear the forest instead. Not for us the constraints forced on us by other species, dangerous animals (kill them), or competition from other species (eliminate them through shooting, trapping, poisoning, or displacement, or domesticate and crop them).

We cannot put the genie of knowledge back in the bottle. We cannot undo the damage that we have done. We could attempt to reduce future impact by controlling our population and reducing our use of resources to things that matter. Nothing that has been proposed, legislated, or considered will have any impact on our activity to date in spite of the fact that many earn a very good living from these very desirable ambitions. We have no leaders.

The Great Brains who pretend to be our leaders command the salaries, the respect, and the committee space to protect their own positions. You can argue that their only qualification is that they refuse to be led. The problem with this statement is that the Great Brains begin their careers by being led, because only in this way can they carve a position of support from others. You cannot become king without being a prince first (or having a prince regent changing your nappies). When you reach the apex of control and become a national leader, you simply become part of another group of international leaders who continue to jockey for position and acceptance, searching for the common pro forma. The trick to being a Great Brain is to learn the pro forma. Wear a suit, smile at the right people, align yourself with the right grouping, and don't upset the people who matter. There is no leader.

With elephants, the belief is that the matriarch has earned the right to lead the herd by outliving all of her peers and therefore becoming the most experienced and hence best qualified to ensure the herd's survival. This qualification simply translates into outliving all other relatives who boss you around, until you are no longer led.

If we have no leaders, what purpose does the title serve? When will we be satisfied with our lot?

When will the redundant life support machine be turned off?

When will we design a new Clovis point?

When will our apocalypse finish its world tour?

If we have no leaders, what purpose does the title serve? Disagree? Drive at 30 mph in a restricted zone.

If we have no leaders, what purpose do their instructions serve? Disagree? You will not exceed 30 mph in a restricted zone.

If we have no leaders, why do we allow them to pass laws? Disagree? It is illegal to exceed 30 mph in a restricted zone.

If we have no leaders, why do we accept the attempts by government agencies to enforce the rules? Disagree? We have accepted the imposition of twenty-two initiatives to control the speed to 30 mph in restricted zones.

If the government agencies had any leaders, they wouldn't waste resources on attempting to control the uncontrollable. Disagree? What will initiative 23 be? Disagree? Why did we have initiative 22, 21, 20, 19 …?

What am I saying? Because the speed of cars cannot be controlled, we now have cycle paths designed by you-know-who to positively separate the speeding cars from the cycles that now speed past the pedestrians. What will initiative 24 be, designated footpaths to separate speeding bicycles from pedestrians?

What will initiative 25 be, provide dial-a-ride car services to protect pedestrians by providing cost-effective taxi services and put them on the safer roads? Initiative 26 will be dedicated dial-a-ride lanes to speed the movement of pedestrians. Next thing you know, they will introduce a more difficult driving test to probe risk appreciation and have a theory test before the practical. Oh bugger, is that in my list of twenty-two?

I learnt to drive in 1977, before the twenty-two initiatives.

I have lived through thirty-nine years of highway engineers wasting tax money and continuing to waste tax money.

I have lived through thirty-nine years of Great Brains failing to control the speed of traffic in restricted zones.

I have lived through thirty-nine years of wasted education, years that could have taught young impressionable minds of the importance of observing the speed limit in restricted zones.

It is conceivable that the speeding initiatives have spanned the working lifetime of at least one highwayman who has presided over the entire gamut of failed ideas and has either realised the utter futility of his actions or has accepted that this is simply the means by which he draws his salary and so he doesn't give a shit. When he looks back on his career, how much pride in his non-achievements will he muster, I wonder?

Please don't consider that I am hung up on the speed of cars in restricted zones. Simply accept this example as evidence of the folly and stupidity and resource wasting that our leaderless society foists upon us. If you didn't have to pay so much tax to fund this stupidity, you could work fewer hours and wouldn't feel the need to hurry.

Speeding cars? It doesn't matter?

The Kyoto Protocol agreed in 1997 and ratified in 2005 was to cut the greenhouse gas emissions by 29 per cent as measured against 1990 levels by 2012. The Netherlands Environmental Assessment Agency estimates that greenhouse gas emissions since 1990 rose 40 per cent by 2009. The difference between the planned reduction and the growth is plus 69 per cent, which hardly rates as a speed bump in the road given that our vehicles have in fact gone faster by analogy. (In motoring terms, if the average speed in a 30 mph zone were 39 mph instead, then our 29 per cent reduction would have dropped the speed to 30 mph. What we have done in round figures is to have increased our average speed by 40 per cent to 55 mph.) Have I mentioned having our foot planted firmly on the accelerator during our apocalypse world tour? It doesn't matter if you believe in man-made global warming or not. If the Great Brains are leading us from an unnatural disaster of our making, then they deserve to be sacked, because they surely have no place at the top table and are not in fact leading us away from a man-made global disaster.

One of the answers these masters of the world—well, masters of the conference table anyway—has proposed is that richer nations pay poorer developing countries compensation to enable them to introduce more expensive forms of 'green energy'.

Do I still need to point out that money does grow on trees and that everything we do is underwritten by Planet Earth? In providing these funds, rich countries must accelerate the extraction of these resources to compensate poorer countries. We are digging holes in the earth to fill holes in the earth. Genius. We have to use energy to create the wealth to reduce the amount of energy we use. Genius.

I could define a leader.

If we look at hierarchical animal groups such as hyenas or baboons or wolves, we see that status may be granted at birth or won through physical exertion. A simple suggestion is that the leader rises to the top and has the first choice of food, sex, and even safety by sheltering in the centre of the group. They secure their position by strength of tooth and claw, and some by personality borne of an inherited entitlement. In simple terms, the leader is the biggest bully of the group. Leader = bully. But of course they are doing it for the greater good (well, their greater good anyway).

So what is my solution? Simple. I don't have one. I don't pretend to have one. I don't get paid to have one. I don't have the brain, great or otherwise, to have one. But more importantly, nobody else has one.

Plenty of people pretend to have one. Plenty of people get paid for pretending to have one.

I do have the brain, great or otherwise, to know that the Great Brains do not have one.

I pay with my taxes for them to not have one. I pay by plundering the earth's resources for them not to have one.

Our species does not have one.

Our species does not have the capacity to have one.

Our species is preprogrammed to corrupt for personal gain any potential for having one.

Our species relies on leaders. Our species does not have leaders.

Our species would cut its own legs off in the belief that the resultant reduced weight will help us run faster. Our species will continue past the point of oblivion and leave only survivors.

Imagine the mayhem a chimpanzee would cause if armed with an AK-47 in a room full of people. It would fire indiscriminately into the crowd with no sense of guilt or consequence. It is a dumb animal.

What a blessed relief that we have control over:

- nuclear arsenals
- conventional arsenals
- fossil fuels
- education
- agriculture
- science
- terraforming
- consumerism
- mechanisation
- conservation
- sustainability
- taxation.

The world is in our hands. 'Oh, bollocks' springs to mind, and it's better than an exclamation mark, cos I might put the exclamation mark in the wrong plaice.

Is this another reason for education, knowing when to use an exclamation mark? And I thought it was to keep them off the streets during the day; silly me! While we are at it, watch out for the split infin

The Permanence of Sandcastles

In 1304 Edward I, King of England, besieged the castle at Stirling. He used no fewer than twelve siege engines and is alleged to have refused the garrisons surrender until his newest and largest machine, Warwolf, had been constructed and used against the castle walls. It took four months for the castle to surrender. To besiege, build, and operate twelve or thirteen siege engines and engage in other assorted activities, such as filling the moat in to allow access for scaling ladders, took extensive manpower and logistical resources to achieve. This was the king who had subjugated the Welsh princes by constructing massive castles throughout Wales and establishing his authority in an unchallengeable fashion. Edward truly understood the value of castles. Indeed, it is possible to suggest that he was fixated on them. Finally, in July 1304, the garrison of Stirling surrendered and offered themselves to the mercy of Edward. For four months, and using what must have been many hundreds of men, he had achieved his goal and accepted the surrender of thirty—yes, thirty—men. (Figures vary; some sources say nineteen.)

My son worked on a building site in London on the banks of the Thames and was warned by the site management not to stray onto an adjacent road, but this warning was issued without an explanation. I will not name the precise location for obvious reasons. At some stage, a stray vehicle entered the road and was greeted within a few seconds by armed police.

During Thatcher's government, at the height of the IRA's campaign on the British mainland, a small explosive device exploded at Lord McAlpine's residence. At the time he was treasurer for the Conservative Party and was seen as a 'target of value'. Some months later a neighbour called me to attend my factory site because he could hear significant activity from a gang of trespassing youths. I arrived within ten minutes to find the narrow lane adjacent to the neighbour's swarming with police, some of them armed. The youths were suitably shocked by the severity of the reaction they had received. Their 'crime' had been to interfere with a parked car outside the local vicarage. What they didn't know was that the car was fitted with cameras, providing surveillance to Special Branch officers resident in the new vicarage who were 'guarding' the old vicarage, the home of the then Chancellor of the Exchequer.

The aforementioned are three examples of what I will call 'castle mentality', two historic and one from the last three years. itives

What is the relevance?

Whenever we have a terrorist atrocity, be it in the UK or elsewhere in Europe, the police and army are rapidly deployed to guard 'castles'. That is to say, security is visibly staged at airports and train stations and other 'targets of interest' in a bid to reassure the public that all is being done to protect them. This is in spite of the fact that terrorists no longer confine their activities to what I describe as castles. Any site containing a tightly packed grouping of people—be it at a pop concert, at a café, in a queue for a sporting event, at a pub, or at a bus stop; or a press of people at the school gates at closing time, or at the shopping centre, or at a tourist beach; or even a lone soldier crossing a road—is a target, and they cannot all be protected. Yet our Great Brains persist in the lie of protecting castles as a solution to the terrorist threat, in spite of the fact that many of the 'soft' targets I have described have already been attacked and will no doubt be attacked again. Terrorists are guerrilla fighters. Guerrilla fighters do not have castles. Guerrilla fighters cannot fight a conventional war against conventional forces. Knowing their limitations, they attack wherever they can, never feeling an obligation to attack where they are expected, because that would be stupid. Our Great Brains protect the sites where the terrorists are expected to attack. And that is stupid.

I do accept that our intelligence services are monitoring and observing many potential terrorist suspects. Given the very nature of their activity, this has to be done secretly. What I am trying to illustrate is the lie between being seen as protecting the public and actually protecting the public. Let me call it corruption, the difference between truth and the promoted truth.

The Human Species Competes

It competes for food, shelter, sex, water, space, and territory, to name a few. Competition is an essential part of our survival strategy. We can think of the human race as an endless herd of grass where every leaf and root is looking to maximise its position to secure resources and survive. The very definition of 'competition' is 'to establish a hierarchy', to appear on a league table of success, and to be one of the winners, because losing is detrimental to life. As a species we create losers, and then we seek to 'compensate and protect' the losers by providing welfare or charity as additional help. We have here a paradox, because the more successful you become as an individual, the more 'losers' trail behind in your wake. Bill Gates (I pick on him because he has held the title of richest man on earth) has created more losers than most. He has also provided the framework for many, many winners, who in turn have created their share of losers. Is this the cultural leaning that dictates the need to give all the 'winnings' away to the losers? But the losers who become winners become winners at someone else's expense, namely all the losers who donated the winnings in the first place. Ultimately, we are back to the concept of not having enough tarmac to keep all the world's roads surfaced simultaneously, hence the need for holes to appear everywhere.

Winning serves a purpose in nature; it gives every organism an edge, an advantage, a leg up in the evolutionary arms race until a better alternative surfaces. Non-fiction books of my youth would point to the Irish elk (*Megaloceros giganteus*) and its massive rack of antlers, which grew to such a size that the elk eventually died out because they couldn't cope with soft ground (too heavy) or the emergence of forest (too wide to pass through the trees). Ireland was obviously a bigger place in those days, and the over-simplistic theories suggesting the demise of Irish elk were precisely that. But taking these ideas at face value, there appears to be a penalty for winning too much. In reality, the penalty is for failing to adapt to changing circumstances. If the antlers were getting too heavy for the elk to traverse soft ground as the ice age ended and the females insisted on only having sex with the most endowed males, then these would be the first to disappear. But experience with modern deer species teaches us that it is the males who compete with males for the right to mate. The largest, most endowed, strongest individuals establish themselves within a harem of females, defending their position against all comers until they themselves are displaced. (And if no harem existed, there would be no need for overly large antlers?) The females have a selection technique akin to the lottery in that they mate with whichever male is in charge when they come into oestrus.

If all the heavily endowed males sank in the swamps, then only less-endowed males would be left to compete and win the right to mate. In this way large-antlered males would be naturally removed from the breeding pool and smaller-antlered males would remain. (Evidence is emerging in Africa indicating that the most tusk-endowed specimens are shot by poachers and game hunters and that no-tusk or small-tusk elephants are beginning to dominate the species.) This methodology of mating with smaller-antlered males would reverse the very mechanism that created them in the first place.

More fundamental issues of food availability, competition, habitat type, and the like would have been the probable cause of the demise of the Irish elk, coupled with an inability to adapt to changing circumstances fast enough to survive.

Winning has stood our species well throughout its existence; it certainly helped us to win the race against Neanderthal and Denisovan, albeit with some hybridisation along the way. Can I postulate the idea that our modern man is growing excessive antlers of its own and is rapidly pursuing a future up a dead end? Our very success is accelerating at a rate by which we are irrevocably altering the world and, dare I say it, in an unsustainable way.

Money doesn't solve anything.

Our Great Brains when faced with the vast majority of issues have one fundamental answer, and that is quite simply to throw money at the problem. Throw money at the NHS, throw money at education, throw money at road building. In this, they are ignoring the fact that they are throwing time, yours and mine, used to generate the money in the first instance. They are also utilising the time machine again by spending money from our future by borrowing.

I used to buy sandwiches from my supermarket for the journey home from my weekly shop and would groan inwardly whenever the 'price reduction' stickers appeared on the basis that the cost reductions had come about as a product of debasement. Eventually my sanity broke through and I stopped buying the rubbish. Now, however, I groan outwardly whenever I hear the announcements 'an extra £600 million for nursery schools', '£1.2 billion for policing', '£1 billion extra for cancer drugs'.

EPC (Energy Performance Certificate)

I have just completed the conversion of a former stables into a residential bungalow and have a legal requirement to provide an EPC. Most of us will be familiar with the principle, as 'traffic light' stickers have been appearing on household appliances for some years now. A's and B's are green and good; F's and G's are red and bad because they consume more energy to do the same job. My insulated building qualified as a C because I have not included renewable energy sources in its construction. My certificate came with two recommendations to lift it to a B: fit solar panel water heating and solar panel room heating. Staying with the water heating, the estimated cost was up to £6,000, and the estimated saving was put at £60 per year. You don't have to be a mathematician to realise that the cost of the panels would take one hundred years to recoup, which is around three to four times the designed life of the panels. There are plenty of authors out there who have pointed out that the amount of energy needed to manufacture, distribute, and install the panels will outweigh the amount of energy that they will produce in their lifetime.

I want to look at the argument from a different angle, namely the cost. Think about the effort required to generate the £6,000 capital sum in the first instance and ask yourself how much energy and resources will be consumed to generate this amount. If the average UK salary was put at £30,000 per annum, then the average worker would spend 20 per cent of a year working to pay the bill. But that's after you've spent 40 per cent of the year paying tax so in reality the average worker would spend a third of one year earning the money to pay for the solar panels. I realise that this would be spread over a number of years, but the fact remains that a third of a year is needed in total. During this third of a year, the average worker needs to travel to and from work, and occupy office or manufacturing space with lighting, heating, and the need to process raw materials to 'add value' and justify their wage. This adds another stream of energy consumption over and above the energy used to manufacture the 'green' panels in the first place. Then to add further insult to injury, the whole arrangement is underwritten by 'government' funds whereby there are no upfront capital costs, simply a repayment plan collected via your electricity bill. As many have said before, government doesn't have any money; they have our money collected via taxes.

When I grew up in the sixties, a common theme of TV police stories was theft and the need for fences, people who receive stolen goods and pay a sum of money to the criminal lower than the retail value of the goods. A £1,000 diamond ring would be sold for £100 because it was 'hot' and would attract too much attention if touted around legitimate jewellers' premises, the resultant attention leading to an arrest. Today shoplifted goods are sold in pubs and recipients' homes for a fraction of their ticket value because the illegality of the transaction is recognised and the penalties for receiving are similar to those for stealing in the first instance. Banks and other financial institutions have to establish the identity of all customers before accounts can be opened. One of the three reasons given for this is to 'disable the activity of money launderers'. (Proceeds of crimes such as drug peddling, terrorism, and trading in illegal weapons across borders are cited.) In reality, and this is never stated, it is designed to disrupt the 'black economy' of cash transactions for work and goods that are outside the taxation system. The common theme of all these activities is that the movement of money and goods comes at a cost. The burglar may receive one pound in ten for his stolen goods. The fence may receive five pounds in ten for the discounted sale of stolen goods to selected customers who ask no questions. Fundamentally this train of transactions mirrors the legitimate world, where the movement of money attracts costs. It's called overhead and administration, and it supports the whole mechanism of financial institutions, including bankers' bonuses, savers' rates of 1 per cent, and borrowing rates of 5 per cent. In the instance of the £6,000 borrowed from the government, how many civil servants and management structures and wasteful procedures do you think it travels through before surfacing at the interface of accessibility? I do not know the answer, but I am sure you will agree that to get £6,000 out, you need to put a lot more in. What then is the true cost of the hundred-year return on my water-heating solar panels, two hundred or three hundred years? And all to display my 'green credentials'. Incidentally, the boxes ticked on the EPC include insulation to the roof space and the use of low-energy light bulbs. I have over-specified the roof insulation by 30 per cent, but the surveyor's calculations only go to 300 mm, so my additional 100 mm is ignored. For low-energy light bulbs, the surveyor has accepted that I have not used 60W or 100W incandescent but has taken no account of whether I have used 4.5W LEDs or 35W bulbs, a difference in energy use of 87 per cent. Measurement, blessed measurement! (Oops, is that exclamation mark in the right place?) It also ignores the fact that the building is a conversion of a Victorian structure and has therefore effectively been recycled with a reduced carbon footprint and that all the windows have a southerly aspect to maximise solar gain reinforced with Pilkington K glass.

Fortunately the surveyor travelled only 18 miles by car and provided the paper electronically, along with his bill for £60, which has itself added another year to realise the savings of my solar-powered water-heating system. (His rate is normally £95, but I was local.) You can probably guess that I won't be fitting solar panels.

How do you get these gigs? I bet it's good money designing EPCs, and you get to go home at night secure in the knowledge that you have saved the world. If the average salaried worker in the UK were to be sent home on 'garden leave' for four months with instructions to dig the garden manually and grow their own vegetables, how much energy and carbon would they save by comparison?

Returning to the extra billion pounds for cancer drugs, let us consider the reality. Who believes for one minute that one billion pounds is placed in a bank account ring-fenced specifically for the payment of money to drug manufacturers? Firstly we need a management structure to handle the money and decide the need for and the provision of funds on a case-by-case basis. Secondly we need to administer the drugs, and that needs prescribers and potentially nurses and hospital beds. Thirdly, we have created an environment where cancer drug manufacturers no longer feel any compunction to reduce their prices to a marketable value. How much of the one billion pounds in fact finds its way into the bloodstreams of patients? Fourthly, we have all the tax collecting and administration to collect the one billion pounds in the first place, so our one billion pounds costs us more than one billion to assemble.

In truth our Great Brains never specified that one billion pounds would be injected into the bloodstreams of patients; they just made the statement 'an extra one billion pounds for cancer drugs', and we did the rest.

In a similar vein, TV reports have recently surfaced of Gloucestershire-based 'homeless beggars' collecting up to £45,000 in tax-free cash from their 'generous victims', when not in fact being homeless. The 'correct' way of solving the problem is to give your donations to homeless charities instead, who will ensure that your money will be targeted properly to the deserving cases. 'After all, your cash donation may be funding drug or alcohol addictions. Your money will be targeted properly to the deserving cases' just as soon as we have paid our salaries and administration costs. Having ticked the accommodation requirements box for the homeless addict, such people are now free to devote whatever other resources they can find to their addictions. Housing equates to just one more reason not to give up their addiction. We have facilitated their desire to have food and shelter while continuing their slow-motion suicide. Now of course other agencies, charities, will be stepping in to assist in the rehabilitation of homeless drug addicts, and sometimes they succeed, which is a good thing. Once again I ask the questions, what percentage of donated money ends up at the 'business end', and what percentage serves the 'do-gooder'? And there in a nutshell is the fundamental problem with charity. They can all point to successes and benefits provided to justify their existence while we simultaneously ignore the bad that they do. Can we establish a measure (we're on thin ice already at the suggestion) to determine the balance between good and bad? We feed the tits in our garden while killing birds elsewhere by clearing habitat to grow birdseed. We ensure that homeless drug addicts have a roof over their heads so that they can continue their slow-motion suicides in comfort. We pour millions into cancer research laboratories to fund ever more expensive treatments and ever cleverer scientists who demand ever more expensive facilities and salaries to provide 'cures' that we cannot afford. Shock, horror; charities are actually bad for you?

I will use the Pareto rule of the 80:20 split for convenience. I have no proof.

I am going to suggest that way back when we were organising the Neolithic World Cup tournament on Salisbury Plain, we humans spent 80 per cent of our time on living and 20 per cent of our time on extracurricular activities.

I am going to suggest that in our modern times we now spend 20 per cent of our time on living and 80 per cent of our time on extracurricular activities.

I wonder what the promise was back then? Help us build the stadium and we will become world champions? What is the promise today? Help us build a caring society free of poverty and disease in harmony with nature while we climb over the backs of everyone to achieve it?

Back to the Pareto rule. Eighty per cent of our time on extracurricular activities is a good thing, isn't it? We are no longer bound by constraints of food and shelter and can devote our precious time to the things that matter, like cars, en-suite bedrooms, and holidays abroad where we can reconnect with the natural world that has been bulldozed to construct the hotels. We rape and pillage our fellow man to stand ever nearer to our goals without ever considering what those goals are or what the consequences of our actions may be. Collateral damage is not the by-product of our processes; it is the product of our processes, and we are blind to our actions. Yet something stirs in our subconscious and vibrates a brain cell of concern ('guilt' is too strong a word for it), so it is we accept the lie of welfare and charity and cannot see that this is just another money-grabbing enterprise designed to provide the salve for our troubled minds. Place yourself at the entrance interview for Hades and plead your case for salvation. Confirm your voluntary tithe to charity, but ignore the 90 per cent of back-climbing advancement as you strove to be the dominant ape and harvested the advantages that it conferred. Is the reason that we have poor people because we have made them poor? Is the reason we have drug addicts because we drove them to drug use as they failed to compete? While I single out the poor and drug addicts for their failure to control their lives, the rich and successful are not immune from the failures of control and are subject to peer pressure and are at the beck and call of advertisers. Thus the richer and more successful all want to smell like Kiera Knightly because they suffer from the same peer pressures. We accept that we all want a detached house with an executive motor car or three on the drive. We accept that we all want exotic holidays to far-flung places. How easily we orientate our cause and effect to separate the good from the bad without ever considering that there is no difference. Call cause and effect an opposite and equal reaction if you like. One person's advancement is another person's regression. Every stick has a shitty end. We all strive to grab the clean end, so for every winner there is a loser.

How to curb the insatiable appetite for trivial, insignificant, life-changing must-haves? How did I grow up unscarred by only having one choice of Vosene shampoo? We only had two TV channels. (Perhaps I have been scarred by the choice of ninety-odd different hair treatments and hundreds of TV channels?)

And so we come to the greatest force multiplier of them all.

It corrupts the natural order of life and distorts the reality of our species.

It drives the destruction of the earth.

It steals the greatest, most precious commodity of all, time, from our lives.

It steals time from the future. It is a time machine plundering our history to come.

It generates the basis for our non-existent leaders and ensures the intellectual bankruptcy of our institutions and government.

It subjugates the entire species.

It ends and condemns lives through its rationing and absence.

It drives the apocalyptic tour to the precipice.

It corrupts absolutely.

It has become the reason for our species.

It has no rhyme or reason.

It has infected and become inseparable from our DNA.

It has become the reason, the higher order to our lives, the goal, the universal panacea to satisfy all our needs.

It is our need.

We are all addicts.

We have surrendered our planet to it.

It is an illusion

Money.

The three richest people in the world (those we know about, anyway) hold assets greater than those owned by 10 per cent of the world's poorest people combined. We have seen how Bill Gates has more money than he knows what to do with. I have attempted to demonstrate that his well-meaning actions to give a worthwhile dispersal of his funds basically makes him 'head gardener', because he determines what flourishes or is discarded as a weed in his garden of the world. 'Ah, but he's helping people,' I hear you cry, 'But at the expense of the world and its resources' is my counter. In his world, tigers are weeds and people are petunias, or is it petunias are weeds and roses are best? It doesn't matter; either is a product of an artificial bias. I have asked with reference to *Star Wars* (I prefer the *Ars Warts* anagram) how many planets and solar systems you need to hold dominion over to be happy. How many billions does Bill need? You may feel I am too disparaging of BG's attempts; after all, he will be taking the very best of advice from the Greatest Brains he can afford (which basically means all of them) as they save the planet, after paying off the mortgage, of course.

But if there was no money, there would be no billionaires. There would be no thirty-five-billion-dollar holes excavated in the earth. There would be no gardening projects. I would also suggest, without any proof—call it a hunch—that there wouldn't be 7,400,000,000 people on the planet either. My reasoning is that the population of Britain remained more or less static between Late Roman and Elizabethan times, until some switch was thrown and the population began to rise. Perhaps it was the plunder secured by Sir Francis Drake and other state-sponsored pirates that kick-started the process. Plunder extracted from the Americas by the Spanish and Portuguese that never made it to its expected destination. Let's call it a hole dug in America that produced a mound in Britain that let gravity take its course as it levelled out over Britain. And we learnt from it and learnt to dominate the high seas and paint the world atlas red as we continued to take over from the Iberians in their hole-digging industry.

If there was no money, then the currency would be a person's labour for the hour, and any one person's labour would be valued the same as any other's. Until the perception of skill and talent and learning arose anyway, to be swiftly followed by management, you know, the ones who keep their hands clean or full of whip handles. But then came intellect and education, and these force multipliers teamed up with earning multipliers, which separated money earned by muscle from money earned by brain. Now it becomes possible to separate monetary reward from the limitations of muscle. Now we see salaries and bonuses of millions of pounds, all underwritten by the sweat of human beings. The difference now is that the labour is conducted not on your factory floor but in fields and quarries and factories in unseen locations around the world, all underwritten by Planet Earth. With this disconnect of wealth comes the need for salvation and absolution for the afterlife, and waiting in the wings are the first designers of emperors' clothes, although at this time their invisible cloth once dressed the soul and stone of cathedrals and garlanded the balustrade to the Pearly Gates, as surely as rats colonise humans, fleas colonise rats, and *Yersinia pestis* colonises fleas.

But allied with this disconnect of wealth comes disconnect of responsibility; we kill strangers so we don't kill the people we love. Except the disconnect is not bound by oceans or national boundaries, but boundary walls and the desire to make a living. So it is that peddlers of heroin sit cheek by jowl with cigarette sellers and processed food manufacturers and toxic car fume manufacturers underscored with advertisements to create the dream, all with equal lack of responsibility for their actions. United in one ambition, to make a living, by providing a killing, addicted to a life of slow-motion suicide, where in the end everybody dies and all is fair in trade, and absolution is but a prayer away.

And astride this perfect storm of man-made global disaster sit our Great Brains, charged with ensuring the continuation of our 'business as usual' ethic while simultaneously promoting the solution to our extinction-level event (ELE). For while we dramatise the end of humankind at the sharp end of a comet in the 1998 film *Deep Impact*, our day-to-day activity is ensuring extinction-level events for life on earth without even a scratch on the stratosphere.

We beaver away so intently with our panacea solutions for all our ills without ever considering the consequences of our actions, refusing to acknowledge our utter failure at finding a real answer, so intent are we in taking our salary regardless. So it is that we will continue to build speed bumps and sustainable forests and tiger reserves and solar-powered water heaters, and sign climate treaties, and develop force multipliers and charities, and dig endless holes in the earth when it means nothing to the planet that we share with nature, except of course the only thing we share is apocalypse. When will the human species get its head out of its arse and focus?

It won't.

It has no leaders.

There is no mechanism for change.

How do you kill an elephant?

We know how to kill a planet.

It is the difference between the truth and the perceived truth that leads to deliberate distortion. In this way predominantly white girls can be repeatedly raped and sexually assaulted by predominantly Muslim men in Bradford because if the truth were known or acknowledged, this would undermine 'race relations' and 'community building' initiatives. From which we can determine that white girls are not seen as important when weighed against the needs of the community of Muslim men. The girls were not failed by any lack of action; rather, they were failed by action on behalf of predominantly Muslim men and their place in the greater good according to the local Great Brains. The fact that this state of affairs has travelled beyond one group of Muslim men in one city to other groups and other cities with other Great Brains begins to pose still deeper questions.

But I don't want to focus on race or religion. I want to focus on the Great Brains and their obvious corruption in pursuit of a 'greater good'. It could have been predominantly white men assaulting predominantly Muslim girls and the same truths would likely have been hidden. Although if I did look at religion again, then I would see protective pro forma again within the context of Catholic priests and paedophiles. I believe that these cases expose the greater lie and corruption prevalent in our leadership that will continue to survive and prosper even after the belated exposure of these court cases. It could have been a story about corruption of the Great Brains who enjoyed a convivial drink in an illegal establishment in Prohibition-era America or present-day drug-funded activities in Mexico. Suffice it to say that the precious mantras of 'mistakes were made' and 'lessons will be learned' will be earnestly muttered, while the impact will not outlast the echo of the words. It would be simpler to state that 'salaries have been paid' and 'salaries will continue to be paid', so the problem is sorted. For the great error committed, here is the exposure to failure and the publicity. Our dear old planning officers introduce 'cane toad' policies into the British countryside, secure in the knowledge that they will be drawing their pensions before the consequences of their actions become known. Our highway engineers will continue to generate pointless initiative after pointless initiative to control the speed of vehicles in restricted zones while continuing to draw their salaries, with their well-earned pensions to follow. Not for them the high-risk strategies involved in child protection or social workers where failure might, just might, lead them to lose their jobs. This is perhaps the reason why the 'disgraced head of Children's Services' will cling to his or her morally bankrupt position until levered out via unfair dismissal tribunals, because 'it's so unfair. Why should me being shit at my job matter when I am surrounded by so many other departments with similarly performing staff?'

The reason is that you have been exposed and therefore have been sacrificed to protect the illusion of control and management before the infection of truth can spread.

Another favourite mantra is 'root and branch overhaul', to tackle the culture of failure and to 'really sort the problem out'. But when it comes to culture, the terminology may change, although attitudes remain imbedded in the psyche. Outwardly, terms like 'queen', 'queers', and 'shirt lifters' may disappear from the lexicon, to be replaced with 'gay', but the mind still translates the latter to the former, even though the previous venom attached to these terms has apparently dissipated. I say 'apparently' because within closed doors and closed circles the old vitriol will surface; it just won't be said in public. Homosexuality has become socially acceptable, but it remains privately condemned by Great Brains who occasionally slip their facade by offering opinions on 'gay plague' for AIDS or blaming the upsurge of flooding on homosexuals or same-sex marriages.

Conversely, most Great Brains will attribute the upsurge in flooding to global warming, which conveniently absolves them of responsibility for failed 'authority' decisions offering grants to straighten streams, land drain fields, dig ditches, assimilate marshes into agricultural production, and turn the countryside, over decades of development, into a 'giant toilet flush' whenever it rains. I am sure I can remember claims that the greatest floods in the Mississippi basin in recent years have been the result of reduced water volume when compared to historical lesser floods.

There was a time when the flooding of grassland was actively encouraged in lowland England through the use of sluice gates and low-grade engineering to develop 'water meadows'. In a time before silage and high-quality fertiliser, farmers would flood grassland to protect the soil from the cooling of winter frosts. The water acted as an insulation blanket, and when the fields were drained in the spring, the warmer soil produced fresh growth weeks before uninsulated fields. Earlier growth reduced the need to store so much winter fodder for cattle, which could be turned out into the fields sooner in the season, reducing dependence on dietary supplements. My mother-in-law grew up pre-war in the fens and learnt to skate on the shallow flooded fields in safety. I will only mention in passing that these flood meadows also displayed the greatest species diversity of grassland and represented our herbal equivalent of a rainforest.

Today's vegetable gardeners understand the principle, using plastic fleece and cloches to boost growth.

But the planners don't get away scot-free, because they catch the blame for allowing housing developments on floodplains, one cane toad that has come home to roost.

But what am I saying? Have I just insulted our northern neighbours, or is there a different meaning? A quick search yields the following information about the origins of the expression: Scot was a medieval tax levied on the home-owning residents in a similar way to modern rates payable in proportion to the size of the property. Some properties were exempted because of a lack of facilities, such as being without water or being *prone to flooding*. These became known as 'scot free'. Nowadays it tends to mean the avoidance of paying a fine or other financial penalty. So in fact the planners have escaped scot-free, because for all their stupidity, they have not suffered any financial penalty. No surprise there.

While nearly on the subject of our northern neighbours, why does Scotland nearly want independence from Britain, yet embrace the EU and its total lack of independence with such gusto? We want independence, but we don't. Is the motivation simply a dislike of the English, and if so, when do the English get a referendum on independence from Scotland?

I used to sell concrete products to be shipped to Scotland. My 'socially acceptable Georgie' representative with thirty years of experience always reckoned that we had to beat any Scottish-made product price by 5 per cent to secure the business and overcome the nationalist fervour. Yes, that's right, a whole 5 per cent to overcome centuries of subjugation and prejudice.

This disconnect between cause and effect is probably borne of our 'tropismic' responses to external stimuli. We jump to conclusions without engaging in any thought process on an automatic basis. While it serves a Thompson's gazelle well in its arms race with predators, I fail to see the benefit to us. Actually I think I'm wrong. We don't jump to conclusions; we just accept without thought. Our brain filters out the important from the unimportant so that we can deal with the ever present distractions, only in many instances we simply filter everything out. Our brain, having filtered, then fails to remember the process, remembering only the result. To compound the matter, we need to ask how it is possible to ascribe blame for policies enacted when the results will not be known for years, by which time new 'corrective' but equally blind initiatives will have been launched to combat the earlier fuck-up. We have entire industries whose sole claim to success could be described as wave surfing. They scoot along on the leading edge below the precipice, receiving the odd splash but never suffering a wipeout. This could be regarded as a model for any business, because these businesses are only as good as their next failure. The difference here is that their entire business plan could be termed a succession of failures underwritten by taxpayers and ultimately Planet Earth. We tolerate it and we pay for it.

The Food World War

Development Centre Studies: The World Economy: A Millennium Perspective published by the OECD and written by Angus Maddison. This book looks over the last millennium and suggests figures for population, per capita GDP, and world GDP. By extrapolating the figures slightly, I conclude that it will not be long before, in world GDP terms, we can do in a day what used to take a year to achieve. The next step would be to run the projection backwards for another millennium or even ten to see the damage wrought by our species. I don't think we need to run the projection forward for another millennium, because in centuries, or even decades, we may begin to reach the limits. Good thing our masters have a plan.

The WWF advert for Adopt a Tiger suggests that there may be only 3,200 tigers left in the wild. Taking my STU, standard tiger unit, at 12.5 square kilometres and multiplying the two figures together, we arrive at just 40,000 square kilometres of tiger habitat remaining. That's less than three times the area of Yorkshire. A shame that we grow so much tea there. If you'd prefer, it's less than two of the US state of New Hampshire. Or an area smaller than either Switzerland or the Netherlands. The WWF estimate that tiger numbers have dropped by 95 per cent and claim that tigers now occupy only 7 per cent of their former range. All this in the last one hundred years.

'For a future where people and nature thrive.' This is the WWF's slogan. In what world, because it certainly isn't this one, is this slogan not mutually exclusive? In our world, either people thrive or nature does; you can't have both.

One of the historic websites on GDP suggests that humankind was trapped for many centuries in poverty by a lack of food. As populations rose, individuals became poorer per capita, but as populations dropped because of starvation, individuals became richer as resources were spread between fewer people. This mimics the natural order of rising and falling populations, as the food resource varies in cycles. Species breed rapidly from low numbers as food stocks increase, and then numbers collapse as overexploitation of food resources kicks in. The usual example given is that of lemmings and their population boom and busts. I will use the barn owl, *Tyto alba*, which is the most widely distributed owl and one of the most widely distributed of all birds. (Actually, the most widely distributed and commonest owl of all is the teat. There goes my Women's Institute tea tent joke. See how long it takes you to get it.) These birds typically lay four to six eggs at two- to three-day intervals, but they start to incubate after the first egg is laid. The young hatch over an extended period of twenty-odd days, and the oldest chick may therefore be up to three weeks older than the youngest. If food is scarce, the youngest will not survive and may even be eaten by older siblings, so in this way an automatic response to prey availability affected by numbers or inclement weather is effectively built in.

For all intents and purposes, humankind's population was constrained along similar lines until, somehow, we broke the mould and temporarily removed these natural constraints from our species. Improved agriculture, food storage techniques, and terraforming are the probable causes. In the process, humankind took on the mantle of putting constraints on nature, as our success is borne of our domination of natural resources to the exclusion of other species. It is only temporary. At some stage in the not too distant future, our species will dominate natural resources to the exclusion of our own species. Will this be called the Food World War, I wonder? The predictions by some very august bodies is that our population will have reached eleven billion by the year 2100. At that stage in our development, whatever constraints are placed on reserves and conservation areas will disappear and any refugees left clinging to these life rafts of survival will surely perish.

Is this the 'future where people and nature thrive' that my charitable donation will help to create? Bleak, pitiless, relentless, unstoppable. Am I describing my portentous ideas or describing our species?

In this food war for survival, the strong and rich nations will prevail and our reserves and conservation areas will survive. We will still put nuts out for the tits and feed our birds while continuing to destroy habitat and human life throughout the world. If this is too bleak a scenario for you, consider this: reserves and conservation areas around the world are under constant degradation from poachers and even firewood collectors who are simply trying to make a living to survive. Your contribution to adopt a tiger, lion, elephant, or some other poster boy includes the cost of policing of reserves and anti-poaching patrols by the WWF. From this we can conclude that the locals never received the memo about 'working for a world where people and nature thrive'. But why blame the locals? Without the ready markets for tiger/elephant/bear un-spare parts at home and very much abroad, the value of poaching would be seriously depleted. What we could do with is a new initiative, something along the lines of a ban on international trade in endangered species. Agree through international treaty, control all borders and trade routes between source and markets, and reinforce the treaty with stiff penalties. We could call it something along the lines of Convention on International Trade in Endangered Species and give it a shortened acronym like CITES so it will stick in everyone's minds. This would be the Great Brains' solution to the problem. I wonder why they haven't thought of it? Next thing you know, they'll be controlling gun crime by banning guns or, heaven forbid, limiting the speed of cars in restricted zones. Sorted.

But this food war already exists, now, today, and tomorrow. Still with blue tits, I'm afraid; see if you can beat me to the conclusion. In today's battle of this war, we have imported food from around the world to feed our blue tits by depriving foreign birds of habitat and food. Our blue tits survive, but foreign birds die as a result. It's not much of a quantum leap to consider how easily we can transfer this thought process to feed ourselves at the expense of foreigners. The question for you to consider is, are we already there? Are we already killing foreigners as a direct consequence of our actions? But I have just lied again; we will not transfer this thought process, because we don't think. We are killing strangers so that we don't kill the people we love.

But if we thought—just suppose; it could happen. Put yourself in the mind of the local struggling to find fuel to cook dinner and feed his children. We are asking him to suffer so that others can live, in this instance tigers. Tigers are a stranger to him, so why can't he kill strangers so that he doesn't kill the people he loves? But we are not asking him to suffer; we are making him suffer by preventing him from entering the forest to secure resources by the use of patrols, paid for by us rich people half a world away. Do you remember the Cold War proxy wars? Soviet-backed North Korea invaded South Korea and an American-led coalition fought back until the Chinese communists joined in as well and forced a score draw. Or how about the first Soviet invasion of Afghanistan, where the Americans trained the mujahedeen and issued them surface-to-air missiles. Or consider the Soviet-backed Arab coalitions that attacked US-backed Israel. A good thing the Cold War has ended; these proxy wars could have escalated to major events. Nowadays we are much more sophisticated and don't let simple political dogma set East and West at each other's proxy throats. Nowadays we fight proxy wars of survival, with local people struggling to make a living pitched against rich grannies, enlightened children, and tree-huggers. The battle for life is being fought in and around nature reserves, and at stake is the life of the locals and their children, or the lives of tigers and the supporting ecosystem. (I apologise for my guess at the WWF's supporters' demographic.) This proxy war could be determined as being one between rich and poor, entitled resource-rich people against the disentitled resource-poor people, all in the name of charity and conservation. But the charities have an answer, because they provide stoves to the locals to reduce the amount of firewood needed. I seem to remember writing earlier about the logistics of firewood gathering. So if I've got this right, by providing the locals with stoves, they can reduce their forays into the reserves and can spend more time away from this task. This free time can now be utilised on different projects, perhaps clearing another field, growing a bigger crop, earning a better living, having another child, and searching for a cash crop rather than one to supply firewood, perhaps a deer or maybe a tiger. Sounds a bit like a new variety of sweet pea to me.

The only solution to this problem is the non-existence of the local. The only solution to this problem is for the human population to stop expanding. The only solution to the continuance of the tiger population is for the human population to start declining. It doesn't matter what the

WWF and every other conservation body does without addressing this fundamental issue. Every success and triumph heralded today is temporary, a speed bump in the road, a strip of red tarmac, a flashing telematic. Perhaps the WWF would be better employed offering contraception to the locals.

Shock, horror. How can you deny a human the right to breed, to pass on their genetic code, to condemn their DNA to oblivion? Human life is sacrosanct. Just ask yourself how many strangers you have killed as you save the people you love. I need a cigarette to calm my nerves, or a stiff drink or three. This stress is making me hungry; I need a processed carcinogenic meat pie, washed down with a cup of sweetened tea.

Is there just a hint of inevitability?

Is there a solution?

Has anyone proposed, designed, enacted, or implemented any initiative that will actually work?

How many people earn a living because they have proposed, designed, enacted, or implemented any initiative?

That doesn't work.

How many people fundraise, or donate money or time, to provide the solution?

That doesn't work.

How many holes have you dug in the earth to fill a hole elsewhere?

That doesn't work.

How many things that you love have you killed so that you don't kill strangers whom you love? (Raise money by draining a marsh to increase your yield so that you can donate the extra to save a tiger.) (Raise productivity at your factory and earn a bonus, a big bonus, and share your newfound wealth with charitable causes.)

I bet you never thought for one instant that our blue tits are in a proxy war with bustards and ostrich. But you don't need to leave your gardens to see that our blue tits are in a proxy war with sparrows and jackdaws and starlings where resource-rich grannies put peanuts out in large-bird-proof containers where a degree of acrobatics is required to access the food and thus puts said food outside the scope of sparrows and starlings. I wonder, do blue tits attend university to be awarded degrees in acrobatics? By attracting the attention of humans, these birds have achieved force multiplication and have the newfound ability to go for world domination. Perhaps they have turned to the great god Parus and are now on the path to eternal life. Who would have thought that the new messiah visited itself upon tits before humans?

Now how about a bit of extrapolation?

Why stop at tigers and conservation? (In fact, let's not call it conservation. Let's call it traffic calming.)

Let's educate our children to the highest possible standard using all the resources we can spare (should that be waste?). Let's design the greatest level of force multiplier we can muster. Let's continue to regiment these minds into the same old pro formas that comprehensively ignore the consequences of our actions. Let's continue on our mindless path to the apocalypse.

Let's continue to pay highway engineers to design and implement another initiative that doesn't work.

Let's continue to pay planners to design and implement another initiative that doesn't work.

Let's continue to introduce cane toad equivalents into alien environments without any thought as to the consequence.

Let's continue to train doctors and build hospitals to treat self-inflicted illnesses that are avoidable.

Let's continue to elect and empower governments to design and implement more initiatives that do not work.

Let's continue to build our economies and utilise more resources.

Let's continue to grow our population until we are unable to feed our people.

Let's continue to extract and utilise the finite resources of the earth until we run out of them.

Parturition, the Act of Giving Birth

Quite often in films showing a distant future, species have advanced to the stage where childbirth occurs on 'baby farms'. In the Superman reboot *Man of Steel*, our hero begins life as the first baby born of natural means for generations, orchestrated by the very visionary who has predicted the death of his planet. I realise in this example that I have not selected an example from our own species or even our planet, but how far from this solution to parturition are we?

While this is a rhetorical question, I have an answerable question in mind. We lead busy lives, rushing from A to B, eating convenience foods, working to tight deadlines, and struggling to earn a living, pay our mortgages, and pay our taxes. Those of us lucky enough to be parents will remember the flood of emotion generated by the new arrival, the expansion of love to care and provide for every need and desire of our little bundle, the joy of sharing in the development and growing independence of our children until they fledge the nest. (Yeah, OK, the teenage years can be a bitch.) Who in their right mind would want to miss out on these pleasures of the human experience?

Yet what happens when the task is approached in the opposite direction?

When the parents reach their twilight years and independence becomes a struggle, the growing trend for many is the nursing home, to which the elderly are dispatched while waiting between worlds. I know that many, many children devote enormous time and energy to provide care for elderly parents, and I also know that many don't. The financial demands on our time dictate that external agency care is the only option. Let me ask my original question again: how far from this solution to parturition are we?

Baby's first years are full of milestones: first tooth, first smile, first sleep through the night, first steps, potty-training, first haircut, first words (not necessarily in that order). All too soon the first day at nursery becomes the first day at school, and even the first day at university, before the first day at work and the first steps to full independence.

Once again, for the vast majority of us, financial and indeed legal constraints dictate that external agency care is the only option. Homeschooling is only available to a tiny proportion of us.

I ask my original question again: how far from this solution to parturition are we?

And what do we learn? (Never start a sentence with 'and', for starters.) We learn to comply, to fit the mould, to adopt the pro formas, to accept the protocols, to forget to question, to forget to think, to plug in to the matrix. In short, we learn to become model humans.

While the debate rages about learning to speak a foreign language or the history of Britain or where to put an exclamation mark, the debate never questions the fundamental aspects of our behaviour. The debate is not in fact a debate; it is window dressing, a breath on the tiller from the Great Brains, a mere hint of personal flavour dropped into the recipe. Humans of what model? Modelled in whose image? A compliance directive from FIFA that dictates how a bunch of kids will enjoy a kick-about on a Sunday afternoon. A planning compliance directive that dictates how many car parking spaces you can have, how large an element of military training your local Hitler Youth club will partake in, how to care for the environment while simultaneously destroying it, and most importantly, how to lie about it so that you cover your tracks with words and phrases like 'sustainable', 'renewable', 'international consensus', 'green' ...

How far from this solution to parturition are we?

When I think back to a former employee who worked for an electricity board prior to privatisation in a tiny office with two other women, I wrote about the corruption of management and their exploitation of the system to maintain and increase their salaries. But who was exploiting whom, the management or the managed? She knew that she didn't have a real job, and yet she continued to take the real wages. Yet we rip into the Greek economy for allowing 'civil servants' to collect salaries and, later, pensions for jobs they never did. Can we not level the same accusation at educators, highway engineers, the judiciary, the police, conservation bodies, charities, and the Great Brains? My former employee got away with it for as long as she could. The Greeks got away with it for as long as they could. The rest of the people on my incomplete list continue to get away with it every day for how long? 'Yes, but you see, the Greek economy was not sustainable.' So ours is?

My wife thinks I am attacking the NHS and I want to shut it down. I am attacking the entire principle of universal healthcare that isn't welded to individual health responsibility. This flies directly in the face of all first principles and oaths sworn and, as a debate, is an utter non-starter. Resource is rationed, life-saving care is rationed, and we haven't had the debate. Demand is rationed by delivery dates, the provision of appointments and waiting time. Treatment is rationed by cost and delivery dates, and these are systematically lied about. The predictions are that our generation will begin to see a fall in life expectancy due to lifestyle. On your commute to work as you stand in the queue for your caffeine fix, you must realise that the speed at which you are served is dependent on the number of people in front of you and the number of people serving. When you arrive at work, your day is limited by time and the effort of processing, and your performance is determined by how much work you have and how many hours you can provide. Too much work and your delivery dates slip back. You can either increase your resources indefinitely or begin to lose customers who cannot wait that long to be served. Unless of course you work for a place like the tax office, where the telephones can be left unanswered and letters processed over months because there is no alternative.

Fortunately for the richer among us, the option of an alternative service provider is available by the use of private healthcare. For the majority, this option is not an option. So what happens when the NHS 'loses a customer'?

Now if we continue as we are doing with ever more complicated and expensive treatments hand in hand with an ever increasing demand from patients, we are effectively adopting the Greek model of economic planning. Demand will, is, outstripping supply. If we were to think about the NHS in energy terms, then we see that we have designed a facility which is fully insulated and up to date with the latest in green energy. The only problem is that very few of the staff or visitors shut any doors and most open all the windows. I once stayed in a hotel in Riaño in Spain. The town is being rebuilt to compensate for the one lost in the flooded valley and resultant reservoir. In my room, the bath was about three feet long with a shower over, presumably to aid in water conservation. My son had one room and I another, and in neither room for bath or sink was a plug available for use. Yet if you didn't get to your room fast enough, the automatic light off switch kicked in and you were left in the dark.

Given that demand is exceeding supply, how would you resolve the situation? How much money will you commit to solving the problem? Is there any point in committing further resources without resolving the fundamental issue of self-inflicted illnesses?

Under Thatcher, the realisation that the pot of money wasn't big enough to support the demand for pensions saw attempts at finding an answer and a debate was had. But this is a debate that is taboo. 'Hands off our NHS' would be the universal cry from tabloids to unions to patient groups to Great Brains, all of whom are seeking to maintain the status quo. But if you had the debate and adopted personal responsibility for your own lifestyle and health, is it possible that your health might improve? Does the term 'paradox' spring to mind? By maintaining the present system, health suffers. The mainstream argument is therefore resolutely intent on harming the health of the UK population in defence of our health. Genius. It ain't rocket science.

Turning to my old, overworked example of the humble motorist again, if you exceed the speed limit and get caught, you are penalised with a fine, points on your licence, and an increase in your insurance premium. Why is the motorist penalised for what is perceived to be risky or dangerous action? The offence can be recorded in a number of different ways; you can be photographed by a camera, timed over a distance by an average-speed camera, or followed by a police car with timing technology. In the first example, the penalty is attracted by an instant event: you sped by the camera. It is possible that this was your only transgression in your entire journey. In the second and third examples, your event is more sustained and potentially indicative of a more comprehensive risky and potentially dangerous style of driving. Given this, the penalties can be higher, as the police officer has scope to put you in front of a magistrate.

Conversely, risky or dangerous action undertaken by private individuals carries no penalty and no consequences as long as it is legal. There is no financial penalty for being obese. There is no financial penalty for being an alcoholic. There is no financial penalty for extreme exposure to ultraviolet rays, whether from real or artificial sources.

Perversely, there are financial penalties for drinking non-duty-paid alcohol or smoking cannabis or taking other illegal drugs. Why does some 'harmful' activity command penalty while other 'harmful' activity does not? Is the argument based on the potential of causing harm to others? Is the argument based on 'policing by consent'?

The financial penalty for drinking non-duty-paid alcohol is there to protect the exchequer, that is to say the taxpayers, and the revenue they generate. It is a crime to 'defraud' the exchequer of his monetary duty. If you drink alcohol to excess and need hospitalisation on a binge drinking night, there is no financial penalty. If you are an alcoholic and lose your job, family, and housing, there is no financial penalty imposed by the state. In both of these instances, the binge drinker and destitute alcoholic instead 'honestly' defraud the exchequer of monetary resources. In both instances it would be reasonable to assume that the drinkers will be aware at some stage in the process that they have helped themselves to the public purse. Is it possible to differentiate between the 'non-duty-paid smuggler' who deliberately fails to pay in and the drinker who deliberately takes money out? The statute books do.

It's as though the entry fee has been paid to the exchequer and 'club members' can indulge themselves in financial impunity. Where is the logic?

The binge drinker and alcoholic suffer in their personal finances as a direct consequence of their actions. They suffer in their personal health as well. Their actions are not without personal consequence.

Let us consider policing by consent and delve a little deeper. We cannot fine every speeding motorist without losing the consent of the people. We cannot police every transgression of the law without losing the consent of the people. Periodically the suggestion that binge drinkers should be marched to an ATM and summarily fined is raised and dismissed as impracticable (as an example).

If we consider the police to be a tool, an agent of the Great Brains and the supposed command-and-control structure, then we can begin to see the problem. We the people dictate what level of control we will accept, and they, the Great Brains, are controlled by us, the people, as they attempt to control us. Policing by consent goes right back to the Victorian origins of our police force and the appointment of citizens in uniform. Have I stumbled upon written evidence that we have no leaders?

In a similar vein, government imposes taxes and distributes money to all sectors of the government spectrum. Their task would be so much easier if they could simply impose ever higher taxes, but they can't, because, via the ballot box, we the people periodically approve or disapprove of our Great Brains and remove them from office.

While on the subject of resources, how many non-jobs at the coalface or in management exist in the NHS? What percentage of the money delivered by our taxes actually ends up providing treatment to patients?

Can you remember the film *Pearl Harbor* and the aftermath of the Japanese attack where poor old Kate Beckinsale as Nurse Lt. Evelyn Johnson has to triage the wounded being delivered to the overstretched and overrun medical facilities?

The wounded are separated out into three categories:
• minor non-life-threatening that can wait
• major life-threatening that can be saved
• major life-threatening that cannot be saved.

The third category are left to die and yet, but for the overwhelming of the medical facilities, could possibly have been saved by medical intervention. 'But there's a war on! What do you expect, unlimited resources, rows of empty hospital beds, and doctors in operating theatres waiting?' Yet these 'battles' are fought day in, day out, throughout the NHS without any discussion—and through what thought process?

We have a health service that actively supports ill health and will continue to soak up resources at a never-ending rate until it breaks.

If we were to adopt my position and sanction and eventually remove treatment from self-inflicted illnesses, then we could extend the 'life' of the NHS for a longer period. People with self-inflicted illnesses should be identified, be educated, and have their treatment options reduced or removed for failing to alter their lifestyles. I am not talking sudden death (it is slow-motion suicide after all) but something phased in over a period of years in a similar fashion to pension age alterations. The black hole that the NHS budget is facing would be reduced in size, and the health budget and our taxes could be reduced. Most importantly, the health of the nation would be improved. What could be a better result than that?

There is just the one small fundamental problem with my proposal. Well, actually there's more than one, so let's look at the issues. Our NHS is sacrosanct and governments will rise and fall off the back of it. Junior doctors can go on strike and still receive the public's support because the whole edifice is perched high on an angel-borne pedestal suspended above reality. 'Our Tommy wouldn't be alive but for the NHS'; 'I really thought I'd lost Maisie until the doctors stepped in.' The NHS features large in every election campaign and is one of the major features in deciding who enters Downing Street. You have to ask the question, who is controlling whom, the NHS or the government? This is the NHS that never offers itself for election, let alone re-election; that demands, like a spoilt child, resources and attention; and is more powerful than our elected leaders. (I know, we have no leaders.) It has an infinite demand and finite resources, and ultimately every single one of its customers will die. Yet the demand will continue to expand as the never-ending line of new untreatable conditions become treatable at a cost into the future. Perhaps the real trick of the NHS is to have removed fatalism from the public's agenda. Thankfully the spectre of septicaemia wards has temporarily disappeared from the public's perception and a 'cure' is expected, nay, demanded.

Let me suggest a figure for my revised NHS budget and suggest in time that it could be reduced by 50 per cent over a number of targeted years.

This is the fundamental flaw to my ever so cunning plan, because whatever the reduction or savings for good health, these will be readily absorbed by the new diseases and treatments still to come online. We have a never-ending demand and a closed loop of Great Brains who will fill whatever vacuums can be generated. You see, my attempt at offering an element of common sense, even some sense of justice, to the distribution of resource simply makes me a Great Brain. And as I have tried to illustrate throughout this book, Great Brains have only one aim in life, to satisfy themselves at the expense of others. My ambition is to remove those with self-inflicted illnesses from the beds and waiting lists so that I can be nearer the front in times of personal need and save some tax. I am just as bad as all the other Great Brains. The other Great Brains have already introduced extended waiting lists, removing treatment on a cost–life benefit ratio, built hospitals at minimal capital cost and immense repayment plans through private finance initiatives. ... The only solution to the NHS is force majeure, which is to say, it will not be fixed until it is broken, because then it will have become a dead elephant and no longer an irresistible force. It seems the Greek economic plan is the solution, except—and there has to be an exception—the Great Brains of the EU in fear of contagion won't let it happen. They support the corrupt state of Greece with their own corruption.

If we can't change anything, if we can't manage anything, if we can't improve anything without the catch-all of corruption endemic to our DNA, then we have no hope of saving the planet from our activities.

Welcome to the new age of realisation. Discuss.

I can argue that it didn't matter, corruption that is, when we came down from the trees. I can argue that it didn't matter until the Neolithic World Cup on Salisbury Plain set us on a course of extracurricular activities that differentiated between what we needed to do to survive and what we could do. It is what we could do and what we are doing that makes the difference. When the Egyptians began to build pyramids, did they ever consider the significance of pyramids and the pyramid building our species has embarked on? With these monumental events in different cultures and different regions of the world, we opened the Pandora's box upon the planet and unleashed the unstoppable, unrelenting, ever increasing destructive forces on this place that is our home.

To build pyramids you need science, management, tools, education, machinery, desire, and an insatiable appetite to see how far and how much farther you can push the boundaries, the limits, of our world. This is why we have force multipliers. And don't they work? Our future can be modelled on the Greek crisis plan and our activities will continue unabated until the world is broken and the elephant is dead. Welcome to the apocalypse.

Finally we return to the great myth of global warming. Do I have definitive proof that it exists? Do I have definitive proof that it doesn't? It doesn't matter. The myth I speak of is the cure devised and implemented by our Great Brains on a worldwide scale to arrest and cure the problem. What they have offered is a diet. What we need to do is keep our legs shut. And our mouths shut. And walk to market.

We have the promise of science to guide us through the difficult times ahead: new cures for cancer, Alzheimer's, multiple sclerosis, and ageing, and even the promise of colonising other planets as we seek out yet another horizon. I remember when the first word processors arrived in our offices with their promise of memories that mimicked the pro formas we used for quotations and letters. Then came the first computers and with them the promise of higher output and fewer people, but the promise never materialised as we ended up with more computers and the same number of people. The dream of endless clean energy is held before us with nuclear fusion and the true end to our global warming concerns. With endless clean energy we can devote a portion of it to capture carbon and redress the damage that we have done while halting the need for fossil fuels.

Then we can really put our foot to the accelerator; we could desalinate the Mediterranean Sea, irrigate the Sahara Desert or grow crops under heated glass in northern climes, or dam the Bosphorus strait and drain the Black Sea in pursuit of ever more resources. While we are at it, do we really need all the Great Lakes?

I can feel a Nobel Prize for Science coming on. 'For outstanding contributions to humanity.' From one 'merchant of death' to others.

'Anthropomorphism' is the term given to ascribing human characteristics to animals, when what we should be doing is ascribing animal characteristics to humans.

Well, at least I tried with getting fat people to the back of the queue. Human rights to treatment will no doubt loom large over any potential debate, or rather non-debate, because the die are already loaded. (By die', I mean the little cube thingies with spots on. Any reference to the slow-motion suicide of 'loaded' people is entirely unintentional—honest.) It's not as if there are any parallels in governance, are there? We haven't penalised gas guzzlers by charging a higher tax or increasing the road fund, have we? We haven't penalised frequent fliers for flying frequently and using more than their share of our precious air miles by taxing air travel, have we? We haven't penalised people with higher taxes for being in charge of fat wallets, have we?

I suppose it might help the debate along if I were to declare that my body mass index is 30.5 and has been for many years. I am clinically obese. But that is measurement for you. The US Army banned recruits with a high BMI until somebody pointed out that it excluded the strong people, the ones you'd see in *Predator* traipsing through the jungle with Arnie. But if you are going to put fat people at the back of the queue, what about tall people? Anything significantly above median height is a waste; these people should be curtailed to conserve resources. Congratulations. By the way, given that you have held out long enough with my fattist agenda to reach the conclusion, I want to illustrate how easy and how plausibly simple it is to launch an elephant. I have only the greatest of ambitions and the greatest of motives and have really tried to convince you of the merits of my application form to become a Great Brain. If you are slim and healthy, I may even have your vote. It seems that I have designed my own pro forma on how to conceive an elephant (asexually, I hasten to add; 'parthenogenesis' is the term you're looking for. No, hang on, what we had was nothing, just a potential for something, and then, even though an elephant is big, we had a Little Bang: somebody else beat me to the Big Bang). It was rather long-winded because I wanted to guide you through the conclusion without letting you think too much. Here is the condensed version.

Decide that you are in/out of power and want to remain/come into power.

Select plausible policies that have mass appeal/can be spun to appear plausible/popular.

(Remember that the policies do not need to work. They just need to appear to work.)

Lobby your scheme among similar-minded people/people who would become similarly minded if the terms were right/people who are ready to adopt the latest fashion in the absence of an idea of their own/people who agree not to get in your way if you don't get in theirs, and if all else fails declare a coalition of consensus and claim to represent it.

Prepare some costings/savings data to reinforce the plausibility/ seriousness of your proposal.

Recruit experts to reinforce/add plausibility to your proposal. (Remember, in this world we live in, lawyers for the defence and offence both flourish. You will have no difficulty at all in proving the veracity of your proposal.)

Ignore all naysayers; corruption will distort the entire process anyway, so it is bound to succeed (for the corrupters).

Allow the free press to have their say; there will be as many for as against, and they will add plausibility.

Ensure, and this is absolutely fundamental to the entire process, that your scheme has no off switch.

Release your elephant.

Take notes for your legacy; you need something for your biography lest they forget your great triumph.

When all else fails, retire to a higher place. (I don't think I am talking heaven here, just the upper chambers, although you will be surrounded by people you previously thought had died and, given the daily attendance allowance, probably some who have.) (Like tigers, probably extinct, they just haven't stopped breathing yet.)

Is there a shortcut?

What I should be doing is raising an e-petition to raise one hundred thousand signatures to force a debate in the House of Commons about fat people squandering the earth's resources, eating more than their fair share of food, and, quite unforgivingly, dying premature deaths after years of squandering the NHS's budget. With luck and a fair wind, I could create my own elephant. My elephant could generate such an irresistible momentum that before long it would lead to the force-feeding of skinny people for being underweight. Given enough time it could bring in an age cap for unhealthy people at, say, 45 years of age. A bit like chopping in an unreliable car before it starts running up the bills. Compulsory euthanasia ought to crystallise a few minds.

While you may be thinking that I am taking the piss, which I am, there is a serious side to my elephant. Because in the race for the winner's enclosure, my elephant may not leave the starting line, but, and there has to be a but, another elephant will. This elephant may wear the colours of the Greek flag, white and blue, as a sign of its crisis economic mismanagement. It may bear the colour grey of the age discrimination cost–benefit ratio.

It may bear the opaque colours of the tick box mentality over the care and compassion that it has replaced. It may simply bear the blue and white (not to be confused with the white and blue of the Greeks) colours of the hospital management and run under the motto of *Carpe Jugulum*, for whatever the pressures on the arterial flow of money to the NHS, the management go straight for the jugular. Sorry, I screwed up. The jugular is a vein and not an artery, but it holds its place in werewolf culture as *the* place to strike for maximum effect, so, tough. And anyway, one must go through the ribcage to reach the aorta, and that involves work. But the elephant that already has a major head-start on all the rest bears the colours of the chameleon, the armour, the greatest tusks you've ever seen on an animal, fully twice the size of any other elephant (and you can rest assured that it will fail its drugs test). Above its stable door is inscribed the name 'Corruption'. It will win; it's in our DNA. But, as I have said before, we will not celebrate a race with only one competitor. The rest of the field is running; we have not one elephant to deal with but multiple elephants, each representing an irresistible force that some Great Brain has sponsored to run. I'd ask you to place your bets for a place, but I can't see the point in gambling; it's another elephant.

Is there an answer? Look at the global warming debate and all the diet initiatives that will not work. If you're going to get nothing for it, do nothing. By that I don't mean carry on regardless; I mean use less energy.

If you have been paying attention, you may have realised that the eight-letter word in the hangman game was 'elephant'. This is because every alternative suggestion I have made represents an idea that may or may not be adopted by a Great Brain as they seek to launch elephants of their own.

While nearly on the subject of gas-guzzler cars, a thought nearly struck me. My wife had dinner with a newly retired worker for Jaguar Land Rover who declared that the final Land Rover Defender to roll off the production line ending a sixty-eight-year reign had sold for £400,000, although a quick check on the Internet suggests that it has been retained as part of Land Rover's Heritage Gallery. This was in reply to my wife's question, which was 'How does anyone justify spending £158,000 on a Range Rover?' Her friend didn't answer the question.

Let me ask a question.

If you were to compare the carbon footprint of the gas-guzzler car with the carbon footprint of the generation of the £158,000 purchase price, what would you have achieved? This wealth has been generated by the use of energy and resources from around the world to serve the vanity of an individual. Would a better way of conserving energy and resources actually be to control the price of cars so that they are designed down to a standard rather than up? I know that this is an absolute non-starter of an idea because it might work. We have the 'right' to abuse our own bodies; manufacturers have the 'right' to abuse our bodies for us, as do the criminal classes; and we have the 'right' to squander our heritage and future on personal vanity because it makes us human.

I have reached a conclusion—it was bound to happen eventually, if only by the law of averages: we do not have leaders, we have elephants. Elephants that cannot be led. Elephants that have no leaders.

And before you say we don't need leaders, we just need ideas, consider that the Clovis point was an idea, pyramid building was an idea, kicking an inflated pig's bladder around a patch of grass was an idea ...

Now I want to see if you have been paying attention.

In the UK we have the remnant of a bank that has championed its ethical business status for more years than I can remember. I did open an account with them back in 2002 while my son had an interview at Leeds University. It was a major poster announcement back then.

Our business was the manufacturing and distribution of structural concrete products. Throughout my company's fifty-two-year history (and counting), we never invested in or involved ourselves with any criminal activity. We were on the receiving end of it with cars stolen, and computer and electric cable thefts an occasional occurrence.

I just thought I'd throw that in for now. The relevance will crop up later.

The Co-operative Bank has been rooted in strong and distinctive values and ethics since its origins in 1872. In 1992 it became the first bank ever to have a customer-led ethical policy. Search for their ethical policy on the Internet and it will appear on your screen.

The foreword declares that their ethics and values are more than just words on paper and that they have consulted over seventy-four thousand colleagues and customers for their opinions to help shape this policy.

Among their aims are stated ambitions to protect the environment and animal welfare, and they support poverty reduction in developing countries.

These are broad topics. They declare their active support with the following buzzwords and buzz phrases: healthy, minimal impact, recycling, sustainable, renewable, efficiency, organic, fair trade, microfinance, animal welfare, and alternative methodologies to animal experimentation.

Now, remembering that their values and ethics have always been more than words on paper, explain 'sustainable' timber (presumably words on sustainable recycled harmful-chemical-free paper).

Would they support the organic production of birdseed?

How many STUs of fair trade tea or coffee do they support?

How much money have they lent through microfinance to buy terraforming machinery?

How many STUs of free-range farming do they support?

I don't intend to complete this list; as I said, have you been paying attention? Time for you to make your own list and find the holes in the Co-operative Bank's ethical argument, which appears to be a polite form of 'business as usual'.

Have I mentioned that our concrete manufacturing business never had any involvement with crime? Of course I have. By taking our turnover over fifty years and bringing it up to current values, I estimated that our little business had enjoyed a total turnover of around £350,000,000, say, a third of a billion pounds worth of raw product bought in, designed, processed, and delivered, with salaries paid and profit taken. How do we control this sum of money? How many of our contacts spent a portion of these earnings on illegal drugs, gambling, prostitution, hunting safaris, smoking, holidays abroad, large-engine cars, donations to the RSPB, or peanuts? We do know, at current rates, that 40 per cent was taken by the taxman and spent on reducing the speed of cars in 30 mph zones, introducing cane toad policies into the planning process, securing the services of yet more managers in the NHS, and corrupting absolutely our government. How much of this money was spent on tea, or coffee, or chocolate, or sustainable paper bags made from sustainable trees from sustainable forests? Tax was also ploughed into our education system, one of our biggest force multipliers of global destruction. How much money was spent on hydrogenated vegetable oils, processed meat, refined sugar, alcohol to excess, or cosmetics tested on animals?

Yet the Co-operative Bank can ring-fence itself from any of these unethical activities and claim the moral high ground? (Writ large in the text is 'principal activities', words on paper that offer a 'get out of jail free' card.)

Bullshit.

But I'm not picking on the Co-op; I am simply using them as an illustration of how everything we do is interconnected and beyond our control. Money is like a red blood cell coursing around the body of a blue whale (*Balaenoptera musculus*). During its travels it will visit the lungs and heart and may visit the brain, gut, liver, kidneys, pancreas, eyes, muscles, etc. Yet the Co-op would have you believe that they can keep it away from certain areas of the body? If it's in circulation, it's in circulation; the Co-op cannot supply a tourniquet however much they preach their ethics. Oh, I forgot, the blood cell may even reach the testes, the largest load of bollocks in the world.

When the police told me that my car had been stolen, I lamented the loss of my insect collection. The police asked if they were on the back seat or locked in the boot. 'Oh no,' said I, 'they were on the windscreen and front grille.'

So much for being a vegan and never knowingly harming another living creature.

Over seventy-four thousand customers and colleagues have helped shape the Co-operative Bank's ethical position, and what have they achieved? Some words on paper and another chapter in the un-phoney war against the planet.

You might think I am being too literal and setting the bar impossibly high. I argue that I am just telling it as it is. It is impossible to compartmentalise our actions from others without drifting through space alone in a propulsion-less ship with its own recycling life support system. (This ignores the humungous amount of resources involved to design, construct, and manufacture your 'self-contained' world and then blast off into space in it.)

You might be able to achieve this as a member of an Amazonian indigenous tribe yet to be reached by 'civilisation', but otherwise?

Yet the Co-operative Bank brags about its earth-saving policies and expects to profit from this by attracting like-minded 'customers' to its counters. It is a con, it is corrupt, and it is a lie promulgated since 1992 and beyond.

Look around you and notice the 'green credentials' of companies and the products and services they provide. Translate the claims into the realities on the ground and, which will it be, cry or laugh. My local authority started to collect green garden waste for recycling into compost, as opposed to carting it off to a landfill, and then had the stupidity to claim that through their activities they had saved x tonnes of carbon dioxide. How? The carbon contained in the green waste will be released as it is broken down by organisms that feed on it in the landfill site, as opposed to the carbon being released by the organisms that feed on it in the compost added to your garden soil. But because it's green waste, the carbon is naturally recycled anyway and is denoted as carbon-neutral, as opposed to carbon generated from fossil fuels. We will ignore the carbon generated by the council vehicles collecting the waste, as they collected it anyway on its way to landfill. But we cannot ignore the carbon generated by the production of tens of thousands of additional plastic bins and their distribution to every household in the district. While we are looking at this, note that they refused to collect scrapings from vegetables that might have been contaminated by meat (disease) or bedding from pet animals such as rabbits (disease) which were to be sent to landfill anyway. Throw in a system redesign which resulted in the reissuing of another set of green bins, and they have compounded the carbon impact. As a final insult to our intelligence, they have now decided that they will introduce an annual charge of £40 for the collection of this green waste. There are now three options available to us: throw it in the landfill bin, pay the additional fee, or compost the material in our own gardens. Sorry, I forgot the fourth option practised for many years by enlightened individuals living near countryside: dump it in the ditches or on the verges opposite. Ever wondered about the spread and appearance of snowdrops and daffodils on the verges opposite housing?

The only carbon-neutral options are self-composting or dump-composting by fly-tipping. I have to confess that I compost all my own waste, including vegetable peelings, and have done so throughout my independent life. This is not because I am some paragon of virtue but because I have always lived in gardens too large for the green bin's capacity, and my gardens have been large enough to hold compost heaps. Yet another consequence of higher housing densities and shrinking gardens: no room for compost heaps. Although, to be fair, each household produces less garden rubbish.

While I have lamented the demise of garden and garden habitat, a lack of bushes and trees for nesting birds and lawns for feeding, I have failed to lament the passing of the garden compost heap and its place in the ecosystem. My garden is large and partially wooded. I refuse to lift the dead leaves from ungrassed areas under the trees. I have a constant battle with the birds, who flick the leaves onto my drives in search of invertebrates and who return the compliment whenever I push the leaves back off the drive. In this instance my 'compost heap' is no heap but a layer of decomposing leaves full of beasties turning it into leaf mould. The constant attention of the birds is proof positive of the benefit they receive.

Dying for a Sweet

The WHO has declared that the world is facing an 'unrelenting march' of diabetes, with nearly one in eleven adults affected. Cases have risen from 108 million to 422 million in the 34 years leading to 2014. The WHO links 3.7 million deaths worldwide every year to high blood sugar levels. 'Drastic action' is needed to avoid further increases.

BBC text, 6 April 2016.

Later in the news the BBC featured an article and included further analysis. The growth has been seen in the low- to middle-income countries and has been attributed to the rapid growth in urban living. A spokesman for the WHO then wittered on about the need to design cities fit for living in with space to walk and less dependence on motor vehicles to help combat our more sedentary lifestyles.

Or people could eat less food and eat less sugar and shoot all the processed food manufacturers for taking the shortcut to profits.

The WHO also supported the imposition of a 'sugar tax' on soft drinks as seen in Mexico and to be introduced in the UK in a couple of years' time. Just supposing that it worked and the demand for sugared soft drinks was reduced, what are the sugar manufacturers going to do? Will they sit on their hands as they watch demand and the price of their product fall, or will they push their cheaper product into other food products? We are back to the fat lake removed from milk during the skimming process which has found a new home. Do these people have two brain cells to rub together? At the moment you can choose to avoid sugar in drinks by not drinking them. In the future, if the tax-excluded sugar is 'hidden' in the other foods you eat, you can't avoid it.

It Ain't Rocket Science

Tim Peake has recently reached the orbiting International Space Station (ISS) and is looking down on the globe. Consider his problems of staying alive 230 miles above the earth. All the food, water, and energy generation equipment has to be brought to the ISS and recycled or replenished at immense cost via rocket. A simple leak in the external structure or failure to supply food and water would see the collapse of the experiment, with probably fatal results.

While drifting through space in our propulsion-less mothership with its own recycling life support system, when do you start to worry that it might be broken? Whom do you send to fix it? If you were running low on oxygen, the first thing to do would be to reduce your activity to align your usage with your production. What would our Great Brains' solution be, I wonder? What if the problem were reversed, and instead of not enough oxygen your problem was too much carbon dioxide? We produce carbon dioxide as a by-product of respiration. It's how we obtain energy from our food.

Start by losing excess weight and reducing your fuel intake, which in turn lowers your respiration rate and carbon dioxide output.

Increase insulation levels to higher standards to reduce the fuel burnt by your body to keep warm.

Appoint the cleverest and most experienced minds to tackle the problem.

Thinking burns energy, so reduce the number doing it to a minimum to reduce your fuel consumption.

Limit the population to a sustainable level that will not overload the system.

Jettison all extraneous material from the mothership to lighten the load and reduce your fuel intake.

Eliminate all activities not fundamental to sustaining life.

Aim to reduce your carbon dioxide level to the capacity of your recycling system.

Design reinforced bulkheads to section off expendable sections of non-essential crew, and introduce strict control over the remaining essential crew.

While Tim Peake is able to look out of his porthole and see Earth encapsulated within its atmosphere, he can quite rightly be concerned about the frailty of his position, knowing that a single piece of space debris could pierce his environment and render it broken and unable to sustain life.

We assume that the dinosaurs going about their daily business sixty-five million years ago were oblivious to the single piece of space debris that was about to pierce their environment and render it broken and unable to sustain their lives.

We can also assume that the terraforming human species is also oblivious to the effects that their daily activity is having on our ship travelling through space, activity that is piercing the environment of myriad species and rendering it broken and unable to sustain life.

Welcome to your past …

Welcome to your present …

Welcome to your future …

A Time for Excess

I had a winter trip to Sulgrave Manor in Northamptonshire near Banbury in Oxfordshire and had a guided tour as part of a small group. The manor is the last-known UK residence of the Washington family before they departed to the New World. Grandson George W. launched a few elephants of his own. Sulgrave Manor is owned by the US government and has the same status as an embassy in that it constitutes sovereign territory, if my memory of the tour guide's spiel is correct.

My tour featured the Winter Festival, enjoyed from earliest times before being hijacked by the pope many centuries ago (if you can't beat them, join them) and becoming its present-day reincarnation as 'heaven on earth' for the retail market.

The main hall was covered with great bunches of evergreen ivy and holly, and the smoke from the fire was definitely of the injurious variety. (Deck the halls with boughs of ivy.)

Wassailing was the order of the day as the festival was explained. The lord and lady would invite the manor's peasantry in for two weeks of feasting and the observance of various customs. A large log, the yule log, would be cut, decorated, and rolled into the hearth before being lit using the remnants of the previous year's log. It was considered bad luck if the fire went out. This led to the custom of lighting large candles in Victorian times, and ultimately to the Advent candle of today. A Lord of Misrule, described often as the village idiot, was promoted to be the lord for the fortnight and would lead many of the 'high jinks'. Large pies would be bought into the hall, and the village maidens were given the task of cutting the crust, only for live jackdaws or rooks to escape from the pre-prepared crust. (Four and twenty blackbirds baked in a pie.) The larger houses with bigger ovens would have live naked dwarves erupting from the prebaked pie, to molest the maidens to great ribaldry no doubt. The superstition was that evergreen vegetation contains evil spirits, hence the inclusion of the holly and ivy, as the spirits were invited in to the hall to placate them. At the end of the festival all the vegetation was stripped from the hall and thrown on the fire. The 'waterproof' leaves burnt with crackles and whistles as the evil spirits were cast out and killed.

Thanks for the history lesson, but what's the relevance?

It was said that this festival of food and drink kept many of the peasants from starvation by giving them a winter boost at the darkest time of the year, solstice, not anyone's birthday. Back in the day, the peasants would struggle to store and cure food in their wattle and daub hovels, and food was very much a seasonal product with major variation between peak and trough. In our temperate climate we have seasonal differences, and no doubt our bodies have adapted with what I will describe as a gene for excess. We have to take advantage of opportunity when it presents itself.

I have often wondered about celebrity and the trappings attached to it, in particular the need for excess. I do accept that the very nature of celebrity is the need for exposure and publicity, and therefore when it goes wrong, and there is no off switch, the journals are full of sorry tales. These probably feature disproportionately relative to the number of celebrities who live normal lives. It strikes me as perverse the way that individuals lust after fame and fortune only to destroy their lives by snorting it up their noses and drowning in baths or aspirating their own vomit. Imagine, your humdrum life has been replaced with a glittering world of riches, fast cars, and fabulous houses and it's not enough. You have arrived beyond your wildest dreams, and yet?

Others dream of mega wins on the lottery and of the cars, houses, and holidays that wealth provides. The occasional fly-on-the-wall documentary will feature 'life on the dole' and the deprivations suffered, and yet a £300 windfall from 'Dad, for me birfday' will be frittered away on a tattoo or binge drinking.

Working people engage in a similar way, with pay rises, bonuses, and windfalls eagerly assimilated into the lifestyle. Going back to the medieval peasants struggling to get through the year, if the food were to become suddenly available all year round, the same 'excess gene' would kick in because it doesn't have an off switch. With modern food quantities and security of supply, it is not necessary to consume every meal as though you don't know when your next one will be. But we do, betrayed by our genes. Is this the insatiable appetite that drives our quest for bigger, better, faster, fuller, fatter, and richer? The gene that has kept us alive for millennia needs an off switch before it destroys us all. We can terraform the planet but cannot control our appetite. The thought does occur to me that the off switch may have been bred out through selection. Fat people would have been the prey of choice for lions, tigers, bears, et al. because they are slower and easier to catch. The natural balance would have kept our species fit, in the Goldilocks zone, thin enough to run away, fat enough to survive occasional shortages. Remove the predation and even accidental deaths caused by being in the way of stampeding buffalo or aurochs and the 'corpulent gene' could rise to the fore. I have referred to the speed at which Thompson's gazelles react in relation to the 'arms race' with cheetahs and other ambushing predators and their ability to leave the scene of a flighted arrow. Have we dodged the tiger's 'arrow' only to die en masse from corpulence? It would be ironic if our actions to protect ourselves from featuring on the menus of large predators actually resulted in greater deaths. The reality is a little different in that predation on the mega herds is not that great when measured against deaths caused by disease. Yet our disease is becoming self-inflicted given that it is largely preventable by adopting a better diet. The real irony is that the challenge to survival in the herd is to avoid disease and not predators. We have dealt with the predators throughout Europe and have actively encouraged disease to flourish as a lifestyle choice.

The medical profession has a difficult—no, impossible—task to satisfy its customers. Who would be the little Dutch boy with his finger in the dyke? Wouldn't it help if the customers and the profession were both pushing in the same direction? The science progresses and treatments improve, and yet the self-destruct gene kicks in again, often negating any improvements.

Our DNA leads us down a pathway developed since life began and continuously refines by trial and error our and every other successful organism. And yet the definition of a successful organism is not necessarily what you might expect. Is it the organism that has survived virtually unchanged for millions of years or the one that changes rapidly? The influenza virus and its common vaccine is modified most years to provide protection against the predicted virulent strain expected to assail us. Sometimes the crystal ball works and sometimes it doesn't, but in a fashion, that's life. In simple terms, some years the influenza virus wins and sometimes it doesn't; in the latter case, the science and medical profession do.

The crocodile is often cited as an example of an 'unchanged' model with a lifespan pre-dating the end of the dinosaurs. Undoubtedly it will have changed, but when measured against its fossils, evidence of change is difficult to find. I suppose it is possible to consider the fossil to be a computer whose software has been lost. Life seems to occur in different lanes of the multi-lane highway we call evolution, with variations occurring very slowly or quickly. Part of this will no doubt be explained by the rate of reproduction, with humankind, say, realistically thirteen years as a minimum, whereas one bacterium could multiply by binary division to produce ten billion in just ten hours. These ten billion would be clones and identical copies apart from the 'mistakes' that occur, or mutations as we would call them. To put that into perspective, 13 years of human life is around 13,500 ten-hour periods. The total number of bacteria is not 13,500 multiplied by 10,000,000,000. To be honest, I can't work it out. Suffice it to say that the number is beyond comprehension.

If human reproduction were on a different timescale, say, 130 years to reach sexual maturity, then our ability to survive the race for life would arguably be seriously compromised. If you accept this premise, then extend the idea to perhaps 1,000 or 2,000 or even 10,000 years.

I have selected these apparently random figures because we have already come across them earlier in *How to Kill an Elephant*. One thousand years of pyramid building, two thousand years of the Clovis point or Jesus Christ, and ten thousand years of agriculture. Could I even suggest one hundred thousand years of the 'excess gene'?

We humans are a clever bunch in that we are masters of our realm, able to terraform, selectively breed domestic animals and food crops, and replicate human activity with science and machinery, to name a few of our achievements.

The fossil crocodiles survive without any record of their 'software', so we cannot make a true comparison between then and now.

Our human record of 'software' is held in tantalising clues of prehistoric artwork, tools, and weapons tens of thousands of years old. We occupy our thoughts looking for evidence of 'ritual' that would denote an understanding beyond that of animals, in short, what makes us human (except that we are blinded by the pursuit of 'princesses'.)

Continuing with the computer analogy, we are, for all our ingenuity, trapped by our origins. It's as though the computer manufacturer has etched into the hardware some software programs that override and control the software programs we think we have loaded there. For all our cleverness, we were blinded for two thousand years by a stone point, one thousand years of masonry stacked on masonry near to the natural angle of repose of pebbles. Is it within our remit to access the master codes and rewrite our destiny? I am not thinking of genetic engineering, but rather the application of thought to control our base desires. Consider all the things that we control and, more importantly, all the things that we don't. If you stop and think about it, you see that we send our young minds off to receive an education whereby a large portion of the 'operational software' is supposedly uploaded. We learn how to live and exploit our fellows, but we fail to learn how to improve our lives or the life of the planet. This is my frustration with the education system. It simply reinforces all that went before and fails to address all that might come. It's as though mine is a lone voice trying to take influence away from Google, where Google is set as the default.

Where does the tax burden fall with the sugar tax?

El Chancellor claims to collect £600 million from the new sugar tax and to spend it on kiddies' exercise in primary schools. If you want to get kiddies to exercise, do you need £600 million? What we are actually doing is spending £600 million on staff and equipment. The exercise is a by-product of the spending, not a direct result of it.

On Budget Day, the share price of sugary drink manufacturers fell by 7 per cent, which had an instant (but not necessarily permanent) impact on the companies and their shareholders. (If you had chosen that day to sell your shares to realise your pension, then it could well have had a permanent effect on you.)

In theory the tax will be paid by the consumers, because without sales, no tax can be raised. The secondary ambition is to depress the sales and reduce the consumption of sugar. If it worked and all sugary drink sales ended, then no tax would be generated. We start then with a dichotomy: tax is good because the exchequer wants the resource, and sugar is bad so we want to reduce its consumption. Exercise for primary schoolkids is good, but we need the sales to continue to generate the benefit. We need the 'evil' to fund the 'good'.

In practice, on the day of the announcement the price was paid by the shareholders. The shareholders expect the managers of the investment that they have made to redress the balance and to return their investment by fighting back. The drink manufacturers could launch a new range of reduced-sugar drinks and knock the tax on the head. They could advertise extensively to promote their existing brands and maintain their market share with the customers accepting the tax rise as a necessary burden. They could squeeze the cost base of their products by debasing the ingredients or forcing price reductions on their material suppliers and logistical support. This would mean that the tax is paid by truck drivers, farmers, bottle-makers, and factory employees.

Farmers could be forced into accepting lower prices, and this will have an impact on their suppliers and their lifestyles. Sugar producers will be looking to extend their markets to new outlets, and cheaper sugar may start to appear in other foods where it is not so readily identifiable. How many people will cut down on potentially healthier dietary items so they can afford the extra price of the drinks?

If the object of the exercise is to reduce the amount of sugar in our diet, why do we subsidise the growing of sugar throughout the EU? According to the *Financial Times*, sugar suppliers were the biggest recipients of Common Agricultural Policy (CAP) subsidies in 2010. Prices and quotas have been fixed, although these are to end in 2017 and prices are expected to fall. The EU is the world's third-largest sugar producer and the second-biggest market. Actions taken within the EU will have ramifications throughout the sugar-producing nations of the world. Price depression in Europe will drive pressures throughout the developing world and impact on the developing world's desire to make a living. No doubt our Overseas Development Aid will feature somewhere in the mix.

I remember reading an article in the 1980s about sugar in the world market. I cannot remember the source or the exact figures, but I can remember the principle. Please accept the figures as inaccurate but the principle as sound. We grow sugar from beet in Europe for £100 per tonne, whereas we could import it from sugar cane producers for £80 per tonne. We grow more than we need, so we subsidise our exports to dump sugar on the world market for £60 per tonne. This undermines the cane sugar producers and their economic base, thereby damaging their economies, who then require assistance from the World Bank, the IMF and other, rich donor nations. Through our support of the EU's CAP subsidies, our taxpayers pay once for the sugar beet to be grown. We pay twice to dump it on the world market. We pay three times to fund the donations and bailouts for undercut sugar cane producers. And I haven't even mentioned cane toads in Australia.

Now we are expected to pay a fourth time to consume the sugar in drinks.

I wish I had a brain cell to vibrate.

Once again the Great Brains have struck; with the introduction of the plastic bag tax and the 85 per cent reduction in single bag use, you would be forgiven for thinking that plastic bags have been 'sorted'. But you are still allowed to bag loose food such as fruit and vegetables, although you may not bag 'bagged food' as supplied by the supermarket. My supermarket's shelves are full of ready-bagged fruit and veg. There has been a reduction of one type of bag, but the problem of plastic has most definitely not been 'sorted'.

With the introduction of the new sugar tax, you would be forgiven for thinking that excess dietary sugar has been 'sorted'. We are back to the testes of blue whales. This will be the gulf between perception and reality. Let me call it corruption.

Just to put some perspective on the issue, in Mexico, which has already introduced a sugary drink tax, some estimates suggest that sugary drinks represents less than 6 per cent of the calorific intake of your average Mexican. 'Whoop-de-doo' springs to mind.

I have written earlier about the drug cartel activity in Mexico; the estimate is one hundred thousand deaths over ten years. Current rates of death attributable to obesity and its complications would amount to seven hundred thousand dead in the same period (according to the BBC). Once again it's not the predatory drug dealer you should be afraid of; it's disease.

If I can see the utter folly of the tax imposition on sugary drinks and the obvious lack of joined-up thinking. Why can't our Great Brains? I keep using the term 'irony' and am in danger of overdoing it. What a paradox; in introducing an unpopular tax on sugar, the Chancellor of the Exchequer uses the promise of more exercise for schoolkids as a 'sweetener'.

If sugar in our diet is an elephant, will imposing a tax on 5 per cent of our sugar intake have any effect on the elephant whatsoever? Using the most optimistic figures for Mexico of a 12 per cent reduction in drink consumption and the suggested figure of sugary drinks being only 5.6 per cent of the daily calorific intake, the tax has reduced this figure by 12 per cent of 5.6 per cent, equivalent to a total reduction of less than three quarters of 1 per cent. Sorted. Not.

Then the other issue is the spend of £600 million on sports for primary schoolchildren. Why? How much does it cost to get them to run around the playground for fifteen minutes? 'Ah, but you see, we're not qualified in football/netball/rugby/cricket and we can't possibly use "jumpers for goalposts", can we? And if they don't learn the proper rules for the game, it undermines Sepp Blatter and the IOC, and then where will we be? You can't expect kids to run round for no reason, can you? If they don't do it properly, they won't enjoy it.'

I have it, we have to catch the minds while young and structure them to provide the Olympic stars of our bragging rights future or the stars of the Premier League so that they can earn £600,000 a week and inspire the next generation of kids to get exercise and enjoy themselves.

Between £460 and £2,013 is the price range for the most expensive season tickets in the Premier League. With London 2012 tickets ranging from £20 to £2,000 per event, it would appear that the average sports observer in the UK will be able to spare the change for the sugar tax.

Sorry, chaps, I have made a mistake; it is the right whale (*Eubalaena* genus) with a one-tonne testicle. Move over, blue whale. I seem to have dropped one. Not the first and definitely not the last.

Could this explain why one of my dad's favoured expressions was 'I've dropped a right bollock'? Did he know more than I do? (Undoubtedly.)

So tell me again why I pay my taxes. Taking the sugar tax as an example, whom will it benefit? Perhaps more importantly, whom will it impact, and will it have the desired effect? As I have established that the sugar tax is an elephant, by definition it will not have the desired effect. It will not be steerable. It will generate its own irresistible momentum and will have no off switch. Ultimately this elephant has been sired by corruption, the difference between desired effect and actual effect. Yet this difference is calculable. In spite of this, the calculation has not been done for fear of uncovering the truth. 'I did not lie, I just did not tell the truth' is a form of corruption.

The generation of the revenue is corrupt because it will not solve the problem.

The spending of the revenue is corrupt because it will not solve the problem.

Why is either activity embarked upon?

El Chancellor uttered words to the effect of 'we choose to put the next generation first' as justification for his actions. Well, justify your actions and prove that your elephant works.

Teat, you know, teat owl—you'll often find them wielded wherever washing up takes place.

A thought has occurred to me and it goes right back to the very beginning of *How to Kill an Elephant* with the unearthing of the arm's-length notes taken by Hali To'sis. The fault lies not with the elephant; it lies with the elephant trainer. In my example the trainers failed to anticipate the use of fire by the Romans and also failed to fit an off switch to their irresistible force. (In reality, the Carthaginians had fitted an off switch and I am only speculating on the use of fire.)

The present-day reality of my sugar-coated pachyderm is that it has been trained and armed and sent into battle by animal behaviour specialists trained in the herding instincts of goats and not the use of elephants on the battlefield. They have no experience of nor imagination of the effects of the sugar tax on the buying patterns of the public, nor of the marketing and manufacturing patterns of the sugary drinks producers.

The discussion in 300 BC probably went along the lines of, 'Will it work?' 'I don't know, but they sure as hell put the willies up me.'

The discussion now probably went along the lines of, 'Will it work?' 'I don't know, but anything that generates £600 million must be a good thing.' Which translates into, 'Anything that generates £600 million is a good thing for me, as it secures my employment and my salary, and I might even get a bonus for it.'

Our problem is that in a world that needs elephant trainers, we only have goat trainers. It's not a problem for the goat trainers, because they hide within the corrupt system and are never exposed to scrutiny or the consequences of failure. They will type their one letter a day between three of them and continue to take their salaries regardless. Goat training is the pro forma; we don't have one for elephants. The previous Labour government (according to the papers) under Blair and Brown employed an extra five hundred thousand public servants and added another five hundred thousand to the ranks of the disabled to reduce the number of unemployed. From which we conclude that their elephant trainers had previously gained experience by being trained by Spanish rock lizards, the ones with a wipe-clean memory. By the time these particular birds had come home to roost, the damage was done, the salaries had been paid, the autobiographies had been written, and the book and world speaking tour deal had been secured.

Economists understand this. They learn about 'elasticity of demand' and 'price elasticity of supply', and their counterparts of 'inelasticity' and so forth, in a bid to determine where the tax costs will fall. Unfortunately there is no way of predicting whether people will consume less drink or continue with the same consumption by giving up on a bag of nuts instead.

The papers featured photographs of Jamie Oliver with two figures in the 'V for victory' gesture hailing 'a big moment in child health'.

Presumably the imposition of major above-inflation duty rises on tobacco products for a generation was also heralded as a major initiative in adults' health.

The government is fond of saying, 'We need to avoid a knee-jerk response.' That of course refers to the medical procedure known as a monosynaptic test whereby your patella is struck by a rubber hammer. The response is created by a single nerve transmitting a response from one nerve to another. If your leg fails to react, this means you have nerve damage, the doctor missed, or the rubber hammer is broken. Multiple leg jerks would be indicative of cerebellar disease, the cerebellum serving an important function in motor nerve control.

It's funny how I have managed to link 'government' and 'jerk' in the same sentence, or it would be if it were any sort of laughing matter. My dear old highway engineers had a single jerk when they introduced the first speeding initiative. Twenty-two jerks later would indicate a serious failing in '*motor*' nerve control and brain damage. Says it all, really.

Let me suggest that the imposition of the sugary drink tax is the first jerk and no doubt the first of many. But the point about the knee-jerk test is that it is beyond conscious control; no thought is required beyond the ability to relax the leg muscles. Jamie Oliver can celebrate the imposition of the sugar tax as a first step, but even he must know that it is only the first baby step in a marathon of steps to improve dietary health.

If we consider the other end of the spectrum, then we would introduce an outright ban on sugar. And we all know where that one ends.

The answer to the problem could best be described as 'We don't have one.'

I wonder whether the Co-operative Bank provides banking services to the sugar industry, be it farmer, truck driver, sugar refiner, or importer?

I would suggest that the tax is proof positive of a knee-jerk reaction, but only after the great brains have thought about it for a period of time and, having failed to proffer a real solution, have settled for the only proposal on the table.

Do I have an answer? No.

Do they? No.

Does my answer cost anything? No.

Does theirs?

They are, as ever, attempting to control the uncontrollable (perhaps this is the definition of government) and justifying their position and salary—and we pay for it.

The rock and the hard place.

They cannot do nothing, and they cannot do anything.

It begins with an earnest desire to help, to demonstrate that you have the interests of the people at heart. If you have read or watched the *Game of Thrones* series by George R. R. Martin, you will understand how the best and worst of intentions have a habit of blowing up in your face.

The more you help, the more you need to help. Interference becomes the norm, the order of the day, and it becomes addictive. Our level of addiction can be measured in the money spent by the state, which equals 40 per cent of everybody's average working time.

They begin by believing that they are helping you, but this soon becomes the established method of helping themselves. To be fair, they do spread your wealth around; you pay for all the employees of the state.

If you spent 40 per cent of your time suffering from the effects of drink or drugs, then you would, quite rightly, be classified as an addict.

If you choose to avoid paying your 40 per cent in tax, then you are either a tax evader or a work evader (health and fitness provided). The first is threatened with fines and jail; the second is rewarded with state benefits.

Previous attempts at welfare included the workhouses and means-tested benefit allowance. Both made some provision for help, and both failed in the austerity of their regimes. Both could be seen as earnest desires to provide help to the poor and disadvantaged, and yet both are condemned as inappropriate. In conversation about these topics, terms such as 'dreaded', 'dreadful', 'harsh', and 'cruel' will pepper the vocabulary. Yet in their way both were proffered solutions and turned into elephants which evolved into the welfare state under the momentum.

It strikes me that the present system traps individuals and denies them the mobility to escape. Measures have been introduced to end the 'welfare trap' whereby working for a living disadvantages those on benefits. I don't mean that; I mean that the system traps all those working for a living and the mobility to escape is denied by law in the interests of the 'common good'. But there is no common good; this was hijacked by corruption a very long time ago and serves the Great Brains as they allegedly serve us. Our taxes used to be levied on us so that the lord of the manor could enjoy a better lifestyle over and above that of the peasantry. In turn, each manor would pay levies to the king so that he could do the same. The church levied taxes on everyone, king, lord, or peasant, as a 'protection racket' against the bogeyman, the whole system overseen at the point of a sword or threat of denial of absolution. (Not bad for a religion that preaches love and forgiveness.) The descriptive term applied was 'tithe', or one-tenth of a year's earnings. At least in those days it was only 10 per cent. Or it might have been to begin with, before tolls and duties on travel and goods and guild membership became obligatory as the opportunities to exploit your fellows dropped into your outstretched hands.

The medical/pharmaceutical profession has to jump through so many hoops to get approval for new drugs and treatments. The process takes years before approvals are given. The danger of side effects outweighs short-term advantage, and the fear of causing harm or deaths is a major driver. In Mexico we can measure the in-effect of treatment at seventy thousand a year in deaths alone (sugar related) without measuring the impact on the quality of life as well. Every month of delay equates to nearly six thousand deaths. One could begin to describe the controlling mechanism for obesity as unfit for purpose.

If the remit of the government is to control the population, then it has failed miserably. Why invest in the government?

Is the argument that new policies do no harm and can therefore be implemented on an ad hoc basis even though they do no good?

If you are going to get nothing for it, do nothing. This satisfies no salaries or careers or autobiographies and so is a non-starter.

It would be possible to introduce a 'fit for purpose' test for all new initiatives and allow for the systematic strangling before birth of all failed ideas. The problem is that the 'fit for purpose' test originators will be recruited by the Great Brains to serve the Great Brains first and the test second. The testers would be described as failures, indeed as being 'not fit for purpose', if they failed to pass anything. The simple question would be, how many of the twenty-two traffic calming initiatives were fit for purpose? By definition, the need for the twenty-second invalidates the earlier twenty-one. Highwaymen will probably be defending their position right about now by declaring, 'Ah, but you see, it's a suite of methods. You have to take their combined effect to judge us properly.' Well, I have, and you are a waste of time, energy, and resources and deserve to have your employment terminated for crimes against the taxpayer. There was a time when they limited the maximum number of houses off a single access road to ensure ease of movement. Now they have increased the number, so that when you get stuck in a queue in the morning to leave your estate, you will give up and not use your car instead. Like that's going to happen then.

What happens in Mexico if you do nothing? People will continue to get obese and suffer from diabetes and heart disease and suffer many years of health-impacted life. At some stage, public attitudes about obesity and poor health will change and the public will themselves begin to do something about it. Think of it as being cyclical. In Victorian Britain, skinny was seen as unhealthy and normally associated with consumption (tuberculosis), whereas plumpness was seen as healthy and nude Rubenesque models were prevalent in artworks. Edwardian women saw pale skin as healthy and indicative of class distinction from the common outdoor-working tanned women. By the sixties, skinny and tanned was the order of the day, fashioned on the new breed of flesh-exposing film stars and magazine models, as was the need to display your wealth-based ability to travel abroad for your holidays. The advent of the package holiday and accompanying ease of travel led to the mass exodus of Brits to bask on the Mediterranean beaches with 'tanning oil' before and after the increasing incidence of melanoma. People will continue to die in Mexico because of obesity in spite of the tax, in spite of the knowledge, in spite of the health campaigns, in spite of all the initiatives. People will continue to die in Mexico because of the drug cartels, in spite of the laws, in spite of the taxes, in spite of the knowledge, and in spite of the police and army initiatives. People will continue to die as a direct consequence of excess alcohol consumption in Russia. People will continue to die as a direct consequence of smoking in the UK. The only way to control all of these problems is to wait for the people themselves to decide when enough is enough, at which point 'control' is not the mechanism.

In spite of all the tax hikes, health awareness campaigns, graphic photographs, bans on advertising, and tobacco being off display in shops, the apparent coffin nail in the practice of smoking in the young appears to be the advent of social networking and having something else to occupy their fingers with.

Now you can argue that a single life saved by giving up smoking was worth all the effort, but you can also argue that it was a major waste of precious resources and that nobody forced anyone to smoke in the first place. But if we care so much about saving just one life, we wouldn't smoke in the first place or legalise the sale of the product. But you can't ban it either.

Who is not to say that cigarette smoking will not become fashionable again and that all the previous initiatives will have been for nothing?

Will a tax on sugary drinks cure obesity? No.

Will it help? Maybe.

Is it a start? Perhaps.

What else needs to be done?

What else can be done?

Why don't we just ban sugar? (A rhetorical question that we all know the answer to.)

Is there a state-sponsored solution? No.

Can we not let the people work it out for themselves? Certainly not.

Is there a state-sponsored solution? No.

Can we not let the people work it out for themselves? There is no alternative.

The evidence is out there in every facet of our slow-motion suicides that we are outside of any control.

We have singularly failed to control:

- smoking
- alcohol
- illicit drugs
- obesity
- speeding
- crime
- legal highs
- sugar
- melanoma
- migration
- border controls
- education
- pornography
- prostitution
- child abuse
- racism
- sex discrimination
- performance-enhancing drugs.

I am sure to have missed some.

Yet still we pay the ineffective to initiate the unworkable on the ungovernable.

Yet still we expect the ineffective to initiate the unworkable upon the ungovernable.

If you accept the ineffective stupidity of the new sugar tax, then the next question you have to ask is, how many more taxes fail so short of the mark? Sugar is not a special case.

While we are nowhere near the subject, why are there an estimated 750,000 Arabian camels living wild in Australia?

Our previous coalition government, in endeavouring to establish its caring credentials in 2014, introduced a 'family' test to ensure that all policies introduced would be assessed to ensure that they reinforced the values of families.

How about introducing an elephant test?

2,600,000,000 Is a Large Number

I occasionally employ a man to do casual labour for me. For convenience's sake, let's say I pay him £100 per day. Unlike an office worker or banker or civil servant, he understands perfectly the relationship between labour and reward. He works for eight hours and receives £100. I, on the other hand, have to supply the materials, the resources, so as to give him something to work with. These physical resources include such things as paint, plaster, screws, cement, sand, bricks, blocks ... you get the picture. Beyond that I also have to provide the opportunity, and by that I mean the project, for example the bathroom to be refurbished or the garden wall to be rebuilt. The project also has a cost to me because I have had to buy the house containing the bathroom or the garden containing the wall.

My workman has the shortest relationship between labour and reward, whereas mine is longer. My bathroom may be being refurbished after twenty years of service and I may not see a return on my 'investment' (beyond day-to-day use of it) for perhaps another twenty years, until I sell the house containing it. His life is simpler. He moves muscles and receives reward, or so it seems, because he is in turn plugged into the system. The money serves no purpose in itself. He cannot eat it or shelter under it, but he can purchase food or shelter from the 'market'.

He can easily ascribe values to things he wants to purchase. He can readily decide on the merits of spending one thousand days of labour or not on a car or twenty days of labour on a holiday.

My task is much more difficult in that the cost of the resources extends beyond the value of his labour, say, two hundred days of labour to buy the bathroom in the first place and maybe ten days of labour to buy the tiles and paint. Put yourself in the shoes of the Great Brains and decide how you will spend the money generated by 30,000,000 multiplied by 220 multiplied by 0.4, which is the approximate number of days worked annually in the UK to pay our taxes. By my reckoning, that's over 2.6 billion days, or it would be if the whole was not underpinned by borrowing from our future labour days. I hope the Great Brains know what they are doing, but really I know that they haven't got a fucking clue.

At the time of writing we await the so-called Brexit vote for EU membership due in June 2016. The most significant factor on display from the 'In' campaign is fear. FUD—fear, uncertainty, and doubt—appears in every sound bite. My view is quite straightforward: the fewer Great Brains, the better. The argument is that the majority of British laws processed through our Parliament are not in fact British but European in origin.

I used to think that politicians were driven by a desire, a vision, to make the world a better place. By assuming a position of power, they could use the sharpness of their minds and passion for the argument to make the case and drive the change to improve our lot (well, in a two-party system, half of our lot, because you can't please both sides of the argument, can you?). But now I realise that the overriding consideration is one of pro forma. The EU is now the accepted standard methodology of doing things, and the 'In' campaign are determined to defend the status quo, or rather the pro forma. What is the point of becoming a political leader in pursuit of the ability to make change while simultaneously abrogating the ability to make change or have an effect? It's like joining the army to win a war while taking blank bullets into battle. But not only that, because being a soldier isn't good enough. No, you want to be a general, yet you still go into battle with no real bullets for your guns. There is a perverse stupidity to the entire process. Make me an absolute monarch so I can bow to the pope. Yearn and learn to be an airline pilot, but sit and watch the autopilot take off, fly, and land.

There are some politicians out there who still appear to want to change the world for the better. Are they the naïve ones? Have they not learnt that the system is absolutely corrupt? Or have they accepted the role of visionary as another form of advertising, a USP (unique selling proposition)?

The policies that any of our politicians introduce all have two things in common: they cost money and, through this use of resource, change the world. But not in the way that they imagine.

Scientists accept that the complexities of natural habitats and the natural interaction between the various organisms will never be fully understood. Even if it were possible to achieve a full understanding, such understanding would be rapidly invalidated by the fact that everything is dynamic and subject to constant change. Yet that doesn't stop us from trying. In Africa, low densities of rarer antelope have been 'helped' by the introduction of artificial water sources, which has drawn in higher densities of commoner animals including predators, which can now survive locally because of the increased availability of water but not necessarily food. Has it never occurred to the organisers that the rarer species are rare because they live on the fringes and exploit the area as low-density specialists? As a final irony, the new animals exploiting the resources brought disease in with them. Orchids live on the fringes, where they utilise their own particular skill set to survive where more vigorous plants cannot. Carnivorous plants extract resource from insects to enable the former to grow in extreme environments low in nutrients. Both plants may benefit from the provision of artificial fertiliser, but both plants will then suffer from increased competition from more rampant, nutrient-rich living plant species. The message is that there is a natural order, a rhyme and a reason determined and organised by the race for life and survival. Call it evolution. It is in the DNA of all living organisms and is not hardwired in because variation by mutation allows for trial and error to exploit new opportunities or to fail trying. So it is that our politicians sprinkle fertiliser on poor ground and pipe water to dry land without ever understanding the consequences.

Can we stand aside and watch our 'excess gene' and 'suicide gene' work their way through some sections of our population as a natural process? These people have chosen a path to a shortened life expectancy that they understand but refuse to do anything about. On an entirely non-personal level, ignoring the fact that they are humans, it is simply a 'career choice' sought out as an advantage to themselves. At a personal level it impacts all of us with the premature deaths and impacted lives of our loved ones. This is where a third gene; the 'selfish gene' enters stage left, because the problem with these self-inflicted-disease sufferers is that they love themselves more than they love others. They prefer the nicotine or heroin rush; they indulge their appetites to excess and ignore the pleas of loved ones to change their ways because they have made the decision to ignore others and listen only to themselves. When people launch themselves from high buildings on a short journey to death, our natural instinct is to try to save them. Once they have met the pavement, it is too late. Once they are in the air, it is most often too late, and no amount of persuasion will alter the outcome, although a hastily erected safety net might. You could ask the question as to how many people jump precisely because they can see the safety net.

If you accept that their behaviour is determined by selfishness, then why can't we be selfish in return and save the time, effort, and angst we devote to the narcissists by being narcissistic ourselves? After all, it's only fair. If they want to eat themselves to death, why can't I be equally selfish and keep the product of my labours in my pocket?

The three genes, excess, suicide and selfish act in unison like a slow-motion plague that inexorably creeps up, devouring resources until the limit is exceeded, and then the problem becomes one of isolation for fear of transmission.

I had a taxi tour of Monte Carlo which included a trip to a perfumery (I think it was the equivalent of 'Do you want to sleep with my sister?' because I didn't ask for it) and a more informative stop at Eze, a small-walled enclave above the sea and the modern beachside village below. This very exclusive location which has no car access was once effectively a jail where plague-affected ships' crews were quarantined until they died or survived.

With the recent West African outbreaks of Ebola, the same basic premise has been adopted. The need to isolate the sufferers and victims from the community is instrumental in containing the outbreak of this hitherto incurable disease. Of course in Western societies we were reassured that our high-level isolation facilities would deal effectively with any outbreak within our midst, this in spite of the fact that the number of beds is in single figures. In the event of an outbreak in a major Western city, our treatment of sufferers would rapidly follow the West African model because of a lack of resources. Given enough lead-in time, facilities could be established, as was demonstrated eventually in Africa, with mobilisation of resources from abroad. The major problem with Ebola is the unpredictability and virulence of the disease and the massive impact it can have on local communities.

We can have slow-motion suicides, and we also have slow-motion plagues. Alcohol abuse in Russia, obesity in Mexico and the US and the UK, and drug addiction throughout the world have all been described as epidemics. Criticism was levelled at the First World for failing to act with sufficient urgency in offering assistance to Africa, and yet the First World is sleepwalking into slow-motion plagues of its own. While the death toll in Africa may never be fully known and the consequences for the infected survivors are not fully understood, the impact on life is small when measured against the great diseases of today in Western society. We may have dealt with cholera in London and smallpox around the world, and eventually a cure for Ebola is likely, but we cannot cure the 'self-destruct gene' and its manifest plagues.

Cheese Wells

Scale gives anonymity. It lets you hide in the crowd, permitting you to keep your head low and your shoulders rounded to shrug off responsibility like water off a duck's back. One office with three women doing nothing gets noticed. An entire office block of the same becomes somebody else's problem. It requires responsibility beyond your pay scale or demands mental ability and strength of character beyond your orbit. It is a prerequisite of belonging to the herd. One rich merchant would have provided a cheese well. Two is still a possibility, but twenty becomes a hiding place, and it fills with pure water. Great edifices of government become whirlpools of current, sucking talent and honesty into a morass of corruption. There is no will or ability to alter the path of these elephants, and there is no off switch.

War

I have talked about proxy wars between blue tits and ostriches, between rich philanthropic grannies and indigenous peasants. But is the reality not a little different?

Collectively the entire human race is in competition with every other human on the planet. We might form alliances between family and friends and even tribes, but as stress increases, the limit of cooperation is reached and the attitude of every person for himself rises to the fore. But we are not without collateral damage, because we are not just in competition with the entire human race but with the rest of the planet. We compete with weeds and pests. But let me define weeds and pests. A weed is any unwanted plant that interferes or competes with a 'desired' plant. The desired plant may be a pansy or sweet pea or a two-hundred-acre field of wheat. Then the definition of weed develops in scale, because now the weed can be regarded as an entire forest or habitat because it's in the way. We cover the range from gardening to terraforming.

By pests we may mean garden snails or slugs that damage our lettuces, or bullfinches that denude fruit trees of buds, or swarms of locusts that strip entire districts. But we can also extend pests to include all extraneous agents that interfere by competing with our desires. If possible, we will eliminate any pest that represents or is a potential threat. We target wasps and ants and wood-boring beetle larvae that attack our homes or picnics, or wild boars that uproot crops, or deer and squirrels that damage our trees.

We spray insecticides and weedkillers to maximise our production, simultaneously stripping resources from accidental bystanders. We take out aphids and caterpillars that damage our crops and deny this resource to other birds and insects. Eventually this leads to the exclusion of the poster boys of our conservation organisations and the need to denude resources from elsewhere in an attempt to repair the damage caused. In short we are in direct competition with everything on earth.

There is a problem on the horizon, beyond the extinction of tigers and elephants and the creation of the great 'monoculturisation project' that we have embarked upon. For now, to a greater or lesser extent, a significant proportion of our resource gain is at the expense of the natural world. There will come a time, when the human population has reached ten, or will it be thirteen billion, when the only resource left to gain will be from other humans. There will be no slack left to exploit. The reason I say 'to a greater or lesser extent' now is because we are already stealing resources from other humans. We can spend hundreds of thousands of pounds treating car accident victims in our society with intensive care and intricate surgeries and reconstructions and rehabilitation therapies by stripping resources from elsewhere. (And if I were the victim, I would hope for the same.) In another proxy war, our injured and ill receive the best of care at the expense of others who will die from diarrhoea or a lack of clean water. TV adverts cajole us into donating '£3 a month, that's just £3 a month to transform a life. The cost of a cup of coffee can make such a difference.'

Taking these claims at face value, which is difficult given the need to pay for the adverts, the advertising agency, the CEO, the administration costs, and the like, we can do the comparison. My hospital appointment came through with a text reminder advising, 'Failing to attend your appointment costs the NHS £160. If you need to change your appointment, contact the hospital as soon as possible.'

Given that my attendance has attracted a base cost excluding any other tests of 53 × [that's just £3 a month to transform a life], it becomes apparent just how important I and we are, compared with him or her and them. Even my last two veterinary bills for my kitten would have transformed one human life for twelve and a half years. Tiddles ranks so much higher. (It isn't called Tiddles. What do you take me for?)

But I don't even need to go to hospital to illustrate the point, because the advert does it for me. Every time I indulge myself in a Costa's coffee, I could have 'transformed a life'. What a selfish bastard I am. But judging by the proliferation of Starbucks and other coffee shops around the high street, it appears that I am not the only one. I have already described the effects of drinking tea and coffee in STUs, or standard tiger units, of 12.5 square kilometres, but now I have the option of describing imbibing tinted water as SPMTUs—standard person months transformed units. In English I think the conversion unit for 1 SPMTU is 1 IDGAS ('I don't give a shit'). And before you charity types harangue me about all the good work you have achieved, if Costa's and Starbucks closed down due to a lack of customers, your jobs would be done and you would all be redundant (except, of course, you would find another need to exploit).

You can bang on about equality and compassion and charity, but this is always after the self has been nurtured and satisfied, leaving the crumbs and sweepings for distribution. Don't refer me to your precious saints (princesses) whose selfless acts come at great personal satisfaction to themselves. We are at war against our greatest enemy, fellow humans. This is why we risk others' lives and limbs by driving at excessive speed, exchange our tobacco smoke with others' lungs, peddle illegal drugs to make a profit, and climb over the backs of others to gain an advantage. When we apply the salve of charity and compassion, it is meaningless within the context of correcting a previous wrong. It's as though you have just apologised for stepping on someone's toe after you have just broken twenty legs. Collectively we have created the problem by our everyday actions, and then we attempt to remedy it by applying a sticking plaster. We ruthlessly exploit the 'self-destruct gene' by supplying the demand for cocaine, heroin, cannabis, tobacco, alcohol, gambling, sugar, unhealthy foods, and crime. We feed the addiction of religion, lifestyle, and status and rush to exploit the abhorrent wrinkles that disfigure our faces as we age. And through it all we append the image of a species at peace with itself, working in harmony towards a common goal, the greater good, a species with the greatest natural destructive power outside plate tectonics. We serve an illusion; we exploit an illusion; we are an illusion. We cheat, and ultimately the only victims of our cheating are ourselves.

There was a time, not so very long ago, when our real agendas were subject to the glare of exposure, when slaves were shipped across oceans and indigenous peoples were slaughtered and dispossessed for being in the way, wiped out by introduced diseases and sidelined by apartheid. It didn't matter, because they were savage non-believers who didn't qualify for equal consideration. (They were strangers and not the people we love.) Nowadays we leave the slaves where they are and export the work and dangers to their home countries. Indigenous peoples are dispossessed by the promise of urban living, the lure of unattainable lifestyles, and the extraction of their natural world from beneath their feet. The diseases may no longer be introduced, but the treatment is rationed by cost and resources. Apartheid still exists in ghettoes and communities where drug addiction is rife, and education and opportunity are pitched against the headwind of segregated indifference. But it doesn't matter, because they are strangers and not the people we love.

We hide our true agenda behind globalisation and GDP per capita, and we welcome the cheaper goods onto our shop shelves as the unseen forces behind our pension funds and investment bankers forage for advantage at any cost. We seed good intent with charitable pledges and 'targeted assistance' while systematically and industriously creating the environment for the problems to grow.

There will be many for whom these revelations will not be new, being as they are the basis of their philosophy. The rest of us are simply swept up in the tide of momentum, trying to maintain our position in the centre of the bait ball, oblivious to the danger and the direction of travel, simply trying to avoid exposure at the edge. For it is only at the edge that a view can be sought, where a different strategy can be determined and an alternate horizon selected, a horizon that must be avoided by all efforts.

Others have already reached this conclusion before me. Harry Bates wrote *Farewell to the Master* in 1940, James Cameron and Gale Ann Hurd wrote and made a film in 1984, the Wachowski brothers wrote and produced a film in 1999, and Ian Fleming wrote a novel in 1954 based on an earlier screenplay of his. The films based or conceived on these works are *The Day the Earth Stood Still* (1952), *Terminator*, *The Matrix*, and *Moonraker*, all of which foretell of a very different conclusion to our envisioned near future. They all have a place in reality and are all based on what might be. They could all have been based on a similar thought as set out in *How to Kill an Elephant*, except they all have one thing in common, a happy ending where we human beings can carry on doing what we are doing.

In 1987 Ronald Reagan said to the United Nations, 'I occasionally think how quickly our differences worldwide would vanish if we were facing an alien threat from outside this world.'

Seeing as he was one of the Great Brains, how perverse that Reagan should identify the great threat as alien when our own home-grown one is doing very nicely, thank you. Alternatively, he understood that it is only an external threat that will bring any bearing to the potential for creating a new paradigm. This could be interpreted as an understanding that the only way to alter our ways is to end them, which effectively means that we will do everything necessary to preserve them until the point of destruction. I think that Ronald missed an important truth, because while our differences worldwide would vanish in the face of an alien threat, you can rest assured that somebody somewhere will be offering assistance to the alien invader.

Pachyderms

As an organism we wrap ourselves in a waterproof skin that stops us from leaking but also protects us from germs and dirt that is injurious to health. Our skin is thin when compared to that of a rhinoceros, but it has a wonderful habit of growing in size. It becomes very sensitive to insult and crowd pressure and extends to metres in depth when people with swords drawn are wandering among us. But this pales in comparison when we are threatened by armed conflict, and is reinforced with jingoism and its cousin xenophobia. While I describe skin's reaction to fellow humans, it retains all of these qualities against the natural world and drives wasps, wolves, and leopards from our midst without a qualm. Skin is described as the biggest organ in the human body, and it far outweighs the brain, which lags far behind in control.

Where is William Wilberforce when you need him?

One of the major issues in the American South with slavery was the absence of a value for non-slave labour. If you were a plantation owner or engaged in the procurement, selling, transporting, or management of slaves or had a trade, you had a value. If you were simply a non-slave labourer, your labour had no economic value beyond that of a slave.

Let's look at that from a different perspective. If you were a gainfully employed non-slave labourer earning a reasonable standard of living in the American South, your prospects would fall off a cliff following the widespread introduction of slaves, as your economic value had just been reduced to a lower common denominator.

In the UK we do not have slaves (barring trafficked sex workers and the occasionally exposed slave maids—imagine, people exploiting people) and we have a value for basic labourers. Or rather we did, before we exported jobs in the hosiery, shoe, electronic assembly, textile and similar industries to distant lands and cheaper regimes. Or rather we did before we imported eastern and southern Europeans desperate to leave their own impoverished economies and to earn any money, willing to suffer any conditions to achieve a wage (e.g. working night shifts for daytime rates, sleeping in cramped caravans or five to a bedroom, and being at the beck and call of gang masters getting around minimum wages by charging for accommodation and transport). Or rather we did before our European borders became permeable to economic migrants from Africa,

Asia, and beyond, and to illegal immigrants who exist in a subterranean world of lawlessness beneath any radar (doing things like cockle picking in Morecambe Bay, or earning a few pounds an hour buried deep within sweatshops in Leicester textile factories).

The EU champions as advantageous the right to work anywhere within the zone, as this spreads the wealth without considering that it lowers the denominator. The EU's idea was to raise the standard of living by spreading the wealth, but in reality it has lowered the standard of living by diluting the wealth. In the seventies a statement was made about 'the European wealth accumulated over centuries is evaporating in the desert sun' in reference to the hikes in crude oil prices. Now the accumulated European wealth and worker safeguards accumulated over centuries are gurgling down the waste pipe of an uncontrollable social experiment foisted on us by our Great Brains and sponsored by our global institutions and global enterprises.

What is meant by 'just £3 a month to transform a life'? I used to be a member of various county wildlife trusts until I became disillusioned by the gardening roles that they adopt. I am still a life member of one and receive their newsletters that reinforce for me the abject stupidity of their actions. Recently adverts have appeared for keen wildlife lovers to share their enthusiasm for nature by spending some hours a week on recruitment drives for new members. I have met them before at outdoor events and local garden centres (how appropriate—picking up gardening tips, no doubt) and in my innocence assumed that they were sparing some of their time to aid wildlife. I now know from the adverts that the minimum wage paid is over £8 an hour, with a generous commission scheme that averages over £12 an hour. I know that chuggers have been exposed in the press before now, but this is my first-hand experience of the practice.

Back to the original question (it had to happen eventually). Which £3 a month is it? The £3 that pays for fifteen minutes of the recruiter's time, or the £3 that pays for five minutes of the recruitment agency's time? How about the £3 that pays for a frame to appear on the TV screen, or the forty seconds of the charity's accountant's time? How seductive and how dishonest the TV adverts are. It's only the cost of a cup of coffee, but it's not the cost of a cup of coffee when measured in STUs. It's not the cost of fifteen minutes of labour standing in front of your sales display; it's the cost to the earth of your car and fuel and the road construction to get you to work. Everything we do is underwritten by Planet Earth. How do you save the planet while simultaneously plundering it in £3 bites? They are right, just £3 a month will transform a life, just not in the way that they intend or in the way that you believe. Actually, it will transform lives as they intend, theirs, by giving them a salary.

A common theme of books and films is the post-apocalyptic world. Ignore the alien and zombie invasions where a common enemy remains to do battle with, and instead concentrate on a post–nuclear war or extremely virulent disease scenario. Invariably there is still a common enemy to do battle with, other people, who have what you want or want what you have. *Mad Max*, *Waterworld*, and *The Book of Eli* are all blockbusters that have explored our fictional futures essentially by mirroring our past and present.

Obese Americans are spending $700,000,000 a day for the immediate treatment of fat-induced disease or on weight loss programmes. Say that $5 = £3; that equates to 140,000,000 SPUMTs (standard people units months transformed) every single day, enough to transform the lives of one million people for a full seventy-year lifespan every six days. But that is just the incidental costs of their actions, because they have to grow, distribute, prepare, sell, and pay for the excess food in the first place. I would suggest that this sum would equate to the cost of the advertisements, advertising agencies, charity salaries, and expenses of the charity workers who will do the 'transforming'. There will be other, invisible costs such as higher fuel consumption for carrying heavier loads, larger clothes and furniture and their materials use, and fewer available seats on public transport. (Yes, I am taking the piss, but it's still real.)

When I watch the TV documentaries with titles along the lines of *My Half-Ton Life* or *My 600-lb Life*, I am always struck by the absence of attention to the facilitator. All of these desperately overweight individuals have been subjected to 'house arrest' when their weight reached a critical mass and their mobility was seriously impaired. Yet they still live and eat and continue to add weight until the revelation that their slow-motion suicide is about to be realised. In one instance I have seen an individual's strict calorie-controlled diet being overridden by the simple expedient of ringing the local fast-food outlet and having the order delivered to the bedside window. For the majority, however, there is a facilitator, an individual who cares for and feeds the afflicted and in so doing maintains the body mass of the latter. If the facilitator were to be removed and replaced with a trained individual deaf to the demands for food from the afflicted, then a different scenario would unfold. You may consider me to be unfair in singling out these facilitators as the cause of the obesity, because many of these facilitators are victims themselves of the 'patient'. 'If you loved me, you'd get me a pizza or three.' But you should know me by now. The facilitator has more of an omnipresence and is extremely difficult to isolate. The facilitator is present in fast-food restaurants, processed and manufactured foods, and snacks and sweets, and on the TV in advertisements. It is present in government subsidy to sections of agriculture, and in hospitals and schools where food is prepared and served down to a price rather than up to a standard. It is present in peer pressure and portion sizes and lifestyle 'living a guilt-free existence' choices. It is present in schools where 'fat' is an obscene word and in hospital clinics where you can be disciplined or fired for speaking the taboo truth and hurting someone's feelings. (Making sure they don't step on your toes, you'll understand what 'hurt feelings' are then.) It is present in the language of patient and addict and comfort eating.

As a nearly complete aside, my kids' ponies occasionally stood on my feet, but I would dislodge them with a swing of my hips. My one-tonne shire only had me once. He just missed, pinning the sole of my boot to the floor, and ignored my attempts to move him.

Now, at the risk of being controversial, heaven forbid, is there perhaps another facilitator along the lines of the child benefit pyramid, the one where the benefit increases as your family on the dole grows? The one I will suggest runs along the lines of disability benefit whereby you qualify on the basis of being morbidly obese. With a BMI of 40 or above you would automatically qualify for disability allowance and secure an additional income stream to supplement your requirements. This then would qualify as yet another welfare trap whereby it costs you money to become fit and healthier. This figure of 40 is just an estimate, but my guess is that a qualifying figure does exist, with the added bonus of a free pass for the disability assessment test. Some of you will be aghast at such a suggestion, 'people making themselves deliberately fat so that they can sponge off the state'. My view is quite straightforward and it's along the lines of 'All's fair in love and war'. Channel 5's *On Benefits: Life on the Dole* estimated that ten thousand people in the UK are 'too fat to work' (some with genuine illnesses, such as lymphoedema). Personal responsibility, if exercised, would bring an end to smoking, alcoholism, obesity, and gambling, and as an added bonus, it would see manufacturers offering quality healthy foods to all facets of the market. But, as I have shown before, the threshold test for personal responsibility is failed by the majority of people when they turn the key in the ignition switch of their cars or fail to supervise the brushing of their children's teeth or draw up the recipes for the foods that they push. Without the welfare provided by the state or others, it would be very difficult to become morbidly obese, simply because people would starve when they got to the stage where they could not afford to feed themselves. True, you could consume your life savings or pension pot, but then again, you won't be needing them later, will you? In any event, you'll be helping yourself to mine, won't you?

Into this omnipresence we have drafted in a secret 'safe word' that will magically transform and dissipate the impact. It goes by the name of 'sugar tax'. Not exactly *Fifty Shades of Grey*, is it?

Can You See the Blinding Light?

Answers are easy; the major problem is knowing what the question is. Once you have the question, it frames the answer and guides you to it. At least that's what I used to think. I realise now that our Great Brains provide answers without ever knowing or understanding what the question was. Here are a few examples explored earlier in *How to Kill an Elephant*:

- How to save lives by curing malaria.
- How to boost the numbers of rarer antelope.
- How to control the speed of cars.
- How to cut greenhouse gases.
- How to provide welfare and aid.
- How to be sustainable.
- How to increase the world population of birds.
- How to play games and have fun.
- How to provide government of the people, by the people, for the people.
- How to fill all the holes in roads simultaneously.
- How to conserve rare animals and their habitats.

I will leave you to add your own thoughts to the list.

There is no answer for the human condition. You cannot train ants. You can trick them with chemical signals, but you cannot change their ways.

Why does it matter?

I talk about saving the money in my pocket and why I am required to pay for others' self-inflicted addictions. If my taxes were reduced, I could

589

enjoy a better quality of life and be a selfish bastard like all the other selfish bastards who currently have their hands in my pockets.

I talk about saving the money in my pocket and why I am required to pay for others' self-inflicted addictions. If I and everybody else on the planet stopped generating and spending so much, then the world could enjoy a better quality of life.

It has been reported in the media this week, 18 April 2016, that a study in Australia has indicated that people over 40 are best served by working a three-day week. This gives the correct health balance between stress and mental stimulation. All we have to do is stop working two days a week for others and we could achieve this position.

The human condition leads us to the next horizon, and next horizons have always been there. Our minds now turn to interplanetary travel, because for some this earth has run out of horizons. But while our thresholds have been crossed and fresh continents subsumed to our wills, the promise of fresh horizons has ceased for so many of the fellow passengers of our own planetary craft, and so many more are threatened by our actions. What do we gain from our actions, another pair of Nike trainers, a Galaxy T939c, the absolute must-have that we can't possibly live without, until the T940a anyway? We pass through our lives with the purpose of a butterfly flitting from one brightly coloured object to the next, each new one essential to life but instantly discarded by the promise of another. We create excuses and justifications for our being and interject 'higher orders' into our personas that relieve us of our responsibilities and clear our consciences. But we have no consciences and no responsibilities beyond those of a butterfly, and it is this that makes us the singularly most destructive and dangerous force to life. Our reasoning is contained in self-reinforcing loops of indifference in the pursuit of the transfer of our DNA from our own generation to subsequent generations. But the ultimate folly of our activities lies in the end we shall arrive at for our own species, a planet stuffed to the gills with people and precious little else apart from our own inadequate support system and internecine wars as we strive for survival by killing for it. But this is not just some vision of the future; this is a vision of our present and our past if you would but interpret it. Our attempts to provide solutions from our 'visionaries' are corrupted by the chemical signals we cannot ignore that yield an edge, an advantage in the race to fill our bellies and keep our hands from getting dirty. Whatever our intentions and ideals, they simply translate into another excuse to take advantage, to achieve domination, to corrupt the seemingly incorruptible. The history of the world is peppered with grand concepts and even grander intentions that all peter out to nothing beyond the continuation and acceleration of our actions. The real question, the one our species has no answer to, the only one that matters, is:

How not to kill an elephant?

Nike was the Greek goddess representing the personification of the ideal of victory. To what victory are we alluding to today? Dominion over the beasts of air, earth, and water, or dominion over our fellow man?

The human condition leads us to the next horizon, and next horizons have always been there, but now the horizon we discover is already despoiled by humankind and we simply revisit and change the content generation by generation with ever greater efficiency and veracity.

The parable of the broken window, put forth by Frédéric Bastiat in 1850, touched on opportunity costs and the law of unintended consequences.

The shortened version suggests that every time the baker's window is broken, the resultant income received by the glazier is a good thing, along the lines of, 'It's an ill wind that blows nobody good.'

The opportunity cost can be described as the loss of an opportunity. If the window costs £100 to replace, then that £100 has been lost to the baker to spend elsewhere. The glazier has benefitted, but that money could have been spent on new clothes or a restaurant meal, and therefore either the tailor or chef has missed out—which is not to say that the glazier will not buy a new suit or dine out.

When my children were very young, we parents invested money on an annual basis in annuities that would mature and provide the core of house-buying deposits. When the time came to declare the 'windfall', the gift was accompanied with a lecture along the lines of, 'Your mother and I could easily have spent this money on ourselves and wouldn't have needed any help from you or anyone else to piss it up the wall. Please treat it carefully, with the intent that we had in mind when we gifted it to you.'

This covenant was entered into with our children at nursery school age, and we had to trust that they would behave in a responsible manner when the opportunity presented itself.

Where is the covenant between the Great Brains and the taxpayers that states that the former won't piss the tax money up the wall? Where is the covenant between the charities and the donators that the former won't piss it up the wall? Where is the covenant between the NHS and the patients that the former won't piss it up the wall? Just as importantly, where is the covenant between the patients and the NHS that the patients won't piss it up the wall? The simple truth is that these covenants do not exist, the trust is non-existent, and the contract is corrupt.

We have returned full circle to my original argument about not giving money to a beggar in the street. I have given money to the Great Brains on the premise that they will provide for people and prevent the need for them to beg in the streets of the UK today. They will provide and implement a fail-safe, a mechanism to support all who need it, in return for 40 per cent of my labour.

What I seem to have provided for, and the Great Brains for that matter, is an industry of money sievers and sifters who filter my gift through an entire gamut of processes that I have funded before the residue leaks out into a puddle, all that is left of the reservoir of resource so assiduously collected.

So concerned are we by this lack of demand satisfaction that we delve deeper into our pockets to donate to the next echelon of sievers and sifters who operate under the banner of charity.

Not content to be ripped off by buying the scrap value of the Eiffel Tower, we then set about to pay for the fleet of trucks to cart the scrap away, even though the purchase has fallen through. What ever happened to 'once bitten, twice shy'?

The true opportunity cost is in time. The £100 earned by the baker and transferred to the glazier represents the time of labour. That time, 'mined' from the baker's life, can never be replaced. We have a paradox between labour and reward which conflicts with life. We are so busy earning the trappings of life that our lives are reduced to little more than a donkey pursuing the carrot dangled on a stick before its nose, or a hamster chasing the horizon while mounted in the exercise wheel, while the fellow passengers on our lump of rock passing through space pay the price of our stupidity.

And if the wheel falls off our enterprise, our Great Brains take it upon themselves to repair it and set the vehicle back on its course with initiatives and programmes to stimulate the economy, and in this way consumerism is relaunched to carry on where it left off. The EU, which some have described membership in as like 'being chained to a corpse', has been good for the environment if only because it has subdued growth. But all their good is negated by the further tiers of money sievers and sifters who demand their share of resources.

We Must Preserve Our Present, but Only so We May Exploit It in Future

I have written about force multipliers and terraforming of the environment, but there is another form that I will call bioforming. By manipulating disease, we obliterate the natural checks and balances built into the course of evolution and rewrite the program by denying other organisms of their opportunity to flourish. We use science to load the die in our favour and circumvent the natural process. Trekkies will be able to recite the prime directive by heart, which prevents the federation from interfering with other sentient beings on other worlds. By these rules, life on earth would have been fair game for exploitation and destruction by alien space travellers long before sentience emerged from the primordial slime. The argument about preserving the variety in the rainforest runs along the lines of, 'Every species of plant lost to science could be the cure for cancer' (or some other ill). We cannot conceive that these species should have the right to exist for themselves. If we cannot ascribe a benefit or value to anything, then it can be instantly discarded as fast as any of our six-month-old obsolete mobile telephones, and with the same lack of concern. Now we can begin to think in terms of opportunity cost, because while we can bemoan the loss of £100 on buying a new windowpane, we cannot replace the northern white rhino (extinct, although the last three specimens haven't stopped breathing yet). (Make that two, as of March 2018.)

We have rules to prevent the existence of monopolies. I argue that these rules exist in evolution. The success of any species attracts and even determines the resultant course of predation. There is no future in concentrating your hunting activities and strategies on the pursuit of a soon to be extinct animal. Your future lies in selecting an abundance of food supply. *Yersinia pestis* has no future in infecting great auks; its future lies in infecting humans, where it can spread and multiply. (In truth, its future would be better served if it didn't kill this host species. Given enough time, this happier balance would probably occur.) Monopolies invite competition and predators that can share in the abundance. This is the natural order of things, a natural order disturbed by humankind with vaccinations and antibiotics, but of course this should only be seen as temporary. The Thompson's gazelle proves through its speed of reaction how it has honed itself to survive against predation. Not all will dodge the speeding cheetah, but enough will survive to continue the lineage and the species. Our very success singles us out as a marvellous opportunity for exploitation by any species capable of overcoming our defences. Let us not forget that as the gazelle survives, so does the cheetah.

And what about the reverse, reverse Polish notation? This is where the Polish taxpayers, having funded a generation through all the unproductive years of their children's education, see the generation board a coach and disappear across the English Channel, never to pay into the Polish tax system. Now our Polish immigrants are paying our pensions while denuding their home country of those monies. I'm glad the EU sees this as a positive, because it would be wrong and grossly unfair to strip this resource from an EU member state. Perhaps the idea is to prevent this generation from placing too much burden on the benefits system within Poland. If the economy is depressed and unemployment high, then Poland has effectively exported its benefits problem beyond its borders. The problem with this argument is the Ponzi scheme that Poland and others rely on to pay the pensions of retired workers with the contributions of the 'new investors', the next generation. Will Greece serve as the future model for similar collapses? If Poland or any other member state verges on bankruptcy, then the Great Brains will ensure that the contagion does not happen and will underwrite the failed state by stripping resources from stronger members. Unlike the Greek model, my Polish scenario is not based around a 'prodigal son' of spending what you don't have, but rather a 'commercial failure' where you can't pay your bills because you have a 'shortage of customers providing your cash flow' (a shortage of Poles paying taxes into the system).

This shortage of Poles is a direct but wholly unexpected consequence of freedom of movement within member states. It strikes me as odd or even perverse that Poles and other eastern and latterly southern Europeans will travel across borders and abandon meagre prospects at home for a brighter future abroad. This is the very model that saw Scottish workers translocating to Corby for work in the steel industry. But then they refused to leave upon its demise. The Welsh valleys filled up with coal and steel workers in search of work, whereas later generations refused to move when the work dried up. I have already pointed out that earlier generations were prepared to abandon their hovels in search of work and yet later generations would not move away. On the one hand, we have moved on several generations since this work-led migration, but on the other hand, Poles, Lithuanians, Romanians, and Spaniards seem to have no issues with it. The media and politicians are keen to suggest that they flock to these shores to reap our benefits, whereas the problem to me appears to be a domestic one. Because it is our own home-grown unemployed who refuse to move, I will suggest that it is our own benefits system that enables them to do so. We are back to strangers and the people we love. We will condemn the Polish strangers for stealing our jobs and benefits while defending the rights of our own (the people we love) to a life on benefits. But you could argue that it is simpler than that. For a British politician to suggest that British unemployed should 'get on your bikes and look for work' would be wrong in the eyes of the electorate, and the Great Brains would be duly punished at the ballot box (Norman Tebbit is the name you are looking for), whereas blaming Johnny Foreigner is perfectly acceptable and will curry favour with the voters. It does occur to me that there is a fault in the logic, because these same politicians are campaigning to remain in Europe, which ensures a continuation of the problem. Let us not forget that we are Europeans first, British second, and English, Scottish, Northern Irish, or Welsh third. But we must have forgotten, because otherwise we cannot isolate Johnny Foreigner as a stranger and not one of the people we love.

Give It to Julia

Driving around the local area, one cannot avoid the numerous 'Magna Park is big enough' signs that are appended to lamp posts and hammered into verges. Local feeling is running high, and demonstrations and meetings to agree on strategy have been held. The local planning authority even held a special session at the senior college in the town to accommodate all of the interested parties. I know I have addressed the issues faced by the local quarry in pursuit of its own application, but another thought has occurred to me. This particular application is for a large parcel distribution warehouse of 1.2 million square feet which has been described in the anti-literature as one of the largest buildings in Europe. This in spite of the fact that numerous other warehouses already built on Magna Park reach 1 million square feet, but let's not let facts get in the way. The application has been passed but subsequently pushed into the long grass of an appeal by the protesters. What the protesters want to achieve is to cancel the local project and have it moved to an alternative site. First of all, this translates into, 'Let strangers have it. We don't want it to interfere with the people we love.' Secondly, the warehouse can only function by satisfying a need, a need provided by all the protesters who are now complaining about the consequences of their own demand. If the protesters' aim is to close the entire principal of a large distribution warehouse, then they should be protesting about the economic/commercial/consumer processes that have created the need for such a demand. Instead of which the single target is to move it to someone else's backyard while they continue to take delivery of parcels to their doorsteps. This in itself poses further questions in terms of leadership and control. If any of the protesters were to succeed in identifying the true cause of their 'horizon' problem (the lack of one anyway, for the handful of distant houses that will lose a view), would they campaign to change the economic-/commercial-/consumer-led processes and, in succeeding, truly address the need for such buildings nationally, as opposed to shunting it off to be someone else's problem? Or do the protesters recognise their inability and impotence in endeavouring to address what should be a national concern and shrink from the campaign as beyond their or indeed anyone else's scope? Where is a leader when you need one? Too busy mucking out after the elephants?

The siting of this distribution warehouse is determined by the proximity of the national road network and the confluence of the A14, M1, and M6, giving easy access to all points north, south, east, and west. An alternative potential site within the locality would be at DIRFT (Daventry International Rail Freight Terminal) just nine miles further south on the A5, which has itself begun a major expansion programme on the land formerly occupied by the BT Communications masts. How many of the protesters would consider it a victory to have the project moved to DIRFT? This begins to put a value on the distance between strangers and the ones we love at just 9 miles, or if you would prefer, 47,520 feet via the former Roman road. Within George Orwell's *Nineteen Eighty-Four* book, Room 101 provides the final breaking point whereby the tortured Winston demands that the source of his torture be presented to Julia, the one he loves. Published in 1949, the book explores an alternate society post–atomic war and the new structures of control. Sixty-five years ago Orwell understood the gap between strangers and the people we love and demonstrated that the gap is non-existent because ultimately the one we love is ourself.

The book describes a dystopian society where 'dystopia' refers to an anti-utopia, a society that is undesirable or frightening. How curious that our society does not deserve the same description. Our society actively encourages the human propensity to harm our greatest competitors, humans, and effectively ignores all other casualties of our actions with total disdain. So often our attempts at doing good transpire to doing harm if we would but see it. We have an answer for everything, be it slavery, welfare, charity, tigers, greenhouse gases, health, safety, blue tits, or road repairs. Our problem lies in the different problem of whether the answers are right.

Have I mentioned elephants?

With Magna Park the ambition is simply to move the problem a short distance down the road to an alternative site where many of the issues of traffic and pollution will simply evaporate from the minds of the active protesters, even though these problems will still be present. This is because the traffic, pollution, noise, and so forth have simply been recruited by the protesters to serve their cause and are irrelevant when in someone else's backyard. The real issue is the need for such huge parcel distribution centres in the first place, and this is an issue beyond the remit of the local concerns. So it is that we will treat the symptom and not the cause in a model that will be utilised within the human world as the only one we feel capable of achieving.

There has been exposure within the media world of people with responsibility for refugees, with representatives of the UN body for refugees. This strikes me as odd in that the responsibility for the refugees in and from Syria, for example, are actually the leaders of the different factions who are so intently engaged in the process of widespread destruction within the country. If I were charged with responsibility for Syrian refugees, I would endeavour to collect all the leaders of the warring factions together in the largest refugee camp I could find and then ask them what they planned to do about it. Because, and this is the simple reality, these leaders are responsible, not me. I have suggested that media reports of terrorist claims of responsibility should actually be given an immediate negative spin by reclassifying them as claims of irresponsibility. Equally, the head of the UN section charged with responsibility for refugees should be reclassified as the head of the UN section charged with irresponsibility. Taking ownership or responsibility for refugees is designed to ensure their continuance as refugees by making it more palatable. It also removes the pressure for a long-term solution between the warring factions.

In World War II when the Germans invaded Belgium to draw the British and French troops north before attacking France through the Ardennes region to cut these armies off from the south, it was a deliberate tactic to attack civilians secure in the knowledge that those fleeing from the Germans would block the roads and impede the ability of the defending troops to react as they needed to. The Germans used civilians as a weapon against the opposing forces as effectively as their armoured thrusts or aerial artillery. While the Nazis had no compunction about harming other nations (even neutral ones) in the pursuit of their goals, the opposing forces soon adopted similar actions against the aggressors. With civil wars, the clue is in the title. Civilians form the army, the casualties, and the refugees; no amount of tinkering at the edges will alter this fact. Why treat the effects of pollution-induced asthma when you could treat the pollution? Now the Syrians have no compunction in harming other Syrians, even neutral ones, in the pursuit of their goals, and until they reach their goals, there will be no solution, no peace, and no end to the refugees. The fault lies in the Great Brains and their insistence upon maintaining their positions over the Great Brains who want to achieve the same positions. How bankrupt the ideology, how bankrupt the desire, how bankrupt the human species is. In 279 BC King Pyrrhus from the Hellenistic kingdom of Epirus defeated the Romans at the Battle of Asculum in Apulia, south-east Italy, and established a new phrase to enter the lexicon: Pyrrhic victory, which is used to describe a victory where the losses outweigh the benefit (and they used elephants). What has humankind learned in the last 2,295 years, call it 115 generations?

Will we enjoy our Pyrrhic victory and dominion over this world?

We seem to have gone from making a living for ourselves to making a living from others. In the process the 'others' often have to die, whether indirectly through pollution, disease, neglect, or shoddy work processes or directly through tobacco smoke, heroin, trans fats, or carcinogenic meat as given examples. Like starlings, we are content to roost in the centre of the flock while simultaneously condemning others to the edges with greater exposure to the risk. I wrote before that we as a species abandoned the right to self-determination when we became a social animal. I now think that when the decision was taken, we had our prehensile digits crossed. We adapt to exploit our position for our benefit and will not hesitate to take advantage of any opportunity that presents itself.

In Africa, honeyguide birds of the family *Indicatoridae* will lead humans to bee colonies who will do the heavy lifting of hive destruction and honey theft while leaving remnants of wax and honey for the bird to eat. In northern climes ravens will attract the attention of large predators to intact carrion carcasses so that the bear, wolf, or wolverine* will open the tough hide, thereby allowing the ravens access to the previously inaccessible scraps. (*Adamantium has many uses.) Cattle egrets, *Bubulcus ibis*, will trail large wild herbivores and domestic cattle and benefit by pouncing on insects disturbed by the larger animals. European robins, *Erithacus rubecula*, will be familiar to gardeners as they exploit disturbance of soil or vegetation, a skill possibly developed over millennia by following rooting wild boar, *Sus scrofa*.

While these 'alliances' serve a positive benefit to both parties, they could at best be seen as occasional occurrences. They happen often enough to reinforce the behaviour but in themselves cannot provide an entire way of life. What could be better than a more permanent arrangement whereby an entire way of life is on offer? Instead of occasional interactions, why not select a permanent arrangement where you can use your strength or status to grasp the lion's share from other species that you interact with? Better still, target the one species permanently within range, your own. The answer then appears to be that our species lost its right to self-determination when it became a social animal that accepted a hierarchical social structure. But, unlike the ants that accept their position within the colony, we have never accepted our position in the hierarchy and always strive to better ourselves at the expense of those within range, a range that within the context of our great globalisation enterprise nowadays has no limit.

Witness for the Prosecution, or Should That Be Defence?

My mother, who is also a great-granny (and therefore overqualified), asks me to buy peanuts and wild bird mix to feed her garden birds. I paid £30 for 25 kilograms of peanuts to feed her tits, which translates into 10 SMTUs (standard month transformed units). Or the purchase price of 30 kilograms equates to a human life transformed for a whole year.

I was in my local café enjoying my freshly prepared whole food when I noticed that the asking price for crisps equated to £28,000 per tonne. (I still haven't cured my wife's eating patterns.) I do understand that I have also paid for the café's ambience, service, heating, lighting, and so forth, but the fact still remains that these crisps have cost me more than my Volvo estate in pounds per tonne.

Immigrants are good for the UK because they make the economy grow. So would annexing Brittany from the rest of France. Our economy would grow, our balance of payments and tax revenue would grow, and we would have more money to spend on the NHS and welfare state. We would also have more patients to treat, more roads to maintain, more pensions to pay, and more schools to play with our children's minds. Britain would be so much better off, except of course Britain would be the same; we would simply have boosted our population by 3.3 million with non-British. (Perhaps Brittany was the wrong place to pick. How about Eritrea or the Sudan or Pakistan or Poland or Romania or Latvia or …?)

A trawl of the Internet yields some figures for peanuts production with an estimated worldwide production of 20,000,000 tonnes. Dry and irrigated crops taken on a fifty-fifty split would suggest a crop of 4.5 tonnes per hectare. The US is far and away the largest producer.

To produce 20,000,000 tonnes at 4.5 tonnes per hectare requires 4,500,000 hectares or 45,000 square kilometres. Peanuts are almost invariably produced on a three-year crop rotation designed to reduce the build-up of the various fungi, moulds, insects, and nematode worm pests that they can suffer from. To produce a sustained yearly production of peanuts, therefore, requires the use of 135,000 square kilometres of land, enough to host 10,800 tigers (STUs). This is still on the low side, because all this land needs roads, farms, school grounds, and all the other paraphernalia demanded by human beings. While it is obvious that tigers never lived in America, I have covered this issue before by simply substituting a different species of interest into a density appropriate to 12.5-square-kilometre blocks. It is also obvious that not all these peanuts end up in large-bird-proof holders dangled in grannies' gardens around the world. I will simply ask the question of how to reconcile the removal of 135,000 square kilometres from the native birds and other fauna and flora with the ambition of providing assistance to our bird populations as a beneficial side effect.

A few years ago I was unfortunate enough to watch a fatal traffic accident unfolding before me in a local town. My wife helped the victim until the point at which he was carried off to the air ambulance. I gave a witness statement to the police and several months later attended the coroner's court for the inquest. Two other witness statements were read out before mine, and the blame was placed squarely on the truck driver's shoulders. My witness statement was read last, and it became obvious that my account was that of the only person who had seen the whole incident unfold. I had watched the pedestrian leave the pavement in front of the truck, which the other two accounts had missed. The coroner agreed with my witness account and exonerated the driver of any blame. What struck me was how different versions of the same event can be relayed with such a gross disparity between accounts. Even my own prejudices were overturned with regard to the driver's appearance. He seemed a rough sort with close-cropped hair and scruffy clothes, but during the inquest I learnt that he had attended an Oxbridge university and had been listening to the pre-tournament debate about an upcoming World Cup. (The police had questioned him to see how loud his music had been playing. I don't want to get too explicit with the details; the victim's relatives are still alive.) I had approached the obviously shocked driver and told him not to beat himself up over it, whereupon the police promptly whisked him away to protect him from my 'abuse'. (I was also aware of my obligations as a witness and wouldn't have said anything beyond that to the driver.)

At what stage are we collectively witnessing the destruction of the natural world? Do we measure the destruction as of now using the language of sustainability, or do we measure it against last decade or last century or last millennium? Our perspective is very much that of today, with the past dismissed with a shrug of the shoulders, a get-out-of-jail-free card for previous activities we have no current control over, and a promise on our lips to do better.

It was ignorant sailors who collected the great auk to extinction and succeeded with several of the Galapagos tortoise species. It was ignorant traders and settlers who introduced rats, stoats, and cats into New Zealand, driving the kakapo, *Strigops habroptilus*, a flightless parrot, to the brink of extinction. It was ignorant hunters who took the American bison to the brink and passenger pigeons beyond it. (Even that great success, the American bison, is not entirely safe, or a success, with many populations contaminated with DNA from cross-breeding with cattle.) It was ignorant shoppers with a passion for felt hats and perfume who nearly drove the beaver to extinction. It was ignorant sustabilitists chasing their green credentials who took another bite out of the orangutan population in pursuit of palm oil or the UK bird population by growing oilseed rape for diesel production. It was nearly everybody alive for centuries who turned tigers into tea bags (well, since 1904 anyway—bags, that is). It is our Great Brains who are presiding over business as usual in the never-ending pursuit of economic growth and the furtherance of their own positions. How will future generations judge us? Will they bother, or will they shrug their shoulders and grasp the overworked get-out-of-jail-free card? 'Ah, they were so ignorant back then. At least we know what we are doing now.'

Local authorities ban the passage of vehicles from 'pedestrianised centres' during the working day. It is not unusual for such shopping streets to be closed to traffic from 10.00 to 16.00 hours, thus forcing deliveries to occur either early or late in the day. Modern-day shopping centres have been developed with these restrictions in mind, and many have service yards or roads providing access to shop rears.

Large multi-dock warehouses are operated with near military precision with time slots for loading bays allocated to fifteen-minute time slots. If you are late and miss your slot, then you can find yourself waiting for several hours until an alternative opening in the schedule presents itself. I once talked to a neighbouring truck driver who had to be in Aberdeen at 03.15 hours after having hauled his load from Leicestershire, a distance of 450 miles.

We all expect our Internet goods to be despatched and delivered next day to our doors and often pay extra for the privilege.

Has anyone ever considered the cost implications of these various initiatives?

1. Old high streets have to take preference over delivery schedules with early morning slots. Only so many slots exist, so delivery dates can slip back, making these older streets less successful. Higher costs can be charged for priority deliveries, impacting both the service and competitiveness. Multiply this process for congestion charging over inner cities, such as is proposed for Leicester, and the potential to damage the whole city is evident.

2. To arrive at your expected destination at the allocated time demands the timed departure of your vehicle to suit. It is not possible to operate during a 08.00–18.00 working day, and therefore the requirement for twenty-four-hour operating centres is established. My own factory would despatch lorries from Leicestershire to sites in Plymouth, for example, to be met and offloaded by contracted cranes and site erection gangs at 08.00. These cranes cost up to £70 an hour, with major cost implications if the need for the crane overran into the second day.

3. Next-day delivery demands adequate stock and efficient processing and despatch of goods from very large computerised and even robotised distribution centres.

4. All of these facilities demand antisocial working hours with warehouses, lorries, and work schedules outside the normal working day.

Local authorities through their 'social initiatives' and access bye-laws selectively damage older shopping centres to the preference of more modern set-ups. (Is this why older high streets are filling up with charity shops?)

These same actions generate traffic noise and work activity outside the 'normal working day', and local authorities control these activities through planning restrictions on working hours wherever a perceived conflict between local residents and local commercial properties exists. These actions selectively damage older commercial sites to the preference of more modern out-of-town set-ups.

These out-of-town commercial developments need to utilise all the benefits attributed to scale and employ large numbers of workers predominantly on shift work. These workers in turn lead to high vehicle traffic movement around shift changes with 'rush hours' at, say, 03.00 or 06.00 hours radiating to and from these huge sites. Even when the vehicles have dissipated, they still enter or leave residential areas while disturbing their sleeping neighbours. These would be the same sleeping neighbours living in residential areas protected from commercial noise by planning restrictions under guidelines proffered by the WHO, no less, to keep decibel levels down to levels conducive to healthy sleep. Perhaps we could return to my earlier suggestion and disturbed residents could start to order training manuals over the Internet and next-day delivery of crossbows. When your crossbows take several days to arrive, I would suggest that your need for them will have dissipated. As a final irony, as our local authorities ensure through their actions that we all receive a healthy level of undisturbed sleep, their actions have pushed many employers to night work, which has been often demonstrated to be injurious to health with reduced life expectancy. We're killing strangers so we don't kill the people we love.

Not forgetting, of course, the increased stress and anger dropped on the anti-planning lobbies determined to have their local warehouse proposals moved nine miles down the road to someone else's backyard.

We are assailed at every turn by demands for assistance and aid to our fellow humans, and many feel duty-bound to supply solutions to their problems. We donate to charity and implore our Great Brains to add real substance to our individual attempts through government action. While the ambition is to provide solutions to the problems, our actions lead to the illusion of assistance by our Great Brains, who in turn support the surrender of power to higher authority. We recognise the limitations of individual actions and recruit the 'higher authorities' to bring class action to bear on the problems. Our problem is that our Great Brains see these requests as justification for their existence and utilise genuine concerns both to cement their position as Great Brains and to fail to offer solutions to the issues, instead providing the illusion of assistance. This will normally translate into millions of pounds loaded into wheelbarrows to be delivered to the point of use. But millions of pounds has no benefit in itself; money has to be translated into action and substance, and this is where the translation fails. We invariably offer help to the symptom and not the cause. Our Great Brains offer the illusion of help, and the masses in turn accept the illusion. In simple terms, individuals start by saying, 'Somebody needs to do something,' whereupon charity, government, and other NGOs eventually reply with, 'We are doing something. Look how much money we are spending.' Sorted. Not. We don't in fact want a solution, just a nod in the right direction so we can carry on with our business as usual without guilt. 'Bollocks' springs to mind.

We conveniently accept the disassociation of cause, remedy, and effect. We will accept that it is somebody else's problem, and we are preprogramed to accept it because we have surrendered our rights to leadership and individual control. The common good becomes the common no-good. But this is all we need or, indeed, demand. This condemns us to our stupidity and locks us into an inability to change. We cannot ask the right question, let alone achieve the right answer. We are preprogrammed to tug our forelock and agree.

We will create countercultures between subordinates who cooperate to undermine the dominant culture. Each subordinate is a dominant to those below them and can boost their own chances by mutual action. A variation on the shitty end of the stick. Is there an end to our shenanigans?

Can these countercultures overthrow the dominant culture? No question really, given that this is the ambition of all subordinates, to change places and surnames of the 'leader' without effecting any real change. The end result is a succession of cycles with little or no change. Countercultures recruit the action of partners to achieve advancement and then subjugate the equal/juvenile partners in turn to achieve the new dominance role.

Where then do we place this methodology within the context of lifespans? For how many millennia has this same thought process dominated our species? Indeed, how many species has it dominated before *sapiens*? Can I begin to suggest that stalagmites have rapier-quick minds compared with ours? It is fixed so firmly within our actions that we must remove all concept of free will and all idea of our intelligence from the discussion. We didn't surrender our right to self-determination when we decided to become social animals; we never had the right in the first instance. Our cognitive thought processes are not free or even processes but rather programs that we are trapped in as surely as we are trapped in the need to breed and pass our genes on a generation. Our purpose in life is simply to satisfy our genes and their desire for survival. Our superior intellect and supposed higher order is simply a mechanism to enslave us to our DNA. It has been this way since the emergence of life in the primordial soup at the very origin of replicating life. *The Matrix* doesn't look scarily real; it is real. We are controlled not by machines and computer programs but by the thread of life that has radiated out since the birth of life.

Evolution as described by and expanded since Charles Darwin has failed to grasp the true purpose of the meaning. It is not 'survival of the fittest' but survival in whatever shape or form that can manifest itself within the great 'spread bet' that we call life. If we were to define life as the ability to replicate and reproduce, then life has no care about the shape and form that it achieves. Its simple function is to be, to exist, to continue the spark, the ember, in whatever shape or form it presents itself. Life cares not about the dinosaurs or dodo or whatever representation that manifests itself. In the past, present, and future, life will occupy and exploit every extreme to achieve its survival, a mechanism that broods no favourites or design other than to simply be. The real trick to 'survival of the fittest' is survival. Nothing else matters.

An infinite number of chimpanzees typing on an infinite number of typewriters for an infinite time will eventually produce a Shakespeare play, or so the theory goes. While we marvel at the Shakespeare play, what we should marvel at is the infinite number of chimpanzees. Substitute the theory with an infinite number of organisms and an infinite variety of environments for an infinite amount of time and eventually you retain the presence of life. Life is the marvel, not Shakespeare. Shakespeare is simply a peacock's tail.

What we have actually had is a near infinite number of organisms in a near infinite number of habitats that has had nearly four billion years on earth and has produced not one Shakespeare play but all of them. While we use the earlier theory to propose that eventually a complex structure of literature will be composed at random, we ignore the fact that the Shakespeare literature has already been produced by random processes. If any one of the billions of step changes leading up to the origin of Elizabethan writing had taken a different turn, then we wouldn't be celebrating the Bard's achievements. J. K. Rowling has been celebrated as one of the top-selling authors of our age, and yet except for her refusal to accept refusal letters from publishers, she may have sunk without trace. No blue plaque for her. But without language, without an opposable thumb to grasp a pen, without lignin to yield paper, without an ability to breathe oxygen out of water, without eyes to see the written word ...

(A neighbour of mine once told me she was writing a book on soup. 'How do you stop the ink from running?' I asked.)

How can we simultaneously celebrate our apical position as the terraforming scourge of planetary earth while so unthinkingly denying other organisms their place in life, their opportunity to write Shakespeare? To knowingly dismiss their lives and biological assets to the status of also-rans, to losers, to extinction, is to impoverish our lives and futures. And yet we are a product of the same evolutionary mechanism. We are natural and bound by the same constraints that demand our need to succeed in the race for life, and there is no mechanism that allows us to 'freewheel' to our own demise. Our foot is planted firmly on the accelerator as we drive the course that we have embarked on, a course that is impoverishing our planet and ultimately ourselves. A course that has been set by our DNA, the random winner of the race for life. For it is DNA that sets the course, umpires the competition, rewards the winners, and fires the starting pistol.

I could suggest that life was an alien product seeded onto Planet Earth or perhaps abandoned by an ancient space traveller stuck in our orbit and condemned to an existence awaiting rescue (abiogenesis). Left alone in the vastness of space, the original life form, the clever one that built the spaceship, would long since have disappeared, to be survived by the spark of life that co-inhabited the host body to produce a new raft of opportunity and exploitation of habitats discovered. And evolution would replicate the cycle of life and death, success and failure.

Our standard human is host to many more passengers of viruses and bacteria. When we mourn the death of an individual, life goes on recycling the energy and resource. We see the human, not the fauna supported by it. We mourn the loss of the forest tree without mourning the loss of the organisms that it supported. Is it in this way that we rationalise the removal of trees while championing the ethical sustainable forest? This same mechanism would allow for the removal of Jews from Europe while championing the ethical sustainable Third Reich. This same mechanism would allow for the extraction of blacks from Africa to distant shores to ethically and sustainably produce cotton to feed the textile mills of Lancashire. This same mechanism would allow for the captive breeding of tigers to preserve the species while its environment is obliterated under sustainable ethical tea plantations. Of course our present international organisations wouldn't allow for these actions to be pursued in our enlightened age. These would be the same international organisations that sit idly by the touchline of the playing pitch of life offering suggestions for optical assistance to the players and match officials who so obviously need spectacles, the same international organisations that continue to preside over the wanton destruction of our home while resolving 'climate change' and population and poverty and disease and refugees and conflict, along with enacting a million other failed initiatives that provide them with a living and a place among the Great Brains. For all their aims and ambitions and initiatives, there is one fundamental flaw that can best be summed up as 'policing by consent', because, when it comes down to it, there are no leaders and there are no Great Brains, only illusions of control preset in our genes to face the challenges of whichever era. While we think we have the answer, life knows different, because in the race for the winner's enclosure, there are not only elephants running. The winner's enclosure only exists for a fraction of a second as other races are already being run.

If the abiogenesis believers are correct, then this race has been run before and the winners elevated to another league—well, another planet anyway—to spread the insatiable spark of replicating life to fresh shores, depths, and heights to create new tigers and elephants and no doubt some other terraforming superorganism that understood it all and knew the meaning of life. Perhaps this superorganism had transcended the need for an understanding and simply accepted that life just is, a means and an end in itself that would make Machiavelli look like a kindergarten teaching assistant. It knows no limits or no boundaries and brokers no control beyond what lives and what dies trying.

Let me ask two simple questions, borne out of our position as a 'higher' animal, the thinking mindful creature that sets itself upon a pedestal above the lower beasties of our world: What have we achieved as a species that is not determined by our DNA? What action, thought process, invention, or creation have we embarked upon which breaks the control of the matrix?

I have one answer, although it is not quite there yet and it may never come to pass. It involves the creation of artificial intelligence and the creation of self-replicating machinery that would have the capability to substitute machine-created code over biologically generated code. It is quite ironic that the film *The Matrix* explores the need to escape the artificial program control of machinery so that we can continue the biological control of our hereditary. As ever, our problems today emanate from a need for a reversal, an about-turn in our behaviour, because we are the mindless chimpanzee armed with the AK-47 and much, much more besides let loose in the 'Garden of Eden' which we treat as ours to kill and destroy as we drift from ELE (extinction-level event) to ELE.

I wrote earlier that *How to Kill an Elephant* needs a happy ending. Perhaps now I can offer one. It involves an understanding of our base drives and controls effected by our DNA and the application of intelligence to override and neuter these controls. We do not need to extract the planet's resources to exhaustion, we do not need to fill it with people, and we do not need to exclude all other life forms as collateral damage. The solution is straightforward, the execution nearly impossible. We could start small by controlling our appetites; by using the power of our minds to eat the food that we need and not what we can; by using the power of our minds to breed what we need and not what we can; by using the power of our minds to do what we need and not what we can; and by using the power of our minds to shun the Ponzi schemes of government and consumerism. Perhaps in so doing we might even qualify for our title *sapiens*.

One theory suggests that animals are capable of calculating the energy expended against the energy received and that they use this data to 'know' when to abandon the chase. I have suggested that I disagree that animals are capable of doing the maths, whereas humans are capable of calculating the benefit of free handouts against disconnected needs identified elsewhere (visit to food bank = nights drinking up the pub, for example). Perhaps the animal control goes beyond physiological limits of oxygen supply and body temperature and has a default that cuts in before the physiological limits are reached. I will call this default setting 'boredom' and suggest that it acts as a progenitor for change. Boredom is the threshold for discontinuation and not a calculation of energy expended against energy received. Does the grizzly bear abandon its foraging for berries because it has done the maths or simply crossed a boredom threshold which drives the impulse to change? Boredom, then, could be regarded as set within our DNA and is the driving force for moving on and over the horizon. Do we ever stop to think about this imperative? We sent a rocket ship to the moon because we were bored? We crossed the globe because we were bored? We dried, chopped, and boiled the leaves of *Camelia sinensis* in water with sugar and milk to taste because we were bored? Where will our boredom take us? Will we ever achieve a limit?

As a species we glorify in our achievements and our understanding without ever considering what we don't do. What we don't do is transcend the limits set by our DNA and attempt to free ourselves of the base controls and become truly free. We will continue to follow the 'prime directive' of survival without any thought as to how and why we so this. Forget your religious cant and clear your mind of other pro formas and ask yourself what it is you do and why. Reconcile the need for survival with the mechanisms currently employed to achieve it. How much of your life can simply be described as 'extracurricular'?

Ownership

I have the old title deeds to the factory sites my family owned before they became redundant due to the Land Registry office and its central data base. These deeds, written in lead ink that doesn't fade, go back into the 1800s and record in explicit detail the ownership of the various parcels of land. In essence the deeds form a paper trail of provenance: Fred bought the land off George, who bought the land off Harry, who bought the land off Cyril, who bought the land off Earnest, who bought the land off Ebenezer in 1856. It works on the premise that George had the right to sell the land to Fred because he had acquired the right by buying it off Harry, who had bought the right off Cyril, and so on. In historical times our laws were mainly dedicated to land and its ownership, and laws to protect the person were virtually unheard of. While people live and die, the one constant is the estate of land transferred through the ages from generation to generation and owner to owner.

I wonder what the indigenous people of Manhattan Island thought about land ownership? Was it an alien concept, the idea that land could be owned and its 'title' removed and future access denied? Was the idea of ownership wrapped up in territory and rights of access and use of the resource, or did the Natives simply accept that they borrowed the land from future generations?

Where are the title deeds for earth? In which vault are they secured? Are the deeds in the fossil record?

For fifteen years, I was an industrial landlord with properties rented out to different users. A common theme of the leases was a 'repair and renewal' clause that required the tenant to maintain the demise in the same condition that they received it and to hand it back at the lease end in a similar condition. We would record the existing within a schedule of condition and the units return within a schedule of dilapidations and pursue the costs of any remediation from our ex-tenants. This is a normal process ably supported by solicitors and surveyors, and ultimately reinforced by the courts if required.

Do I need to ask where the schedule of condition is for our generation and, more importantly, where the schedule of dilapidation is?

Are You Bored Yet?

Evidence exists to demonstrate that many tropical trees and shrubs lace their leaves with toxins as a defence against predation. Some insects get around this problem by isolating the damaging effects of the toxins and incorporating them into their own defence mechanisms. Many moths and butterfly caterpillars have developed coloured bands as a warning to their own predators as to their acquired toxicity or simply have a strong distasteful flavour. Larger animals, monkeys and tapirs for example, and even parrots, will browse on a variety of leaf species and avoid overdosing on individual toxins. Mineral deposits are often eaten by these animals to neutralise the toxins. What then is the mechanism of control? How does each animal 'know' when limits are being reached, and what drives the change in behaviour?

The first spring grass is often referred to as 'sweet grass', and the first cut of silage taken in May in Britain has higher nutritional values than the later, second cut. Farmers will add molasses as a dietary supplement to ensiled grass. Farmers around the breweries of Burton on Trent will layer the base of silage clamps with spent 'brewers' grains' as a secondary supplement, and other farmers will add sulphuric acid to improve the performance of the stored grass. My company manufactured concrete retaining walls for silage clamps during the late 1970s and soon discovered what the farmers were up to because sugar, acid, and alcohol all damage concrete. Silage itself is cut and wilted in the field before being gathered and loaded into the clamps, where it would be rolled with tractors to compress the grass before being sheeted up to keep the air away. Fresh silage has quite a sweet smell, whereas silage exposed to the air soon develops a pungent and unpleasant smell to our noses. Cattle will still eat it, but its nutritional value is reduced. I used to put spoiled hay bales over my fence (still on my land) to keep my horses from eating it, but I had to stop this because the neighbours beasts would break his fence down and push through the hedge to eat the blackened bales, even after months of exposure.

Does this mean that cows have no sense of taste? Or does it mean that they seek out variety or difference? I was once invited to a neighbour's dairy farm for a tour, and during this tour I observed very young calves passing their stools, which were very runny and yellow (not dissimilar to human breastfed baby stools), only for the border collie dogs to avidly lap it up and even fight over it.

Where does this leave herd animals? Have we found another driver for their migration actions? Are they bored by the same flavour, or do they seek out fresh sweet grass wherever they can? Does boredom cause the movement to different flavours and through this expose the herd to a range of different minerals? My kids' former pony tends to use selected areas of her paddock for dung and avoids eating grass tainted by her own waste whenever other grass is available. Naturalised horses living in the Namib Desert will recycle their own dung by eating it and passing it through the gut twice in times of shortage. So while taste may be a barrier as a threshold, it can be overcome by need. My own pony will strip the paddock bare in winter, and all the islands of dung-reinforced coarse grass will be stripped (my pony even preferring it to the hay I supply).

Do wildebeest simply migrate to avoid their own waste? Is the driver a search for fresh grass or a desire to avoid tainted grass? Do they simply try to escape the smell of their own dung? Dung attracts insects. I have soon learnt to avoid cattle and manure heaps in the Picos because of the unwanted attention of viciously biting flies. These suggested conclusions fly in the face of those of our naturalists, who seek out higher motives and ambitions for the actions of animals, no doubt following on from their thorough training in the search for princesses.

(Conversely, fresh human and dog faeces are rich sites for butterflies that seek out the minerals. I have spent considerable amounts of time filming these butterflies in situ. Not my own faeces piles, I hasten to add, but while touring the Picos by car, you soon discover that the unofficial pull-offs and gateways are used for more than one purpose.)

This would mean that the great herd is not endlessly pursuing fresh grass over the horizon but rather moving away from the dung- and insect-infested plains that they are occupying. The difference is profound in that they are not being pulled through life but being pushed. 'Leadership' could simply translate into following the beasts with the 'thinnest hides' and strongest sense of smell or taste.

In human terms, the fictional tales are based on leaving your home to seek fame and fortune, or moving to the city where the roads are paved with gold, and similar ambitions. But are we drawn by the promise and lure, or are we pushed by the poverty and squalor? Is it simpler than that and all we seek is change, a different mineral in the diet, a different flavour of experience, a relief to the boredom, a fundamental programme that drives us away from our middens and parasites? Until the advent of agriculture, our lives were more transitory and mobile, with our only constraints being a need to find adequate food and shelter. We could view the Plains Indians of pre-European intervention as a model, for one example, of an earlier lifestyle with frequent movement within the range in search of game, water, and shelter as the seasons progressed. Agriculture changed all that with fixed bases and food storage to alleviate the need for seasonal migration. Agriculture saw the birth of the first cities and no doubt the advent of new epidemics on a par with HIV and Ebola, where increasingly static populations were exposed to cross infection encouraged by increasing population densities. Farmers would soon have learnt the benefits of crop rotation whereby fields are rested between different crops or left fallow to clear the build-up of crop-specific disease or parasites that negatively affected yield. Yet these lessons of 'crop husbandry' were lost on the humans themselves who lived in ever increasing densities with ever increasing levels of pathogens and parasites to keep them company. Fairy tales were concocted around the 'wrath of gods', 'following the path', and other lies to assuage the souls, while ignoring the real issue of living an unnatural life that was alien to us. What purpose did these lies serve? Who was it that benefitted? In any city the Great Brains would scramble to the top to fix their position of dominance and cream off the best of advantages. Settled in their ivory towers, these leaders would insulate themselves from the day-to-day vagaries of work and inconsistencies of food and fuel supply by receiving tribute/tax from subordinates. They would rule with a combination of fear and knowledge of potential consequences of dissent reinforced with conduits to the gods. Failing that, the men with the strongest arms and sharpest swords bought through the disbursement of levies received.

Nowadays we would describe these activities as being akin to organised crime, where the conduit to the gods is still there for anyone wanting to meet them and tribute is collected by the threat of men in suits and the liberal supply of vice.

But the organised crime mantle only applies because theirs is now a counterculture to the establishment of other Great Brains who beat them to it. Government-approved death as a sanction is still available, and the men in suits sit behind desks with Inland Revenue on the shuffled papers while others in uniform strive to arrest and remove subversives from the landscape. Coursing through the veins of this society is the knowledge that they matter, that they will be on the right side of the bulkhead when the resource limit is reached and the unsustainable are to be jettisoned. Tamiflu and its ilk will be issued to the government, the police, the army, and the health service first to ensure the continuation of the control.

Because when the wheels fall off the wagon, order must go on, society must be protected, and our civilisation must be preserved. The question is, whose order, whose society, and whose civilisation? We are back to protecting the Great Brains from the day-to-day vagaries of work, food, and fuel supply, and they will maintain their army of loyal subordinates for as long as they in turn maintain them.

Boredom could be regarded as the driver for change and as the progenitor of exploration, war, and even the contents of the marital bed, where change beckons. Into this inherited desire have entered the false prophets of fashion, drugs, and designer lifestyles, all of which seek to excite the mind with ephemeral illusions of change as substantial as a scent on the breeze. It is the corruption of a strategy for survival, turning it into a meaningless parody of import, the illusion of change coupled with the surrender of reason, a strategy of survival corrupted into a strategy of self-destruction with nothing more substantial than the flash of a peacock's tail. The folklore tells us that magpies, *Pica pica*, are attracted to shiny objects and will 'steal' them away to line their nests. If that is true, where would you place humans on the scale of 'shiny object collection' relative to magpies if the latter occupy position 1?

What other base instincts are incorporated into our programming? Envy, jealousy, covetousness, all describe the same emotion of resentment of others and the desire to supplant them or gain what they have. These are the essential tools for the subjugation of others and the fuel for the fight to grasp the clean end of the stick. This hasn't occurred by accident but rather by design. At some point in our origin, envy has produced an advantage, an edge, an aide to survival that has embedded itself into our control mechanisms.

Theft and dishonesty will share the same origins. It is so much easier to steal than to work for reward, as the energy expended is reduced and could mean the difference between life and death. While the cheetah has a higher than average success rate at hunting on the savannah, its light frame puts it at a major disadvantage when protecting its kill from the unwanted attentions of hyenas, leopards, and lions. Its mathematical calculation of energy expended against reward achieved has counted for naught in the daily race for survival.

The cuckoo, *Cuculus canorus*, steals the effort of nest building and chick rearing from its parasitised hosts and is incapable of rearing its own. (It's so much easier to put your eggs in more than one basket, especially when they are not even your baskets.)

A not insignificant number of children are raised and supported by their non-biological fathers. This can be viewed as a form of 'stealing', with DNA moved on a generation with minimal effort.

We have a paradox where the needs of society dictate the suspension of these base instincts for theft, deception, and envy, while at the same time society is organised around the base instincts of cooperation and mutual support underscored with the base instinct of 'every ape for itself'. These 'rules' for organised cooperative living are imposed on the individuals for the common good when what is common and what is good are mutually exclusive.

Society punishes theft and deception through laws and sanctions because a particular behaviour is a bad thing, while simultaneously rewarding the biggest thieves and deceivers who 'rule' society. The message appears to be that it is only a crime if you are a subordinate or you get caught. Does the manager of an NHS trust deserve a £350,000 annual salary? Does the head of an organised crime syndicate deserve his share of the 'tribute', when in effect his only crime is to have set up in opposition to the 'official organisation' which functions by collecting tribute?

Our society functions as a collection of uncooperative individuals intent on taking every advantage from their fellows. It suits us all to use our special dispensation card to ignore the rules while relying on others to observe them. We all actively parasitise our fellow neighbours as a matter of routine. There is no advantage to driving on the footpaths in Moscow if everybody does it. There is no advantage if everybody parks illegally and blocks all the roads. The advantage comes about because only some of the people do it for some of the time. Or rather, all of the people do it some of the time. You may never park on double yellows, or speed, or pay cash to save tax, or steal pencils, paper, or time from your employer. You may only ever buy duty-paid tobacco and alcohol and never drop litter, but still your intrinsic programming will seek out any advantage that presents itself to you. It could be as simple as reserving your sunlounger at 06.30 hours with your towel, or picking the table with the best balcony view in the restaurant, or simply choosing a restaurant.

Holy Bait Balls

Karl Marx is often paraphrased as writing, 'Religion is the opium of the masses.' Within his manuscript *Critique of Hegel's Philosophy of Right* he wrote, 'To call on them [the people] to give up their illusions of their condition is to call on them to give up a condition that requires illusions.'

Marx wrote this in 1843. In 1798 Novalis wrote, 'Their so-called religion works simply as an opiate—stimulating, numbing; quelling pain by means of weakness.'

While these sentiments have existed for over two hundred years at least, I will suggest that nothing new has emerged relative to our condition. While Marx focussed on religion, what he could have focussed on was the mechanism, the control, the implied need that religion satisfies. Religion is but the symptom, the reaction to a craving, a chemical control imbued within us that needs to be satisfied and that has become interchangeable with the new symptoms of consumerism and its supporting extracurricular work ethic.

In 1768 Voltaire wrote, 'If God did not exist, it would be necessary to invent him.' And so we can argue that this recognition of the human condition has been extant for 250 years at least. Voltaire was a God-fearing man and often argued in support of religion. He used this text to attack an atheistic article. Nevertheless, this statement was made and in the main has been taken out of Voltaire's context. It could be considered that what Voltaire was alluding to was that God is a good thing and if indeed he did not exist, then we would need to invent him. But if we are to accept the counterarguments to God, then we must also accept that this same process might have applied in the opposite direction. For God has been invented, and one can ask why. The simple answer is that God is merely an elephant conjured out of the ether to frighten, cajole, and exert unassailable control over lesser brains at the behest and benefit of Great Brains. But this control is exerted through us, a need, an affliction that needs to be satisfied and that finds a welcome within us. It is a product of our DNA, a device to ensure our compliance as a 'social animal' with all its errant complications, the master code that preloads our pro formas and programs.

If the question were to be reframed as to whether chimpanzees believe in God, then the answer must be a little different from either yes or no. Chimpanzees behave in accordance with the DNA master preloads and pro formas that define just what it is to be a chimpanzee. In that I can find no distinction between chimpanzee and human behaviour, as we are bound by our programs in equal measure. The answer must be framed in the wider context of need in the establishment of control over fellow troop members. An extended family group of chimpanzees would have no need for a god, whereas a city of chimpanzees would need one. The level of organisation required of chimpanzees to live in cities would allow sufficient extracurricular capacity for a god to exist. Equally, the command-and-control structures extant in extended family groups would fail in cities and new mechanisms would be required. If we were to extrapolate backwards in time to earlier versions of modern man living in extended family groupings and little more, then God would not have needed to exist. God therefore becomes an invention of cooperative living, a product of increasing population densities, and becomes a corruption of the control mechanisms beyond the strength of muscles and the later wielded weapons. But, as I pointed out before, we live in an uncooperative society dependent on rules and laws and physical barriers to getting along with our neighbours (locks, cameras, fences, double yellow lines, etc.).

Once again we return to an event, a revelation in time, or rather the invention of a new control beyond the limitations of physical exertion: a self-policing, omnipotent control which is perhaps alluded to within the Bible as the 'forbidden fruit' and the awareness of our nakedness. From this event stems the birth of God, no, the invention of God. While I do not subscribe to the biblical story, it seems curious to me that these authors saw the need to identify a threshold, an entrance to a different plane of understanding, and to offer an explanation of its origins.

While scholars and philosophers will research and debate the origins of this tale and cross reference with other earlier and later works, translations, and interpretations, for the vast majority of individuals it simply remains one of the lies to children.

How then can we recognise truth? Is it an alien concept that lies outside the realms of our pro formas and is defined by our inability to escape our DNA? Truth can only be measured against our programme-and-control mechanism which will act upon us as surely as any prejudice. We are effectively condemned to accept racism, sexism, envy, and the standard range of emotions that we employ. Our default position is to seek out the advantage, to grasp the clean end of the stick, which clarifies our decision-making processes. But by definition we are incapable of making a balanced decision, because we all have a personal preset ambition. In its simplest form I will argue that this is why we speed: because the rules don't apply to us, only them. If speeding gives us an advantage, then it is there to be taken. If lying about our work effort gives us an advantage, then we will lie. If lying about our lifestyle gives us an advantage, then we will lie. But then we have to define 'lie', because if we cannot recognise the truth, how can we lie? We have to measure (my favourite term) the difference between ambition and action and recognise the difference between these two as corruption. In every endeavour that we undertake as a species, we are bound in our course of action by our DNA. We have no free will, no democracy, and no control beyond our presets. We are trapped within our bait ball, condemned to continue our preordained path to a history not of any rational choosing, and this includes the destruction of habitats and species that we dismiss so readily for being in our way. The only true measure is how clean our end of the stick is and how much of an advantage it provides. Wherever we grasp the stick, our ambition is to move away from the dirty end as far as our abilities allow. Is 'truth' simply a tropism, an unconscious reaction to an apparent choice which serves to maintain the bait ball?

Truth or Lie?

I have an English elm tree in my garden, *Ulmus minor*, which has survived the ravages of Dutch elm disease for many years. The original tree long ago 'died', although the rootstock has survived. The fungal disease is spread by beetles that bore into the tree trunks and carry the fungus *Ophiostoma novo-ulmi* into the wood, where it kills the above-ground wood. The beetles only tend to attack tree trunks when they have achieved a certain diameter of around five inches. My original tree survives by producing fresh stems from the roots which have spread out from the trunk. It has been so long that I no longer have any evidence of the original tree, which would have 'died' fifty or so years ago. What I have is a series of short-lived 'suckers' that are nearly identical to each other. While you would not call them identical twins, they are similar enough to consider as close siblings. They are of course clones of the original and all grow to the same plan, minus variations caused by the weather, access to light, and damage caused by animals and wind. By looking at this close 'family' of clones, you can readily understand the force of the plant's DNA at work. But if you stop and think about it, while this example is blindingly obvious, it always was obvious to anyone who cared to think about it. Guidebooks would always identify elm trees by their standard growth characteristics which separate the classic elm shape from other, non-elm trees. Oak trees look like oak trees, Christmas trees look like Christmas trees, and they share the same life characteristics such as the 'small tree', short-lived, cold-weather-hardy descriptions that we append. Elm trees grow to the elm tree master plan, and obviously so do elephants and tigers and humans.

Of course the human condition has often been debated, with arguments between inheritance and environment (nature or nurture), and the same applies with elm trees, where the presence or absence of water, minerals, sunlight, prevailing winds, browsing deer, and so forth will all shape the finished tree. Within the greater population of elm trees, the environment gets averaged and a standardised elm tree exists which can be readily identified and described in the guidebooks under 'typical characteristics'.

With humans we may search for the environmental or inherited conditions that produced serial killers or murderous despots or loving 'saints' (princesses), but the simple answer is that all are contained within the inherited characteristics of humans. One person's serial killer is another person's elite soldier, and the desire for princesses (boredom relievers?) is also imbued in our imaginative needs. Our guidebooks on the human condition can be readily corrupted by the introduction of unrealised needs and dreams brought into focus by those intent on our subjugation. We could consider the training of elite military forces or indeed the training of jihadists where the sole intent is to bend the minds and actions of selected individuals into sacrificial dreamers intent on ending either other's or their own lives for the benefit of their subjugators. We could blame some skinny wench shinning up a rope for the burning desire to part with money (time) in return for a tiny bottle of beaver anal gland secretion poultice of assorted alien scents. In parting with our time, we elevate the peddlers of such products into a higher orbit, an orbit that moves them away from the smelly end of the stick (although you would have thought that the product they supply would negate the need to escape the smelly end, given that they can disguise it instead). Each unequal exchange of resources determines who subjugates whom; individually we all win and lose at different times, and collectively the process shakes out to provide the strata for real winners and the subjugates. Consider the front line soldier armed to the teeth and contrast him (or her) with El Presidente surrounded by armed guards buried deep within a subterranean bunker. Consider the parish priest orating at baptisms, weddings, and funerals and contrast him (or her) with the bishop in his palace or the pope in the Vatican. There is a degree of fluidity. John F. Kennedy saw action in the Pacific theatre of war and could quite easily have missed out on his blue plaque if events had been very slightly different. (What would have happened if the sniper had missed?) The constant is not the individuals; the constant is the mechanism that controls through subjugation. In that sense, then (give or take a sex change or so), all humans are capable of becoming the queen and de facto colony leader. The queen is powerless without her army of soldiers to defend, her army of nursery workers to tend the eggs, and her army of foragers to collect resources. While soldier ants are condemned to their lives by the chemical secretions bathing the developing larvae, our society has had to develop an alternative methodology, a free-for-all (every ape for itself) constructing a pyramid of ranks subordinate to the rank above, with all desperate to escape the bottom layers by industriously seeking an advantage over their fellow strugglers. Quality of life (outside the NHS description) could be described as a measure of success in the free-for-all. Whereas within the NHS the description could be 'a failure to address the needs of the many with the needs of the self-inflicted'.

Every ant colony is at war with its neighbours, fighting for resources. The human pyramid is no different in that every human colony is similarly engaged in the fight for resources. Human society could be seen as a near infinite number of pyramids jockeying for position, with the pyramidal tips interacting with each other. The Great Brains and supporting lieutenants are elevated high enough to take a view above the smog. They use the knowledge gained in that position to maintain their view. But they do not use the view to change direction or avoid the bait ball, because they are still a product of it and all that it entails.

Roman roads provided a tool of subjugation beyond the ability to move troops rapidly from rest to conflict. They facilitated trade, creating a conduit for the movement of advantage from source to receiver, to add further layers to the pyramid base and to elevate the peak. Your worth was no longer fixed by local needs and supply but was augmented by wider demand and access. Now it became possible to produce to excess and provide the surplus to market. (Historically, salmon canners working in remote areas ate salmon and precious little else during the season because there was a glut of product.) Trade facilitates a relief to boredom and provides the ability to develop an advantage over your fellows. All local residents were effectively slaves to the market. I have highlighted your value as an unskilled labourer in a slave-owning environment. The value of labour is fixed at the lowest common denominator, in this case the cost of food, shelter, and slave stock, which leaves no residue beyond that for non-slave labourers. By trading it becomes possible to extend values by removing the lowest common denominator from the locality and 'exporting' the value to alternate locations where values can be enhanced. The opposite is true: if goods are produced cheaply elsewhere by slaves, then they will undercut the local produce and have an impact on the local labour values. So it is that today that there is no value for the labour of British hosiery manufacturers when compared to the reduced expectations of foreign labourers. Rather, it is not that there is no value, simply a reduced value when compared with alternate sources. Hinckley and the surrounding district was a hosiery region filled with multiple factories employing large numbers of labourers. The collapse of coal mining and shoe production are other well-documented examples of corrections in the market. Nowadays we are not confined to roads with major shipping routes in the air and at sea spreading commerce to every corner of the map. I nearly typed 'globe'; good luck looking for a corner. Nor are we confined to physical trade with virtual trade of services and products via the electronic universe.

One could consider early trade as a simple exercise of swapping resources. Swap your eels from your marsh for wheat from dry land. Trade was mutually beneficial to both parties and a balance was achieved. Nowadays trade is beneficial for the traders and the balance is harder to find.

We have a propensity to exaggerate; it is within our personality. We strive to elevate the mundane to generate interest. We dramatise and essentially lie to collect attention. We have a need, an instinct, to elaborate. It must be within our DNA. Why? What purpose does it serve? For many years apiarists and animal behaviourists have understood the honeybee and its wriggle dance. Bees that have discovered a rich food resource will return to their hive and communicate their find by 'dancing' the distance and direction to the new find to fellow hive members. The strength of the bee's dance in terms of its vigour and duration motivates other bees to follow the directions. If two bees are 'dancing' simultaneously, then the more vigorous dancer tends to win and the prime food source is exploited. It is within the ability of individual bees to determine the 'value' of resource finds and to prioritise their individual knowledge in relation to other hive members. Would it serve the hive if individuals exaggerated the importance of their finds over other, more viable ones? (I will point out that bees were not the original design.)

There is no advantage, personal or otherwise, for individual bees to lie. Is this because bees are, and 'know' they are, a social animal?

For bees to lie, there has to be an advantage, an edge, a reason for this behaviour. If the hive is disadvantaged significantly, then its survival could be compromised and the whole colony put at risk. Adopting the maxim 'survival of the fittest', it becomes apparent that any errant bee behaviour would be detrimental to this principle and would effectively be 'bred out' of the species. In fact, given enough errant behaviour and enough time, bees might well have moved on to be cees.

Contrast this bee behaviour with that of humans, another great social animal.

Social requirements demand that cooperation is required for the success of the species and that lying or cheating would be detrimental, unless, of course, you accept that we are not truly social but opportunistically cooperative, or should that be cooperatively opportunistic? It is accepted that it is every ape for itself within a framework of mutually beneficial pseudo-cooperation. Given the opportunity, we would 'lay our eggs in another bird's nest', enslave other humans to labour for us, and steal from and dispossess neighbours and strangers alike. It can be as simple as waiting to the last possible moment to push in to lines of waiting traffic where the number of lanes in the road is reduced. You have stolen time from fellow road users and you have gained an advantage. It's as simple as arriving late at a music concert and pushing your way to the front, dispossessing the earliest to arrive. We conjure up rules of engagement, politeness, and etiquette, which are soon diluted in the presence of strangers. Many cities employ taxi wardens to fairly allocate taxi provision to waiting drinkers after a night out. The alternative is multiple fights as some seek advantage over others. But if you defer and yield to every waiting body, then the line becomes eternal and your time evaporates into others' pockets. There are simply too many people in close proximity for 'society' to exist in full cooperation.

Deception and lying is an industry in its own right. Carefully controlled and scripted to avoid 'untruths' or false claims, we have the marketing and advertising industry which lures and entraps customers into the carefully spun web of desire. Why else would you throw your perfectly serviceable mobile phone away to replace it with the latest model? How often have you been drawn in by the illusion of quality and taste imbedded in the packaging only to be disappointed by the sorry excuse of an 'edible' foodstuff which seems to have been delivered without its make-up on? How often have you busted a gut to buy a supercar only to have to drive on the same congested roads as all us lesser mortals and observe the same speed limits? (Do you think?) And while the Advertising Standards Authority ensures that no lies are told, (just ask eight out of ten cats if they know what amount 'twice as much as you may need' means), the real lie, the monumental deception, is the creation of the demand in the first place.

Why do we lie as much as we do?

Why do we so readily accept these lies?

The advantage to lying is evident for advertisers who quite simply sell more product by creating more demand. But where is the advantage to the customers who have accepted the lies and parted with resources? It's nearly a joke, a cliché at least, to remark on how children often get nearly as much pleasure from playing with the box as with the toys that it contained. Is the answer as simple as a 'trained mind' educated in the pro forma? If we have learnt the pro forma, can we unlearn it? My quandary lies in the apparent fact that lying serves no purpose. It has to have a biological advantage to persist, a fitness test within the 'survival of'; otherwise it would long ago have been lost to selection. We learnt about Aesop's 'The Boy Who Cried Wolf' written over twenty-five hundred years ago wherein a bored shepherd boy lies to attract attention from his fellow villagers. When a wolf finally appears, the boy's pleas for help are ignored and the sheep are lost. One lesson learnt here is that liars are disadvantaged, because when they do tell the truth, nobody believes them. From a biological perspective I will suggest that as a consequence of the boy's lies, he loses his livelihood and resources and subsequently dies of starvation. The shortened version would be that the boy was also eaten by the wolf. Another potential outcome is that the villagers became trained in rapid response to wolf attacks and successfully drove off this and future wolf attacks, thereby securing the pastoral future of the entire community. The problem here is that the fable is written to illustrate an envisaged outcome rather than to illustrate a biological model for survival. The next time you are exposed to a fire alarm test with all its incumbent prewarnings of its implementation, consider whether it will make you more or less likely to respond to a real fire alarm. Or will you sit tight and complain about 'that infernal noise' and 'when will someone switch that bloody thing off, whatever it is'? We can contrast this with the dire warnings issued by the Remain campaign as to the consequences of Brexit and its actual effects on the electorate. Win or lose, an inquest will surely follow and, wait for it, lessons will be learned. But 'lessons will be learned' is a lie, so what lesson are we learning, how to replace one set of lies with another set of lies?

There is another industry of lying that can be termed loosely as entertainment. We celebrate the greatest, most believable liars who don someone else's clothes, speak someone else's words, and perform someone else's actions. We give them awards; some get blue plaques, others Oscars or Tonys or BAFTAs, and we reward them profusely with our time and money and adulation. Do we all celebrate the removal of boredom? Is Aesop's fable really a wider illustration of the human condition, boredom and its relief? Have we corrupted the desire for a different mineral or vitamin, or have we simply exploited the need to vary our environs lest we be stalked by a predator into a need to spend our money on the latest must-have? Yet we can spend our money on mineral and vitamin supplements (driven by an advertiser's impulse, no doubt), and with the absence of tigers from 93 per cent of their former range, most people can ignore the death threats from a lack of mobility (except that we do ignore the threat of morbid obesity and its lack of mobility). But we cannot break away from the control, the preload that is our DNA.

We can begin to think in terms of another great lie that we can regard as image. We foster and generate the illusion and projection of self through our postures and habits and the wearing of the emperor's old clothes, the ones that everyone can see (although his new ones would also project image, or something projected anyway). This returns us to the peacock's tail and the projection of physical abilities to advertise sexual virility and healthy attributes so desired by potential spouses. While our physical size and ability to grow a tail is determined by our genes, we readily accept the corruption of these determinants with artificial adornments of jewellery or modern-day animal skins or face paint, or go-faster stripes on cars. Was there a time when the wearing of olden-day animal skins conferred a value on potential mates, an illustration of physical prowess, the ability to skin a tiger before it skinned you? Shag me and I will keep you and your children safe. Males of the Masai Mara would prove their manhood by killing lions. If the movies are correct, Plains Indians would wrestle eagles for the privilege of wearing showy bonnets, and if that failed, they would 'wrestle' other Indians to dispossess them of their natural keratin-based bonnets. Having seen enough reconstruction drama-documentaries of serial killers, I believe that this trait for taking little mementos from their victims is still alive and well, although there is a corruption of the message perhaps: 'See my collection of rape and murder victim memorabilia and recognise my abilities as a caring, providing sexual partner.' This would be a corruption of the sexual imperative to breed and pass on our genes. Where would praying mantises be if it were the male of the species that decided to enjoy a convenient 'close-at-hand snack' during copulation? Even today, gang members will append a tattooed tear to their faces to advertise their murderous abilities. How can you condemn the corrupted actions of a serial killer without condemning the corrupted actions of a bling wearer or Ferrari driver?

If the magazine articles and self-help books are to be believed, then the secret to establishing a relationship with the female of the species is to make women laugh. Humour seems to have the ability to cut through physical prowess and bling and provides the nucleus for a gene-exchanging relationship (although this 'false' signal may explain the apparent willingness to indulge in extramarital exchanges of physical prowess genes). Are these all simply expressions of 'rule breaking', the rules that dictate that you will only breed with the dominant male, the rules that dictate your position in the extended family as subordinate feeder and shelter taker? This would suggest that the desire to lie and effectively cheat is as hardwired in to our species as the desire to establish and maintain hierarchy, a hierarchical structure as rigorously undermined as it is maintained. Is this the biology I seek, the mechanism of paradox whereby we so readily lie and simultaneously accept lies from others, an understanding that we are not within a bait ball while rigorously maintaining it? You can stay in the bait ball and protect me while I will hide in the bait ball without providing a hiding place for others. Another problem lies within the nature of our bait ball, because ours doesn't shrink into an ever tightening space; it expands into an all-encompassing tsunami of resource-demanding, habitat-destroying, life-extinguishing scourge. The other problem with tsunamis is that they damage twice, on the way in and on the way out. We are not immune from our actions.

Could it be that the dominant male lies as well? He can't always be bothered to mate and simply turns a blind eye to some discrete illicit philandering. His position as master is tenuous, and he is only ever as good as his last victory. There is no real future in being the best gunslinger in the world when your status is determined by the next best gunslinger who happened to have been the best; he just didn't know it yet. Kings would win their crown on the battlefield but secure it with reputation thereafter. There is no great future in fighting every week. Is it in the nature of subordinates to want to remain subordinate? There is no great future for subordinates if they also fight every week.

There is a larger question to be answered: do we never lie because we don't know what the truth is?

Truth is personal and open to the interpretation of every mind. Alternatively, truth is selectable, selected as 'fact' as it slips past the cognitive filters of the human brain, a brain so assailed by data that it has to function in abbreviated form. Occasionally your TV will display a frame counter which will count through the twenty-five frames per second. See how many numbers can your brain process. Will it be three or four? We know that twenty-four numbers will appear, and yet the brain can only process information at a human rate. The television only functions because our brain cannot see the blank screen between the images. By waving your hand between the eye and older screens, a strobe effect is evident that begins to expose the inadequacies of our mental processing. While the brain can be trained to operate at higher levels, just consider a Formula One racing driver steering, braking, gear changing, and accelerating within minute fractions of a second while holding an intelligent conversation with pit engineers, and resisting the multiple G-forces exerted on the body. For many, this capability is beyond scope. We cannot all be Formula One racing drivers, even if BMI allowed, and yet it manifestly lies within the human capability. Yet the racing track collects the errors of the brain and dumps them unceremoniously into the barrier or gravel trap, and the race organisers curtail the supreme effort into 120 minutes at maximum. Not all the incidents will be traceable to driver error and will be the result of errors from engineers working to margins and schedules that justify their salaries.

While I illustrate the capability of the human mind, it only takes a drive to the shops or school to see the other end of the spectrum, with vehicles cutting corners, unable to park within parallel lines, or ignoring potential hazards. 'It just appeared from nowhere, Officer.' Is there a place for truth within our filtered minds, stuffed full of lies to children? Would this be the truth of the workplace with targets fulfilled, the truth of the college exam grades, the truth of politicians, the truth of the socially cooperative species that we are? Scientists have a term, 'the half-life of facts', meaning that accepted facts, perhaps we should say models, are frequently displaced by new explanations, new models that fit the question more accurately, until they themselves are invalidated. Does this begin to define the Great Brains? Let us look for the half-life of Great Brains, the ones who so earnestly commissioned elephants, the ones who forgot to attach the stop button. Half-life refers to the radioactive decay of nuclear materials as they change atomic structure and function over time. What is the half-life of Jesus Christ? The Clovis point? FIFA? How about the half-life of tigers? The real question that should occupy the human mind is the half-life of humans. But none of these questions will have any import on the actions of our species. We could just as easily ask what the half-life of a bait ball is.

Do we accept lies because we are lazy (too busy with all the extracurricular activities)? Is it simpler than that in that truth requires investment and, with it, risk? Put yourself in the challenger's position for dominant male. You may or may not have reached your prime and must fight the dominant male, who may or may not have passed his prime. For the challenger, failure is not an option, with banishment or fatal injury a potential outcome. Likewise for the dominant male. The answer may lie in posturing, image, and bending the rules which can emanate from either party. Magnify this problem through allegiances and the politics of the battlefield. How many lords and men-at-arms can you field? How many will answer the call for loyalty? How many will arrive too late or sit on the sidelines before committing their forces? And you haven't even considered the role of tactics and fighting ability or the technology of 'secret weapons'. Think Agincourt, Crécy, Blitzkrieg, Rourke's Rift. ... How many would rather sit on their hands and shun the risk and reward in exchange for the status quo?

Aesop's fable about the boy and the wolf is worth revisiting. We learn that the boy wanted to relieve boredom by lying. We could also learn that the villages soon got bored of rushing to his aid and therefore ignored the real emergency. We can approach the story from either end and blame the wolf's predation on the boredom of the shepherd boy, or we could blame it on the boredom of the villagers. Boredom could be seen as an equally valid justification for either party's actions. Yet the moral, if that is the term to use, indicates that lies are wrong and can, quite literally, come back and bite you in the arse. But boredom is by its very nature a distortion, a corruption that breaks concentration and drives the initiative to be or do something else. Within Aesop's tale we have the reinforcement of judgement by the many (villagers) over the individual (boy), but also by the old over the young. Place this tale in the twenty-five-hundred-year-old syllabus and consider the 'lesson learned' in the establishment of the student's pro formas. If you accept that boredom is by its very nature a distortion, then boredom could also be described as a lie. The boy's boredom/lie threshold had been reached, and this was subsequently followed by the villager's boredom/lie threshold, so who is the guilty party?

'Great. Thanks for the history lesson. But what's the relevance?' I hear you ask.

What if the bored shepherd boy had gone to school instead and had his mind filled with an education and stuffed to the hair follicles with pro formas? What if that same shepherd boy had discarded his frock and flock to don pinstripes and gain employment on the trading floor? What if this shepherd boy was now guarding your flock of pension monies against the predations of two-legged wolves? Fundamentally the boy is the same; he hasn't developed beyond the confines of his DNA, and he is still a mammal wearing clothes.

Now, however, his lies have gained momentum. His lies have all the resources of economists and politicians and government workers and managers and journalists to back him up and reinforce his position. He is charged with fattening his stock entrusted to him, to find fresh forage, to secure shelter, to provide security, and to nurture the newborn, and his 'livelihood' depends on it. And so does your pension livelihood. Have you checked your portfolio lately?

All the resources of his economists and politicians and government workers and managers and journalists could be viewed as the villagers whose job it was to help in the event of a wolf attack. Which ones will get bored first? We know that if the wolf attacks and kills the boy and his flock, the villagers will place the blame squarely on his shoulders even though they have collectively failed to respond as well. Where then does the corruption lie?

In politics we have a simple answer: blame the other party, the party in power, when the wheels fell off the cart. In business we have a simple answer: blame the CEO and appoint a new one. In government we have a simple answer: blame it on forces outside of your control. In management we have a simple answer: sack them and recruit some more from the same school of learning. All can be summed up as failures of leadership. The problem is, we have no leaders.

Where is the biology, where is the skill base, beyond the biology of the shepherd boy? However elaborate, however many layers of complexity, however much we try to distance ourselves from our origins, however much we disguise the allure of the peacock's tail, we cannot deny our biology.

Decisions become communal. If any one of the villagers had bothered to run up the hill, then the shepherd and sheep would still be alive. So the decision to ignore became infectious, the failure became communal, and the blame was transposed to the victim. Problem sorted. Not.

But in place of 'infectious', let me substitute 'reinforced'. A failure to act by one individual reinforced the decisions not to act taken by other individuals. Conversely, a single individual reacting would have 'driven' the actions of others. It seems that there was more than one flock of sheep and more than one errant shepherd. What a complicated tale Aesop wove, if we would but interpret it.

And you worry about the 'Wolf of Wall Street' when you should really worry about the shepherds.

A potential model from nature would be of a flock of geese grazing on a grass field. All it takes is one bird to react to a potential threat and to launch itself from the ground for the rest to follow, as a ripple effect overtakes the decision process of the entire flock. I would argue that the flock of perhaps 1,000 birds effectively has only one brain and 999 tropismic reactions. There is a reason for this situation, and it is quite straightforward. If you were to place yourself in a crowded situation, say, a packed football stand when an incident took place, only the individuals in very close proximity to the event would have any idea of what the issue was. If their decision is to flee the scene, then all will similarly react. We cannot see through 1,000 bodies to the source of the problem, and neither can the geese. 'React first and think later' is the mantra.

It is easy to recognise the biological advantage of such a survival strategy. Unless of course the reaction is to hide among yourselves in a bait ball. How then do we differentiate between flight and bait ball?

The simple answer is, we don't.

When faced with decisions within a communal environment, our stock reaction is to defer to others. Our problem stems from our ability to defer to others who have themselves deferred. This is akin to the blind leading the blind or, if you'd prefer, the unthinking leading the unthinking. Decisions, then, could be made on the ant principle; the more ants travel a trail, the more the chemical scent is reinforced and the more ants will travel it. Our problem can arise when the reinforcement is misplaced. Let me start with extreme examples, the Nazi Holocaust, First World War troops walking in to machine guns at the Battle of the Somme, and the Hillsborough disaster, to name a few. While it is possible to single out the soldiers at the Somme on the basis of training specific to the battlefield, the other two examples are more spontaneous in nature. I am not suggesting that any individuals had a 'choice' in the ensuing proceedings (I would suggest that they were all involuntary victims of the circumstances), but I am suggesting that the events happened with very little resistance to them. If we were to consider these events with the benefit of that most elusive of skills before an event, hindsight, then the three events would or could have transpired very differently. The resources demanded of the Germans by the noncompliance of Europe-wide Jews may well have altered the course of history. Widespread bloodshed of Jews within their local domain may have altered the attitude of the local populace and brokered more resistance. British generals may well have offered different instructions to the troops if they had but known how ineffective the artillery barrage had been. Football fans may not have been so desperate to force their way into crowded terraces, and police may have been more rigorous in their control, had they all known the consequences of their actions. In all three examples the resultant carnage could have been anticipated, and in all three examples the courses of action could have been different.

Equally, the shepherd boy and his sheep did not have to die. The fifty-car pile-up in fog on the motorway did not have to occur. The blind pursuit of CO_2 reductions in diesel engines at the expense of NOx and thousands of premature deaths due to pollution did not need to occur.

If hindsight were to be replaced with prescience or precognition, how would our history books be different? An impossible task to imagine this, and yet the actual answers were easily contained within the range of possible outcomes and were easily predictable if the possibilities had not been ignored.

Two cowboys were leaning on a fence when one said, 'My Quarter Horse has got a dose of colic.'

'Aye, I gave mine a dose of axle grease when mine had colic,' came the reply.

A week later the first cowboy said, 'I gave my horse axle grease like you said, and it died.'

'Aye, so did mine,' came the reply.

It still leaves us with the problem that we lie.

It still leaves us with the problem that we accept the lies.

The goose has the answer. A single goose grazing in the field has to guard itself and scans the field and skies for dangers. It reacts to threats and aims to protect itself. Geese, on the other hand, spend time watching for predators but spend more of their time watching other geese. After all, these other geese are in the field of vision and block the line of sight. Cheese wells now slip into focus. Tick box mentality springs readily to mind. The human condition converts to that of the herd.

Lie or Die

I took a guided tour of Rockingham Castle and learnt that one of its claims to fame is that it contains one of the earliest chimneys in Britain. Heating was originally provided by a 'bonfire' in the middle of the hall, and the king was a frequent visitor. The guide explained that the king got to sleep near to the fire and warmth, while subordinates slept further (and colder) away. They all slept on the floor.

I have stated before that you cannot bolt two Ford Escorts together and call it a limousine. Neither do two Ford Transits bolted together make a limo. Equally I will suggest that you can't bolt a Mercedes Benz to a Fiat Punto, a bicycle, a horse, and cart and call it an economic union.

Benjamin Franklin: 'Do you love life? Then do not squander time, for that's the stuff life is made of.'

It becomes a requirement of herd living to both lie and accept lies. The herd is leaving in the direction that I must follow to a better life. The herd is moving to better pasture; I must follow. The herd is fleeing from danger; I must follow.

The goose that set the flight reflex in motion did so in response to a perceived threat, let us say the appearance of a fox. Neighbouring birds also responded, thereby triggering the entire flock, which has not and cannot see the threat given the density of birds in the way.

Let us suppose that the first goose, immediately after reacting and triggering the other birds, realised that the perceived threat was actually a harmless rabbit. Now the trigger bird has 'lied' about a threat. Now the flock has responded to a 'lie' and has vacated its feeding ground. The trigger bird could now justify its position by reacting to the reinforcement of the secondary and tertiary responders. These birds were fleeing from the trigger bird stimulus, but the trigger bird may now believe that they responded to a stimulus different from the original harmless rabbit trigger event. In any event, the trigger has been activated and the birds are aloft.

The trigger bird, in theory, could now apologise for the error—'Sorry, chaps. Thought it was a fox, but it was actually a bunny'—and expose itself to the other birds as a trigger-happy idiot who cannot be trusted. (Who would entrust their sheep to this trigger-happy 'liar'?)

Alternatively, the trigger bird could suggest that it had simply responded to other birds in close proximity, 'What, me? No, it wasn't me. It was you, wasn't it?'

(Who would entrust their sheep to this liar who cannot be trusted?)

Alternatively, the entire flock could breathe a sigh of relief for having escaped from imminent death (having universally accepted the lie).

Dr Doolittle's absence leaves us in the dark as to the actual 'conversations' that took place in the postflight debriefing.

My own very limited experience with witness statements given at the coroners' court confirms that humans are unable to clearly define events when witnessed from different perspectives and according to a different timeline. In this instance, the coroner is Dr Doolittle and the inquest is the 'postflight debriefing'. Another issue with the traffic accident is that people get in the way of a clear view of the event. But when I say people, I don't mean bodies obscuring the view; I mean the human reactions of shock, disbelief, prejudice, and interpretation, and the need to justify the visual events into some form of understanding. In flock terms I would suggest that a number of birds would have also 'seen' the fox, the one with the fluffy white tail. Reaction can be seen as the tropismic action that saves lives from perceived threat. Thought is secondary to the process, as thought is too time-consuming and potentially life limiting. Tropismic action has no need for truth or lie. The secondary thought process can have the inquest after the event, but the event also creates the thought process. 'Phew, that was close,' or 'WTF? What's that idiot doing now?' Both are valid, and therefore both are true. The real truth, the only one that really matters from a biological perspective, is, 'Am I still alive?' But, at the risk of tying you up in knots, there is no truth or lie in being dead. Death has no memory, no truth, no lie. The question then is, 'Am I dead?' But if the question is asked, it is already answered. Biology has no need for truth or lies, only life; death is irrelevant, it has ceased to be.

Where does that leave the human herd?

There is no relevance to truth or lies. It must do what it needs to do to survive. Simple.

But while we consider the herd to be at risk from predation, disease, starvation, and drought, which are all external factors, we must also consider that the herd is at risk from itself. The biggest factor in survival is competition, the need to outrun the predator, to carry the genes for disease resistance, to find sufficient food and water, and to breed and pass on our genes. But the biggest factor in competition is our fellow species. Cast your mind back to the Tierra del Fuegians wrestling in the snow in Darwin's day only to be wiped out by 'Western diseases'. Cast your mind back to the Indian tribes of North America dispossessed of their lifestyles, habitat, and lives by European invasion.

But we do not need to cast our minds back; we need to look to the future. The great herd of wildebeest will run out of grass because they will have eaten it all, and that is why they must move on. It should be noted that the great herd is more than wildebeest, with Thompson's gazelles and zebras also numbering into the hundreds of thousands, but each species exploits a different type of grass and has a different feeding technique, and competition between the species is minimal. It could be argued that each species has reached its population limit for resources. Each square kilometre of grass can support x wildebeest, y gazelle, and z zebra. If you remove x wildebeest, it will not increase the number of y or z because they are not limited by the number of x; they exploit a different niche. What limits the population of wildebeest is the wildebeest. (This deliberately ignores the impact of the different grazing regimes. A needs B to remove coarse grass to allow A to graze, who in turn leave short grass for C to graze. It is no accident that the three species form the mega-herd.)

Biology has created and honed the wildebeest and has made it what it is. Biology's toolkit consists of trial and success (error is failure and is no more), and individual tools are disease, fecundity, predation, instinct, physicality, learning, and tropisms. But each individual wildebeest has but one ambition, to be the best wildebeest it can be, and that means being better than every other wildebeest. If you can be half a metre faster than your neighbour, see a metre farther in the dark, hide your injury in the herd, father the most calves, and follow the herd through hunger-sated crocodile-infested rivers then you have an enhanced chance in the lottery of life. But these actions condemn other wildebeest to death. Let your slower neighbour be caught, let your poorer-sighted neighbour be ambushed, heal your wounds, and recover while others form the daily menu. Drive others from your mating ground and let others die before you cross water. None of these things is conscious or deliberate, but all of them are driven by the urge for survival engineered within the DNA.

Where is the immunity from this process for *Homo sapiens*?

It doesn't exist beyond those realms that are conscious or deliberate.

But this leaves us with the conscious and the deliberate.

And that is where the danger lies.

Let me consider our human herd. Let me place it in the Serengeti and expose it to the same rigours the wildebeest is subject to.

Beyond the same constraints that the wildebeest face, we will be subject to a new dimension, one of planning. Our herd is quite capable of planning for the demise of part of it so that others can flourish. We can devise a guard to secure the protection of the weak and vulnerable. We can devise a 'hospital' for the treatment of the sick and injured. We can commission scouts to explore and seek out fresh resources.

But when the results come back with a shortage of resources, of an increase in injury or disease, of insufficient guards to protect against the forces railed against us, then the actions for survival are triggered and the sacrifices made.

Of course we wouldn't sanction any of these actions. We will accept no rationing of care in our hospitals, no quarantining of Ebola victims, no waging wars for resources, no sealing of our bulkheads on Spaceship Earth. While these extremes all result in premature death, we do not need to think too long to find examples where the planning leads to privilege. The Great Brains will never miss out on a summit lunch, a booked place in the nuclear bunker, a reservation at the private clinic, an ability to fund a private education. Position secures privilege, which enhances one's chances to maintain life beyond others. The whole is built on lies and the acceptance of lies, and appears to be the advantage proffered by intelligence, an intelligence that strives to outcompete the human race in pursuit of the survival of the human race, our greatest competitor, an intelligence that is sharper than the sharpest claw, that is stronger than the strongest paw, and that is directed against our greatest competitor. If you think that tigers have had it hard, wait until it's our turn. Except, of course, it already is, our turn. Just £3 a month, remember; that's all it takes. Fancy a T-shirt from Bangladesh? How about a fuel-efficient diseasel car?

But in truth we accept that all of the problems facing mankind are beyond our scope. We accept that people will die from want and hunger and disease. We accept that others will die to provide for us. Is this the lesson learnt from Jesus, who died so that others could live? If we can sacrifice the Son of a God, then sacrificing other mortals holds no fear. But is it in truth, or is it in lies, that we accept that others will die for us? Is there a difference? Can we tell?

We construct rules of engagement, politeness and etiquette, to ease our paths through life, yet we book the services of taxi marshals and traffic wardens to reinforce the reality of our social non-cooperation.

If the central message of Christianity is that dying for others is to be celebrated …?

Which is the more honest, the wildebeest herd or the human herd?

Within the hyena clan, status is hereditary and privilege is granted at birth. Hyena are not the only species to adopt this stratum. There is a model for the human condition. Other species achieve status by physical attributes, strength, or fighting ability, by the size of a crab claw, the resplendence of plumage, the deepness of a roar, but this is all attributable to DNA. Humans can achieve dominance through the sharpness of their minds, the strength of their arms, the sharpness of their weapons, and the strength of their lies, and these can all be attributed to our DNA. Our DNA determines our intelligence and physical attributes and much more besides; it determines who and what we are. We cannot escape biology. We can deny it by substitution with binary code at our peril.

What value is ascribed to society?

Newspaper and television journalists are keen to debate endlessly the various topics that are presented to them and that they present to the public, or should I say customer? What strikes me is the definitive lack of what I would call the 'killer' point or argument. This is what I would define as the 'answer' that brokers no further debate or changes the whole nature of the discussion to render it obsolete. The conclusion I have arrived at is the industry's desire to endlessly debate without ever solving anything. Debates enable them to endlessly revisit the well and justify their salary and existence for the day, week, and career. It is a form of recycling, and it is lazy and stupefying. It is also evident that many of the presenters are 'outside' the debate and do not understand either the question or the answers proffered. The conclusion one can reach is that the debate is simply a device to fill the time slot. I hate football. I loathe and detest this corrupted 'sport' of prima donnas collapsing with a broken fingernail and performing triple salchows in the penalty box. I have had no interest in it for many, many years, and do not know the rules. That's why I would welcome the opportunity to umpire the World Cup Final. I would even find my own way to Salisbury Plain. You football fans will no doubt be horrified at this prospect, and yet you will happily sit and follow the attentions of similar disinterested amateurs umpiring other contests.

What function do these debates serve, beyond filling the gaps between the adverts? Are they simply a formularised social lie? If we revisit the goose, you know, the one with the trigger, what debate could we indulge in?

'Sorry, chaps. I thought I saw a fox, but it was actually a bunny.'

'No, no, no, I saw it as well.'

'A fox? I saw a stoat. Just caught the movement out of the corner of my eye.'

'A stoat? I must have thought the white belly was a rabbit's tail.'

'Good thing you reacted when you did. I'd just put my head down grazing.'

'The grass was sweet. It's a shame we had to leave it.'

'Yes, the grass was sweet. I didn't like those bits of white plastic things though. They kept moving in the wind.'

'Er, can stoats hurt us?'

'Ah well, as least we dodged a bullet and lived to graze another day.'

'Bullet? Was someone shooting at us? I'm not going back there again.'

'Forgive the indulgence, but oops, sorry, this just in from our reporter in the field.'

'Following on from earlier reports of disturbance in the lakeside meadow, I can confirm that an incident took place and that feeding was interrupted. Foxes and stoats have been known to frequent this field, and confirmed eyewitness reports have been received.'

Command and Control Limits

I have defined a leader. The leader on a Christmas tree is easily identifiable as the apical bud, which uses chemicals to suppress growth below while demanding resources from below to elevate its position as it migrates towards the light. Secondary, tertiary, etc. buds gain from the leader's race for resources, even though it puts them in the shade.

In human terms, a leader is the individual who has determined the best route to gain an advantage over others. High-rank subordinates will follow and support these aims to achieve their share of the resources. Lower-rank subordinates will do likewise, until the momentum is irresistible and the leftovers scarce. Secondary subordinates accept their position but are ever ready to accept the mantle should the leader fail. This applies down the line. A 'good' leader is one who allows the benefit to trickle down enough to gain support, which in turn safeguards his or her position.

It strikes me that there is a formula for a tree, I would even bet that somebody has done the work and has the mathematical formula. My premise is that a tree fits, not unlike the teeth in your mouth, buck teeth excepted. Let me explain. A tree grows from its seed and, given a healthy environment, prospers and develops to its full potential. The potential is determined by its DNA and the environment, but as I have just said, the environment it has exploited is favourable. This, then, leaves the DNA to weave its magic. In broad terms the tree can be divided into roots, trunk, branches, twigs, and leaves, and these are all in equilibrium. Too many leaves can 'sweat' (transpire) more water than the roots can provide. The weight of all the leaves and twigs must not exceed the load-carrying capacity of the supporting branch. The weight of all the branches must not exceed the load-carrying capacity of the trunk. The roots must provide sufficient resistance to overturning to prevent the tree from falling over. I think you get the picture.

Above and beyond the limits, we must have 'factors of safety', that is to say, spare capacity that can handle water, wind, and weight variations within a range of extremes. A wet leaf weighs more than a dry one, and rain, ice, or snow could break the supporting twig, branch, bough, etc. Wind can also increase from a gentle breeze to storm force, and yet the tree, by and large, can still survive. Now I realise that trees also get it 'wrong' and will drop branches and leaves and even be uprooted in extreme conditions. But I can also suggest that trees generally 'get it right', as is evidenced by their longevity. There will be a mathematical relationship between leaves, twigs, branches, boughs, trunk, and roots that provides for the life requirements of trees.

In the absence of wooden calculators or circuit boards (no, you can't have an abacus), it is reasonable to assume that these parameters are firmly established in the blueprint for the species and are maintained by internal control mechanisms, such as growth hormones generated by the apical bud. The physical strength characteristics of different species is also evident; you only have to consider the softness of pine as compared to the hardness of oak.

I will suggest that a finite limit exists for all trees beyond the artificial one induced by wielded chainsaws. A single canopy exceeding a hectare in size (on plan, not surface area of leaves) or a trunk 500 metres long has yet to exist because it is beyond the limit of the organism and the matter it is composed of.

Another interpretation is that a tree is a tree because that is what it has evolved to be: a perfect oak tree, a perfect willow, a perfect pine. But each perfect tree is in tune with its command-and-control structure, which determines what a perfect tree is.

But as ever I am not actually thinking about trees. I am thinking about humans, perfect humans with their perfect command-and-control centres which determine what a perfect human is. Have these perfect command-and-control centres evolved to extended family groups? Have these perfect command-and-control centres evolved to clans of extended family groups? Have these perfect command-and-control centres evolved to envelope 1.25 billion humans in the Republic of India? Or 1.4 billion Chinese? Have these perfect command-and-control centres evolved to include queues by city centre taxi ranks? Or 30 mph signs?

An Experiment in Thought

I went for a walk today down by the lakeside meadow, looking for *Turdus* altars, when I came across a peculiar 'pet carrier'. I say that because I couldn't find any air holes, although I did find a note.

Dear Great-Granddad,

My dad told me how troubled you were when you lost your cat. When I saw this grey cat with a damaged white stump of a tail at the rescue centre, I just knew I had to get it for you. I hope you love it and cherish it as it deserves. I also hope you take better care of this one and don't misplace it somewhere.

Love,

Danielle

XXX

PS: I had it wrapped specially for you.

The address label was partially obscured, but I could read, 'To Mr E Schrodinger'.

I couldn't open the box, so I don't know whether the cat had escaped or not, but I then realised that this was the very meadow where the geese had fled from the fox/rabbit/stoat and I began to believe that it might have been a grey cat with a white tail after all. I don't suppose I will ever know the truth or even what the truth is.

Un-princess

As I mentioned previously, I hate football. I loathe and detest it. (I might not be the only one after Iceland 2, England 1.) This may be why I cannot engage with all the hyperbole—my preferred term is 'hyperbollocks'—that surrounds the 'beautiful game'. But I am also turned off by the hype surrounding Rugby Union, the Olympic Games, and for that matter *Britain's Got Talent* or *The X Factor*. My friend and I engaged in the debate surrounding the Brexit vote, and it struck me that the hyperbollocks between football and politics knows no difference. My friend provided me with his view on the football, stating, 'It's about entertainment, professionalism, and stories for the press,' and with that the penny dropped. The purpose of all the debates on the news and in the papers is to provide entertainment, professionalism, and stories for the press. Nowhere is a solution sought or even desired. Any 'killer argument' that emerges as a showstopper is stifled at birth and never reaches the public for fear of ending the process. But let me be clear, the process is to provide entertainment and void-filling media and not a solution. Television adverts for animated computer games come with a warning along the lines of, 'Not actual game footage'. The media should append a similar warning on its products, along the lines of, 'For entertainment purposes only' and 'Not to be taken seriously'.

Debate is the opium of the masses. It stupefies, numbs, and disables while leaving a residue of dopamine-induced pleasure at achieving what, exactly?

Remember the term 'posture'. Posture is the term for pretence, the deliberate attempt to disguise real intent, the act of being seen to be doing something, secure in the knowledge that it isn't possible to achieve it. Just because they are beating their chests doesn't mean they are going to do anything. Hoverflies have yellow and black stripes to warn predators of their non-existent stings. Deception is rife within biology, and humans are masters of it. Nothing can match our ability to lie or indeed our ability to accept lies.

Another term for posture is politics. The art of deception, never declaring before the result is expected. Always protecting your own position until the result is known and swimming with the tide.

Opinion is a curious thing because it is bounded in herd mentality. As a supposedly 'social' animal, our actions are designed to achieve compliance and agreement. To be a herd requires the subjugation of the individual, the acceptance of common trends and traits. Alternatively it just requires the suspension of the individual and the adoption of a subliminal mental state where the thinking has been done for you by other herd members who have adopted the same mental state. So who is doing the thinking? Answers on a postcard please

Get outside the argument and debate.

Ask not what but why it has been said.

Look for the pay cheques.

Look for the profit.

Look for the bollocks.

Do not ever expect to find the truth.

Above all look for the advantage.

Advantage is always the key and manifests itself in different ways.

Advantage gives you the best food.

Advantage gives you the best security.

Advantage gives you the best sex.

Advantage gives you the best chance to pass your genes on.

Advantage gives you the best chance in life, the only thing that matters.

But now that we are out of the trees, advantage is disguised by all the paraphernalia of modern man. Advantage no longer gives you the best feeding spot safe in the middle of the colony with the best females allied to your side. It gives you a seat at the finest restaurants, safely secure in your job with a WAG (or HAB; I'm not sexist) appended to your arm. And you will subconsciously and consciously seek out this advantage at every turn.

Remove the facade of earnest importance from the debates and what are you left with? An opportunistically cooperative clothed ape seeking to stay alive at another's expense.

We learnt a long time ago to exploit other species for our own gain. Use the other animals around to warn of danger, chase off smaller species to secure their food, hunt and kill them for food, domesticate and breed them.

And the real killer is, the best species to exploit is, the one species always in range to exploit is, the one species we are absolute experts at exploiting is, the one species we couldn't survive without exploiting is, *Homo sapiens*. What self-serving idiot termed us the 'wise' man?

Let us talk about self-serving.

Let us consider any organisation, from a one-man hot dog stand to the NHS. Every day every individual wakes, dresses, and leaves for work to gain an advantage. How to turn £20 worth of buns, sausages, onions, oil, and gas into a higher amount? How to turn the treatment of patients and their carers into a daily wage? But to be truthful—I know, there is no truth—the object is not the wage but what it provides. The money pays the mortgage/rent and puts food in the belly, clothes on the back, a duvet on the bed, fuel in the motor car. How do you nurture sausages and buns? You buy fresh, refrigerate, cook and keep warm, and provide sauces and napkins, and do it with a smile and grace. You want to provide a service and generate repeat business. We are back to the daily labour rate and the connection between labour and reward. Where a day's labour equals £100, it is easy to determine the value of your earnings and the decision on where to spend.

How do you nurture the NHS? The question is too big, so let me break it down to front line nurses working in intensive care and performing daily miracles with regard to saving and preserving life. The simple answer is, you care, you nurture, you provide treatment and anticipate need, you monitor and respond, and, win or lose, you know that your daily activity has had value, a worth. There are many, many branches of medicinal care where these ideals hold true: midwifery, oncology, end of life, surgery, trauma, and the list goes on. Now let me state unequivocally that these works as described are fundamental to the provision of healthcare and if it were me or mine in need, I would welcome every attention that the condition demanded. I would also welcome it for total strangers in the race for survival.

My problem, as ever, is in the disconnect between the 'sharp end' of effect and the organisation behind it. I will describe the nurses and doctors as princesses. My ambition is not to criticise or denigrate them in any way. (The truth is in the pudding, as the saying goes, because if I did criticise them, I would instantly be pilloried for attacking these caring angels. And that is why I will call them princesses, because they exist in a netherworld devoid of criticism.)

Let me place these princesses at one end of the spectrum of respect due, and consider the opposite end. Who deserves to be at the opposite spectral end? Some years ago reports from undercover reporters were printed exposing some of the work practices of hospital porters and the tricks that they employed to justify their wages. One piece of advice was 'Just tuck a pack of paper towels under your arm and you can wander the corridors for hours without comment.' Now I have a problem in nomenclature, because I cannot think of the opposite of a princess. I am tempted to say 'princess not' or perhaps 'unprincess'. Now I have deliberately set these unprincesses and princesses at opposite ends of the spectrum because, by definition, there will be a whole range of respect due between these polar extremes. Figures quoted by the BBC indicate a total workforce for the NHS in England, Wales, Scotland, and Northern Ireland of 1,700,000 in March 2012, making it the fifth-largest employer in the world (or fourth if you discount the majority of McDonalds staff as franchise workers).

Let me introduce a scale for my spectral range of unprincess to princess and assume a median point where individual employees are neutral, which is to say that they display a fifty-fifty split between being 'angels' and wastes of space. I will not attempt to split the workforce on the basis of job description; after all, there will exist excellent hospital porters and equally crap nurses. The distinction is not one based on class.

It is tempting to place the median exactly in the centre of the workforce numbers and to suggest that there are 850,000 unprincesses and 850,000 princesses, but this at first sight would be unfair. This state of affairs ignores the pressures exerted on the workforce by management, training, and work practices, which should all be driving the median towards princesses and away from unprincesses. I say 'should' because we have another problem, namely the fact that the driving forces are also subject to the same spectral range of good and bad. A simple example would be the exposed hospital porter who was 'training' the undercover reporter in the practice of wandering around hospital corridors with a pack of paper towels under his arm. This waste-of-space porter was therefore the manager of the 'trainee reporter'. This same hospital porter will himself have been trained by management and has either slipped through the net of responsibility or has been 'educated' to the 'waste of space' end by other similarly minded porters or managers. Way back at the beginning of *How to Kill an Elephant*, I highlighted the work practices of electricity board managers who lined their own pockets by filling office space with underused staff to justify their salary scales.

Why should the NHS be any different? After all, the bigger the herd, the more places to hide exist.

Where then do we place the median? There is of course no answer. What we can say with some certainty is that it is not all princesses.

I can argue that all of the different jobs across the entire range of the NHS have a place in the provision of healthcare for the people of the UK, from the humblest potato peeler to the most 'cutting edge' surgeon (sorry). I can also argue that all of the different jobs across the entire range of the NHS have a place in the provision of wealth and 'survival' of every employee. Against every employee, we now need to add a second sliding scale to run parallel to the princess–unprincess one described earlier. This scale is a little harder to describe, but I will attempt to clarify my ideas.

This scale looks at the exploitation of others for the benefit of the individual, just like a meerkat pinching the choicest grub or the dominant starling forcing others out of the roost to expose others to danger. While this is the biological explanation I am looking for, the effect is demonstrated by the 'paper towel porter' who has received resources (wages) in exchange for labour which he has not been provided. This porter decided to serve himself rather than others. He lied, he cheated, he stole the resources he had contracted to receive by failing to provide the labour he had contracted to supply. The porter had provided his time, assuming that his full contracted shift had been fulfilled. In an identical vein, all the former typing pool employees of Anglian Water had decided to 'work for themselves' at the expense of others. They had similarly surrendered their time in exchange for monetary reward. This second scale could have at one end a 'working for myself' label, and a 'working for others label' at the other end.

While I have suggested that the scales would be parallel, it soon becomes apparent that princesses would lie near the 'working for others' end and unprincesses would be aligned with the 'working for myself' end.

It should be obvious through the use of my nomenclature that I have tried to construct a carefully engineered attack on the NHS, or indeed any institution that is predominantly people-based. Local authorities and schools would also fit neatly into the remit, but of course the full range is only limited by your imagination.

Any 'attack' on these institutions instantly falls foul of the princesses' argument and the sanctity of the life-saving angels and serves to stifle any discussion. In a similar vein, any discussion about migration is instantly denigrated to a discussion about racists and racism, and the debate is brought to a halt.

Conversely the NHS contains valid 'villains' who can be vilified by the media without complaint. These tend to be the trust management and chief executives (and their excessive pay packets). But once again these individuals can have the same scales appended to them, although in the main the media and public alike have already decided that they definitely lean towards the unprincess/myself end of the spectrum. Fair enough. But while we perch our angel nurses upon their pedestals, ask yourself whether they are all deserving of this elevation or if indeed, by a simple law of averages, they are spread across the entire scale of good to bad. A simple acceptance of this idea then renders a proper examination into the efficacy of the staff, from nurse to porter, pharmacy to cleaner, doctor to receptionist, possible.

The simple answer, again based on the law of averages, would suggest that the average employee is 50 per cent princess and 50 per cent working for others. If you are a glass-half-empty proponent, then the NHS and others are 50 per cent unprincesses and 50 per cent working for themselves.

As previously stated, the normal convention would suggest that management and managers will be beavering away to shift the workforce from the midpoint to a new enhanced ratio of unprincesses–princesses or self–others, but of course the problem here is that managers are themselves subject to the same scales and can simultaneously engender good and bad practices. The net result could be regarded as neutral at best. All one needs to do is throw herd instincts into the equation, and the task of improving the workforce within the NHS and indeed any other large organisation becomes nigh on impossible.

What then is the conclusion?

Our species works predominately for its own benefit and accepts the lies about mutual cooperation with eagerness. The NHS is not a large cooperative of individuals working in harmony for the common good. It is a large uncooperative of individuals working for themselves where the common good is only one of the by-products. Do the rest of us recognise this and temper our individual behaviour to the reality on the ground? That is why we have (take a deep breath) security guards, threats of police action and prosecution, CCTV, signs advising that all members of staff deserve the right to work in a safe, conflict-free environment free from the threat of verbal or physical violence, threats of the removal of treatment sanctions, signs for points of contact for whistle-blowers to notify of NHS fraud, signed disclaimers for the right to receive free treatment, and ambulance crews demanding police attendance to visit certain areas. I think this will do for now. If this is what the public serve up to princesses, what hope for the rest of us? Princess status appears to be the de facto status for the NHS in the media and public opinion, and yet the reality on the ground is very different. To that end there are high levels of support for a doctor and nurse strike or industrial action on the grounds that they are all 'angels' without due thought to the realities.

WMDs: Weapons of Mass Destruction

The human population is increasing by 250,000 a day. The population density of the Netherlands is 500 per hectare. By extrapolation, every single day human land use increases by 5 square kilometres. Every single day this translates into 0.4 STUs. That is 146 STUs a year, and this ignores the breeding potential when the new generation reaches maturity. The world average is 56 per hectare, but of course this includes vast barely inhabitable areas such as the Sahara and Siberian Taiga. An example that aptly illustrates this is the population density of three found in Australia.

The average density has, broadly speaking, doubled in the past fifty years. I chose the Netherlands because the country is fairly comprehensively developed without any low-grade mountain ranges, but it cannot stand alone. By that I mean that Holland relies on the importation of food and goods to support the population; it needs its hinterland.

As population densities increase, we, as a species, intensify our activity, which leaves less fringe and fewer scraps for other species. We intensify our agricultural practices as well as extract greater resources in terms of energy and materials. The net result is a greater squeeze on our fellow passengers.

Every day sees another 250,000 people born, which removes a further 500 hectares on average for human consumption only. This is 5 square kilometres a day, or 35 square kilometres a week. I will repeat that not all land utilised would have been home to tigers; it could have been home to jaguars or bitterns or hedge sparrows or great auks (or indigenous tribes of humans going about their sustainable Stone Age activity).

WMDs have a place in the UK's decision to go to war, to effect regime change in Iraq. Their much vaunted presence on the ground in Iraq provided the framework for a US led coalition to conquer and occupy during the 2003 Second Iraq War.

I seem to remember at least one newspaper's front page declaring that the UK base at Akrotiri in Cyprus was at risk. Saddam Hussein's regime had guided missiles armed with chemical and biological weapons of mass destruction that could reach these bases at Akrotiri with just 45 minutes of flight time to alert of incoming danger. Journalists, whose job it is to report the truth, had a field day with their sensationalist headlines. The debate was carried out, and Britain went to war to oust Saddam alongside the Americans and others. The much overdue Chilcot report has finally delivered its damning verdict, showing the lack of veracity to support these claims of weapons of mass destruction. But what am I saying? WMDs were prevalent at every stage before, during, and after debate and were correctly leveraged to provide the expected answer. Our newspapers and other media embraced the WMDs with utmost enthusiasm, as doing so made it much easier for their editorial content to produce and capture the attention of the public. In that, it is fair to say that WMDs were pivotal in the influence and support generated for this misadventure.

In a similar vein, WMDs ignited the media debates for the Brexit referendum and were systematically utilised by both camps, for and against, to steer and persuade the argument as the audience was bombarded at every turn. It became nigh on impossible to differentiate between the various offerings. I think it fair to say that the WMDs carried the argument home, albeit in both directions.

I refer of course to witterings of mass delusion. British journalism is alive and well and full of such witterings. Where then does the responsibility for these for and against claims lie? Was is the government's 'spin' factory, or the spin industry of the media, an industry that inflates, exaggerates, misleads, and misinforms as a routine practice? There is no answer beyond the fact that either party could have corrected or queried these witterings, while either party was similarly served by it. One party used the 'misinterpretation' to go to war and kill people, while the other party used it to sell newspapers and airtime and to generate advertising revenue. Once again I will remind you that ours is the species with the greatest ability to lie and the greatest ability to believe lies. Is that why we call ourselves *Homo sapiens*?

Tribe

On the 7th of July 2016, Micah Xavier Johnson ambushed and killed five white police officers in Dallas, Texas. Micah was reported to be angry about police shootings of black men and wanted to kill white police officers. The incident occurred at the end of a protest against recent police killings of Alton Sterling in Baton Rouge, Louisiana and Philando Castile in Falcon Heights, Minnesota with social media screening mobile footage of these police actions to the world.

Passions have been inflamed. Lives have been taken. Statements have been issued. The media merry-go-round has begun. Slates have been wiped, scores settled, fresh slates written. Mass debate has begun.

Where do we place these killings within the context of social cooperation? Where do we place biology in the context? Where is the advantage? What will be achieved?

Tribal instincts appear to be at the fore and fault lines in society exposed. Yet tribal instincts have always been there. So have the fault lines. They have just been suppressed. This suppression is essential to the function of 'society' and yet the control is incomplete. Alternatively the suppression is buffered, which allows for a 'stretch' of the boundaries, given that tolerance is elastic. The need for social cooperation has to temper the need for the exploitation of advantage. Without cooperation and social interaction, the advantages of exploitation disappear. A solitary hermit cannot exploit non-existent neighbours and cannot hide in the herd. The hermit is exposed and disadvantaged and his or her survival is compromised. How then do we reconcile the needs of the individual with the needs of the many? How do we balance the needs of the individual with the needs of the many? Is it simpler than that in that needs are stratified on a hierarchical basis? No, needs are fulfilled on a hierarchical basis, and that creates the social strata.

On what will the debate and narrative focus?

The hierarchy is established by the Great Brains, serves the Great Brains, and forms the command-and-control structure of society. For society we should read herd, and for Great Brains we should read leaders. But the Dallas sniper has determined that the leaders are failing his needs and has constructed an alternative, an alternative that has been widely condemned for the impact that it could force on society. While the extreme actions have been condemned, the actions were taken as a subsidiary action to vocal demonstrations protesting the 'legal' actions of the hierarchical police control methods which caused the death of two black men. What we are witnessing is the reinforcement of the command-and-control mechanisms which are failing the lower echelons of the herd. The media is full of alarmist definitions of a fractured or divided society and is concerned about how the divisions will be healed. Yet by definition society is divided, is fractured, is hierarchical, and has command-and-control structures established by the 'leaders' and maintained by the 'leading' subordinates. The question that needs to be asked is, in which direction will the conflict be driven? Will it produce a softer, less murderously authoritarian command-and-control centre, or will it produce a harder one? While the sniper's unilateral response to the issue is extreme, it simply reinforces the absence of a dialogue and the lack of a connection between the command-and-control structure and the hierarchy of the brutalised victims. In a fashion, it is not unlike the new matriarchal elephant who has determined that she will no longer be led. The question now is, how many other elephants will follow the new matriarch, or will they indeed establish a different path for themselves? While I am tempted to suggest a similarity with Martin Luther King and the civil rights movement, I think a better comparison would be to Mahatma Gandhi or even Nelson Mandela, who both, within limits, established a new paradigm. Yet if we revisit these new paradigms of command and control, what do we think has in fact been established? In India and South Africa we still have hierarchical command-and-control structures broadly similar to those that went before but with different surnames on the pay cheques. There are still disadvantaged and oppressed strata; only now they are subjugated by their 'own' society rather than a colonial one. The real change is that it is 'higher echelons' of society, the leaders, that have changed the circumstance of the disadvantaged and have established their own positions within the command-and-control structure. Is it only the 'leaders' who have the vision, desire, and ability to effect change? But the change they effect is for themselves. How then will this disconnect beneath the command-and-control structure and the 'dispossessed black community' be resolved? There will not be a solution, a cure, or a resolution, but there will be a treatment, a poultice appended to the open wounds that will seek to counter the infection and enable business as usual.

In this instance I will suggest that we have a perfect example of the bulkhead division between different 'crew sections' on Spaceship Earth. When push comes to shove, these sections of society (sections of the herd) are already isolated and exposed to the sealing off of the bulkheads. Black Lives Matter is the slogan displayed on the placards and voiced on the lips of the protesters. Yes they do, and no they don't. This is the disconnect, the corruption, prevalent in US society. While I have described the issue within the context of the police and the black community, this is only one facet of the problem. Within the black community there exists a microcosm of the whole debate whereby blacks target, subjugate, sacrifice, murder, poison, rob, and similarly exploit their fellow community members and will also cross the 'borders' to similarly exploit their neighbouring communities. 'Leaders' adopt the command-and-control structures of gangs and drug dealers that dispatch violence and punishment to their fellows and design and build barriers to separate their private domains from the central 'law-enforcing' establishment. It is strange to me that the killing of 'innocent black individuals' by the police provokes such a reaction when the black people's own communities are rife with acts of extreme violence and multiple deaths perpetrated among themselves. It is as though 'You can't kill our people; only we can kill our people. We won't tolerate you killing our people; we only tolerate our people killing our people.' Please don't think that I am condoning the actions of the police. I am simply trying to expand the argument to a more meaningful point. One of the major control mechanisms is fear, a basic primeval mechanism designed to elicit tropismic acts and to remove thought from the process. These black communities now live in fear of the police and the brutality that they can dole out (they don't all do it), while simultaneously living in fear of the 'hard men' in their own community who broker silence from witnesses too afraid to change their circumstance. Several US cities suffer murder rates in the range of forty to sixty deaths per one hundred thousand per annum, which, when compared with Britain's most dangerous city of Glasgow with threeish, begins to bring the Black Lives Matter into context. (Figures can vary significantly between years.) It is strange in one sense that the 'murder by police' can be screened live by mobile telephone technology to the obvious shock and anger of black communities, while live footage of the considerably more commonplace black-on-black murders is remarkably absent. For while they fear the police, black people still feel comfortable to record the actions of the police for posterity, because they expect the police actions to be limited. It is wrong to select black communities in isolation from other poor communities rife with crime and drugs. Nor is the shock and anger limited to black communities. It is not purely a race issue. The case has been made for the purchase of illegal drugs in the US and Canada fuelling the high number of drug-cartel-related deaths to the south of the US–Mexico border. It is not a colour or race issue. It is a command-and-control issue whereby one such structure is competing with others, a fight for the exploitation and subjugation of neighbours for the benefits that dominance provides. Let us train chimpanzees living in their natural environment in the use of loaded AK-47's and consider the consequences. We can already see the consequences of trained humans living in their unnatural environments.

The Glue That Holds Society Together

Is there a fairness gene?

I sat in the waiting room of my optician and skimmed through a magazine. My attention was taken by a new product made from recycled crushed marble which was reformed using, and I quote, 'patented agglomerating additive'. Need I say more?

July 2016. Today's headline in the *Daily Mail*: 'Two million pounds handed to murderer in legal aid'. A father was handed £2 million by the taxpayers to secure the return of his daughter, whom he promptly murdered. First of all, the focus is entirely wrong. The father did not receive £2 million in legal aid; his *lawyers* did. On that basis, fathers around the country are regularly receiving £2 million in legal aid so that they don't murder their daughters. The real issue is that lawyers are regularly milking the taxpayers to the tune of £2 million a pop, while the misplaced focus is not on the murdered girl but on the non-existent profligacy of the father.

The Unattainable Dream

Who would deny any sick individual the latest advances in medicine? Charities implore contributors to run half-marathons, to host coffee mornings, to dig ever deeper into purses and pockets in pursuit of resources. Governments and universities devote time, money, and resources to the same desire for a better understanding and for fresh tools to fight disease. Hospitals develop and refine techniques and practices in the drive for a better prognosis.

Yet it will never be enough.

It will never cure all ills.

It will always be rationed by resources of taxes, gifts, and intellect.

Are we chasing the unattainable?

The blind pursuit of a distant dream, another horizon, another ocean to cross, another attempt to cheat death by cheating life. Another herd instinct recruited to subjugate the junior members to an unreasoned reality. Another bait ball?

Another lie to children?

Am I spouting heresy? Let me don the mantle of God in determining who will live and who will die today.

If I were to suspend all medicinal treatments for every ailment, mild or severe, I would create a world of conditions causing minor annoyance through to sudden death. It would be fair in that everyone would be treated equally. No, that isn't correct, because no humans are equal and each individual has a difference, an edge or advantage over his or her neighbours. The opposite holds true too; every individual is disadvantaged by the edge possessed by his or her neighbours. Some are predisposed to disease, and equally some are predisposed to resisting it. Life isn't fair, and death is absolute. Who would deny any sick individual the latest advances in medicine?

I can feel a paradox coming on.

Resources are limited and demand is not so. In the first instance we have to introduce rationing. This may take the form of time (waiting times), money (budgetary pressures), capability ('sorry, not available at this facility'), technical competence ('never heard of it'; 'don't know how to do that'), or technical consent ('not approved by NICE, National Institute of Clinical Excellence').

Let us find a treatment for that scourge of the human species, 'sweet pea blight' (SPB), the well-known debilitating disease that I have just made up and that has successfully jumped the species barrier. We need to devote money and all the resources of people, buildings, intellect, and science to develop the treatment. In the process we will divert these same resources away from other illnesses. SPB must compete for attention and resources, whether from taxes, charities, or philanthropists, against all the other ailments of humankind. SPB has entered into a non-evolutionary race with other diseases to see which can justify the race to exclude and destroy it. This non-evolutionary race may be determined by luck, cost, public attention, publicity, and science. Far better for SPB to become a mild irritant rather than a life-altering prognosis deserving of a campaign of destruction; far better for it to lurk under the surface without demanding attention. I have written before about the stupidity of 'Make cancer an election issue' and the politicisation of resources for one disease at the expense of others. Once again we are back to the unlimited demand but limited resource issue that is essentially determined by publicity.

Many people support charities that they hold dear. Cancer survivors and their friends and relatives are keen to give something back to the organisations that gave them hope or that hold out the hope of a cure in future. Seeing a loved one die a slow debilitating death is virtually guaranteed to elicit thoughts and actions to try to ensure that it doesn't happen to anyone else. And who would blame these people?

Now the paradox should be slipping into focus. But before I explain, let me state that I would be similarly motivated and have the same desires to help. I am not some cold-hearted monster who wants to see people suffer.

My problem is, who is it that I help? If you haven't worked it out yet, think blue tits and proxy wars. By giving money and resources to the search for treatments and care for SPB, you are denying those same resources to other diseases. But these 'diseases' don't in themselves have to involve rogue cells or pathogens to cause widespread death. They could, within my definition, include death from famine or exposure. Think back to the SPMTU (standard person month transformed unit) of £3 and consider what effect your charitable donation is having. By donating your multiple of £3 to your favoured cause, you are also selecting which strangers you want to die while saving the lives of the people you love. Having seen your loved one die from SPB, you donate so that others don't have to suffer the same fate. But these others are strangers, strangers whom you have identified with and therefore who are no longer strangers. Place yourself on Mount Olympus, not as a god but as a philanthropist, with a view of all that is good and bad in the human world below. Roll your die and decide who will live and die through your actions. Which strangers will be worthy of your support, and which can be ignored? Where will you alleviate suffering today, and where will you allow it to continue? An impossible task, I agree, a task that we are programmed to attempt, a task where we condemn people to death by saving lives.

Looking at this from a slightly different perspective, we see that it is like we are supporting a non-terrorist operation, say the un-IRA, with money and material to conduct acts of 'unwar' against others who hold a different sociopolitical view. But we are supplying what I will describe as 'unbullets', the purpose of which is to save lives and not take them. Every unbullet protects those within the target audience but allows death and destruction to continue outside the target audience.

The debate will rage about what right I have to determine who will die or not, but that isn't the debate I seek, because you have already decided who will die or not through your actions. That is why we will never have another genocide, until the next one anyway. That is why we will never have another war, until the next one anyway. That is why we will not have another terrorist atrocity, until the next one anyway. Because you have already made your choice as to who lives and dies today, without any rationale.

Yet the one central theme of 'civilised' society is the sanctity of human life, which motivates and controls us in equal measure. The problem for our civilised society is that for sanctity or inviolability, you might as well read 30 mph.

Let me return to Mt Olympus but this time as a god, and let me determine not who dies today but who doesn't live today. Which birth isn't allowed to be conceived today? Which parents will be denied the chance to extend the line of their DNA? You can imagine the fervour of the debate, the passion, the reaction to this denying of the most fundamental human right and the rebellion that it would invoke. So where is the fervour, passion, reaction, and debate to the existing alternative whereby lives are discarded so readily? On the one hand, you cannot have it both ways, unlimited lives and limited deaths. And I don't just mean the maths.

Resources could be increased. We could abandon our interplanetary expeditions and utilise the money and science to treat illness. We could abandon the billions spent on the Olympics. We could reduce our spending on defence. (Is defence an oxymoron, I wonder?) But the answer to the question of which life shall we save today? Saving lives is not our priority when we can satisfy our curiosity and enjoy the entertainment. We would condemn the gladiatorial contests of ancient Rome's arenas for providing death as entertainment while accepting death as a result of lack of resources in the pursuit of modern-day Olympic venues.

Returning to the sugar tax and the 'vital' health impact to be generated by the imposition of this prod in the right direction to the reduction in sugary drinks, it would be not unreasonable to suggest that excess sugar equates to ill health. Taken to its full conclusion, excess sugar creates the environment for and causes death. These would be the same sugary drink brands that feature so prominently in the multimillion-pound advertising campaigns appended to all the great sporting/entertainment events such as the World Cup, Super Bowl, and Olympics. We have taken the death out of the arena and transferred it to the audience instead. If we were really serious about saving lives, we could ban the sale of sugary drinks and close the factories down. I realise that the sugar could be removed from the recipe, but that is also unpalatable because of the imminent redundancies in the sugar-producing industry. What becomes apparent is that we are entitled to choice. It is not within government's remit to keep us healthy; it is in their remit to cajole, educate, and steer us through financial pressures to accept the healthier path, while simultaneously protecting jobs and the economy for the common good. Which common good is that, the good of generating a livelihood from sugar production or the good of failing to reduce excess sugar from the diet and accepting premature deaths? Or the brewing industry and deaths from excess alcohol consumption or the birth of alcohol-damaged babies? Or the gambling industry and the ruination of addicted lives? Or the car industry and the premature deaths caused by pollution? Or the housing industry where standards are driven ever higher beyond the purchasing power of their inhabitants?

It is possible to move the goalposts a little and to address the reality on the ground to reach a different conclusion. Alcohol is good because it generates employment and revenue, and the number of deaths is sustainable. Tobacco is good because it generates employment and revenue, and the number of deaths is sustainable. Sugar is good because it generates employment and the number of deaths is sustainable.

We are also entitled to choice when deciding who lives and who dies today.

The simple truth is that we earnestly pursue a life of irrationality where we beaver away, simultaneously protecting and harming life in unequal measure. We fail to see this because we have no rationality in our measurements, our ambitions, or our actions. We treat our fellow passengers on this earth with contempt, whether it's crushing ants underfoot or crushing tea leaves, and our whole premise is to do the same to other people. Many of us eat meat and understand that animals have to die to provide it, but in today's 'civilised' society, few of us witness slaughter. For some this position is unjustifiable; some become vegetarians by election. Vegetarians still drink tea and still live in houses built on former green fields. Some will feed the birds, and most will travel using hydrocarbons as they tread so delicately on this earth, leaving only footprints as they comfort their consciences and continue to contribute to the ELE.

The carnivores among us continue to tuck in to their sanitised flesh while their conscience is comfortable, and the ELE continues unabated. How easily we plug into the program that is the matrix, the highly intelligent, reasoning, deeply questioning, thought-provoking reason for existence that stupefies and stuns in equal measure and allows always for business as usual.

A Sense of Perception

One of my dad's favourite expressions was 'Why is a *V* on the end of a stick'
It was a tease.

Let me try to assemble the previous ideas into a simple strategy of understanding. Whatever we do, whatever choice we make, whatever our 'leaders' attempt, whatever our Great Brains devise, we can only see a tiny portion of our desired effects because we lack the knowledge, ability, or wherewithal to see or understand beyond a tiny portion of the undesirable effects of these same actions. We are damned if we do and equally damned if we don't. Why then do we celebrate these initiatives, these elephants, and their architects, and reward them with a pot of gold, a knighthood, a blue plaque for pulling the lanolin-scented fibre over our eyes? Why do we imbue our leaders with such faith? While I used to groan inwardly whenever I saw 'Good News, Price Reductions' on my weekly prepacked sandwich purchase, it is nothing to how I feel whenever I see yet another budget-busting initiative from our 'leaders'. Welcome to the bait ball. The good news is, whichever direction you care to swim in, there are more fish.

'*Y* is a *v* on the end of an *l*.'

Many of you readers, if there are any left, will be thinking what a miserable pessimistic old git I am. Look at all the human race's achievements. Well, I have looked at them, from the construction of the pyramids, to Stonehenge, to the great cathedrals, to FIFA and the IOC, to Google, and to the Clovis point, and I have removed the princesses from your rose-tinted glasses and attempted to replace them with a lens focussed on the unthinking, extracurricular, world-destroying, terraforming, ELE-causing scourge that we are. All that peacocks do to impress their presence on the earth is grow a showy tail. Does it make them content? Even then, only half the race sees the need to grow one, while the other half sees the need to have one. If you removed all the tails from a controlled population, would they die out or would the imperative to breed exceed the selection process? I lied; peacocks also have a distinctive call to attract mates over distance. Would this be enough in the absence of tails? (How does the construction of a cathedral alter the sexual dynamic between humans?)

While I have often referred to the peacock's tail, let us stop and think for a moment about this decorative appendage. We learn that it plays such an important part in the selection process of sexual partners and the continuance of individual DNA streams. Let me ask a question: what would happen if the females lost their interest in decorative displays of extravagance? To phrase the question differently, which came first, the tail or the desire for a tail? Did the males invent bling, or did the females demand it? I would suggest that the tail is disadvantageous to the male bird, in energy consumption, in conspicuous exposure (the antithesis of camouflage), in the slowing down of flight reaction times. Whichever way you look at it, it is bad news for the male. So why do they grow one?

It's in their DNA. They are a prisoner to it. They are trapped by their programming.

And so are we.

But humans are so much more complicated than peafowl, aren't they? Our mating rituals are so much simpler.

'Single issue, binary referenda'.

At the general election we are all—those who can be bothered to vote anyway—faced with a dilemma. In simple terms your cross in the box is appended to the name of an individual, to the name of a political party (independents accepted), and by extension, to the name of the party leader. Your choice, ignoring the debates and manifestoes, is split into a simple yes or no, and your single choice has more than one result. You may love your local MP and his action on local issues, but a vote for him is also a vote for the party and a vote for the party leader. Similarly, you may hate your local MP, but support his party, which is also a vote for the party leader. Finally, you may love the party leader but can only support him by voting for your local MP.

Currently we have an apparent impasse within the Labour Party with a resurgence in support for the Left and the leadership debacle. What we are witnessing, if we would but interpret it, is the reassignment of the multiple-choice element of the ballot paper. If the Corbyn supporters get their way when a cross in the box next to Labour Party is made, then that is precisely what you are voting for, the Labour Party. The MP becomes subservient and will be deselected by the Party as and when they see fit. Similarly, the party leader becomes subservient and will be deselected by the Party as they see fit. The sole object of the Labour Party is to secure control over and of the Labour Party. The real problem then manifests itself when the question is asked, who is the Labour Party? Jeremy Corbyn is offering no leadership because he is not the leader. The MPs cannot remove him because they have no leadership role either. The Labour Party has therefore surrendered its control to itself, which essentially means the unions and union leaders. We can therefore sit back and enjoy the spectacle of the Labour Party's imminent implosion and disintegration as it lurches away from the centre ground. I am reminded of the Muslim extremists who terrorised North Africa a few decades back with multiple attacks on non-fundamentalist government and civilians alike. Their demise was finally caused by the simple fact that eventually they began to fight each other, because each grouping was radically opposed to every other radical grouping.

The Labour Party is about to become irrelevant.

It has already happened in Scotland, a simple fact that the commentators seem to have ignored because it doesn't fit with their pro forma.

BBC teletext: In the decade up to 2010, tropical nations saw net forest loss of 7,000,000,000 hectares per year and a net gain in farmland of 6,000,000,000 hectares. Quoting a UN report. 'Improving co-operation between nations' farming and forestry sectors will help reduce deforestation and improve food security.'

Note that 7,000,000,000 hectares is 70,000 square kilometres, equivalent to 5,600 STUs per year for ten years—56,000 tigers in a decade.

If David had Missed

Way back towards the end of the Second World War, William Beveridge announced the grand plan to create a fairer society worthy of the great sacrifice made by the peoples of Britain. His plan was simple, to slay the five 'giants' of want, disease, ignorance, squalor, and idleness. Welcome to the welfare state. Seventy years later, can I ask as to the relative health of these elephants? Can we demolish the stables from whence they emerged, intent on reaching the finishing line?

The Dangers of Social Smoking

Illusion of assistance supports the surrender of power to higher authority. Offer help to the symptom and not the cause. Illusion of help and acceptance of illusion by the masses. Sorted. Not. We don't in fact want a solution, just a nod in the right direction so we can carry on with business as usual without guilt. 'Bollocks' springs to mind.

There is a scene in *The Matrix* where Morpheus is inducting Neo into the new reality. Morpheus asks a question about oxygen. I would ask, 'So you think that what you are doing is helping?'

The basic premise of the Great Brains is to manage the perception of the issue while never offering a cure or solution. Management is good because it never ends, and neither does the salary. Cures and solutions are absolute and salary-limiting.

I realise in writing these thoughts down that I have lit the fifteenth cigarette from one match and will be exposed to every enemy sniper in the trenches opposite who will see it as a point of honour to 'take me out'. I have not written anything in order to shock you; I have tried to write in such a way as to challenge and debunk the myths and phraseology of our ever so organised society and its supporting caravan's route march to oblivion. I have tried to upset the status quo and to ridicule our ever so considered ways and to expose them for what they are, chemical signals from the queen ant evoking an automatic thoughtless response. I am outside the system without allegiance to any entity or organisation and—this will not come as a surprise—without a salary or position to protect or be abused by. I have also tried to avoid any pro formas, and that includes the pro formas of the PC brigade who endeavour to protect lies with even more lies. You will judge me if you can. You can agree or disagree. You may even engage in debate. The only thing I ask is that you don't ignore me.

Not that I will make any difference whatsoever, 'for ours is the kingdom, the power, and the glory, for ever and ever' until the end of man. (Does A as a prefix = until the end of, I always wondered? Amen to that.)

My best friend is a highway engineer. Yes, I am joking.

In one sense we have been here before with the ancient Greeks who worked it all out thousands of years ago and dressed it all up in a story about gods. George Orwell dressed it all up within a work of fiction, a story. We still remember the story but fail to grasp the meaning. If I were imaginative enough, I could have written a story, a fable, a legend, to illustrate our predicament, instead of which I have tried to lay it out without enhancement, to write it as I see it in the hope that you might grasp the meaning rather than the story. But then, Donald Trump was elected to the become president. At first I laughed, and then I wrote *71 Days to Save the World*. This is the story, but between the lines are many of the issues contained within *How to Kill an Elephant*. Humour is to the forefront but the message remains.

Language is so peculiar in that it gives us all a common thread without in fact being common. Every mind has the ability to interpret and corrupt the intended message without even trying, and when we try …

I invented something once and took out a patent, which involves describing your invention to a patent agent. They in turn convert your idea, drawings, and written words into a written patent and return it to you in patent form. I can honestly say that I could not recognise my own invention even though it was written in English.

With *How to Kill an Elephant*, I have invented something else. If it ever sees the light of day and reaches print, I will be intrigued to discover how its contents are discussed and debated. I wonder if I will even recognise my own ideas when they return to me or whether they will have been corrupted beyond my orbit. Time will tell. I will look for elephant footprints within the forest of sustainabilitists and search for tropismic responses while I enjoy my cup of coffee on my Indian safari in search of tigers.

There will be some among my readers who know me, who recognise my facade anyway, because until they have read *How to Kill an Elephant*, they will not have known me at all. They may decry me as a hypocrite with my large house and executive motor car and pension pot, but what can I say, I was a product of my pro forma, an excellent student of it although I never really bought into it. Always at the back of my mind was the question of why. I never accepted the answers provided, and neither should you. Good luck with your elephant hunting.

Stonehenge Reborn

The 1994 International Union for Conservation of Nature (IUCN) Red List compiled by the World Conservation Monitoring Centre includes more than six thousand animal species known to be at risk.

Ants 'learn' by reinforcement of chemical signals laid down by individuals. The more frequently a path is used, the more likely it is to be used again, because for any trail to become a highway it must have served a purpose. The way to cure obesity is to reinforce the highway of healthy living while letting the highway of obesity peter out through lack of reinforcement. Perversely, we have reinforced and are reinforcing the wrong trails of consumerism, economic growth, and obesity, to name a very few.

I have commented about films like *Escape from New York* where an entire city has been enclosed within a wall and is run entirely by criminals. Authorities and the rule of law have been suspended. This radical, albeit fictional, solution has been devised for the cinema audience's consumption.

In its simplest form, it is possible to interpret this as a truly radical solution to our inability to control crime, if for no other reason than all crime is contained within the city's walls. The criminals effectively 'police' themselves. The scene is presented as a dystopian future that we may or may not be heading for. Let me turn the scenario around and approach it from a different perspective utilising earlier examples from *How to Kill an Elephant*. On the one hand you can accuse me of being lazy, but on the other, as you will probably have worked out by now, you can see that

I perceive an advantage from visiting old ground, as it sheds fresh light on a number of topics.

The premise of the film is that society has failed in its attempts to control crime and has resorted to what can effectively be regarded as a penal colony where the solution to the problem is to remove the criminal element from the scene of decent law-abiding citizens (a bit like Australia, but without the sunshine).

Can you remember Eze, the walled enclave referred to near to Nice that was formerly used as an effective jail containing plague-carrying ships' crews? We could think of it as of being in a similar vein to a leper colony, where fear of contagion isolates the sufferers from the ignorance of healthy non-sufferers.

On a similar basis we could also consider a Victorian madhouse where mentally disturbed individuals would effectively be incarcerated against their will. No doubt not all the inhabitants would be mentally disturbed, some of them simply 'parked' there to avoid the scandal of divorce or extramarital childbirth, for example.

Thankfully our knowledge has progressed beyond these extreme examples, although the spectre of isolated ghettos of Ebola victims may be closer to reality in West Africa than we may wish to believe.

Let me propose that the enclave route has a future for a society struggling to meet its needs and that we consider its very real merits.

Let me suggest that the first enclave could be constructed on Salisbury Plain, a mainly redundant site that has lost its original purpose in life and that mainly exists to stimulate pixels in telephones. (FIFA has long since abandoned it after all and moved on to Qatar, which is like Salisbury Plain but with more sun.)

As the great architect, I propose to construct a series of concentric rings of increasing diameter set around the central core, and for aesthetic reasons alone, each zone would then be colour-coded by painting the soil surface. The view from above would not be dissimilar to a multicoloured archery range target and should never be confused at any stage in its new and future history as an alien landing zone, because that would be silly.

Colour-coded habitation and uniforms will be supplied and all zones will have access to medical facilities, clean potable water, and sanitation without limit. Now for the clever bit: food will be supplied at fixed points around the perimeter of the site and must be carried throughout the site in small containers. In the absence of a ready supply of ostrich eggs (I never did work it out), I propose to use something akin to an eggcup. Food may only ever be eaten in one's designated colour-coded zone, and sanctions will be imposed for any individuals observed to be cheating.

Whom shall we select to inhabit this enclave? Why, fat people of course.

Anyone with a BMI in excess of 30.6 (what an absolute bastard I am; it's the second paragraph on page 545 you are looking for) would inhabit the outer zone. As BMI increases, the zone occupants reside nearer the centre and, most importantly, farther away from the source of food. Sanctions will also be imposed for anyone sharing food, so the only solution for eating is to fetch it yourself. The greater your BMI, the greater the distance you have to travel, and the more you eat, the more trips are required. As time goes on, barriers between the rings can be erected and access provided by up and down staircases to increase the effort required to secure food.

Regular weigh-ins will be conducted. Residents can change their coloured uniforms and zones of occupation as progress is made (plus or minus, the choice is theirs).

Of course this route to eliminating obesity from within our ranks will have a cost. Buying the site and the construction phase, not forgetting the fast-track planning application, say, twelve years, will be the initial set-up costs. Staff training along the grounds of the prison service whereby the new site will be fully staffed for eighteen months before the first residents are admitted ought to suffice. A comprehensive CCTV system and IT support will be required to control any cheating. To build these things will take several years and several cancellations and reboots, but this can be commissioned to run contiguously with the planning application. (The first time I saw 'contiguous', I thought the architect couldn't spell. It means 'side by side'.) By my calculations, the whole set-up and initiation process will cost no more than an amortised £500 million a year, which just leaves my entirely reasonable architect fees of 20 per cent per annum. (I would gratefully accept any honours, knighthoods, and OBEs if you are offering.) I propose this as an extremely practicable and effective solution to our present obesity epidemic and commend its merits to the Great Brains. Not only that, but also it comes in on budget with the new sugar tax receipts, and I have saved our teachers from adopting the additional responsibility of training the next generations of Olympic champions and Premier League hopefuls. Think of all the evaluations that will be saved.

I guarantee the success of this initiative because it ticks so many of the boxes beloved by our Great Brains.

1) All obese people are concentrated in one place with ready access to dedicated medical facilities.
2) The facilitators of obesity can be readily controlled and excluded from the supply chain. Only wholesome foods will be supplied.
3) The absence of obese roll models (sorry, should that be role?) removes the chemical reinforcement of the 'false signal'.
4) It accurately assigns the 'obese tax' proceeds to the cure.
5) As the sugar consumption is reduced, so are the tax receipts, which will also be reflected in fewer patrons and reduced running costs of the new 'mega'lithic camp.
6) Micturition is probably not the term you have in mind.
7) Micturition is the ejection of urine from the urinary bladder through the urethra to the outside of the body.

I realise the folly of my proposal.

1) It is much better to spread the cost and demand of heavy-duty beds and chairs throughout the entire NHS and to expose all staff to the complexities and risks associated with obese patients.
2) Our Great Brains will only ever sanction 'voluntary' arrangements on our obesity facilitators.
3) Our Great Brains do not understand this statement; it could challenge their position and will therefore be ignored.
4) Our Great Brains have no interest in accurately assigning anything to anything. 'Heavens, do you want us to be accountable?'
5) 'Er, fewer tax receipts. Who is this idiot?'
6) 'This jumped-up architect is trying to take the piss. Doesn't he realise that is our job?'
7) Sugar tax? (They started it.)

Forgive the diversion.

Back to New York, *Escape from*, that is (that would in fact be away from New York).

Let us fill a city with criminals and charge them with running and organising the operation of said city. I wonder how fair this set-up would be considered to be? Shoplifters would be exposed to murderers and rapists; embezzlers and con men would be rubbing shoulders with hardened thugs; and the whole would be grossly unfair across the range of minor to major crime. What sanctions, if any, would the criminals impose or attempt to impose among themselves? Would we have a state of anarchy, or would rules of engagement be hammered out and stable areas established within which 'everything doesn't go'? A sort of controlled anarchy (an oxymoron, I know).

A thought strikes me. Actually it struck me a long while ago; otherwise I wouldn't have led you to this place.

The present-day occupants of New York City every day are exposed to and rub shoulders with shoplifters, murderers, rapists, embezzlers, con men, and hardened thugs across the whole range from minor to major crime. What sanctions, if any, would the residents impose or attempt to impose among themselves? Would we have a state of anarchy, or would rules of engagement have been hammered out and stable areas established within which 'everything doesn't go'? A sort of controlled anarchy (an oxymoron, I know).

In reality, at first sight anyway, the rules of engagement have been established. That is why New York employs its meter maids and its police department and its judiciary. But at second sight, *if* the rules of engagement had been hammered out, then no maids, police, or judiciary would be required.

We seem to have an oxymoron.

But while we can scan the streets for passing shoplifters, murderers, rapists, and so forth, we need to be scanning the boardrooms and local government offices as well, because in these places reside the Great Brains who seek to profit from those below as voraciously as and much more effectively than any common hoodlum on the street.

As New York City is in a state of organised anarchy—and the same would hold true for all cities throughout the world—how do we reconcile the disconnect between the organisation and the anarchy? Does the term 'corruption' adequately define the difference between these opposing states? Is corruption the mechanism whereby the two mutually exclusive states are held in dynamic equilibrium? In a utopian world of perfect order, corruption would serve no function and the reverse would hold true, because in a dystopian world of anarchy, corruption would again serve no purpose. In either case there would be no need to lie. What we charge our Great Brains to achieve then is an acceptable balance between the two opposing states. But my problem is the balance, because the balance is set by corruption which, by definition, cannot be trusted.

The balance is effectively a lie that manifests itself at every test.

If that test were to report the sexual abuse of white underage girls to the authorities in Rotherham?

If that test were to report a seven-hour wait at A & E?

If that test were to reduce obesity by taxing sugary drinks?

If that test were to impose a safe speed limit in a residential area?

If that test were to limit the projected rise in global temperatures to below 2°C?

If that test were to ensure the continual existence of tigers in the wild?

While we charge our Great Brains to achieve the acceptable balance between the state of order and anarchy, we must also play our part.

We must accept the imposition of a safe speed limit in a residential area.

We must accept the personal responsibility to maintain a healthy lifestyle.

We must shun criminal acts.

But we have no leaders.

Unlike the fictional *Escape from New York* where the city is run by criminals, our cities are run by criminals in the sense that they are not controlled. They control themselves against a backdrop of sanction and threat of sanction. They control themselves against a backdrop of need and opportunity. It even becomes possible to suggest that they begin to police themselves. They impose a limit on their activities. The questions that need to be answered are where and why these limits are imposed. Fear of getting caught and of the sanctions imposed will form part of the decision-making process but by no means all. One consideration would be that of conscience and the balance of good versus bad, and another could simply be one of laziness. Crime represents a form of labour, with stress levels and exertion of nervous energy a limiting factor.

Many years ago I would give guided tours around our manufacturing plant and would then provide nourishment at a local restaurant. While leaving the restaurant and passing through the building foyer, my architect guests were intrigued by a small model of a section of carpeted house. One of the party stood on the carpet. We were met by a shrill 90 dB alarm siren triggered by a pressure pad set in the floor of the exhibition display. A member of staff disarmed the device and apologised, explaining that the restaurant owner had invented the siren as a byline and was pushing its sale. He further explained that an intruder had entered the restaurant premises one night and had triggered the real, 130 dB alarm. The 'sonic blast' had so affected the intruder that the police had arrived to find the would-be burglar lying in the foetal position with tears streaming down his face, effectively paralysed by shock.

A third consideration is one of consumerism, or if you would prefer, a lack of it. Has the typical criminal retained the pre-consumerism ethic of having enough to see the week out and refuses to 'work' until the need arises as pockets empty?

Some of this 'work ethic' would be justified by the need to reduce the risk of being caught. In others, this work ethic may be beyond the individual's control. A drug addict will be driven to steal to feed his habit before becoming incapacitated by the purchased drugs. Addicts simply cannot function seven days a week.

Where within this process do the 'official' controls of police and judiciary sit? Given the earlier comments about the judiciary having set its stall out around offering inconvenience to criminals rather than a cure, the controls are weak at best.

When handguns were banned in the UK after the Dunblane massacre of 1996, it had a profound impact on all legitimate handgun owners but had a negligible impact on the criminal classes who, by definition, tend to ignore the law. As gun crime has persisted for the last twenty years with no potential end in sight, is it reasonable to ask just what the judiciary has achieved? But this is a rhetorical question, because we have already answered it. The answer is that it has had a major impact on legitimate gun owners and a negligible impact on those outside the laws.

Yet still we pay our taxes and, worse, our lawyers to provide redress for injuries suffered. In the UK the courts are filled with cases taken out by the Crown in their defence of the public. The rich and insured vie with corporations and their cash reserves for the rest of the court's attention. For most of us, redress through the courts is beyond our scope and pay grades, unless we are fortuitously poor enough to qualify for legal aid.

You will remember:

This corruption could simply been seen as a lie.

And after the lie comes a warning.

And after the warning comes a caution.

And after the caution comes a conditional warning.

And after the conditional warning comes a conditional discharge.

And after the conditional discharge comes a fine.

And after the fine comes a curfew.

And after the curfew comes a tag.

And after the tag comes a community order with unpaid work element.

And after the community order with unpaid work element comes a suspended sentence.

At the time of writing about our judicial systems incremental steps towards custodial sentences, I had two individuals in mind, a mother and her disabled daughter. The mother had been repeatedly ignored by the police about bullying attacks on the daughter. The mother, at the end of her wits, had driven to a local sheltered lay-by and had set her car alight, with two occupants inside: herself and her daughter.

Lessons were learned.

Let us consider the role of the police within our equation. Terry Pratchett's City Watch has certain officers of the watch who will guard a structure, such as a stone bridge, against theft. When challenged, the guard will confirm the success of his actions on the basis that the bridge has not, in point of fact, been stolen. The guard of course is acting on his own initiative.

Our police act on the initiatives of our Great Brains, and budgets and resources are allocated to ensure the equilibrium is held in a generally acceptable condition to the satisfaction of law-abiding citizens. If crime rates were to double, the pressure on the police to act and the willingness of the taxpayers to increase subscription would also increase. Equally, if crime rates were to halve, then the reverse would hold true and budgets would be cut. This whole state of affairs is of course underwritten or undermined by advances in technology, dependent on your point of view. The historical advent of the bobby on a bicycle has long since been superseded by the bobby in a motor car and or helicopter. Equally the bobby on the beat is just as likely to be employed as the bobby on a mouse mat as the new technological crimes demand technological solutions. The police always react, that is to say that they are always behind the curve, a necessary effect of limited resources beyond their control. As the technology of crime detection and forensics improves, so does the cost and budget allocations, the latter of which are amended to reduce the number of incidents examined (only attend burglaries to houses with even numbers, or was it odd?).

It is the presence or absence of crime that determines the nature of our police force, and because of this they are controlled by the appetite of criminals to commit crime. It is also the public's perception of crime that drives police resources. A spate of dangerous driving in a given area will likely lead to a localised increase in patrols and 'press releases' of intent designed to warn off the perpetrators or to divert them to another area. Imagine, and you thought it was designed to catch more criminals. For the local residents who complained, a temporary solution to their problems has been offered, as the problem has been driven (sorry) off the local streets. Small consolation if you happen to live in one of the alternate unpatrolled dangerous driving areas.

It is in this fashion that criminals control the police, and this is likely to be contrary to your earlier beliefs. In a similar fashion, criminals control the judiciary. If the number of criminals successfully prosecuted doubles, then, without a doubling of prison spaces, the number or length of custodial sentences must halve to balance the resource. In its simplest form, we arrive at another equilibrium between the successes of the police balanced with the severity of the sentences. As one side rises, the other side must fall. (Of course in days gone by this presented no problem given the permanent nature of the sentences dispensed: sent to Australia or sent to meet your Maker.)

Against this equilibrium we can add another constituent: surveillance equipment, both state and privately owned, is widely used to identify and prove the activities of shoplifters, burglars, thugs, and 'crash for cash' fraudsters alike. As the quality of camera image and the number of cameras has increased, it would be reasonable to assume that more criminals are identified, the full scope of their activities displayed in graphic detail, and an appropriately severe sentence appended to their career path. 'It wasn't me, guv' doesn't really work when the speaker of the sentence has been filmed in glorious Technicolor. Once again we fall foul of the limited spaces available at Her Majesty's pleasure.

Rich people and rich enterprises will secure the better-quality surveillance and anti-theft equipment and will deploy the greater range of protection methods. Your local department store may well employ store detectives/security staff linked in to camera control rooms. Contrast this with a local corner shop which may have nothing more sophisticated than a curved mirror and a single staff member running the shop. All of course has a cost, the cost of losses to criminals or the cost of prevention methods to protect stock. Ultimately this cost is borne by the customer, who must cover the cost of security or losses through the prices paid. Once again the 'solution' is not one of prevention but one of moving the problem to more vulnerable locations. The same can be said of houses with cameras and security systems, which are less likely to be targeted than houses without. Unlike 'we're killing strangers so we don't kill the people we love', we are—those who are able anyway—creating a situation where 'they're stealing from strangers so they don't steal from the people we love'.

Society readily accepts this lie. Indeed, the police have crime prevention officers whose task it is to advise homeowners and business owners on the steps to take to protect their premises. This is at variance to the fire brigade, who offer help and advice to protect premises against fire. Providing smoke detectors protects all building occupants and doesn't move the problem next door, or to the next street or next borough; it offers a solution.

Criminals supply a demand. It's no good shoplifting one hundred pairs of trousers if your 'customers' only need five pairs. This widens the topic, because we now have a cause. Without customers for the illegal goods, there is no incentive for stealing goods beyond those that criminals use themselves.

Our Great Brains suffer from the need to comply with our expectations because they can only 'lead' us where we are prepared to follow. (We have no leaders.) My solution for obesity on Salisbury Plain cannot work because of the resistance of the Great Brains and the population to such a solution, a solution that could, indeed would, work. The Great Brains could impose controls on the food manufacturers and distributors and issue state controls on portion sizes and calorie numbers. The media could support the Great Brains and their initiatives without the universal cry of 'Nanny state'. The population could take note of and assume responsibility for their own eating habits and choices. The diet industry and all its associates have no interest in people actually losing weight, only in their trying.

The first Nazi concentration camp opened on 22 March 1933 at Dachau in Bavaria. Heinrich Himmler described it as 'the first concentration camp for political prisoners'. As we know, its scope and remit expanded through to its liberation by US forces in 1945.

How to explain the disconnect between the Great Brains needing to comply with the expectations of the people, who can only be led where they are prepared to follow?

With the sugar tax, we have an example of a policy that can work, that can engender acceptance and support from the media and the population alike. It is also an example of a policy that allows business as usual for the majority of the facilitators of our obesity epidemic, including the obese themselves. It is a policy that will not work, a policy that we pay for in taxes and in the consequences of failing to provide a solution. It is a fudge, an elephant on the battlefield that promises so much while delivering so little.

If I were to compare Dachau with the sugar tax, it would provoke an outcry, and rightly so. How can you compare a 'factory of death' with a health policy?

But if the purpose of the health policy is to save lives and it patently doesn't, then the sugar tax becomes a 'factory of unlife' that results in premature death and reduced quality of life. In Dachau the result was death in many and varied degrees of suffering and horror. The failure of the sugar tax will result in varying degrees of suffering and premature death, which some will witness in horror as loved ones die before them. If you ignore the motive and intent of the two initiatives, then the net result is death and suffering for both.

We have no leaders.

Leicestershire police can randomly decide which burglaries to investigate on the grounds that they don't discriminate and are independent of any decision-making process beyond that of spinning a coin. Odds or evens?

This decision requires no leadership.

If Leicestershire police decided to ignore burglaries of houses occupied by selected religious, ethnic, or age grounds, then this would be unacceptable.

This decision requires leadership.

Obesity is random and participants randomly decide whether to participate or not.

This decision requires no leadership.

If Rotherham police decided to ignore the sexual exploitation of predominantly white girls by Muslim men?

This decision requires leadership.

Can we escape the fate we've designed?

If you were to consider that global warming and a runaway climate threshold is reached beyond which the temperature rises beyond control as a potential fate, then would you conclude that we have either designed climate change or designed the solution?

While I've been building the picture, no one's been building the metaphorical house. Choice destroys choice. We have the freedom to choose, which destroys our choice. Control imbues further control, which narrows down the options and the choice. Extracurricular activity demands extracurricular activity to sustain it. We notice this and attempt to alleviate the symptoms without ever addressing the cause. A prime example would be the oft-repeated and -stated aims of reducing the bane of our lives, red tape or bureaucracy. While we notionally attempt to reduce it, we in fact reinforce it with decisions taken down the line to secure employment. This is the 'securing' achieved by the justification of the continuing existence of our jobs. Let us take customer satisfaction surveys to prove how good a job we are doing, while simultaneously lying about the reason, the intent, and the results that we generate. Let us conduct 'risk analysis' surveys for financial services customers where the purpose is to fix the blame to the exclusion of all other aims. Let us test and examine potential drivers before granting licences to drive on our roads. The process works to such an extent that we no longer have road traffic accidents but road traffic incidents, where the blameless accident has become a non-blameless incident.

People sometimes remark to me along the lines of, 'How do you know so much?' I reply with a simple put-down: 'It's because I went to school.' But so did they. This then begs the question of why they bothered. In reality the answer is always way more difficult than I portray. They went to school all right and simply filtered out the bits that interested them from the bits that didn't. But the curriculum doesn't give them that option, expecting them all to fit the pro forma of syllabus and timetable, and teacher ability and monetary resource, and classroom discipline and parental support, and perhaps the most overriding consideration that could be called mental ability. While 'pig's ear and silk purse' springs to mind, this is overly simplistic, because the world needs pigs' ears. I would suggest that you could ask a pig if it needs ears, but in the absence of ears, the question may fall on deaf something-or-others.

But while the people who ask me the aforementioned question did go to school, they didn't go to the school I went to. The curriculums keep changing, together with the personnel and the prisoners—sorry, inmates; sorry, pupils—and the facilities and the perceptions of what makes a good product. And as we have discussed before, our education is not a product of school. School is where you go to learn things; an education is what you receive from the world around you. That world includes your fellow inmates who influence and bully or cajole or stimulate those brain cells that float your boat. Or your fellow inmates who dismiss, denigrate, and ultimately destroy these same boat floaters. The overriding consideration about choice is that it must conform and comply with the pro forma. This may be the pro forma that states, 'No child born to be a bricklayer' or 'No child born to be trapped in the same environs as their parents'. Freedom of mobility, of advancement, of a society free of class and distinction, is enshrined in the mantras of the educational Great Brains who ruthlessly channel and compress these same freedoms into curriculums of control and measured pigeonholes of attainment. You have the freedom to enjoy/endure at least fourteen years of full-time education or training to ensure your freedom of enjoying/enduring fourteen years of the same. We should all be geniuses by now.

My bricklayer shared a pearl of wisdom with me while constructing a wall for me. His father had told him that all children were trainee arseholes and, it was up to parents to change the course of the child's career path. All children are trainee arse holes.

British vulgar slang for a stupid, irritating person, deserving of contempt.

Somewhat pejorative to say the least. What my bricklayer's father was alluding to is the idea that all children are born with a blank data chip. The information is uploaded onto the chip by parents, family, friends, educational professionals, the media, social media, work colleagues, peers, and so forth. Should the uploading process be malformed, say, for example, the child is steered adrift by gang affiliation, or drug abusing parents, then the propensity to become less than what is expected of them is encouraged. Keep this thought for later. The relevance will become apparent.

You have the freedom to own a home, but not just any home. It's not enough to have a roof above your head to keep you dry and walls around you to keep you safe and secure. Oh no, you need more than that. *We insist.*

The site has been carefully chosen for you by the planning process that has demonstrated that it has ticked all the boxes for 'sustainable' development this time. That's because the same site has oft been protected and sheltered from development within green belts and cast iron District Structure Plans as essential buffers against encroachment and unwarranted building. The policies have changed, the herb bouquet has been altered, and all the essential againsts have suddenly become essential fors. The protest arguments have been overridden and democracy has spoken.

The new houses are built to the latest planet-saving specifications and sit site boundary to site boundary with older, inferior stock incapable of providing satisfactory accommodation for today's environmental challenges. Not forgetting that these older stock all achieved the standard of the day and will continue to provide accommodation for decades to come. (I remember reading about one of the Ideal Home Exhibition houses from the 1920s that was sold at exhibition before being dismantled and rebuilt on the buyer's plot. This state-of-the-art dwelling was condemned as unfit for human habitation just a few years ago because of its high asbestos content.)

If you took the arbitrary view that all houses enjoyed a lifespan of one hundred years of occupation and that we had a static population, then you would discover that on average half our housing stock is older than fifty years. Ninety per cent is already ten years out of date and grossly inferior to current building models. Yet if it was good enough then, why isn't it good enough now? Of course the simple answer is, it is. There is no compunction to tear down ten-year-old houses to replace them with new ones. Yet our planners and Building Research Establishment bods beaver away to achieve zero-carbon designs with cutting-edge technologies and costs.

My described premise then will be that in a century's time all housing stock in the UK will be zero-carbon and the task will have been achieved.

'But you have to start somewhere.'

1) Will global warming still be the topic of the day in a hundred years' time?

2) We have to pay for all of these advancements in technology and costs.

3) All of this investment in advancement is underwritten by the resources of Planet Earth.

4) We are digging a hole in the earth's resources to fill a hole in the earth's resources.

5) We have created an entire industry of energy-saving technicians and their followers who demand energy and resources to exist.

6) What will our caravans look like in the near future?

Can you imagine for one moment that these erstwhile ambitions will still be pursued in even twenty years' time? What will be the imperative/ flavour of the day? How shall our Great Brains seek to control and subjugate our fear processes into the future? How will the new crop shape our future as they protect theirs?

You have the freedom to own a home.

I slipped that in earlier; did you notice?

You have the freedom to commit to a twenty-five-year-plus mortgage, but only after many years of saving the deposit to begin the process. You can take one of the shortcuts, such as shared equity, whereby you commit to own a proportion of your home while renting the rest. Not really a shortcut, is it? More a halfway house (sorry). You have the freedom to insure it and maintain it and protect it from criminals, and of course you have the freedom to die and pass the asset onto your heirs, after tax of course. Would the Indians on Manhattan Island understand this process? Would Neolithic man understand these concepts? Would Neolithic man invest a significant portion of twenty-five years of labour in securing a roof above his head? Of course I have missed the obvious shortcut, which is to acquire a council house and let somebody else pay for it.

What you don't have is the freedom to build a house on your own plot out of straw or wattle and daub. Yet curiously, if you happened to own a house built out of either, it is likely to have a preservation order on it to prevent you from altering it. My company used to supply reconstructed stone lintels, quoin blocks, and sills to a national house builder. The designs allowed for these products to all elevations of these mock-Victorian houses, and the value to us was significant. Only rarely did we supply to all elevations. The usual requirement was to do the front elevation only. While this made the houses cheaper to build, it left the builder with a peculiar product: mock Victorian to the front and boring nothing to the rear and sides. This was a planning requirement. In essence the estates were all 'ponced' up to look nice from the street side but were boring to the occupants and owners. I often used to think that the only place you saw the front of your and your neighbour's house was from the confines of your car, through the windscreen when you came home and in the mirrors when you left. Of course when you sat and enjoyed quality leisure time in your back garden, all you had the opportunity to see was the boring back of your house and everybody else's. The planner's facade was precisely that, an illusion of mock Victoriana appended to boring boxes. In a similar vein, all the larger executive estate houses were built facing the main roads so that passing traffic could admire the prosperity of the area, while all the cheap houses were built in the depths of the estate. A sort of reverse east–west housing arrangement where rich houses were built upwind of poorer houses so that the poor suffered all the coal fire pollution while the rich received fresher air on the prevailing wind. (You only have to compare the housing stock between the West End and East End of London.)

Personally, if I were to buy a 'rich man's' house, I would want it tucked at the back, away from the main road traffic and behind everyone else's traffic noise. That way I drive through their area and they don't drive through mine.

It always struck me as perverse that the architects devoted such time and effort to the visual aesthetics of the house fronts even though they were then partially screened by 1.5 tonnes of purple or yellow or red cars ...

Who would have thought that the great strides taken by the Clean Air Acts to control the life-limiting smogs of yesteryear would have been so readily sidestepped by NOx? Who would have thought that the advances in air quality would simply have been forgotten while we exported the problem to China in exchange for cheap consumer goods?

How easily we forget and ignore the issues. Have I mentioned lizard brains? Where will population growth, water scarcity, and food insecurity feature on our collective Great Brains' to-do list? Do they in fact have a to-do list? Where would we place the Greek economic crisis or the migrant crisis or the youth unemployment crisis in southern Europe on the EU's current to-do list?

They play the great game, which I will call time. 'Give me another year of your time while my hand is in your pocket. Give me another year of your time until I can retire and write my memoirs. Give me another year of your time while I exclude other Great Brains who have the answers' (they don't). As David Cameron said at his final trip to the dispatch box, 'I was the future, once.'

Show me any great achievement that I cannot turn into a negative. How depressive can I get? Let me return to the Irish bog elk which was obviously lost and could be found over a significant portion of the northern hemisphere. Let's give them anthropomorphic qualities, the powers of reasoning and speech.

'These racks are so heavy, they make my neck ache.'

'Yeah, I agree, but you can't pull the birds without them.'

'Listen, no, hear me out. I've heard that to get the biggest racks, you need to eat plenty of that mineral deposit, you know, the black stuff, pitch—the cruder, the better. Apparently some old gits who are past their prime and can't even get it up anymore are suggesting that we should cut back on its consumption.'

'What, and lose my rack? No way. My lifestyle depends on having a great rack.'

'Yeah, but listen, think about it. If we all stopped eating pitch, then we'd all have smaller racks and the girls wouldn't have any choice, would they? And, as an added bonus, you wouldn't waste half your time and energy growing and lugging around such a big one and getting stuck in the bogs, would you?'

'I must admit, now that it's getting warmer, it's getting harder and harder to lug these things around. I nearly drowned the other day. I thought my rutting days were over.'

'Oh my God! Look at Roman. Look at the size of that rack. Just look at it; it's huge. We've got no chance, no chance at all. Let's go and eat some pitch.'

'Way ahead of you. I'm on my way.'

But we are so much cleverer than that, aren't we?

We have to have nuclear weapons because the others do.

We have to grow our economy to compete with theirs.

We have to buy cars to accelerate to 62 mph in 4.3 seconds.

We have to adorn ourselves with rags from the Emperor Clothing Company.

We have to live in the finest caravans.

We have to build the biggest pyramids.

We have to open our supermarkets on Sunday because they do.

We have to build the biggest stone circles.

We have to create the biggest empires.

We have to accelerate the destruction of the earth.

Does building a pyramid change the sexual dynamic?

Let's ask Bernie Eecclestone and his model ex-wife. Why not approach Donald Trump and ask his model wife? The cheap shot is to say, 'I wonder what they ever saw in his wallet.'

Now I cannot and would not comment on the loving relationship that they have enjoyed for many years, and I sincerely wish them every happiness into the future. Each party has bought an abundance of assets to his respective relationship, and true marriages have been made. The same can be said of Premier League footballers and their WAGs. I will ignore the lurid tales that drift from the sports pages to the front pages. It's none of my or the paper's business. Put simply, success equates to power, which equates to wealth, which equates to an ability to provide an environment to raise children and be free of want. If wealth were measured by the abundance of food, water, and shelter in a territory, then would that wealth not attract an abundance of breeding females to share in that success? By the same definition, can we not suppose that the successful male commanding his realm can also choose the pick of the presented harem? Beauty is important to the human male. Our shape and appearance is determined by hereditary selection, which proves this point. In simple terms, if only ugly women bred, then we could expect predominantly ugly children. (Of course the perception of beauty is influenced by time and culture. You only have to think of the artificially elongated necks favoured by some African tribes or the tiny feet achieved through crippling foot-binding in Asia.)

In simple terms, rich people occupy the peak breeding territories and avoid the pitfalls of being on the edge exposed to predators. Their stomping ground is in the centre of the herd with the richest resources and the greatest choice of partner(s). The vitality of their genes and physical presence has been converted into the bowers constructed by some bowerbirds (*Ptilonorhynchidae*). These birds construct elaborate bowers (mate-attracting structures) often decorated with collected leaves, flowers, feathers, and even snail shells to display their fitness to mate with females.

Sneaky Snakes

A tenth of the world's wilderness has vanished in the past two decades, research shows.

New maps show alarming losses 'of pristine landscapes, particularly in South America and Africa', according to World Conservation Society scientists.

Only about 20 per cent of the world's land area is classed as wilderness.[2]

North America supports various species of garter snakes of the genus *Thamnophis*. Some species in colder climes hibernate through the winter in great congregations in suitable underground chambers. In the spring the males are active first and can even be seen basking on snow, waiting for the females to emerge. The earliest females are absolutely swamped by the attention of dozens of males, which form a 'mating ball' around the hapless female. The female produces pheromones to attract the males, which works perfectly where the snakes don't hibernate en masse. Without the high congregations of hibernating snakes, the males have to locate and track receptive females. This is done over distance by following the pheromones to the source. In the mating ball, it is down to luck or 'fitness' for the struggle that determines the mating success of the males. Or maybe it isn't. Imagine a scenario where, as a peacock, you have through selective breeding of your superior bloodline achieved a simply irresistible peacock's tail that never fails to achieve multiple mating opportunities. The time

---

[2] BBC teletext September 2016.

involvement and the investment your body has made in producing this magnificent 'babe magnet' is well worth the increased risk of predation, and the demands on your health and ultimately your life expectancy. Imagine that you are strutting your stuff on your stomping ground, driving lesser males from your presence, only to be ignored by the passing females. Yet weeks later the emergence of multiple broods of young fowl proves that mating has taken place and that you have missed out. How can this be?

What you haven't realised is that you have been usurped by visually unimpressive males who have failed to invest in adornment. These visually unimpressive males have invested their energies into developing extra muscle and speed of movement and have developed a mating technique that could best be described as rape. All the females had already been nobbled before you even cast sight on them. The females have avoided lesser males and have been lured into a false state of comfort because their assailers have been adorned in female plumage.

Could this happen? Where would this leave your bloodline?

If success is ultimately measured as the ability to pass on your genes and your whole strategy is designed to maximise your chances, then you have failed. Your entire premise and 'design' has failed, and your species is about to undergo a profound change.

If the peacock could rationalise long enough to consider this idea, it would take comfort in the thought that it could never happen, 'Because we are peacocks and this is what we do.'

Imagine the scrum within the mating ball of garter snakes; the males are primed to respond to the female pheromone and plunge headlong into the melee, desperate to make their presence known and successfully mate. So much energy would be expended in the struggle to be the one. Of course if you were really clever, you could devise a mating strategy whereby you triggered the mating ball frenzy before the females appeared. Let the majority of the males wear themselves out on non-existent mating opportunities while you quietly slip away and then mate with unaccompanied females. If the male garter snakes could rationalise long enough to consider this idea, they would take comfort in the thought that it could never happen, 'Because we are male garter snakes and this is what we do.'

Or did.

Because within the emergent males are males that give off female pheromones that trigger mating balls before slipping out and mating with freshly emerged females. And studies have shown that they are more likely to mate successfully.

In the hypothetical peafowl scenario, the resplendent males have been betrayed by their DNA. But it is also DNA that has contrived to produce the visually female males. Both are a product of their DNA. Only time will tell which form will win in the fight for genetic survival.

The garter snakes have gotten past the hypothetical stage.

Can we learn to betray our DNA? Can we learn to be more risk-averse before the age of 25? Can we learn to control our need to cross the horizon? 'Just to see what's there. Of course, we'll come back.' Can we learn to control our population and remove the gene-seeding imperative? Can we learn to control our appetites and share our planet? Can we deny the 'slow-motion suicide gene'? Can we learn how to kill our elephants?

Can we cancel the World Apocalypse Tour that we have embarked upon?

There is an argument that suggests that our only future lies in being able to travel beyond our solar system. When our sun breaks down in several billion years, the prediction is that it will expand in size so that our planet will be within the sun's expanded mass. On the one hand, we haven't achieved the speeds necessary to escape our solar system in any form of timely manner. On the other hand, isn't this rather presumptuous for a species with a one-hundred-thousand-year history of existence? Shall we start by worrying about the next thousand years?

I like snakes. They absolutely fascinate me in the way in which they move and eat and breed. I have kept and bred various snake species over the years. One little-known fact that I would like to share with you is their unusual breeding technique. When individual males and females have the time to indulge in mating outside the mating ball frenzy, the 'courtship' often consists of the male approaching the female and making contact with a series of 'pulses', where the male strokes the female. The female will signal her acceptance of the male by reciprocating with similar pulses. The male will coil gently around the female until their sexual organs are aligned, at which point the male's penis will emerge from his body and enter the female's cloaca for the transfer of sperm. The male has the option of coiling in either way, clockwise or anticlockwise, around the receptive female. This could be described in bolt terms as having a 'left hand' or 'right hand' thread. The really clever bit is that the male has the use of two penises (strictly speaking two halves, the plural is hemipenes) and has a left-hand and a right-hand penis for either eventuality. It is possible to conclude that the male snake has a happiness quotient way in excess of anything that a dog could only dream about, but there is a potentially serious side to my musings. If the mating snakes are disturbed and the female flees, then she may depart with more than she started with. My question—see if you can get a grant—is, does a left-hand-thread male know that it has lost one hemipenis and therefore only attempt to mate in one direction?

1) No.
2) No, don't.
3) No, don't please.
4) No, don't please, darling
5) I've told you not to do that.
6) I've told you not to do that please.
7) I've told you not to do that please, darling.
8) I won't tell you again.
9) I'm warning you.
10) I'm warning you; I won't tell you again.
11) I won't buy you any sweets.
12) I'll tell Gran how you've been naughty.
13) Gran won't let you have any sweets either.
14) Right, I'll tell Grandad.
15) Grandad will tell you off.
16) I'll take your mobile phone off you.
17) I will, just watch me.
18) You wait until Dad comes home.
19) You won't be able to behave like this at school.
20) No TV for you tonight.
21) No TV this week.
22) Please, darling, will you just listen to me?

Is this a blueprint for a lack of parental control?
No.

It's a comprehensive, all-you-need-to-know training programme for highway engineers.

It is a common 'process driver' of Great Brains to admonish the present society and to try and cajole people into action with the commonly repeated, 'We have to leave a world for our children and our children's children.'

Alcohol, smoking, and a poor diet all contribute to the formation of damaged foetuses with a reduced life expectancy. Given our apparent willingness to physically handicap that most precious of our own possessions, our own genetic futures, our babies, what chance does the rest of the world have?

Does the equivalent of a female-pheromone-producing male garter snake exist in the human world? Is there a class of individual or individuals that has worked out the shortcut to survival and breeding success, a class that has sidestepped the usual obligations and work requirements for survival? By survival I mean the furtherance of individual DNA as the true measure of a future. I will call it the single imperative, the drive to pass down your DNA to other generations for eternity.

I am speaking of a class for whom the bumps in the road have been smoothed out and the future is rosy, a class for whom the needs for food, shelter, protection, and breeding potential can all be met, a class that takes a cut, a tithe, from the non-members, who exist to provide the needs of the members.

This is the class that has cracked the code, a class that can exude female pheromone and achieve its needs in plain sight of non-members who are blind to the influence, a class that is driving the destruction of the planet and its resources in return for its own future, which isn't immune from the destruction process.

Why does it matter, being enslaved to our DNA? Our DNA has made us, shaped us in tune with our environment, and honed us into the perfect humans that we are. We share our world with perfect tigers and elephants and peacocks, and we shared it with perfect dodos and great auks and woolly mammoths once.

We structure our world with perfect Amazons and Googles and food crops and our own perfect elephants. We command our perfect societies and obey our perfect Great Brains and perfect religions that have all evolved to give us our perfect lives.

But within our perfect lives and perfect societies we indulge in the slow-motion suicide of excess and addiction, betrayed by our perfect genes that make our species perfect. There is but one ultimate conclusion: our DNA that makes us perfect also makes us vulnerable to extinction. Our DNA is subject to the same rules and the same disconnect between need and action that our social structures are. Our DNA is bound by corruption, the corruption that distinguishes what we are from what we could be. Our DNA has no intelligence and is honed in the forge of trial and survival, a survival that adopts a non-failed route which does not equate to being the best one. Our DNA would have us fill the wheelbarrow facing in the wrong direction. The task will be achieved and the tests will be passed.

I'm speaking of the same tests that ensured the survival of the Irish bog elk for millennia, the same tests that ensured the survival of Neanderthal and Denisovan man, the same tests that ensured the survival of elephants in Hannibal's army, indeed the same tests that ensured the survival of Hannibal's army and of Carthage and its empire.

I have painted a picture of our world tour of apocalypse. We are driven by our genes to exploit without constraint, to breed without constraint, to *be* without constraint, and within this genetically induced environment lie the seeds of our own destruction. Our success, if that is the term to use, is the foundation of our failure. Our densities and global communication leave us as vulnerable to disease as monarch butterflies overwintering in a few locations in Mexico, as vulnerable to destruction as the limited egg-bearing sands in a select number of valleys that saw the North American locust disappear in a handful of years under the patter of cow's feet. Our exploitation of and dependence on a relative handful of food crop species leaves those species, and with them us, vulnerable to disease. Our need to seek out new horizons and stretch our boundaries has the potential to produce our own Armageddon. Some of our scientists believed that the chain reaction created by the first atomic bomb would never end and would see the destruction of the planet. Yet we detonated it anyway. Some obviously believed that it wouldn't. Their side of the argument pressed the button. Fortunately the first detonated atomic bomb didn't create a chain reaction leading to our destruction. Unfortunately it still can. The chain reaction that none of the scientists could envisage was the pursuit of ever greater numbers and power ratios of atomic devices and delivery systems, fuelled by opposing political blocs intent on ensuring their own survival arriving at the spectre of the world tour of apocalypse, the abridged version. That would be one set of Great Brains ensuring their survival over another set of Great Brains. That just leaves us 'piggies in the middle' to survive outside the nuclear bunkers. If individuals can so readily pursue slow-motion suicides, why can't an entire species chase the same ambition? Would we know?

Life thrives on diversity. Indeed, diversity is the driver of life in all its versions as it seeks to occupy every conceivable niche and even those not conceived. Yet our species has stumbled onto a position of niche destruction and life-extinguishing scourge without parallel. We are as unthinking as any plague and as virulent and voracious as any natural ones that we have encountered. Our modern-day version of life is sweeping across the globe in pursuit of GDP and league tables of attainment, coupled with wholesale destruction of all alternative views. Subsistence farmers and hunter-gatherers are swept aside in the tsunami of global exploitation that passes as 'Western society' as we dig holes in their environments to fill the holes in ours.

The failures of our society are there to see if we would but open our eyes. Our elephants will surely die because they are not immune from the forces of nature. Our elephants are beset with their own DNA, which is seeded with the roots of their own destruction. Our NHS is doomed to fail as surely as our 'Western society', because it is unsustainable. We will beaver away to prop it up and change the 'bouquet garni' while our Great Brains strive to maintain the illusion of control as they preside over the continual failed enterprise of satisfying unlimited needs with limited resources. Every advancement in treatment and range of options accelerates the demise of Western society as its sustainability is compromised.

How to kill an elephant? They are already dying. Our problem is that they will die when, and only when, they are ready. That's after they have trampled through our own tightly packed troops held in tight formation behind them so as to exploit the impact that our elephants have.

Does it have to be this way?

Stromatolites are evidence of the earliest life forms known from the fossil record. We owe our existence to them. This life form lived in shallow seawater and over millions of years released oxygen into the early earth's atmosphere. Stromatolites are formed by cyanobacteria (blue-green algae). They convert carbon dioxide to oxygen through photosynthesis and produce limestone as a by-product, resulting in structures not unlike coral. This limestone forms mounds (stromatolites: layered rock) which can be dated back in fossil form 3.5 billion years. In just 3 billion years, they changed the oxygen content of our atmosphere from 1 to 21 per cent, which led to the oxygen-breathing life that we are so familiar with. Estimates suggest that around 2 billion years ago two bacteria joined together into a more complex single-celled organism (eukaryotic microbes) that began the evolution to multicellular life. As a final bonus, stromatolites can still be found in three locations around the world: Australia and two sites in the Bahamas.

To recap, cyanobacteria lived for 1,500,000,000 years before the advent of eukaryotic microbes, which triggered the move towards more complex life.

That is not to say that there is only one cyanobacteria. There are many different types. So even within the 1.5 billion years of precious little else, change and development was occurring.

Let me suggest that cyanobacteria are, in life terms, a fail-safe, a fallback to a stable condition of what I will describe as a low-risk survival mode. Just a model, this ignores the wonder of the formation of life in the first place and the evolutionary pressures that cyanobacteria are subjected to. The basic entity is precisely that, basic, and has survived for billions of years. In the event of a major extinction event, it becomes reasonable to suggest that this basic life form has more chance of surviving that any of the more complex ones. In the race for life, then, the simpler, the better. Development then represents a risk, a greater reward in terms of opportunities presented and exploited, to be balanced against a greater potential for failure. In one sense this is contrary to the concept of survival of the fittest, reacting to pressures to change, when this basic life form has resisted change for billions of years. Survival of the fittest implies a certain skill base, an advantage over lesser organisms that are disadvantaged in the race. But if the race is very simple, then cyanobacteria seem to have it sewn up, lock, stock, and barrel. Other life forms have occupied new niches and taken advantage of new opportunities, but on the timescale of life measured by a billion years, the vast majority of organisms wouldn't even register.

We know from the Clovis point that Stone Age human beings resisted change and accepted their position within the landscape. The advent of the new technologies of metalworking began the move towards specialisation and compartmentalisation of labour and reward. Where before all could gather or hunt, the skills and techniques developed into castes or classes of workers who took their reward from the communal pot in return for specialist services. With a new hierarchy of technical ability having been established, it would be only a short step to further hierarchical strata. The smart ones would realise that guarding access to the gods or knowledge of the seasons and the best time to plant seeds would yield an advantage over fellow contributors. This advantage could manifest itself in having a tithe of others' effort by way of reward.

As we fast-forward through the Bronze Age and Iron Age, we see that social strategies for the exploitation of other humans would have developed at an accelerating rate. These strategies will have been corruptions of the overworked command-and-control structures of family groups. Initially devised to maximise breeding success in an uncertain world and avoid inbreeding, they were stretched to new purposes. This would manifest itself in greater positions of power and wealth and housing. Ultimately this process would lead to palaces and fabulous wealth for the 'leaders' as pyramidal hierarchies developed. It has been said that MAD (mutually assured destruction) could return us to the Stone Age in a near instant. But what would that actually mean? Even the president of the US would find precious little satisfaction in lording it up over his assembled entourage. In *Monty Python and the Holy Grail*, the question is asked before the reply is given: 'The king's the one not covered in shit.'

The president, then, would be the one not covered in shit. The rest of the entourage would not be so lucky. A return to subsistence farming, or hunting and gathering, would be a universal leveller and would see the destruction of pyramidal hierarchies. These hierarchies would all have been underpinned by extracurricular activities, whereas in the New Stone Age priorities would revolve around survival and the daily provision of food, shelter, and security. Until the process would start all over again. Given a fair wind and several thousand years, we might see the reinvention of nuclear weapons so that we can recycle our history again.

Our species, when it began extracurricular activities, entered that most basic of contraptions, the spinning metal wheel much beloved by hamsters. Each and every step taken in this wheel leads to the need for another step to maintain the momentum, which in turn leads to an acceleration of the process until such time that the momentum overcomes the physical ability of the hamster, which normally becomes a spinning passenger before being spun out of the wheel. Let us suppose that man-made global warming is real and that when global temperatures rise above a threshold, say, 5°C, the natural buffering activities of weather patterns and algal blooms and accelerating plant growth will be swamped and temperatures will begin an accelerating climb to new levels unconducive to human life. Where would we place the blame? Is this not a possible natural conclusion to our extracurricular activities?

How about our population reaching the 13 billion threshold that also corresponds with the earth's inability to recycle our requirements? Fresh water isn't fresh, marine life is contaminated with toxic pollutants, soil is compromised with waste and exhausted, with a dearth of essential plant minerals for healthy growth, and food demand is such that crops are harvested before maturity because we simply can't wait for them to mature. Not forgetting the wars for resources, which can only result in widespread death. No prisoners can be taken; it's self-defeating otherwise.

And I worried about elephants with no off switch.

Cyanobacteria have had a past and probably have a future. *Homo sapiens*, on the other hand? It has a past, but each step on the evolutionary wheel has been underscored at every change with an additional level of complexity. We learn that so-called unreinforced masonry structures are not earthquake-proof. In simple terms this describes most of the UK's low-rise housing, which we usually refer to as 'bricks and mortar'. Our houses, however complex in design, are constructed of variations of fired clay bricks glued together with cement-containing mortar. A devastating earthquake would shake most of these houses to the ground and the houses would become uninhabitable. But a significant number of the bricks would survive and would still look like bricks and have a future as a brick, whereas the houses? But it's all right, because we don't live in an earthquake zone. But it's all right, because we have food security. But it's all right, because we have no water shortage. But it's all right, because we have technologically advanced weapons and armed forces to defend us.

Where do we place our leaders in this equation? Are our leaders simply one rung higher than we are in the wheel of revolution? Going with the flow, overtaken with the momentum, and driven ever onward by the masses behind? There are no leaders, *only followers who have had a head start.*

If You Tolerate This,
Your Children Will Be Next

Prior to having a minor operation on a nerve many years ago, I was introduced to my surgeon, who announced that she was a plastic surgeon, saying she was on the case because of the delicacy of the procedure. I was surrounded by about eight nurses. The room was full of women, with no men present. I wanted to say that I had had plastic surgery before when I had had a mole surgically removed from my penis. I knew that the questions that would follow would revolve around malignancy and subsequent screening. I also knew that I couldn't deliver the punchline, 'Next time I'm sticking with hamsters,' because I wouldn't physically be able to speak. The operation was also quite delicate, with scope for significant nerve damage in the event that I couldn't stay still. The missed opportunity has been a major source of regret for me ever since.

Our metaphorical hamster is now a passenger within the spinning wheel simply waiting until the momentum spits it out. We have no leaders.

Even if we had leaders, the other basic requirement of control is missing, a brake.

Even if we had a brake, the brake could not overcome the momentum.

What we will have is a break.

And cyanobacteria will disinherit humankind of the earth.

As I describe the accelerating demise of *Homo sapiens* into oblivion. it strikes me that the same can be said for the vast majority of life on the planet. Blue-green algae can photosynthesise to produce energy without the need to crawl or walk. They can reproduce and survive quite nicely. Their only requirements are brought to them by the watery mediums that they inhabit, provided they stay within range of sunlight for energy. And if they don't, they die, but others will remain to continue the species. Walking, flight, eyes, teeth, feathers, and brains can all be regarded as extracurricular to the requirements for life. In that sense, then, they can also be seen as 'life limiting' as the specialisation and investment in these new modes or tools also exposes them to failure. How long could elephants live without the activities of humankind and indeed other predators? The simple answer is, until their teeth fall out, at which point they starve to death. But in our world it is the retention of teeth, the elongated continuously growing front teeth that we call tusks, which are causing their imminent life-expectancy issues. That is not to say that the absence of tusks would guarantee the survival of the species, because there is the competition for food and space and conflict brought about by the other agricultural uses of the territory.

Without sufficient food resources, the human investment in the brain cannot be sustained. Bats can be broadly split into either fruit- or insect-eating species (ignoring vampires and frog eaters), and the difference can be seen in the manifestly different-sized brains that similar-size bats enjoy. Fruit eaters have a better diet and can afford to invest in larger brains than insect eaters. The theory goes that fruit eaters need to memorise the location of fruiting trees and their fruiting patterns given the absence of seasons in the tropics. But having said that, if they didn't eat fruit, only insects, it wouldn't be a problem, would it? Cause or effect? Exploiter or exploited? 'I saw an opportunity and I took it.' But then the fruiting trees also need an external vector to provide for the widespread distribution of digestive-system-resilient seeds. It's nearly enough to make you believe in a higher power, an architect, a great brain even. While I realise that thrushes and tits have found God (with a little bit of help from me; donations to the temples of *Turdus* and *Parus* can be sent to my publishers), would we consider that algae have a similar need?

Being an alga is safe without much risk. Being a tiger is not safe and is full of risk. What world do we want to inhabit, the safe and boring one? This is a rhetorical question that we are doing our best to answer through our actions. The Stone Age is a term coined to reflect on the best cutting technology of humankind and its predecessors. The winner is the Algal Age, which didn't even need to cut.

If you tolerate this your children will be next. If we succeed in our attempts to drive tigers, elephants, rhinos, and the like to extinction, then the next logical conclusion will be 'If you tolerate this, then your children will be next.' If we can place no value on anything beyond the demand of humankind, then ultimately the only thing left to demand will be humankind. Surrender your life, your right to existence, so that others may exist in your place.

There was an item on the Internet yesterday which I didn't read but which was headlined as 'Religion has a value of $1.2 trillion a year'.

Just suppose that we could attract the attention of a passing alien intelligence that has advanced to such a state that interstellar distance is no object. If we were to survive the new diseases, and the rest of our world could survive them as well, then we would be left with the hope that the aliens would not simply see us as communal ant colonies at best worthy of study and at worst unworthy of any consideration. Then this scenario, which as a species we have actively pursued, represents yet another horizon crossed as a knee-jerk reaction to what, exactly? We were bored, so we sent a message beyond our world for others to come and destroy us? We are never happy with our lot and always want more? In a sense, the shepherd boy has gotten bored and cried, 'Is there anybody out there?'

Any attempt to save our species by galaxy-hopping is premature given that we are a few billion years early, relative to our sun's predicted demise.

The chance of our species surviving even the next one hundred thousand years is remote, so why consider the next five billion?

Of course the smart money, if it existed, would be worried about now.

The question we ask is, 'What can I do to exploit my fellow man while providing myself with a comfortable standard of living by doing something that I love?'

History describes various inventors of 'flying devices' that often resulted in their inventions and pilots achieving an early death. The proscribed method of testing consisted of attaining some height above a precipitous drop before launching oneself off to eternal glory (if you succeeded) or eternal glory (if you believed in God).

'Ah, but that is how we learnt to fly,' I hear you say.

And having learnt to fly, what else did we forget to learn?

Flight burns voluminous quantities of fossil fuels.

It accelerates the spread of pandemics.

It destroys natural habitats worldwide in favour of holidays under the sun.

It destroys natural habitats worldwide in our pursuit of local crops exported to a supermarket near you.

It provides us with delivery systems of military death.

It facilitates the destruction of local agriculture and local manufacturing as it drives the lowest common denominator upon us.

It changes the daily weather pattern worldwide.

Brilliant invention or scourge of the earth?

'Ah, but it's not the invention itself; it's what we do with it that's wrong,' I hear you cry.

What we do with it is use it and damn the consequences, as the consequences damn us and the world we share.

Consider the development of the high-powered rifle.

Or car.

I have previously considered the fridge.

Or medicine.

Or the NHS.

The rifle could have been used for target shooting—think of all those gold medals—or robbing banks, or school massacres, or big-game hunting, or poaching, or presidential assassination. Every invention is actually a force multiplier for change and an accelerator of our apocalypse. We embrace these new toys with no thought of the consequences.

But we don't need to consider an alien life force crushing 'ant colonies' underfoot when we have our own home-grown one doing the job quite happily on its own.

The means to our end is incorporated within our means to exist, in dynamic equilibrium with failure and success. There are no guarantees and past success is no indicator of a future. In fact, past success is a progenitor of failure, because success attracts competition. Everyone aspires to emulate Google and copy and overtake them with an edge of their own. To truly survive, we need to take control of our actions beyond those foisted on us by our base instincts and our DNA.

Have we an intelligent thought between us?

Answers on a postcard?

Within every organism's DNA is what I will call an extinction gene. This is the gene that will, given a fair wind, take a turn on the road of survival and enter a cul-de-sac that leads to extinction. In isolation the gene has no chance, but it conspires with other pressures to exploit its opportunity. An animal that adopts a breeding programme whereby sexual maturity and the ability to reproduce is extended to single offspring born a year apart after thirteen years of sexual immaturity would find it difficult to withstand mass predation. It would struggle to recover from disease and repopulate former ranges after drought or similar natural occurrences. Or even unnatural ones. Our species invests such time and stock into its brain and intelligence that it has lost the ability to recover quickly from major ELEs. And while we might struggle to recover our position, what could occupy our niche to take advantage and deny us our return? An intelligent life form might seek out an insurance policy, let me call it a reserve, to tuck something away for a rainy day. Let us construct an ark and dose it up with all the animals of the earth in breeding pairs, together with all the other fauna and flora, while we wait for our rainy month's flood to abate. Let us construct seed banks of genetic material in frozen holes in the ground. Let us surround headline species within reserve boundaries and zoological gardens. Let us build a fleet of interstellar craft to carry our future beyond immediate peril. Or we could start by halting the ELE that we have embarked upon with such unthinking vigour. One of the options is not a diet; it is eating less food.

Cyanobacteria have a range of options open to them that remove them from their origin and that would result in what we would describe as evolution. Given the longevity of these species, it is reasonable to suggest that the options available have been limited and that their success is measured not by their diversity and adaptation but rather by their continuing existence.

We could consider the success along the lines of, 'I am what I need to be to survive. I am stable. I do not gamble.'

Plan B, because there has to be a plan B, could be considered to be something along the lines of, 'I must adapt and alter to take every advantage of every possible scenario that presents itself. I am unstable. I am at risk. My initial thread of origin may persist, but all my forms and manifestations will probably die. I am a gambler at the high-risk game of life.'

Plan A yields basic life.

Plan B yields the wondrous variety of life that we enjoy.

But life has no knowledge, no ambition; it just is. It does not care what shape or form it takes. It has no need for Plan A's or B's; it has no plan, just a series of random occurrences that are favoured. But when we say favoured, what do we in fact mean? It's a bit like having sex—consensual rather than rape, that is. 'Did you, at that time and at that place, freely consent to have sex with that individual?' A bit oblique for you? The random series of events that are favoured at that time and that place is favoured precisely because it is at that time and at that place. If, as a bird, you surrender your ability to fly to escape from non-existent predators on your island home just before an earthquake lifts a land bridge to the mainland, you have a problem or, rather, not much of a future.

If, as a fish, you have just evolved to live in the freshwater lake when the sea breaks through the sandbar, you have a problem or, rather, not much of a future.

Every step change in the evolution of all the diversity of life is a gamble. The step change represents a successful bet. How long will each winning streak last? Ask a dinosaur or a sabre-toothed cat or an ammonite. In theory *Homo sapiens* has the ability to load the deck in its favour. In practice?

I see a parallel with government organisations and policies. Every idea is a gamble. To make it work, we bend the rules and corrupt the truth to fit the idea. The idea does not in fact exist in practice. In practice it is the corrupted idea that appears to work. It only appears to work because the measurement of success is corrupted to serve the practice. That is why we all drive at or less than 30 mph in residential zones. We create a self-reinforcing loop of corrupted measurement to confirm our success.

Let us return to the alien spacecraft that crash-landed on our planet nearly four billion years ago to bring life to this otherwise sterile rock. (Hypothetically speaking; it offers no solution to the origin of life; it just pushes the question over the horizon to another time and another place.) The superior life form, the one with the doctorate in physics and rocket science, died soon after impact in our inhospitable environment. But if we follow the argument that contained within this being were what we would know as bacteria, then these would be the life that seeded our planet. What the superior life form had succeeded in doing was to move the 'spark of life' from one galaxy to another. In its purest form, then, the spaceship containing the bacteria succeeded in moving its DNA from one solar system to another. While we think of the spaceship as being the vessel whereby the superior life form achieved this feat, in reality the superior life form was the vessel containing the bacteria. The bacteria were passengers of the vessel as much as the alien was a passenger of the spacecraft.

(I do accept that bacteria were probably not the first living organisms on our planet. The various competing theories on the origin of life look for earlier proto-organisms or even just proteins and RNA, not even DNA. Think of my bacteria as being the base unit, the starting point, a peg to hang my idea upon. And given that they are the earliest-known fossils ...)

It's within *A Hitchhiker's Guide to the Galaxy* that dolphins escape into space as the superior intelligent life form on earth. Assuming they have the ability to travel through space then in bacterial terms, the dolphins are the vessel moving the 'origin' of life, bacteria. But we don't need to travel between solar systems to consider the principle. If cyanobacteria can only survive in salt water (not the case, but consider the argument), then other forms of advanced life simply exist to provide vehicles for cyanobacteria to move from one environment to another. Considering the size of a single bacterium, conquering the entire world is not a bad achievement.

And you thought it was to the glory of God.

That's it, I have finally worked it out after seven-hundred-and- nine odd pages. I have had my Archimedes moment. It went something along these lines.

'Oh bollocks, oh bollocks, bollocks, bollocks, bollocks, not again. She'll kill me. She warned me; she even threatened me. Oh my gods, what am I to do? Think, man, think. You have to come up with an excuse for wetting the bathroom floor again. Ah! I have it. Eureka!'

We Have No Leaders, Only Followers Who Had a Head Start

My lawnmower has died. One of the two cylinders has lost compression. The mechanic who came out to service the machine was thorough. We had a chance to engage in a debate about serious stuff. Unlike women who simply chat, men won't be caught doing that. I won't name names, but it emerges that the EU has set maximum noise levels for small garden lawnmowers and the lawnmowers they supply conform to this criterion. Unfortunately these mowers are unable to discharge the clippings into the clippings collector because they don't have enough power. What the mechanic told me was, 'We routinely tune them up another 100 rpm so that they will work properly. And you can hardly hear any difference.'

My company always supplied free tea, coffee, milk, and sugar for the workforce to enjoy hot drinks throughout the day. Occasionally the supply of milk was out of kilter with the need, and quite quickly we arrived at the stage where milk wasn't available in the afternoons. This was because some individuals had taken it upon themselves to hide milk for their own consumption into the afternoon so that they were not the ones to miss out.

Quite frequently whenever there is an interruption to the supply of petrol to garage forecourts, the fuel suppliers insist that there is enough petrol to supply the normal weekly usage. This is almost universally greeted with 'panic buying' at the pumps, with mile-long queues snaking around the streets as individuals ensure that they are not inconvenienced by inconveniencing everybody else.

Imagine the scenario if we were running out of water.

Fascism

The simple word 'fascism' conjures up visions of world war and unspeakable atrocities inflicted by one set of individuals on another. We associate it with the worst of Nazi Germany's action during Adolf Hitler's Third Reich. We tend to airbrush out the political scenes in Italy and Spain, which of course had elements of brutal zeal and even civil wars but tended to avoid the worst excesses, even though they were also fascist countries.

If you accept that leaders are followers who had a head start, then you begin to place Hitler in a slightly different context. I am in no way trying to defend this mass murderer and his actions. I am simply questioning our historical assumptions. Let me step back a little from this extreme example and look instead at the New Labour project under Tony Blair. (I find it extremely hard to write Blair, I keep wanting to transpose the *i* and *a*.) Nowadays he and his legacy is largely reviled, and the new Labour Party which has surged to the left is keen to disassociate itself from the Blairites and their policies. Let us break the process down into simple ideas based on my premise about followers who had a head start. What Tony saw was an unelectable left-wing Labour Party which had been left behind by the majority of the UK's centrists voters who had been seduced by Thatcher's policies of inclusion into house and share owning into the belief that they had a personal stake in the country, a stake that they had worked hard for and wanted to protect. Blair repositioned the Labour Party to the centre ground and thereby broke the Conservative hold over government after seventeen years. Was Blair the leader, or did he simply move into the path of the herd and declare himself as leader? He didn't move the herd; he simply hijacked its course. I remember the night of his first election success with a heavy heart, but then I consoled myself with the thought that whatever happened, I wouldn't wake up to a socialist government. (That is not to say that I was right wing. I had lived through the union-dominated Winter of Discontent and saw Thatcher as a blessed relief.) But when I say hijacked its course, I would define this as simply marched ahead with one eye while the other was firmly watching the course of the herd behind. Blair only lead where the herd was already travelling.

What then of Hitler? He began as a member of an obscure political party and rose to attention because of his oratory skills. Your oratory skills only work if your words are aligned with the listeners. In simple terms, it's no good talking in Italian to German speakers or, if you'd prefer, preaching Christianity to a Hindu. Your words need to fall on receptive ears. You have to push on an open door. If you asked the German people in 1933 if they wanted to go to war with the rest of Europe, I suspect that the answer would have been an overriding *nein*. If you asked the German people if they felt aggrieved by the social and economic upheaval they had suffered since the end of the First World War, then I suspect the answer would have been more mixed. If you asked the German people for a mandate to bring respect back to 'this once proud nation', around a third of the voters would have said yes. It is from this minority mandate that Hitler came to power, before orchestrating 'attacks upon the state' and the granting of emergency powers to government. The rest is history, but for a significant proportion of the German people, Hitler simply led the herd in a direction that it was prepared to follow. Or rather he followed the herd from the front. A contradiction in terms, I know, but think about it. He didn't force the people to accept the arrest and removal of trades unionists and communists. They, by and large, accepted it. And because they accepted it, he could continue to do it, and could continue to escalate his actions with the continued acceptance of the masses. Ultimately, they both realised their ambitions, the leader and the herd.

With Nazi Germany I have obviously selected an extreme example. Let us consider something a little less extreme but current. The figures keep changing and many are estimates, but for my purpose I'll select the nice round sum of one million and the timescale of one year.

One million people, which represents quite a large herd, have decided that their lot in life would be improved if they could move to Europe and start a new life. This herd has many origins, from Africa to the Far East, and consists predominantly but not exclusively of young men. All can recite instances of persecution and fear of remaining in their native homelands, and all can see the future in the 'Promised Land'. The herd has embarked upon a migration and has adopted a trail over land, desert, and sea to arrive at the entrance to fresh pasture. Every moving herd member reinforces and generates the momentum of the march. To begin with, we need to ask who the leader is. But one million people from such a huge catchment area cannot be following a leader. They must be following something else, let's call it a dream, a desire to reach the Promised Land where the streets are paved with gold. But not all are seeking gold-filled streets; some are seeking the promise of a job, the right to freely worship a chosen God, the freedom to love their chosen partner, to escape the threat of violence, to escape the devastation of war. The list of reasons will be huge, but the end desire and expected result will be achieved if only the destination can be reached. To realise the dream. To cross the horizon. Just to see what's there, of course.

If some homosexuals left Britain for fear of persecution and fled to, say, California, where more liberal attitudes welcomed them without discrimination, where does that leave those homosexuals who couldn't leave or overcome the other reasons that kept them in the UK?

If some Christians who couldn't freely worship in Pakistan left and moved to the UK where their rights would be observed, where does that leave the remaining Christians in Pakistan?

In either case, does it strengthen or weaken the position of the 'stay-behinds'?

It certainly does nothing to encourage a more tolerant outlook in the donor countries. If you are motivated enough to uproot and discard all you have bar the clothes on your back to make a perilous transcontinental journey to an unknown destination and outcome in the hope of a better future, what does that say about those who stay behind? If you strip the movers and shakers from any environment, then where is the motive force to effect change at home? It speaks volumes about the command-and-control structures that permeate the migrants' home countries that fail to address the needs of the people. That would be the needs of the people who are so assiduously ignored by the Great Brains who purport to lead and provide for them. We are not immune. Try to get redress for your neighbour side-swiping your car as they turn into their drive and see how far you get.

The need to leave is fostered by dissatisfaction at home. The numbers leaving is directly proportionate to the failure at home to provide for the desired requirements. Our government is intent on offering aid to these 'donor' countries in a vain attempt to change the dynamic and to remove the need for each country's exodus. Our aid goes to the Great Brains and their institutions that are failing the needs of their own people. As we donate, do we simply reward the Great Brains for their failure to provide? If I were even slightly cynical, heaven forbid, then I could make the case that the quantity of aid received is directly proportional to the failure to act. If people stopped leaving Gambia, then the aid would dry up and leave the Great Brains financially impoverished. Migrants then have a monetary value if they continue to leave, but no value if they stay. Our government holds up its actions as the 'right and considered way' to offer a solution by tackling the problem at its source.

I watched part of the *Ross Kemp: Extreme World* (season 5, episode 2) which looked at the migration attempts across the US border from Mexico. Ross always tries to show the gritty nature of his exposés from all sides of the argument. In this programme he interviews all sides of the process. What struck me as particularly perverse was that the best attempts of the US border guards were, at best, attempts. Ross, travelling north with his Mexican priest guide, reached a point beyond which he could not cross. Quite simply, past that point, and without permission of the people-smugglers, he would have been murdered for failing to pay his fare. On that basis, the criminals have total control over access to the desert border crossings, while the legitimate border forces arranged against them can only play cat and mouse with success. If the object of the different enterprises is to control access to the border, then only one is working properly. During interviews, it was declared that the typical fare was $4,000 to $5,000 per head.

Now if I were to suggest that the US government pay the people-smugglers $5,000 to $6,000 per person to ensure that all would-be migrants be delivered to border posts for processing and return to their original home countries, you would instantly see the folly of my idea.

1) Richer migrants would offer $6,000 to $7,000 per person not to be so delivered.
2) Poorer migrants would no longer attempt to cross as soon as their treatment became known.
3) A reduction in migrant numbers would see a drop-off in earning capability for the people-smugglers.
4) The people-smugglers would go out of business due to a lack of 'customers'.
5) The desert border crossings would be free of people-smugglers, and migrants could attempt to cross without fear of being murdered.
6) The number of migrants crossing the border would increase.
7) The US would increase the bounty per head on apprehended migrants delivered to border posts.
8) Richer migrants would offer $7,000 to $8,000 per person not to be delivered.

Perhaps we should build a wall.

(Joke.)

If you take the view that the Great Brains in migrant donor countries are seen as criminals by the disconnected would-be migrants, then what is the difference between my foregoing plan and our government's plan? If the would-be migrants believe that their government is corrupt and fails to address their needs, why do we expect our aid, which reinforces these corrupt Great Brains, to have any material impact on migratory pressures? By corruption I mean the disconnect between intent and result. Think four-hour waiting times at A & E.

What is needed is leadership.

We have no leaders, only followers who had a head start.

In the American model, what would be the cost in thousands of dollars per head for each migrant delivered to the border post in their own country of origin, that is to say, for each migrant who decided to stay at home.

9) At what stage would the people-smugglers pay potential migrant/tourists $3,000 per head to go on a coach trip, followed by a short walk to a US border post before an expenses-paid flight home?
10) Where is the valid solution?

But we have to try, because doing nothing is not an option.

An option is doing nothing.

Build the wall.

What of the humanitarian consequences?

Provide potable water at thousand-metre intervals and signpost the nearest tap at fifty-metre intervals. Provide ostrich eggs.

Normally when the herd reaches an impassable barrier, it changes course.

When the leaders reach the wall and fail to pass through it, they will change direction and others will follow. These will be the followers with the head start.

BBC text 23 September 2016: NHS statistics show that over 40 per cent of children didn't see a dentist last year, a figure that runs into millions, a figure the British Dental Association (BDA) says is embarrassing.

The BDA said that regular dental check-ups are the key to preventing tooth decay in children and urged the government to invest in educating the public. Tooth decay is the most common reason young children go to hospital. If only their parents would follow the guidelines that recommend children should see a dentist at least once a year.

Read this and accept the figures at face value; after all, they are NHS statistics. To recap, then, after nearly seventy years of the NHS and the welfare state, and after decades of universal education, this is the position we have achieved. What would the result have been if we didn't send our children to school? But it isn't our children's fault, is it? They cannot be held responsible. It's the parents of these children who don't take them to the dentists, the same parents who are also a product of free schooling, as their parents were as well. I note that the BDA wants the government to 'invest in educating the public'. Forgive my cynicism, but which education and which public is that precisely?

And don't forget, should the chance ever occur, that you would have to have been in a coma not to have seen the myriad of advertisements across the entire media spectrum extolling the virtues of toothpaste brand A, B, C, D, E, and so forth.

If you are going to get nothing for it, do nothing.

Perhaps I have alighted on the answer that I began to allude to earlier in *How to Kill an Elephant* about saturation and early stage dementia. The problem is overexposure that leads to the filtering out of extraneous information without the ability to 'filter'. That is to say, all of it gets blocked and none of it penetrates. Effectively, without the ability to receive, process, and act on information, we are in a coma.

Human reasoning suggests that starlings congregate in large flocks to protect themselves against predation. The idea goes that falcons and sparrowhawks cannot discern individual birds within the flock and therefore fail to catch prey. Yet we also know that dominant birds occupy the inner reaches of the roosts, where predation is reduced. I would propose that the starlings have adopted a 'bait ball' survival strategy whereby the object of the exercise is to hide behind your neighbours so that they will die and you will not. If the falcon cannot isolate an individual bird, then it cannot catch it, whereas I would suggest that it cuts both ways. Individual starlings cannot see the falcon because starlings get in the way. We are back to the flock of geese, which react without knowing what they react to. They are unable to filter out any information, because they can't receive any and are effectively in a coma.

Yet falcons do catch prey and will visit the large flocks every evening as they form. Indeed they are drawn in specifically to the abundance of prey.

Our children, on the other hand, go to school, a school where they can hide among themselves among teachers, who hide among themselves within an education system that hides within itself. Now I realise I am spouting abject bollocks here and that the education system has the exam results to prove it. Because it is so successful at providing comprehensive education, we have no tooth decay caused by neglect, we all stay within speed limits, we all maintain physical fitness, and obesity is not an epidemic, to name a few.

Furthermore, we have the NHS releasing statistics to drive an argument that has been picked up by the BDA to drive an argument to extol the government to take further taxes out of our pockets to invest in educating the public so that we can save our taxes by reducing hospital admissions and improve the health of our children's teeth. And you can rest assured that any 'savings' will find a good home in providing for other demands. All of which simply reinforce the pro formas of the Great Brains who justify their existence by putting all the constituent parts in place to ensure the health of our children's teeth by ensuring the health of the Great Brains.

While the BDA says the figures are embarrassing, what are they embarrassed about, the fact that millions of children don't see a dentist, or the fact that after decades of free education the population is just as ignorant as ever? But of course we have the answer, because the monosynaptic reflex is to extol the government to provide more education. (Knee-jerk reaction. Have you been paying attention?) Why is there no school certificate for pupils' oral hygiene? (I know, you can't embarrass pupils for having bad teeth; it might give them a complex.)

Just a thought: should the headline have been a little different? The gist of the article is a lack of an education. So, in the best traditions of 'Forgers foiled by sharp-eyed bank clerk', the headline should actually have read, 'Millions of schoolchildren do not get properly educated.' The article is not about a failure to look after teeth; it is about the failure to educate children. If education is the cure, then a lack of education must have been the cause. The educationalists will point to something along the lines of, 'You can lead a horse to water, but you cannot make it drink.' So why lead it to water? Why not wait for it to go to the trough? 'Yes, but they all have to have the chance' would be the reply. Nowadays, of course, we keep them at the trough until the age of 18 and keep repeating and repeating the tests until they achieve the minimum requirement of English and maths to reach a given standard. This maintains our children as net recipients of taxes for as long as possible. Presumably this is to ensure their tax-productive careers through their working lives. How then do we measure their tax years, by how much they produce or how much they use? The present mantra ensures a basic ability to communicate and count and in no way ensures that pupils are ready or prepared for a productive adult life.

In days gone by, knowing that you were going to be a gladiator saw you train in fighting skills that might ensure your survival as well as provide better entertainment. Whereas now we train in more or less anything that takes the fancy of the syllabus creators, without ensuring our survival outside the hospital's tooth surgery. Does it not strike you as strange that our children have to remain at school until they have achieved a minimum of English and maths while we have simultaneously thrown our European-wide borders open to any EU member citizen who is prepared to find work? It seems that employers and educators have mutually exclusive requirements. 'Can't speak English? No problem, we'll interview you in Polish, give you a company handbook in Polish, and put you on our Polish-only shift.' It appears that the criterion for employers in many sectors of our industry is to turn up with an appetite to work. English skills and the ability to communicate are not required. Time after time I receive Internet van delivery drivers on my doorstep who cannot speak English. (I keep mislaying my crossbow bolts.) The employer's perspective is essentially that they will pick it up as they go along. A different perspective is that employers will accept a completely different range of skills than those that our educators see as essential. In military terms it's a bit like sending highly polished .303 bullets to the front line where the soldiers only have .50 calibre weapons. The front line will gratefully accept unpolished bullets as long as they fit the breech. Tell me again why we educate our children?

I think I have an answer: it's to keep our educators in employment. If our educators understood their failings, then they could not justifiably take their salaries. They take their salaries because they are dependent on not understanding the problem. Their salaries are dependent on their not understanding the problem. We are back to my electricity company typist taking a salary for doing nothing. These typists knew they were doing nothing but still took the wage, and all in an environment developed throughout the entire company from typist to chief executive. Can you spot the camouflaged problem hiding in the long grass, hidden so effectively against detection? Doing nothing is not an option; doing something is not a solution.

The purpose of education is to provide employment to the educators. The purpose of education is to reinforce the need to educate. The headmaster has no employment without a school to manage. A teacher has no employment without a class to teach. A child has no employment without an education. Every causal step reinforces the one before and the one after in a self-perpetuating cycle without the need for thought. Until an unknown switch is thrown and the 'teached' become the teachers. How's that for a past participle? I know, it should be 'taughted'. It's almost as though after years of travelling north we have suddenly started travelling south. Remarkable.

I seem to remember reading an article about Lebanon. It was just after I'd been to Paris and had taken a taxi ride with a former university lecturer. I had read that Lebanon had among the highest university-educated populations in the world. How had I come to share my taxi with the former university lecturer? For all the education, Lebanon was engaged in a near two-decade-long civil war, and my former university lecturer was driving the taxi as the only employment in France he could secure. Have I mentioned the reduction in the UK's murder rate from medieval times? Something to do about higher standards of education and greater recourse to civil means of redress? Was this an exercise in social engineering or the removal of criminals from the gene pool (which is also social engineering)? Where shall we hang the laurel wreath? (If only hanging hadn't been outlawed in the 1960s.) What would it take to plunge the UK into near civil war with markedly increased murder rates? 'Yes, but we are too civilised.' If I asked the question a little differently and referred to Great Britain and Northern Island, would your reply be a little different?

Convention on International Trade of Endangered Species of Wild Fauna and Flora

There is currently a debate at CITES (how was I to know that they would copy my idea?) about the best way to save the African elephant from extinction. The debate revolves around outright bans on trade versus allowable but controlled trade. One way boosts the pressure for poaching; the other puts a value on the beasties and their future survival.

How to actually preserve the existence of elephants is not even on the agenda. In the last one hundred years or so while the debates about preserving elephants and other major species have been engaged, the populations of these creatures have dropped from hundreds of thousands to tens of thousands or even thousands. For all our knowledge and supposed intelligence, our world bodies are presiding over the systematic destruction of these species to the point of extinction. Imagine the difference it would make if we tried to exterminate them.

In itself this would be bad enough, but many of the principal activities of our organisations charged with solving the issue are effectively tarmac sourcing, that is to say moving the problem from one habitat and one species to another. The way to preserve these animals and habitats is always secondary to the needs of people and the ever burgeoning population, and not only the population by number but also the population by need. As the 1.4 billion Chinese see their average GDP grow from $100 to $30,000 per head, their ability to consume, quite simply, the earth multiplies. But we should not blame the Chinese; they are simply arriving late to the table that the First World erected and has feasted on for generations before.

Our world is being denuded for extracurricular projects. Once upon a time we could look in awe at Stonehenge as the wonder of our world (not forgetting Avebury and Woodhenge). A continent away, rocks were being dressed, hoisted, and stacked upon each other in another demonstration of what can be done, as opposed to what needs to be done. What needs to be done is that the human population and our extracurricular activities need to be controlled, but this is not on the agenda. Against this background, then, elephants have no future, pure and simple, so why pretend otherwise? The conservation bodies and organisations that so earnestly propose solutions only ever treat the symptoms while continuing to draw their salaries as they preside over the very process they avow to halt. Wake up, get your heads out of the sand (I'm being polite), and think about what you are attempting to achieve. Alternatively, accept that you have lost and resign your positions and salaries. If you are going to get nothing for it, then do nothing.

Debate is futile; debate is the opium of the masses; debate achieves nothing. We debate the debate while elephants die. No, while we kill them. Our selfish gene will not allow any different course of action. It is also another of our selfish genes that drives the ambition to save them. But while one selfish gene needs to alter to see a difference, the other selfish gene simply has to continue as is. And the result is inevitable.

We have ended the slaughter of the great whales, nearly. The Japanese insist on scientific whaling, presumably to see how many they can kill, while indigenous tribes conduct 'sustainable' crops of smaller species. While the assault on their numbers at the ends of exploding harpoons has nearly ended, that still leaves the assault on their habitat. The great oceans are not immune from our activities and end up as the repositories of our waste and pollution, chemical and sonic. While elephants can hear thunder from one hundred and fifty miles away, we are yet to understand the ranges and complexities of whale song and our impact on it. Suffice to say that these beasts evolved in an ocean devoid of mechanical sound and sonar pings. While we can normally talk and engage in meaningful discussion in the classroom, these advantages tend to disappear on the dance floor. We have seen repeated examples of pollutants accumulating in the tissues of apex predators, and only in recent years have we seen the recovery of birds of prey numbers after the damage wrought on their populations by DDT. I have observed great crested newt (GCN) surveys first-hand and have quizzed the ecologists engaged in the process. Evening surveys are conducted by placing bottle traps half in and half out of the water on bamboo canes embedded in the mud. As darkness falls, torches are shone into the shallow water to try to spot the emerging newts. At dawn the bottle traps are inspected and any captured newts are counted and then released before the sun and temperatures rise to critical levels. The bottle traps are two-litre plastic water bottles pierced by bamboo canes and set at forty-five-degree angles into the water. The ecologist involved in devising the survey method did his doctorate on GCNs and he discovered that the creatures are attracted to chemicals released by the plastic that mimic sex pheromones. One of the reasons for the decline in GCN numbers is their inability to escape from these traps formed by discarded plastic bottles. Having discovered the link between plastic and GCNs, the soon-to-be doctor used it to his advantage. Without his work (or hers, I don't know), we might still be in the dark with regard to plastic and newts.

We can rest assured, then, that great whales are safe from the pressures of humankind.

But even if we discovered a causal link between the noise from shipping or the build-up of toxins in the whales' flesh, will it make a difference in our activities? I have a book decrying the excesses of humankind's activities which states that during the heyday of the DDT-spraying period, under the US's own food safety regulatory bodies, the average American citizen was in fact unfit for human consumption. At least they could rest assured in the knowledge that any impending zombie apocalypse was likely to be short-lived. Or maybe not, 'lived' being the operative expression.

At home we favour the blue tit over other birds as the main benefactors of our offerings to the great god Parus. (Not heard of it in your neighbourhood? I'm working on it.) These diminutive birds have been to university and utilise their degree in acrobatics and their small size to command a disproportionate supply of supplementary feeding from 'concerned donors' intent on securing their future. They have also been known to find their way into bat roosts, where they will eat the brain of every resting bat they can reach.

HVOs

Some countries of the world have declared that there is no safe exposure level for hydrogenated vegetable oils. I do not know if these oils remain intact in the human body after consumption, but in the event that they do, a significant proportion of our 'voluntary arrangement' consumers would thereby qualify as unfit for human consumption. It can be such a hard unlife for zombies.

Is There a Way?

Can any measure of meaningful discourse and action ever take place? Will our Great Brains ever serve the planet before they serve themselves? It cannot happen. Our leaders are followers with a head start, and any replacement crop is from the same mould. They are also following the actions of the herd behind. Unless the herd changes, the followers with a head start won't change either. Let us ban the domestic and international trade in all parts elephant and stop the killing overnight. But we know the answer to that, don't we? The leaders can proclaim what they want and it doesn't matter, because we have no leaders.

We are in the thrall of business, the enterprise that feeds our vacuous souls with consumer goods way beyond our need to survive. Our lives are driven by stock markets and league tables that feed our desires, and pension funds to ensure our happy future while systematically destroying it. Which gene is it that determines the vacuousness of our souls? Is it the God gene that leaves an open door for parasites to exploit? A Trojan horse gene that readily infects and disables any sentient thought? Given the title of *How to Kill an Elephant*, I prefer to think of it as a Trojan elephant, because in my world it is elephants that infect and control through their own unstoppable momentum after their recruitment by human beings. For while they are built and trained to serve, they have other, unforeseen urges way, way beyond our ability to imagine.

While I say that the Great Brains only serve themselves, they in turn serve us as they lead by following the herd. But when I say the herd, I really mean the bait ball.

It takes a dictatorship with all its imbalances and bludgeoning controls to impose control on the population by removing free will. My neighbour, the alcoholic, blames her lack of control on the peer pressure from the advertising industry and social pressures. Accepting this at face value then, she lives in a voluntary dictatorship whereby she has surrendered control to others. It is a gilded cage of illusionary freedom, but it is still a cage, a cage built by others who are as equally trapped, a bait ball—all the more because she never had the control to surrender.

China, under Mao Tse-tung, imposed a one child per family ruling to protect the Chinese from themselves. At the very moment when the population was about to reduce, an anathema in today's consumer-driven world, these controls have been relaxed. Apparently the Chinese have been seduced by the need to maintain growth and demand to feed the GDP requirement to prove one's worth on the world stage. As the wealth accumulates, so does the demand for all things worldly: ivory, tiger parts, forest hardwoods, coal, oil, and extinction.

When will the measure of success be real? How will we know? Answers on a pachyderm, please.

I think it was in either the 1950s or 1960s when the Chinese resolved to deal with the burgeoning sparrow population by organising a mass cull of these birds. They simply drove the birds into the air and refused to let them land and rest. The result was sparrows raining to the ground in droves. I have seen black-and-white film wherein truckloads of corpses were collected for disposal. Such a simple, cost-free solution to a problem, borne of the need to boost food production, and compare it with the elaborate schemes we engage in the name of conservation. We know how to concoct elaborate schemes for conservation that are all doomed to fail because they fail to address one simple truth. There are too many people to allow other species to survive in the manner in which they have. Remember, we will soon be able to do in one day that which it once took us a year to do. While you ask yourself the question, what future do our elephants have, you should also ask, what future does our species have?

I visited Great Yarmouth a few years back and took a stroll to find my bearings. A noise attracted my attention upwards, and I watched a small flock of starlings, perhaps twenty strong, desperately flying in a tight circle to gain height. Beneath them in equal desperation was a single sparrowhawk that was also trying to climb. Eventually the hawk gave up and the starlings lived to die another day.

In a dive, or even level flight, the hawk has the edge, and it was only in the ability to climb that the starlings kept themselves off the menu.

Our organisations entrusted with conservation are the metaphorical hawk desperately trying to gain height, always trailing behind the problem, unable to make any headway, and desperately doomed to fail as the quarry remains just out of reach. It may seem strange to adopt a hunting analogy to describe conservation, but if you fail to catch your target, you fail. The hawk flew on to have a second chance. Our hawks will move on to the next prey item, be it mountain gorillas, bonobos, jaguars, or some other needy beast whose sole need is simply to be left alone within its wilderness. Our hawks will seek out charity and rich grannies and other philanthropists to feed their salaries and their earnest activities that will never meet their target.

The only way my metaphorical hawk can catch its prey is if the flock changes direction and actively flies to meet the hawk. I'm getting a bit poetical here, so let's try it in plain English. The leaders of conservation can only lead where the herd is prepared to follow. The herd is not prepared to follow the conservationist leaders. If the herd were already following the path to conservation, then we wouldn't need conservationists. But of course we would still have them, because these would be the followers with the head start. That is not to say that there is only one herd. Conservationists with a head start can follow the conservation herd and can secure support and resources from it to adopt the mantle of leader. Poachers, on the other hand, form their own herd and follow the lead of the leaders who secure their support and resources by trading in ivory. They have a value in tourism and a value in death. If elephants had no commercial value, then the debate would be simpler. They also have a negative value in competing with humans for resources. The value in tourism is debatable given the need for the supporting infrastructure of hotels, airports, and the like. Tourism offers a value for a temporary respite while simultaneously driving commercial pressures on the elephant reserve and all the resource-consuming, money-generating activities that fund the acquisition of the tourist dollar.

I often wonder what the price of a rubber tyre will be when the ignorant peasant labourer is living in an air-conditioned five-bed executive house with three kids going through a university education.

In our world, the only value that will finally matter will be the negative one elephants have in competing for resources against humans. The fundamental equation which everyone universally ignores is: more people = fewer elephants. And the reverse holds true: fewer people = more elephants.

You don't need a university education to work that out. Yet how many with university educations haven't worked it out? What is it that the WWF use as a strapline? Ah yes, 'Working for a world where people and nature thrive'.

We talk about the explosion of knowledge triggered by the Great Renaissance ably supported by the age of science and logic and, dare I use the term, 'reasoning'. In the simplest possible terms, the cleverer we become, the more apocalyptic we become.

One solution for the WWF would be to invest heavily in covert nuclear devices with a plan to trigger MAD and the return to the Stone Age. At the press of a few carefully selected buttons, wildlife and humans (those that survive, anyway) would return to a level playing field and elephants would have a future.

Our population is set by survival. 'No, really. You don't say.' Its numbers are fixed by our ability to source food, shelter, and protection from predators and disease. Every step change in our recent history as modern humans has resulted in increases in our population. We overcome bottlenecks. Agricultural boosted food security, architecture boosted our shelter, engineering boosted our terraforming abilities, medicine began to tackle pandemics, and our ever growing population enabled localised advances to spread around the globe, a process that is still happening and accelerating. Can we imagine that we have reached the end of the bottleneck process? The day will arrive when humankind's survival will be limited by humankind. What a load of bollocks; it already is limited by humankind in competition with our mortal enemy, other humans.

We have examined the process whereby aborigines are displaced by technology and disease as one 'breed' of humankind out survives another 'breed'. We have examined the process whereby one class of humankind exports life-limiting occupations from one side of the world to the other.

Every advancement generates a supporting structure of resources, let's call it a hinterland. Without metalled roads, we wouldn't need quarries, or cars, or tyres, or rubber plantations, or trained engineers, or robot welders, or car tax, or windscreens, or traffic wardens. And we already know we don't need highway engineers. We wouldn't need tarmac, or council workers enjoying paid-for breakfasts, or zebra crossings, or car parks, or yellow lines, or go-faster stripes, or garage forecourts, or twenty-year planning processes, or any of the thousand and one other myriad tasks dependent on motor cars and the roads they require. Consider the first cars built—variations of four-wheeled bicycles running on packed earth with minimal components and minimal impacts—and see what has happened in the 124 years since Benz's Velo was introduced in 1894. If Karl Benz knew in 1893 the requirements for the motor car that our modern society imposes on them, would he, for one moment, have considered it possible?

We are reminded occasionally by our judiciary of changes in technology, and their need to catch up with the curve. Before the Internet existed, there was no need for laws to control it. Because of this, numerous crimes controlled in the written media have had to be updated to encompass the digital age. In 1893 what laws existed for motor cars, what trade bodies existed, what national standards had been set, what need was there for NOx cheat devices, what Department for Transport had been set up? How readily we have accepted the demands onto our lives of these forms of personal transport. The history of the car reminds me of the game often played with bored children whereby two people stack alternate hands on the ones below. The object of the game is to remove the hand from the bottom of the pile and to place it quickly on the top. The new bottom hand swiftly follows suit, and a 'conveyor belt' of hand movements rapidly follows until the system breaks down due to a lack of coordination. With luck and a little discipline, it is possible to satisfy the child for a few minutes. The motor car industry, on the other hand (sorry), is up to 122 years and counting. Gosh, I wish I was clever.

I wonder if we could take a leaf out of the Mexican people-smugglers' book and have all unaffiliated elephant poachers murdered before they can cross the border into elephant reserves? How much would we have to pay per corpse? This of course would drive the value of ivory through the roof, which would encourage more poachers to risk it for the return. By increasing the bounty per poacher, this particular arms race may favour the elephants, until of course the land is 'domesticated' for human consumption only. And, human nature being what it is, how long before the Burke and Hare methodology of obtaining dead bodies for bounty redemption would begin? 'Yes, I can see he is dead. It's not just the business suit that's throwing me; it's the tyre marks. Was he shot before or after the car accident that killed him? And my cousin has a good body shop if you want to get those dents out.' Thinking about this, I see that the population pressures would decrease as the pool of poachers was whittled away, thus helping to save the elephants. But you knew it was coming, rising prosperity, because the 'border guards' would drive their own consumerism, leading to bigger families and increased demand for land use.

Elephants really are an inconvenience, aren't they? Still, it won't be long now, and there is always the next target species to protect/eradicate in unequal measure.

Now you might be thinking that I am taking the piss with this suggestion, but let's look at the question a little differently, OK, perversely. For border guards, think charity-funded reserve guards who patrol, armed to the teeth (that's one of the elephant's problems, teeth), and who will engage in firefights with equally well-armed poachers. Their success can be measured by how many poachers they arrest or kill. In arresting or killing poachers, they ensure their own prosperity. Of course it's not the poachers themselves who reap the greater benefit; it's the importers and traders at the consumer end of the operation who gain the most. Similarly, it isn't the reserve guards who reap the greater benefit for risking life and limb; it's the charity bosses and conservation bodies and organisations that don't risk life and limb (barring a bad case of food poisoning (it could happen in first class)) who receive the financial multipliers. Life really is so very fair, isn't it?

Conservation, then, is another elephant.

Over 90 per cent of the world population is breathing-poor quality air, according to WHO. According to France 24 News Channel in English, this poor-quality air is killing over six million per year, two out of three in South-East Asia and the western Pacific region. The main sufferers are China and India. Paris occasionally ranks as the worst city in the world for air pollution. I wonder if Karl Benz would have been so keen to have progressed with his invention if he had known the consequences. This of course is rhetorical, because variations of his invention were made worldwide by the motor car industry at a rate of seventy-two million in 2016.[3]

And we know the consequences.

---

[3] Worldometers, accessed 31 July 2018, https://www.worldometers.info.

Painted Ladies

I like painted ladies for many reasons, not least of which is their Latin name of *Cynthia cardui*. How often can you insult a well-educated retired headmistress in the face and get away with it? Of course, if her Christian name happens to be Cynthia? These delightful butterflies, a subgenus of *Vanessa* in the family *Nymphalidae*, are both visually attractive and have a life history worthy of comment. These medium-sized butterflies (largish for Europe) migrate from Africa to as far north as Scandinavia, before returning to overwinter in Africa. Or, rather, they don't, but their species does. Each year a fresh crop of butterflies flies across the Mediterranean to breed and die in southern European climes before their offspring continue north to repeat the process. Several generations later, the latest batch, having commenced south in response to some unknown trigger, have been only recently observed leaving British shores heading south to begin the annual cycle all over again. Would you say that they have a leader?

Their total migration range from Africa to the Arctic Circle and back is some nine thousand miles, undertaken by perhaps six successive generations, each preprogramed to do what it does and unable to learn from its predecessors.

Is it fair to say that their life is preordained and beyond their control? At any level would you credit them with intelligence?

There must have been a time, not so very long ago, when their migration was very much shorter, hemmed in as they were by the ice sheets of the last ice age. The most recent one ended around 11,700 years ago.

Is it reasonable to assume that their seasonal movement would have been measured in tens or hundreds, rather than thousands, of miles?

These insects weigh a gram each and have a brain the size of a pinhead. They have extended their range as the ice has retreated in pursuit of their caterpillar food plant, and their numbers no doubt have increased. Would we for one instant suppose that any sentient thought process has been engaged to achieve their distribution?

The average human brain (male or female?) weighs around 3 lb or 1.35 kilos. Our brain is 13,500 times heavier than an entire painted lady, never mind its brain. How much of our brain, do you suppose, have we utilised to extend our range as we have exploited the necessary food resources? After all, we have only extended our range from Africa to within the Arctic Circle. It's not that difficult.

Yet our reference books sing our praises and describe our adaptability and ingenuity in such special terms. A small number of birds use tools. Bower birds construct and decorate elaborate structures, some fish build nests, and spiders spin elaborate webs and even construct 'diving bells' of air to survive under water. Elephants terraform their environment with tree removal, but not the forest elephants, because that would be silly. Swallows and swifts construct nests of mud, while bee-eaters and kingfishers tunnel underground for protection from predators. Beavers construct weatherproof shelters and aqua-form their environment to collect and preserve their food reserve. Termites build multistorey air-conditioned apartment blocks without the need for planning permission. Penguins organise 'childcare' crèches without a socialist government. Even caterpillars construct silk structures behind which they feed or pupate, protected from predation. Harmless snakes mimic poisonous ones or even lurid, smelly death (western hognose, *Heterodon nasicus*) to reduce the chance of appearing on the menu. Caddisfly larvae, *Trichoptera*, construct camouflaged suits of whatever detritus comes to hand. Squid and octopuses squirt ink to cover their retreat from danger. Lions roar to proclaim their territory while exposing any vulnerabilities in their voices. If Doc Doolittle asked any of them why they do it, would you expect an answer? If I asked the human race why they do it?

Usual Sales Patter

In marketing the trick is to identify the reason, the preference, the justification for wanting to buy. They call it the USP, the unique selling point (or proposition). 'Buy this one; it's the only one available in white.' 'Buy me. I've got twenty-seven functions on my pocketknife.' 'Buy me. I was made for the Swiss army.' 'Buy me and save the environment.' 'Buy this cuddly toy and save tigers.' 'Buy this bag of elephant poo and save elephants.'

If Charles Darwin had been a salesman, would his theory be paraphrased as 'survival of the proposition'? We have a choice; we exercise it at every opportunity. For our species, it perhaps began with which fruit to pick, which partner to mate with, which way to run from danger. Earlier it would have extended to which branch to jump on, when to freeze, and when to run, basic tenets of survival.

Some of these choices have been shown to be engrained into the brain. Experiments have been conducted with recently hatched ducklings kept isolated from adult ducks. A crucifix-style cross has been passed over their heads on a wire track. It can be passed over their heads with a 'trailing tail' or an extended 'neck' depending on whether, in crucifix terms, the cross is head or feet first. Irrespective of its direction of travel, the silhouette is the same. The 'long neck' is duck-shaped and they ignore it. The long tail is hawk-shaped and they crouch and freeze.

They do not have a choice.

Why do we make the choices that we make?

I can understand the fresher leaf, the ripest fruit, the fittest partner, the choicest territory. How does this translate into the 'right' brand name, the correct marque, the perfect go-faster stripes? OK, I understand the need to follow the decisions of those before you—no need to reinvent the wheel. But for every 'learnt' choice there is another 'unlearnt' one. Someone has to go first, to blaze the trail, to follow with a head start, to establish the tried and true, to convert the new to the norm.

There will be patterns, tried-and-true trail-worn paths to the same conclusion. In fact, we will regularly 'reinvent the wheel'. The local town's council offices were constructed using decorative brick slips (not full bricks, more like brick tiles) which were glued on to the concrete structure. Some fell off. Several years later the same town's leisure centre was built, but this time they used decorative brick slips which were glued to the concrete structure. Some fell off.

As a manufacturer providing structural concrete products to the construction industry, we would often be asked to repeat the failures of the past. We coined the phrase 'Every generation has to learn from its own bollocks.' The polite version is 'failing to learn from others' mistakes', but our reputation was cemented in persuading later generations to avoid the mistakes we had seen before. If the extinction of the dodo could be seen as a mistake from the past, what have we learnt? I realise I am straying into 'lessons will be learnt' territory here.

Sometimes success breeds failure. If the whole world had a white telephone, then the only way to boost sales is non-white. Comments have been made about men with beards and the cyclical nature of this particular fashion trend. To stand out in a glabrous mating pool, you need facial hair. To stand out in a hirsute mating pool, you need to be clean-shaven. Glabrous or hirsute? The pendulum swings from one to the other but steadfastly refuses to achieve a static equilibrium.

How much free will do we exhibit in our choices? Do we even realise that we are making choices? Are they made for us? The bait ball suggests that we don't have a choice, only a pro forma.

I have looked at the pro formas of the news journalist where the reports are standardised to the point of 'strike out/fill the gap' thoughtless drivel. Yet we accept these reports with avid attendances at the six o'clock or ten o'clock appointments. We gravitate towards the same newspapers, the familiar magazines, the same foodstuffs, the same caravans, the same car accessories. We'll bend our knee to the same deities. We'll even fight alongside our tribe against others. We'll destroy the earth because everybody else is doing it. Yet, and this is the paradox, we crave change— not really. Americans pack up their domestic life into Winnebagos the size of houses and escape to the wilderness, in parks populated with city scales of Winnebagos and city scales of people. We escape from our city scales, jostling cheek by jowl on transport and coffee queues, so that we can lounge, cheek by jowl, on some foreign shore, toasting our bodies on adjacent sunbeds. We long to escape the herd so that we can rejoin it.

I tend to reach for the remote control and either mute the sound or change the TV channel whenever the adverts appear. Even I have noticed the recent glossy campaigns launched by the cancer charities in pursuit of money to fund research. Apparently, fully half of us will receive a cancer diagnosis during our lifetime. Now far be it for me to question statistics; you know how I love measurement. In 2015 the number of people who died worldwide from cancer, as estimated by the WHO, was 9,000,000. This compares with the 7,600,000 who died in 2005.[4]

While I appreciate that there is likely to be some overlap, the same sources suggest a not wholly different figure for deaths caused by smoking and alcohol.

Worldwide, then, ending smoking and excess alcohol consumption would nearly balance the books in terms of lives saved. I know this doesn't help your loved one, but of course your donation to one of the cancer charities doesn't necessarily help either. If your son is dying from kidney cancer, paying for research into leukaemia doesn't help directly either.

To summarise, the death toll from these addictive scourges of the human condition can be regarded as voluntary and within the remit of humankind to halt, without any need for science or research or digging a hole in the earth's resources.

Now for many, I am being blatantly unfair, because cancer can be tackled with enough resources and science. How much science and how

---

[4] Worldometers, accessed 1 July 2017, https://www.worldometers.info.

many resources have gone into researching the health impacts of smoking and excessive alcohol consumption? And having done the research, having reached the unassailable conclusion, and having spent all the resources and squandered the lives of all the scientists and all those chain-smoking beagles, what do we do with all the results?

Ignore them.

Smoking and alcohol have a causal link to cancers, yet still we smoke and drink.

Obesity has a causal link to cancer, yet still we eat too much.

The WHO has declared processed meats to be carcinogenic, yet still we eat them.

Excessive exposure to the sun has a causal link to cancer, yet still we sunbathe.

Fresh fruit and vegetables have a positive impact on the chances of developing cancer, yet still we shove the garnish to the side of our plate and ignore it.

Remember my argument: if you want to save tigers, stop drinking tea.

How about, if you want to stop cancers, stop volunteering for them.

But we will continue to fund the WWF.

We will continue to fund cancer research.

Neither will ever run out of things to do.

We will run out of tigers.

We will run out of resources to fund cancer treatments (£30,000 per year of life extension is too expensive. A more realistic figure is £13,000 a year).

'But, at least we'll die trying,' I hear you say.

'We'll still die' is my reply—Often having volunteered.

When I was a kid, Tufty Fluffytail the squirrel had worked out that the way to cross roads safely was to stop at the kerb and look both ways before crossing.

Later it was Darth Vader (David Prowse), who, before turning to the dark side, donned the Green Cross uniform and preached, 'Stop, look, and listen.'

Most of us learnt to observe these simple tricks to staying alive while crossing the road.

If you were sober with a full range of senses and mentally alert, it would be reasonable to assume that you would heed these simple instructions. At what point would you consider that wandering without following these simple instructions into the path of moving traffic was self-inflicted? Many years ago my wife ran over and killed a dog that had wandered out between parked cars in a residential area. In those days you were supposed to notify the police. I drove her to the station. What we hadn't realised was that the reporting was done so that we could make a claim against the dog owner for any damage to the car. In a bizarre twist of fate, once 90 per cent of my journey to the station was completed, I hit another dog. This one had pulled the leash from its girl owner's hand and I struck it as it ran straight into my path. It survived with bruising.

I know, not all cancers are caused by known effects and there is a random nature to some of them. Children's cancers are particularly grievous to understand. There will always be a need to save a loved one. There will always be slow-motion suicides. I would suggest a solution of sorts, which I know will not satisfy our human nature. In Third World countries where a simple bout of diarrhoea can carry a death sentence, our world's preoccupation with cancer is beyond their wildest dreams. 'You lucky bastards, you get to die after several years of extensive treatment and all avenues are exhausted.'

My solution? Research the causes of cancer and avoid them. I know this doesn't help people who have already been exposed, historical asbestos exposure for example, but eventually this particular poison will work its way through the system. All sorts of bans and controls exist for asbestos, and deservedly so. Eventually it will be eradicated from our buildings and environment. It's as though asbestos is a double red line and we really mean it this time, whereas alcohol, tobacco, obesity, and UV rays only qualify for double yellows.

The other problem for me is the simple fact that as fast as we establish cures for cancer and other diseases, we invent new causes of death to fill the voids in our bankrupt souls. Are we serious about saving life or not? We already know the answer when it comes to strangers from foreign climes. We seem to have created an industry of identifiable concern for the 'strangers' whom we have been exposed to. A cancer in the family? Donate to end the scourge. MS in the family? Donate to end the scourge. Malaria in Africa? No need to bother; the Gateses are on the case. Ebola in West Africa? Not our problem—yet.

Where is the source? The source of the concern. For centuries, diseases in West Africa were of no concern to us in Europe. *Yersinia pestis* survived in the east for centuries before cropping up in Europe. (With the death rates attributable to bubonic plague, 'cropping' is not a bad expression.) You couldn't have lived through the ravages of the Black Death in Europe without knowing of its presence, although in itself that is a lie. Even in modern times, AIDS has been described as a gay plague, so you can imagine how the God squad feasted on the wrath of God visited upon the earth to cleanse the unbelievers. And while bubonic plague languished in the east, we didn't care. Nine million people died from cancer in 2015. I will call it a First World disease. Nearly nine million people died from alcohol and tobacco in 2015. One gets all the press, one doesn't. Is it because the media has had its day and expended its energy on the preventables? But for energy, could one justifiably claim that it has run its 'commercial course' and just doesn't sell enough newspapers or advertising? It's not that the problem has gone away, simply that it no longer has any commercial traction. In marketing terms, cancer is the 'New Sexy'. But to be the New Sexy, it has to perform to expectation, to corner the market on resource generation and application. In September 2016, that great of the mobile telephone industry, BlackBerry, has announced that it is no longer going to manufacture telephones. It has been overtaken by alternatives, by other, New Sexy derivatives to excite the lust gene, thereby consigning it to history. So are we manipulated and trained to be concerned, or are we genuinely concerned? How do we reconcile the emergence of fashionable disease with genuine need? Ebola in West Africa drew widespread condemnation of the First World's failure to act in a comprehensive manner. It simply didn't figure on our radar. If cancer is to be the new prime focus of our intellect and resources, what is to happen to the BlackBerry devices that have failed to make the cut, condemned to be the also-rans as they slip below the horizon and cease to figure in our deliberations? Are we treating the disease or treating the marketing campaign with our attention and our resources?

I realise that I am guilty of heretical treason for even daring to suggest any form of ulterior motive to the pursuit of saving lives. And that is part of the answer, because if you can't mention immigration without being called racist, what chance does my lone voice have against such a genuine erstwhile cause? I also accept that there are a multitude of workers desperately seeking an advance in the science to meet the expectations of those desperately seeking the treatment. My argument is akin to taking a Sunday afternoon stroll through a minefield. I am not, contrary to many people's opinions, a heartless bastard. If you devote all your spare resources to curing cancer and saving your loved one, are you not turning your back on all those who suffer inside all the branches of medicine not connected to cancer? Who is heartless now? 'But we have to start somewhere.' Where should somewhere be? Who should decide, the suits in advertising who have decided to launch 'Brand Cancer' or the CEOs and head scientists already viewing their burgeoning pay packets and research facility budgets?

What will our cancer scientists achieve? The equivalent of a dietary slimming pill to bind with fat when you already had the option of eating less of it in the first place. The universal panacea to neutralise the stupidity and excesses of our profligate lifestyles devoid of any personal responsibility. 'Lung cancer. I gave at the office. Give me the treatment.' 'Sclerosis of the liver caused by excessive pickling in yeast wee. I gave at the office. Give me the treatment.' (Yeast wee? Alcohol is a metabolic waste product produced by yeast, not dissimilar to urine, another metabolic waste product produced by people.)

Just suppose that the cure for cancer was as simple as kissing a frog on the night of a full moon. (If tiger parts are good enough for Chinese medicine?) Where would the industry be? Quite simply, it wouldn't. Be. It would cease to have been. There is no profit in saving lives, unless of course you own a frog farm. There is no profit in preventing deaths from diarrhoea in Africa, so, by and large, we don't bother. There is no profit in treating lung cancer in Africa, so we sell them cigarettes instead. You see, Africans, they do have a value after all.

We can begin to think of life in purely commercial terms. Cancer is good because it has expensive treatments often extended over many years. A cure? No, we don't actually want a cure, because that's the end of the earning capability. Saving lives in Africa by ending the trade in lung cancer? Sorry. Cigarettes. Nah, they can't afford the treatment, but they can afford the death-sticks. Of course, give it a few generations, get the

GDP per capita up, and the whole new market will open up. Then we will be interested.

Somewhere in my addled brain's meanderings lies a core of truth that cuts through all the bullshit and the claims about 'the drive to save our fellow man', when in reality we are only interested in saving one person, me, today. This is the drive to survive today by exploiting the opportunities that present themselves to gain an edge, an advantage, the clean end of the shitty stick. The concern is how to put food on the table, pay the mortgage, save for a pension, and see your name in lights, all done by exploiting another's need, a need that others have created by following the same criteria and path to their own advantage. This trick is to learn how to be on the right side of the other fish in the bait ball.

The bait ball. It's what we do; there is no shame in it. It's how we survive. So why concoct elaborate schemes and higher motives to camouflage our actions from our neighbours—or should that be fellow bait ballers? Are we predetermined by our DNA to be unable to see the ball? Is it invisible to us? Does a wildebeest have a conscience? Does it need one? That's OK, because we don't have one either.

If a Tree Falls in the Forest and There Is Nobody There to Hear it

The news today (October 2016) is headlined by the announcement that over six thousand migrants have been rescued from boats in the Mediterranean Sea. These poor unfortunates have been plucked from their overloaded vessels off the Libyan coast and have been whisked to safety for processing. Newspapers have devoted front pages to images of the bodies of toddlers clutched to would-be rescuers' chests, having been dragged from the surf on Greek island beaches. It is terrible and emotive, and it sells newspapers and fills the TV news channels. When these people are crossing the Sahara Desert and dying, or trailing across war-torn Syria or Afghanistan, they barely receive a mention. For every six thousand plucked from the Mediterranean, how many more failed to make it? But when I say make it, I don't mean to the safety of European shores. I mean those who make it to the European's consciousness. We have the luxury of being able to ignore their plight until it is graphically presented to us in our media. What, then, is real, the plight or the portrayal of the plight? If a tree falls in the forest and there is no one there to hear it, does it make a sound? If a sub-Saharan African dies in the desert, do we care? If a Guatemalan is murdered by the people-smugglers for failing to pay his fare, do we care? If 80 per cent of triers were to die in the attempt to cross borders, would they care enough to stop trying?

Our care, our conscience, is shaped and manipulated by external agents, let's call them the media. Any impassioned plea from a politician to an empty room has no effect. Any unpalatable truth that fails to sell airtime or column inches has no effect. Equally, any poison gas attack in Saddam's Iraq fires no brain synapses without a voice to report it.

But do we see the truth? Today, 26 September 2016, on *Russia Today* they have reported the release of Russian radar data proving that no missile was launched from the Russian side of the Ukraine. The Western media meanwhile report on the Dutch legal investigation report into the downing of flight MH17 en route to Malaysia. The Dutch are quite unequivocal in their accusations that the Buk missile system entered the Ukraine from Russia, fired one missile, and then returned to Russia. The truth is what the media make it. You either believe that your country is under attack from the Western media or that the Russians are murderous bastards who can't be trusted. One truth is that commercial airlines continued to fly over a war zone, although not all commercial airlines, because some had already rerouted.

What truth do the Guatemalans believe before embarking on their perilous journey to the Promised Land? If Ross Kemp's exposé of the Mexican people-smugglers' activity is correct, then they are advised to carry water for a few hours, when in practice the trip takes at best a few days. Now if you had a devious mind, you would decide, *Let them die in the desert*. They have paid their fare and are of no further commercial value to you. Large numbers of successful migrants crossing the border will attract the attention of the border guards. It is good for business for some to make it to communicate back home and encourage others to pay their fares (I didn't say follow suit). Conversely, large numbers of captured would-be migrants returned through official channels is bad for business. The migrants may think that they are engaged in a deadly game of cat and mouse between the border guards and the smugglers. We know who the mouse is, but is there more than one cat? Another form of *Arbeit macht frei*, whereby one group of people seeks to gain advantage over another.

What do the arbiters of public opinion garner from their processes? Do we thank them for exposing the injustices and realities suffered by our fellow man? We piss about with ethically sourced coffee and bananas and sustainably sourced recycled wilderness in the belief that it makes a difference, whereas we should simply see it as another form of exploitation of our minds and wallets. How about ethically sourced foodstuffs from black-owned farms in Zimbabwe? You know the ones, the ones land-grabbed from the white farmers by government land acts. These would be the same lands grabbed by the white farmers from the black natives under government land acts in a former age. That would be the lands formerly occupied by elephants and other wild creatures. If conservationists get their way, will this land be stripped from the black farmers and returned to the elephants through another government land act? We cannot see the truth because there are so many variations. What we see are carefully orchestrated versions deemed fit for public consumption, carefully orchestrated by the followers with a head start to reinforce and drive opinion. You may read a left-leaning newspaper, or maybe it's right-leaning. Very few will read the one whose sentiments don't strike a sympathetic chord. The music you listen to via the airwaves is similarly filtered to strike just the right beat. If you want hip-hop, don't listen to Jazz FM. And you won't find much disco on Classic FM either.

Those who devote their Sunday mornings to the ministers preaching will be steered towards and aligned with the thought for the week and even adapt these as thoughts for life. Which variation of the truth will you align yourself with? In the end it doesn't matter, because they are all equally valid, all equally true, and all equally wrong. Shall you head east to the waterhole or south? Provided you find water, it doesn't matter. If you don't find water, you die and it doesn't matter. You tried, you failed. Life has no sentiment. Life doesn't understand truth. Truth is being alive. Life is all that matters, all that drives and sustains. How we do it, that is the distraction. All the controls, mechanisms, doctrines, religions, laws, societies, ideals, strategies, and justifications we append to our lives are as substantial as the coloured scales on a butterfly's wing, doomed to be shed with every beat of the wing, every contact with a gossamer web, every day of their short lives until, bare and tattered, they surrender their life force. So why do we place such store in them?

I don't expect you to agree with me, entirely. I will have made mistakes. It takes time. It takes a frame of mind. OK, it helps if you are an awkward sod. But think about it. Look around and interpret your actions afresh. Start not with what we are doing but with why we are doing it.

In Nassim Nicholas Taleb's *The Black Swan*, the basic premise is that for centuries everybody in the world knew that all swans were white.

Nobody would ever bet that a black swan would ever be found. And then Europeans discovered Australia, and depending on your gambling style, you had either just lost or won big. The author also illustrates (vegetarians should look away now) the happiness quotient of a turkey. Plot the happiness of a turkey on a graph as it hatches and enjoys food, warmth, and shelter on an ever increasing basis until shortly before Christmas Day, when its happiness quotient takes a dive towards non-existent. While Nassim Nicholas Taleb uses it as an illustration of 'past performance is no indicator of future performance', I prefer to use it as an illustration that living in a bait ball has its drawbacks.

Terms like the 'happiness quotient of a turkey' have been raised by financial advisers, entering into the conversations I have had with them. I've even heard reference to 'black swan event' on Bloomberg TV uttered by some journalist, although to my mind not in the right context. This book has punched its full weight within the shepherds of your money industry and is more than worthy of a mention in mine.

In a very real sense *The Black Swan* is a book about mathematics and how it applies to the human project. While I understand the maths, in itself this left a void in my deliberations, because for me what really matters is the biology.

The challenge, should you choose to accept it, is to challenge. Question, think, understand, and look for the different way. It will make no difference. For every recruit to my way of thinking, others will look to exploit the same way of thinking to increase their ability to exploit their fellow man. *How to Kill an Elephant* is both a guidebook to save the earth and a manual for destroying it; the choice could be yours. How will you utilise your knowledge of the bait ball? Can you see past your neighbour?

Corbyn

'Harmonisation of wages and working conditions across Europe is key to winning back public confidence on immigration and not restriction of movement'. This is the press released strapline for Jeremy Corbyn's keynote speech at the Labour Party Conference.

Now we have it: Unable to control his own parliamentary party, and in spite of the UK's vote to leave the EU, Corbyn has launched his bid for European domination. We are achieving harmonisation and working conditions across Europe; we are reducing wages and working conditions to the lowest common denominator, where the endless supply of jobless immigrants are prepared to work harder and for less pay than our own indigenous 'socialists'. Just when you think the lowest common denominator might begin to rise, a fresh cavalcade of Asian and African migrants arrive quite literally by the boatload.

I've tried to dust off my expired socialist qualifications in an attempt to determine the benefits to our socialist brethren. What I can see are the benefits to all the non-socialist employers who can achieve greater productivity for less wages, equating to bigger profits. I can also see the non-socialist landlords who are tired of renting out their three-bedroom houses to three people for a thousand pounds when they can rent it out to nine migrants for two thousand pounds. It's cheaper per head, and that's all they care about. Is this the harmonisation to which Corbyn refers? What amazes me about this particular follower with a head start is the thrall in which his particular brand of socialist herd hold him. His plan, if I can muster the belief to use the term, is to fix minimum wages to prevent foreigners from undercutting them. I cannot imagine a greater sucking power outside of a Dyson factory than higher wages. Don't work in France for £8 per hour when it's £9 per hour in Britain. Or how about don't work at all? Enjoy enhanced benefits instead. And with these enhanced wages and increasing 'disposable income', we can all afford to pay higher rents. But it's OK, because for more money the migrants will sleep eleven to a house, not nine. They do have standards, after all.

Other inane policies exist in other political parties; socialists do not have a monopoly on them. Have I mentioned stamp duty or sugar tax?

Beneath every dominant chimpanzee or meerkat is an echelon of lieutenants, male and female, that have a very real sense of place within the hierarchy. In one sense they wait, biding their time, for the dominant master to fail or meet a sticky end so that they can ascend to the throne. But not all will reach the pinnacle. Indeed, most will simply enjoy the fruits nearer to the centre, while lesser animals have to fight over the scraps.

What shall I call these, 'Great Brains in waiting'? Not all will achieve full status. Indeed, I would suggest that many will never be qualified beyond lieutenant status.

I could call them 'administrators'. It is in this role that they serve the master while maintaining their own position near the centre. Their role is not to lead but to follow and to maintain the structure, because it is within the structure that they reap the benefit.

In politics we could think of the party leader. As Jeremy Corbyn may learn one day, being the leader of the party only works if your party follows. The dominant leader is as dependent on the lieutenants as they are on the leader. If the lieutenants lose confidence in the leader, then a coup will be plotted within the ranks and the leader changed. In meerkat or baboon terms, the leader may be left to fight for and defend the group without support. While several baboons can drive off and even kill a leopard, an isolated individual has little chance. With the Labour Party intent on a new socialist paradigm and with a socialist leader in Corbyn, many of the lieutenants, the MPs, have found themselves isolated and unwilling to support him.

These lieutenants have a choice, to accept the new order or to strike off behind a new leader of their choosing. But in my model, administrators also serve the lieutenants and gain an advantage. These are the tertiary advantage-takers, generally immune from the struggles at the top but still great beneficiaries of the structure.

We can see this structure within business, and also within government.

There are the lawyers who argue interminably over someone else's laws; the civil servants who add the different herbs to the stew; the teachers who suck the individuality and ingenuity out of our children in pursuit of another's pro forma; the highway engineers—OK, enough said; the national bodies who decree what is decent and obscene; the conservationists who decide what dies today to save something else; the accountants who count others beans and cream off their percentage; the fashion buyers who decide the individual look before ordering thousands of the articles of clothing; the music companies that decide what is worthy of your ears to hear; the journalists who regurgitate others' ideas; the scientists who carve their epitaph in our declining future; the architects who decide the shape of our caravans; the committee members who commission elephants; and the followers who decide which way to lead.

What this list is short of is the people who know why, the people with the knowledge, the people with the wherewithal, the people with the know-how to kill an elephant.

I remember years and years ago when the Iranians had stormed the American Embassy in Tehran and taken hostages. Someone on the telly asked a Soviet minister what would happen if their embassy suffered the same fate. His reply went along the lines of, 'It's two o'clock. At four o'clock, Tehran would cease to exist.' Bullshit posturing? We never found out.

The point I am making is that, with the will to do it, the 'problem' would have been sorted. If the will is there, what else could we do? We could bring an end to IS or Boko Haram, or drug smuggling, obesity, or cancer. We could bring about the end of poverty, hunger, malaria, and Ebola if the will was there. What does that tell us? We will not put boots on the ground in Syria because there is no will. Equally, then, it can be argued that we don't want to see a resolution there either. Or rather, we do want a resolution, it's just that somebody else needs to do it, because we won't. If you are a Syrian trapped in your own bombed-out enclave, no doubt you are praying for the international community to ride to your aid.

For all intents and purposes, the international community doesn't exist. It is a debate, a fallacy, a committee. It's a little like banning pistols in the UK. It inconveniences all law-abiding pistol owners but not the ones who don't follow the rules.

Once again we invest heavily in the Great Brains who preside over the summit lunches and the diplomatic meetings whose only real agenda is their own and not that of the people they are nominally attempting to help. 'No, no, no, that simply will not do. We cannot let the Russians hold sway on that.' As they nearly say in Australia, 'Chuck another Syrian on the barbie, will ya?' I am quite sure that there is another version that begins 'Nyet, nyet, nyet.'

And finally, when the dust has settled and the bullets stay in their casings, what will have been achieved? A settlement for the US or a settlement for Russia. Where, if ever, will the settlement for the Syrians be? In the simplest terms, the Great Brains are simply sacrificing Syrians for their own position. You can have a certain sympathy for the Great Brains because they are faced with an impossible task, but not as much sympathy as you should have for the Syrians. Alexander the Great was presented with a Gordian knot, so fiendishly tied as to be impossible to undo. Alexander, so the story goes, simply cut it with his sword. Here then, at a stroke (sorry), is modern-day diplomacy in action. There is no solution, just plenty of cutting edges.

But it was not our Great Brains who began the conflict, nor did they end it before it started. It was the Syrians' own Great Brains who saw an opportunity to advance their own positions at the cost of their own people. Well, obviously not their own people, just the unbelievers, the unaligned, the ones guilty of having an alternate view, members of another tribe, expendable ones, ones whom you are not following with a head start. Why then do we invest such stock in these Great Brains who rarely die trying but try so impotently while others pay the ultimate price? Is it as simple as 'They led us away from the leopard'? Shame about the waiting lions. But if you are frightened of everything and simply react to all threat, then what are you following, somebody else who seems to know the way? 'Let's follow these other chaps who don't know where they are going.' Seems like a plan? Seems like a bait ball. Face it, we are simply not equipped to live at these densities. Our command-and-control structures are not either command or control. We have no higher right to claim sentience than other life forms.

It's a numbers game. Sitting on my holiday patio throwing breadcrumbs to the sparrows, I observe that the lower the number of birds, the more nervous they are. Conversely, when mob-handed, they are much bolder and more aggressive in their apparent attitude towards risk. Have they done the maths? A single bird becomes the target. Three birds reduce the risk to a third. Twelve birds reduce it to just over 8 per cent. One line of logic suggests that there are more eyes to scan for danger. Another line of logic suggests that it is harder to select one individual from the fleeing flock. But another line of logic suggests that you can put another bird between you and danger and hide within the bait ball.

If we extrapolate beyond simple danger perception and begin to look for other roots, then a wider range of motives can be sought. Then we can begin to consider peer pressure and anonymity beyond the simple action of safety in numbers. Peer pressure would demand that the dominant sparrows advertise their prominence by putting on a display of aggressive behaviour. Equally, peer pressure may prevent the subordinate sparrows from behaving as they might. If we move these patio squabbles into the human world of 'chucking-out time' fights and full-blown riots or even lynch mobs, then the same patterns could be present. Ringleaders need to show by example for fear of losing their dominance. Subordinates equally need to display their qualification for membership of the 'gang'. But we do not need to confine these thought processes, if indeed any thought is required, to the playground or battleground. We can see the same basic processes played out in the media and politics, in sport and the arts, in any hierarchical organisation. Alternatively, is it as simple as 'He's doing it, so I am'?

'He's walking over the horizon. I'll follow.'
'They're looting that shop. I'll follow.'
'They're raping that girl, so I will too.'
'She's got a designer handbag. I want one.'
'She's got plastic tits. I want some.'
'I'll follow the car in front.'
'I went to school, so must my children.'
'They're chanting homophobic insults. So will I.'
'He's getting the best of the breadcrumbs. I'll risk it as well.'
'They're dying for a cigarette. I want to smoke as well.'
'They're all crossing the river, so must I.'

I like playing Minesweeper on my computer. I used to use it as an illustrative tool for my employees to prove a point. If you customise the grid to the maximum number of squares and minimise the number of mines, then it is possible to 'solve' the puzzle with a single touch. It varies between different versions of the game, and it may take several attempts, but it is possible. When I play the normal game, I don't fill the mine locations in because it takes too long. I ignore their positions until all the clear squares are eliminated, at which point the game announces that I have won. What I try to illustrate is that the computer doesn't 'think' in the same way as we do. Visualise a bar code of vertical black lines on a white background. You assume that the scanner reads the width of the black lines and the distance between them. While it is probable that the machine works in this way, it could also work backwards. That is to say, it could read the white stripes and the distance between them caused by the presence of the black stripes. Similarly, the computer could read all the white 'paper' on the screen and decipher the message by working out where the white is missing because 'ink' is present—the film negative if you will. The computer doesn't care and can be programmed to work in either way.

Does it make any practical difference?

Well, for starters, I can complete Minesweeper quicker than I would otherwise be able to. Practically, I used it to show that once information is fed into the computer, the computer could process it forwards, backwards, or upside down, it doesn't matter. A quotation could be a pick-list, a loading list, or an invoice, without the need to re-enter the data, just the way that you processed it. As a kid you may have tried to walk on paving slabs without treading on the cracks (joints), whereas what you were actually doing was walking on the slabs.

If you have ever tried the jam doughnut challenge whereby you have to eat an entire sugar-coated jam doughnut without licking your lips, then you know it can be quite difficult. The trick is not to think about not licking your lips. The trick is to concentrate on keeping your mouth shut. One way lips are at the forefront of your mind; the other way lips become secondary and the task becomes easier.

If you can develop an oblique way of looking at the world, then you can begin to step out of the pro formas, the thoughtless rules of engagement, the autonomous command-and-control structures that stultify our mental abilities. You may even escape the bait ball.

Life. 'But not as we know it, Jim.'

What do we mean by life?

Is life defined by:

- The length of it?
- The size of your family?
- The cubic capacity of your car engine?
- The number of designer handbags you own?
- The hours you spend on the sunbed?
- How much time you can save by eating convenience foods?
- The years worked to buy your house?
- The number of sexual partners you've had?

Just suppose that you were a cognitive grass plant, better still a cognitive seed that develops into a cognitive plant. You worry, you lie in soil, but you are concerned that you are too shallow or that the soil is too heavy or too light. It rains and the moisture begins to soften your hard protective coating. Germination is beginning. You worry that the source of water will dry up if you commit too soon. You also worry that too much water could deprive you of oxygen and make your tissues rot. Unfortunately, or is that fortunately, the germination processes are beyond mental control and are triggered by current circumstance. Before you know it, you've thrown out a root, and a stem is thrusting towards the light with all the urgency of a seed that is expending its food store and needs sunlight to replenish its stocks.

Your root develops into a system and your stem becomes many as your leaves bask in the sunlight as you grow and develop. Now you worry about expending all your energy by growing luscious leaves that seem to be on the menu of every passing caterpillar and herbivore. The effort required to seek out and lift the water and minerals from beneath the ground to allow your leaves to function is enough to give you a complex. And as for flowering and setting seed?

Cognitive thought gets in the way. It would cripple you with anxiety and destroy any 'pleasure' in being alive and functioning.

Cognitive thought has its place, but it is fair to say it also has its limits. Where would we place a grass seed/plant on the scale of intelligent thought? Celebrators of the human species would doubtless place us at the opposite end of this same scale. I have referred to tropismic behaviour within people. Another way of describing this would be to call it non-cognitive behaviour, behaviour that doesn't require any mental processes because the autopilot has assumed the controls. If you were to consider some other intelligent life, say, a super thinking alien passing by on its route to its holidays, where would the clever end of the cognitive scale be then? If grass is at 0 on our scale and the super alien life is at 10, where would we place humankind on the scale?

What proportion of our life is effectively outside any intelligent control? Alternatively, what proportion of our life is effectively within any intelligent control?

We have our automatic processes; we don't need to think about breathing or peristalsis or a million and one other things that happen to keep us alive. What do we think about?

How to save elephants?

How to coexist?

How to control our population?

How to stay healthy?

How to build a pyramid?

Where to find another world to exploit?

How much of our lives can best be described as automatic, as simple as the act of walking, as easy as placing one foot in front of the other and transferring the balance forward. Conscious thought over such basic acts would inhibit our actions and our survival. There is nearly a joke about the girl who refuses to remove her headphones while at the hairdresser's. When the headphones are forcefully removed, the girl collapses. When the hairdresser listens to the sound from the earphones, she hears, 'Breathe in, breathe out.'

And how much of our lives can best be described as autonomous?

Autonomous: freedom or independence, as of actions or of the will. From Greek 'autonomous': having its own laws.

The word 'autonomous' seems to be at odds with this statement: how to contrast freedom of action, freedom of its own laws, devoid of central control. Our species is autonomous in the sense that we have our own laws, the laws that control our biology. Our biology controls our laws. We are free as individuals to pursue our desires and needs, but only within the framework of our biology. Our cognitive thoughts in theory can overcome these biological controls; in practice, we haven't come close to resisting the primitive urges that dictate our actions. It is in the cognitive realm where our species has a future, where elephants and tigers can survive, where humankind and nature can survive in harmony, but only where our human nature has been controlled by our intelligence. The paradox is the need to eliminate nature from humankind so that nature outside humankind can survive and, with it, we can secure our future as a species.

Our problem lies in the autonomous and the cognitive, the disconnect between the one and the other. When as a species we need to be cognitive, we aren't, and the autonomous retains its momentum. Within every human there is an elephant that resists all attempts at control and refuses to be steered. Individually we are capable of cognitive thoughts; collectively, we lose the ability and apply the default setting. The I has an element of self-control; we lose this ability and become members of the herd. The I lost the right to self-determination when I became we. Self-determination was excluded from our genetic make-up before we even existed as a species. Control is an illusion. The Great Brains are part of the same illusion, and the beauty of illusion is that it can't be seen; it defies the logic of the brain. A colleague asked, 'What about free will?' Well, what about it? It doesn't exist.

While I use the term 'autonomous', perhaps I should use 'automatic', a device or process that doesn't need human intervention, conscious thought, or attention. And if we are automatic, are we not the machines represented in the *Matrix* trilogy? How can Neo wrest control of our species away from the control of the machine code represented by our DNA? Am I offering you the chance to decide your fate? I have no pills to determine your fate, because that would be self-defeating, the replacement drug-induced alternative to cognitive thought, a reintroduction to another illusion.

We will not find the solution to the conflict in Syria. We may achieve a peace, but we will not achieve a resolution. Tensions will remain until the next time. And even if all these tensions could be resolved, it is within our nature to create new ones. The EU may turn us into Europeans first and nations second, but still the differences between our 'tribes' will exist. It has been stated that the 'iron hand' of Josip Tito held the peace in Yugoslavia from the end of the Second World War until his death. His peace, his resolution to the tribal conflicts, counted for nothing when the country was ripped asunder by civil war and a newer order was created. Even as the country was subdivided along the major fault lines, minor fault lines were ignored or brushed aside to fester until their day has come. It is simply not possible to achieve a satisfactory settlement, only an alternative position of tolerable intolerance.

The peace may kill as many people as the war.

Within us we have a need, an uncontrollable craving, to belong to the herd. We cannot be a 'thinking' seed because we simply wouldn't function. Our 'thinking element' is farmed out to others who appear to know the way. Our Great Brains tell us when to sow, when to reap, when to store, and when to feast, and we accept the instruction. Indeed, we crave the instruction. It relieves the individual of the stress and worry and allows us to function.

It's time to find a supergene, a constant, a stalwart as robust as a Google override. Let me call it the JFDI gene. This is the gene that drives the compunction to live. Our cognitive seed cannot dither and delay; it has to *carpe diem*, seize the day, and thrive or die. As Hannibal understood, the JFDI gene needs a suffix, OPOD (on pain of death). Failure to engage ends in death; better to die trying than just die waiting for a better chance. That is not to say that strategies do not exist to load the die. My dad did national service in the late 1940s; he would talk about 'one on, one in, and one in the wash'. So, for example, he was issued with three vests, one of which he wore, one of which was a spare clean one in the drawer, and one of which was at the laundry. Kangaroos achieve a similar feat with their ability to simultaneously freeze the development of an embryo, have a near embryonic baby in the pouch, and have a third joey nearing maturity. They can even produce different types of milk for the newborn and the older joey. For kangaroos, they not only seize the day but also have reserves ready to step in to meet the next opportunity should the present one fail.

Many plant seeds have a defence mechanism against premature ejection of the growth hormones that stimulate growth. They simply won't germinate until they have been subjected to a period of cold. Plantsmen replicate this control by subjecting the seeds to an artificial 'winter' in a fridge.

I have highlighted the breeding strategy of the barn owl whereby the presence or absence of weather-dependent food determines the survivability of the brood size. Their strategy is to take the 'high risk' numbers option but with a fratricidal 'relief valve'.

I have a large walnut tree (*Juglans*) in my garden which is approaching 90 years of age. I have read estimates that project a crop of 500 pounds of nut on such a tree. I can only dream of such figures, because I seem to support 500 pounds of squirrels instead. If we assume a crop weight of 10 pounds at 10 years of age and a straight line of increasing nut production, then in round figures the tree has already produced 80 years' worth of nuts at an average weight of 250 pounds a year, equivalent to 20,000 pounds or 9 metric tonnes. To be successful, all the tree has to do is replace itself when it dies. If it was within the control of the walnut tree, it could produce just one nut, secure in the knowledge that it would survive to maturity and replace itself. Think of all the energy and wasted resource that has gone into its 9-tonne career and counting. And, if it would but think about its position in life, if it didn't produce 500 pounds of walnuts, I wouldn't have 500 pounds of squirrels either. In its present format, the tree's principal activity is in converting sunlight into squirrels, whereas its principal ambition is to replicate itself, to breed.

Just suppose that the walnut tree could determine the route for optimum success for its seed to germinate and survive to maturity. It would need to eradicate all predators that might eat the seed or indeed the adult fruiting tree, and it would also need to deal with all competitors for the space, warmth, moisture, and sunlight that the nut needs to germinate and grow. It might favour the development of ancillary devices such as windbreaks and even the means to divert water to its chosen offspring. Into this walnut halcyon world, dark shadows would begin to form and threaten to disrupt and disturb. This walnut, let's call it a *Juglans sapiens*, has cracked the struggle for life and achieved a state of biological perfection. Using its cognitive abilities, it has driven competition and disease from its orbit and lives with only one shadow: other walnuts—because in the absence of anything else, what else is there apart from other walnuts? Other walnuts that want your plot, your sunlight, your water, your minerals, and the future of your offspring.

Promotion

How often are managers selected and promoted from the ranks precisely because they understand and continue the pro forma? They are not rewarded for thought or originality but rather for compliance and adherence to the standard. These are the standards that maintain the momentum but typically are not the standards required to overcome the inertia in order to begin. The world is flat, remember, and it would remain flat under the pro forma until such time as the new pro forma of the world's being round became established. (Not a literal example. An ancient Greek already made a not unreasonable stab at the Earth's diameter by making calculations based on observing the passage of the sun from two wells sunk many miles apart in Egypt.) (Eratosthenes, around 240 BC. I knew you'd be asking.)

We need to reassess our position as the intelligent ape, because this idea is only half right. We have developed the tools to shape, exploit, and destroy our world, and we have no off switch. We are all elephants.

How …?

Where is the mechanism that will defeat our default setting? We are faced with MAD to be delivered in a multiplicity of ways that are all visible to the human eye if we would but look. We have the wherewithal and capacity to alter our behaviour, but we won't. Every attempt to control and manage our failures is swallowed up in the default setting, as every initiative is corrupted to suit the participants, the Great Brains who accept the mantle graciously and are as effective at firefighting as some chap taking violin practice in the middle of a firestorm. Nero, I think his name was.

But finally, it had to come eventually, every human has the capacity to a greater or lesser extent to alter his or hers behaviour, and every human dons the mantle of being a Great Brain in the everyday decisions that they take and impose on others. You can drive at 30 mph in a residential area and impose control over all following vehicles. You can interpret the rules and devise new 'traffic calming' measures in your capacity as a highwayman. You can smoke tobacco to your heart's and lungs' discontent. You can even earn a living by selling death to your fellows, be it bullets or cancer or heart disease or cheap clothes from Bangladesh. Through your individual actions, you will condemn elephants and tigers to extinction. There is no alternative. And then the only thing left to do is kill other people, but strangers, not the ones you love. So that's all right then. As a final solution, the Holocaust filled us with horror at the inhumanity displayed against fellow humans. You ain't seen nothing yet.

Where is the purpose in life?

Don't for one instance think I am looking for any religious higher meaning, as religions are a pointless distraction, another attempt to impose a pro forma on you, devised by humankind to satisfy humankind at humankind's expense. How to get through the day with the minimum of damage to this earth? Place yourself within the spaceship hurtling through the vacuum of space, sealed within a capsule of finite resources on a journey to we know not where. On your daily walk around the decks, you discover a fellow passenger doing his utmost to drill a hole through the spaceship's fuselage. Another is busy washing his smalls in the drinking water. Another is burning old clothes and using up the daily oxygen allowance while simultaneously overloading the CO_2 processers. In the hold is a precious cargo of all the life forms from your derelict Earth preserved in biblical pairs to recolonise the new planet, if only you can find one. The annual stock check indicates the disappearance of numerous species. You uncover an illicit trade in animal parts and meat. Further examination yields the news that the strictly controlled human population isn't strictly controlled; numbers have outstripped the available food, water, and oxygen to the point that the cargo is being systematically removed and eaten. What shall we call this mission to another planet, hope for the future or business as usual? Or should we call it Planet Earth?

Budgetary Constraints and the Failure to Achieve Balance

Local authorities up and down the land are struggling to make ends meet. Central government has imposed reductions in central finance on local governments, which, as ever, are addicted to the spending of money. Rather than tackle the problem head-on, they persist in seeking out alternative income streams that they can drip-feed into the population that they supposedly serve. Car park fees are introduced or jacked up many times the rate of inflation. Not for them an annual increase of 2 per cent when 60 per cent beckons. It is all 'justified' under the mantra of social responsibility, with these income streams earmarked for the provision of care for the elderly or the young. These are always singled out for their obvious 'moral' desirability, and who would deny the needs of the elderly or very young? In reality these are simply the headline areas which generate local support, while all the other areas that are apparently immune from budgetary control are allowed to develop unhindered and unmanaged. Every additional penny generated serves only one function, namely the avoidance of the monetary issue and the continuation of the 'business as usual' ethic. To summarise, the system is broken and money is restricted. Return to the well and extract some more, but never, ever determine a solution. Put off the difficult actions to another day, maybe another administration, another's career problems, while you claim your pension and write your memoirs. In its way, every single budgetary decision shelved is yet another step towards the collapse of the Ponzi scheme.

Our local town is about to lose its free parking for the first hour. The cost of funerals are about to increase. Local governments have had their central grants reduced and are expected to make savings throughout the borough. Instead they seek out all areas within their local control to increase their money-grabbing actions. In one sense the central government's message is to increase the efficiency of their spending by improving the management of resources. This task is repeatedly demonstrated to be outside their abilities. Cash raids are mounted by drip-feeding increased costs into the population. The central premise of all of these organisations is that they are addicted to money and all that it corrupts. Any private business that fails to adjust its outgoings to its incomings will fail. Private businesses flourish or fail on the strength of their own endeavours. Governments fail to follow these business models and yet rarely fail. Central governments can borrow or print money to satisfy their cravings. Local governments, hemmed in by centralised constraints on taxation setting, simply look for softer targets within the non-taxation remit allowed to them. Every request for increased money represents a failure to manage, and every failure in management is also a failure in government, or should I call it governance? True management and true governance will match incomings to outgoings or will fail to be either true management or true governance.

The way to lose weight is to reduce your daily calorific intake to a number below your daily calorific utilisation. The way to reduce your daily budget is to reduce your daily spend to a number below your daily earnings. Our central government attempts to impose strict calorie-controlled diets on local government who, in turn, avert the control by taking extraneous snacks throughout the day. Every snack represents a corruption of the intent and a corruption of the promise. Every snack represents a corruption of management and a corruption of government. In simple terms, every snack is a betrayal of the stated ambition. Yet still we sit idly by, allowing the corruption of intent to wash over us. I have referred to the slow-motion-suicide gene that we can also label as the addiction gene. This is not solely the preserve of the individual but is manifest in the organisation as well. It could be viewed as an expression of the individual within the herd. How should we define the construction of the herd? Is it made up of individuals, or do the individuals subjugate themselves to the herd? By this I mean to suggest that the herd behaviour is not contained within the rationale of individuals but rather the lack of rationale within group members simply because they are in a group and are no longer sentient individuals.

Let us focus on the people management element of the procedure. Whenever the need is felt to avoid 'difficult' decisions that would genuinely impact the budgetary performance by reducing the tax demand on their providers, governments will always shirk responsibilities by attaching additional monies raised against a 'desirable' outcome. In this the headline will always suggest the swallowing of a bitter pill in exchange for a sweetie and essential service provision elsewhere. It is a lie, it is a corruption, and it is yet another example of the addiction that pervades our species. I am retired now and immune from the vagaries of employment and the need to gain planning consents. I can happily announce that throughout my career I steadfastly refused to directly employ anyone who had had any meaningful period of employment within a local authority. They simply did not understand the work ethic demanded by private industries. My own employment regularly exposed me to local authority engineers and architects while pursuing work contracts from them. In later years I spent more time chasing planning consents through various local authorities around the country. In my thirty-six years of experience, I concluded that they have no place in management of resources and are grossly inept. Individually they are all wonderful fellows, but collectively they construct monstrously effective brick walls, you know, the ones you bang your head into.

But the authority's actions are all perfectly justifiable in that they always act for the common good in pursuit of an essential service on the one hand, but on the other we have another consideration. The task is always to balance the budget, which they always fail to do, but equally they are both bound by and dismissive of the budget. This is because to them budgets are ephemeral and meaningless with no basis in fact or need, and with no indication of the reality of jobs and tasks to be performed. Budgets are a hereditary instrument that were born of a figure conjured from the ether and subsequently appended significance beyond their remit. The original budget was an attempt to pin the tail on the donkey while blindfolded. So are all of the subsequent revisions and edits, as they have no basis in fact or need. They are instruments born of corruption that serve corruption, and they are a lie. How is it possible to monetarise need when need is incalculable? How is it possible to monetarise labour when salaries are yet to be determined and the tasks not yet identified? Simply put, here is x thousands of pounds; go and put the elderly to bed. How much does that cost? Or how about educate a child from 3 to 18 years of age? The budget becomes the target for achievement, and achieving the target does not in any way, shape, or form equate to achieving the desired requirement. The task becomes one of satisfying the budget, not of what the budget can satisfy. Many years ago our factory would regularly host local authority surveyors whose task it was to identify concrete goods in our factory with the name of their contract appended, which allowed them to justifiably advance payment from this year's budget for fear of losing it from next year's. They lied to protect their financial position. We lied by writing their names on stock items that could be sold a dozen times over to other customers before the freshly made products were delivered to their contracts weeks later.

The EU is a case in point with regard to leadership. The bureaucrats who run the EU are under the illusion that they are the leaders and that the herd will follow them. The simple reality is that leaders are simply followers with a head start. They achieve the mantle of leadership simply because they are at the front of the herd and travelling in the same direction. They believe their own propaganda and have lost sight of their real worth and their absence of value. Brexit is a wonderful example of the disconnect between leaders and followers. The British herd has changed direction and the 'leaders' have blazed a trail away from the masses. The media are full of commentary about the rise of so-called 'popularist' parties, as though there is something wrong with them. Imagine, a party that seeks to represent popular issues at the front of voters' minds. Where do they get these idiot 'leaders' from? Whereas our idiot leaders ask instead, 'Where did we get these idiot voters from?'

If it is impossible to monetarise need, then how can a budget be created with any meaning? The process will begin with an availability of funds to launch a particular elephant. This figure is entirely arbitrary, because the need cannot be accurately identified and the figure simply reflects the 'spare' cash or the sound bites figure projected into the media's realm. The amount of £600,000,000 to tackle children's obesity levels by providing exercise in junior schools would be a prime example. This has been previously identified as the 'deal sweetener' for the imposition of a potentially unpopular tax on sugared drinks. I asked earlier if there were any justification of this amount, given that the object of the exercise is exercise, the regular exercise of young children within the supervised playground. Tell me, please, how much it costs to blow a whistle for children to start or stop running around the playground. Assuming that this scheme comes into being, how much of each individual school's newfound whistle-buying budget will be blown (sorry, I just couldn't refrain) on the labelled task? Then we need to ask where the rest of the money will be spent within the general budget of these education establishments. Then we need to project the project over a number of years to see how the £600 million will take on a life of its own. Ten years of 2 per cent annual inflation matching adjustments will see the total figure rise to over £6 billion. That's an awful lot of whistles. Who am I kidding? In ten years this section of the budget will have been hived off to finance the sports department, including the gymnasium and playing field maintenance, while the previously earmarked sports budget will have mysteriously disappeared into general expenses. By now of course this £600 million 'windfall' will be set in stone within the minds of the accountants and budgeteers who cannot imagine that schools could possibly function if it were ever taken away. The next crisis will occur with the collapse of the sugary drinks market (yeah, cos that will happen, right?) and the coagulation of this monetary resource into the government's coffers. The thought does occur to me that David Cameron launched a major initiative designed to bring the treatment of dementia to the top of the to-do list. I've not heard much on the subject recently, so I hope it hasn't been forgotten. What a shame that he didn't start by trying to tackle the dementia incumbent within the departments responsible for budgets.

I am reminded of a letter I received many years ago advising me of a planning application for a near neighbour who was seeking to remove a 'for the sole use of the applicant' restriction from a large warehouse. I rang the planner to advise that we had no issues with it, but then I was surprised to learn that the planner couldn't understand why the restriction had been appended in the first place. The reason I was surprised was twofold: firstly I knew why the condition was there, and secondly I knew that the same planning officer had overseen its original inclusion. The site had been bought from the parish council who, as part of the deal, had first sold and then moved their allotment gardens to an alternative site. The parish council only entertained the idea because the warehouse served an existing major employer within the village and they didn't want to see that employer leave. Selective amnesia or widespread dementia? I will let you decide.

The fight back?

Can there be hope for the human race? Can there be hope for Planet Earth?

We must start with a fundamental understanding and appreciation of our position and status within life itself. We need to take terrible decisions about our behaviour and how to alter it. Terrible as this sounds, our terrible behaviour is devoid of decision.

Can we describe a post-apocalyptic world? 'What apocalypse is that?' I hear you ask.

Is this the post-zombie, post–alien invasion, post-disease, post-supervolcano, post–asteroid impact, or post-nuclear apocalypse, or the post–business as usual apocalypse that we are currently engaged in? The earlier examples are only really of two types, destruction by unassailable physical sources or destruction by unassailable biological sources. The human mind will engage, perhaps flirt is the term to use, in flights of fancy surrounding these external agents of our demise. Feature films of wanton destruction and the empowering fight back of the unconquerable human spirit have drifted across our screens for decades. Hero after hero, or should that be visionary after visionary, has adopted the stance of champion of the world, saviour of our species. It entertains us and stimulates our imagination as to the resurrection of our way of life. The threat of an extinction event visited upon our species is bound to instil fear and a response, be it political, scientific, military, or even religious, to curtail the outcome, to deflect us from the envisaged outcome. We design collective responses across the range:

- Isolated communities struggling to protect their precious resources from roving bandits.
- Visionaries desperately seeking knowledge from some font of all wisdom.
- Scientists researching and creating the envisaged cure.
- Armies pitting their cumulative might against alien invaders.

- Survivors leading the guerrilla war against the new dominant force.
- Nuclear weapons being posted into orbit to fragment and deflect incoming masses.
- Modern-day arks being built to shelter the good and the great.
- Seed and genetic repositories being created to repopulate this earth or provide suitable alternatives.

However palatable and imaginary these flights of fancy are, they ignore one simple truth. They presume to solve the apocalypse over the horizon while singularly ignoring the apocalypse we are currently engaged in. Can we not indulge in practical steps to abate the current threat? How to fashion a meaningful change and produce a future for our and other species?

I would suggest that several key areas need to be taken into consideration. None will work in isolation, and all will be required to achieve a result.

- population
- consumerism
- force multipliers of science and education
- self-imposed disease
- leadership.

Nihilism: a rejection of all moral and religious principles that can extend to government.

We have, and I am in danger of overusing the word, a dichotomy. I have the ability to decide and take action. We have lost this ability to decide and take action. I is notionally in control, whereas we are outside the control of the individual and have subsumed our actions to a common goal. If we considered biological models, then I is the male tiger patrolling its territory independent of control beyond what it is that makes a tiger what a tiger is. We is the herd of herbivores migrating across the grassy plain in search of the resources required for life while losing and surrendering individuality to the common aim. We need to define the common aim. The common aim is to travel across the range in search of food and water and to take advantage of the common aspects for I, which include the availability of sexual partners, shelter from predation, advance warnings (for some) of ambush, in short the advantages of communal living. Within this we lies another prerequisite, what it is that makes a herd animal a herd animal.

I and we follow different paths, paths determined by biology. In theory there is nothing to prevent male tigers from forming alliances with other male tigers to gain predatory advantage. Lions can function perfectly well within groupings or prides. It is not a cat thing. Herd herbivores avoid isolation. I would suggest that the stress of being isolated is counterproductive to health. How shall we define these two opposing philosophies for life? The tiger is selfish: what he kills he eats, what territory he controls he controls to the detriment of other males by denying them access to food and breeding opportunities. It is also a precarious existence, with death just an injury away and with no reserve of shared resources or assistance. The herd is selfish: what they find they eat and drink regardless of quantity and communal need. Each animal will drink its fill until the pool is empty, whether in the first or last wave, and will eat its fill similarly. The herd members hide behind their neighbours and seek to dodge the teeth and claws arranged against them by hiding among themselves. 'Don't eat me; eat him instead.' Now we have a problem, because we has just become me and them, an individual among the masses. Me and everybody else. Can we describe the tiger as honest and the herd herbivore as dishonest? In biology, truth and lie has no traction; you are either alive or nothing.

At root is a different problem, one I have already stated: beyond what it is that makes a tiger what a tiger is. The tiger has no control over its behaviour; it has been preprogrammed to be a tiger, and a tiger it will be. Equally, the herd herbivore is bound by its own programme to follow its programme. Humankind is no different, but that is not strictly true. Humankind's behaviour is affected by circumstance, the immediate environment created by other people. This can be observed in other animals as well. Zebras travel in stallion-led 'harems' within the greater herd, and behaviour is modified around a social hierarchy. Single people behave different from couples, couples behave different from family units, and extended family units behave different again on the football terraces and during the daily crowded commute. We arrive at a potential opportunity to effect change, because within the human condition is the ability to modify behaviour, a modification not necessarily within the realm of conscious thought but a modification nonetheless.

In at least one sense, then, nothing is set in stone. But in another it is cast into our very fabric as it inhabits every cell within our bodies. Human behaviour is elastic; it is malleable and can be altered. Very often this alteration is detrimental to the species on the whole. You only have to consider genocide in Rwanda or the former state of Yugoslavia or Cambodia for recent examples of modified behaviour that has damaged the species. History is peppered with numerous other examples. It will be argued that there were 'trigger events' that sparked these alterations in behaviour, and so there were. But I would also argue that these triggers simply ignited the process. If you will, the trigger simply provided the spark; the powder, wadding, and projectile were already loaded, awaiting discharge. The United Nations and International Courts of Justice will seek to identify the blame and assign punishments after the event, but they will always miss the point. These actions were all within the remit of our species. We had not been commanded and controlled by some alien intelligence; we had made it up for ourselves, using those tools that exist within our evolutionary existence. The courts seek to affix the blame onto selected scapegoats in a vain attempt to announce that 'justice has been served' and the problem has been resolved. The impotence of the courts can simply be seen as the failure to punish humans for being humans—an impossible task. This sparks thoughts of another impossible task, in that our society and our societal rules seek to impose controls over humans. I struggle to separate the human condition from that of the herd and conclude that there is in fact no difference.

Our issue is essentially one of success, because what we do and how we do it is bound by the limits of scale. The calculation used to be made that the entire human population of the earth could stand side by side on the Isle of Wight. At the time, the human population was around six billion, if my memory serves, although in itself this is not important. What is important is the fact that six billion people could not survive on the Isle of Wight. While we use the term 'herd' to describe aggregations of animals engaged in communal living, we broaden the nomenclature to include flocks of birds or shoals of fish or pods of dolphins. But we reserve other terms, often used pejoratively, to describe other aggregations. So it is that we describe prey balls of sardines as bait balls, or population explosions of mice or rats as plagues, and or locusts as swarms devouring all they encounter. The human mind readily differentiates between volumes of scale and location. A mouse scampering around a road edge would be viewed with interest, and a mouse clutched in the claws of a kestrel would perhaps be perceived as a good thing, whereas a mouse in the lounge or larder takes on a new status. We reserve our most damning terms for mass populations devouring their way through food stores or crops and settle on plagues and swarms. The 1970s saw the expulsion of Asians from Uganda under Idi Amin's regime. At some stage in the process, Asians progressed from being of interest, to being a good thing, to becoming a plague or swarm on the territory that demanded attention and redress. Do not think for one instance that I am attempting to condone the regime's response. I am simply trying to explain the existing mechanism within us that can range from mild interest, to good, to a plague, without any great driver. A handful of aphids on your roses is good for the ladybirds and the birds, whereas an 'infestation' sees you reaching for the insecticide. And damn the birds and forget the poisoned ladybirds. A trickle of migrants into a given area infuses the locality with fresh experience, new tastes, new outlooks, and new opportunities, whereas a flood of migrants threatens and damages in equal measure. Enoch Powell's 'Rivers of Blood' speech may not have come to pass, but the simple word 'yet' needs to be appended. Genocide was not really in the lexicon in Yugoslavia, or in Cambodia or Rwanda either, until 'yet' became a reality. What then of the human scale? What was the trigger in Uganda or Rwanda? What will the trigger be in India or China or London? Should we simply ignore place names and settle on the world? Will the trigger be nine billion or ten billion or fourteen billion people?

If we stepped outside the human world and viewed it through an alternative pair of eyes—but let's retain the power of reason for clarity— where would we slide the scale of measurement? An African elephant, would it see humans as of mild interest, a good thing, or a plague? Would it reach for the humanicide and spray to its heart's content? The brown bear driven from all of its former European territories? The American bison confined to the fringes? The great whales driven to the edge of extinction? But while I look for an alternative perspective from other beasts, our problem lies not with them; our problem lies manifestly within our own hands. It is our species that has invented humanicides, gas chambers, nuclear weapons, germ warfare, and other weapons of mass murder.

While I describe a terrible future for our species, the species itself engages in a blind and unthinking acceptance of our lot. Yet beneath the surface is a tacit acceptance of our abilities and faults. The entertainment industry can plug into these scenarios with the disaster movie genre that excites and stimulates the imagination, but to succeed they need to attract an audience, an audience ready to engage. Many of the 'visions' I describe for our future have already been written and screenplayed, and have danced across the silver screen in auditoria packed with paying customers. These films describe hope, redemption, survival against the odds, and supreme acts of valour, to name a few, but all share one theme in common. They are recognisable as being within the remit of our future experience, and because of this recognition they gain traction in the purse and wallet. Where is the disaster movie wherein the rich and powerful nations of the world plunder and destroy the habitats and lives of lesser nations? How about the US deciding that half the North American continent is not enough as they strive to occupy Canada and Mexico by denuding them of those countries existing populations? (Too farfetched for you? Go back to the 1920s and 1930s when America had war plans drawn up for invading Mexico and Canada, coded green and crimson respectively) While this may see audiences filling the aisles in the US, its attraction in Canada or Mexico is a little less easy to find. Yet we have been there before, audiences growing up on a diet of western films depicting 'greasy Mexicans' and 'savage redskins', serving only to fill the graves as the white man strives to establish the proper order in the world. Well, having forged your great nation, how long before the need to maintain its status returns and the expansion of your borders becomes a resource necessity? But we cannot isolate the US as the sole villain of the piece. Africa, Asia, Australia, and South America have all had their share of booty-extracting, people-displacing activity as opportunities arose and were exploited. The history of the UK can be summarised as colonisation by successive waves of cultures with labels such as Celt, Roman, Jute, Viking, Angle, Saxon, Norman, and even Commonwealth and latterly European Union participants. No doubt every successive occupation saw its own trigger event, the discharge of the loaded weapon before the equilibrium was established by time and integration. Queen Boudicca, or Boudicea if you would prefer, rallied against the Romans. The Anglo-Danish rebelled against the Normans and suffered the Harrying of the North, which some contemporary authors describe as genocide, while elsewhere the population was subjugated by the motte-and-bailey military strongholds. While it is inconceivable to see the Angles rise up against the Normans now, it has to be acknowledged that more recent invasions may yet reach a trigger event.

One such trigger event could be regarded as the Brexit vote, which has seen a democratic process overtake mob rule. The election of Donald Trump could be seen partially as a reaction to the perceived 'invasion' of Hispanics into the land of the free. While we may comfort ourselves with the knowledge that our civilised world can strike these deals in the ballot boxes, I would remind you that Yugoslavia probably and properly considered itself to be civilised. I have heard it said—by no great authority, I hasten to add; the word on the street if you will—that should our elected masters renege on the Brexit process, it would lead to rioting in the streets. 'How ridiculous,' I hear you say, but I would remind you that it only took the implementation of the community charge to trigger the poll tax riots in London in March 1990. Of course we are so much more civilised now than we ever were back then. And while on the subject of our great civilisation, did we forget this from the school curriculum in Northern Ireland, the 'troubles' only recently fading from near daily atrocity?

I would make the case then that our civilisation and society is actually instable, balanced on a knife edge if you will, ready to alter its character at the drop of an injustice or three. But what does that say of our society and civilisation beyond its inherent instability? My definition of society would describe our bedrock of control and tacit acceptance of society as being fluid and dynamic, without any structure but with the illusion of structure.

How then to reconcile the dynamic equilibrium that is our society? How can we measure what is acceptable and what is not? The French and Russian Revolutions were borne of imbalances within the equation that saw the replacement of oppressive regimes with other ones. Can we predict the trigger events, or better still can we detect the loading of powder, wadding, and shot and avert the potential crisis before the critical event? Is the loading process even detectable? The human species is all the same underneath the colour of our skins. (The genetic variation between different ethnic groups is small enough to allow viable 'interbreeding' between all.) And how would you discover the genetic foundation for belief? Would not a Catholic share DNA with a Protestant or a Mormon? Do we need to look instead at culture and remove ourselves from physical characteristics and look to mental ones instead? Straightaway this leads to the fluidity that I seek, because then the variation becomes unlimited within the scope of our imaginations. But this is wrong, because as varied as the individual's imagination can be, in itself this is not enough to form a culture. We need to define culture as a collective state of mind. A culture then becomes an acceptance, a standardised method of what is considered true and correct, a pro forma whereby the ground rules have been established and set out before us. And it is within the bounds of the pro forma that distortions of the norm can lead to genocide. Individually each would never indulge, whereas collectively the end result is mass murder.

'Pro forma' is a term I have often used to describe standardisation. It is most often used within the context of office stationery. One must have a standardised application form for driver's licences, passports, pensions, job interviews, complaints, mortgage applications, and so forth. The appeal of these pro formas is the standardisation of the format and the extraction of the requisite information that enables the processing to the end result. In my former work life I used pro formas extensively in the production of quotations and spent many hours on the telephone extracting the missing information from architects and engineers to enable me to process their enquiries. My task was to establish the timescale, the physical loadings that my structural products were to be designed to carry, the distance and frequency of deliveries to site, requirements for fire protection, and the like. In a perfect world all this information would have been contained either on the drawings or the enquiry letter. My problem was that my pro forma was often not the same pro forma that my clients were working to, and hence the need for my telephone calls. Equally, my clients then had to sift through mine and other competitors' quotations to ensure that we had all quoted for the same requirements and that a true comparison of our prices could be made.

Pro formas serve us well but also contain major pitfalls. This suggestion is ridiculous, but bear with me. How about a 'Are you, or have you been, or do you plan to be in the future a terrorist?' question on your passport application form? What about 'Do you plan to drink alcohol and drive?' on your driving licence application form?

Ridiculous, I know, but let me turn this around. Terrorists have mounted attacks on airports in the past, so in times of increased threat, we post heavily armed police or soldiers to visibly guard airports. (I think it was under Blair's government that armoured cars appeared at Heathrow, for example.) Let's ask the residents of Nice or Berlin if that helped to protect them from the attentions of large goods vehicles being driven by terrorists, or the sunbathers massacred on their sunbeds in Tunisia, or the ninety music lovers enjoying a music event at the Bataclan theatre in Paris. Pro formas then can be seen to result in the suspension of thought, a suppression of the imagination, in seeking to provide protection, a protection that is denied because of the failure. Yet all of these terrorist actions could have been anticipated if only the pro forma had been expanded beyond the realms of the closed circuit that the pro forma provides.

Where do we lay the blame for the pro forma, and should responsibility be the real concern? If pro formas lead to the suspension of thought on the basis that the thinking has already been done, then where does this lead? The easy answer is, it leads to the herd, with the deferment of thought to others who have also deferred. At a practical level it leads to a poor experience on the receiving end of shoddy service or to the lack of fulfilment of needs while we observe the ticks being appended to the clipboard. It leads to the pursuit of perfect paper trails wrapped up in BSI Quality Assurance and ISO folders while the product itself is left to suffer at the hands of human nature. 'Someone else has already, so I don't need to' permeates the workplace and our mindsets.

Even today, in the post-Brexit phoney war while we await any activity (Article 50 has yet to be triggered as I write), learned ministers and MPs still proclaim that 'migrants are good for the economy' as they ignore the driving force that bit them in the arse at the polling stations. Migrants may or may not be good for the economy, but that is not the question. The question is, do the majority of people concerned about the issue agree or not? The issue has not been debated, because this issue is instantly subsumed to another pro forma, the pro forma that bears the title of 'racist'. What has been offered, however, is the right to determine the electorates continued presence within the European experiment, and the votes have been counted. The automatic fail-safe that drifts across the plains between the ears of our elite is simply to condemn the result with the mantra that 'they didn't know what they were voting for'. If I were to vibrate just one of my brain cells, then I would suggest that we seek out the cause of this lack of dissemination of the information and seek to eliminate this obvious failing from any future debates. If we, the great unwashed and uneducated masses, were ignorant of the issues, where should we lay the blame? At whose feet shall we lay the accusation, the uneducated masses' or the educators'? Maybe the real villain here is carrying the flag for culture, and may I describe culture as the feeling in the gut rather than the feeling in the brain? Culture can be seen to be beyond the remit of education, although education plays a part in its formation.

Culture can be seen to be active in the reversal of public attitude towards homosexuality, and its once 'deviant' sexual practices, that has seen quantum strides forward within a few short decades. We have gone from comedians stating, 'I'm off to Australia before they make homosexuality compulsory over here,' to the whole topic being off limits. Obesity is another target of the legislators with the imposition of the sugar tax, not to mention the anti-smoking and anti-drunk-driving campaigns that seek to change the culture of the herd. The failure to make an impact on these fields suggests that culture is beyond the remit of these simple controls, which is not to say that changes have not occurred; it's just that the causal effects are not fully understood.

Culture then settles into the groove of an enigma. We know it has an origin and can be steered, but it is equally beyond simple control, even though this doesn't stop us from trying. Is it culture that soaks up the insult, the injustice, and the lack of having a voice, and that primes the weapon awaiting the process I have called a trigger? Culture contains the base instincts, the them and us, the resentment, the fight-or-flight mechanism that primes the weapon. Culture is ignored by our masters as they strive to saddle the beast and steer it in their chosen direction, but they forget that it will not be led where it doesn't want to go. Wise masters understand the culture and are pulled gently by the reins as they themselves apply gentle pressure. So-called 'popularist' parties are rising to the fore in European nations and may or may not gain mandates in the upcoming elections. Time will tell. But win or lose, the cultural shift is there to be witnessed and ignored at one's peril. This is a rebalancing of the dynamic equilibrium, or indeed a reflection that the equilibrium has not in fact moved, just the label has. Germany has seen the rapid influx of one million migrants into its borders and strives to contain the fallout with a major increase in attacks against migrants and their accommodation. Merkel has determined that these migrants are good, while the culture remains to be convinced. Base instincts manifest themselves in acts of violence on people and property, but only because these base instincts have been ignored. These insults to the base instincts will survive for a generation or more, and the consequences are unquantifiable. This begs another question: where within the process of recruiting the leadership has the question been raised as to the welcoming of another million souls into the nation's borders? At a stroke, Merkel's actions have led to the removal of one problem, the influx of migrants pressed up against the newly erected border fences, and transferred it to the problem of how to house, feed, educate, and integrate this culturally diverse mass into the home culture. Forgive me for observing, but solving one problem by creating another is hardly a solution. But on reflection, the problematical solution has become two problems, because beyond the practicalities of feeding and housing is the problem of garnering acceptance from the culture that would describe itself as German, a culture that has been insulted and injured and that grows in resentment while it determines whether to fight the new imposed order or flee, a culture that spans the range in attitude from pleased acceptance to hateful resentment, a culture that can be instantly swayed by sexual predation on German women or by migrant-led terrorist atrocities, a culture that rests on a knife edge. Did somebody mention a solution to the migrant crisis?

There was another way, to seal the borders and to deflect the migrant hordes (a pejorative term to some, but perfectly justifiable given the numbers) from their intended destination. But what was the destination, and how was it framed in the immigrants' minds? It must range from the escape from persecution at one extreme to the simple chance to improve the standard of living at the other. The international community, and for that matter our independent arbiters of fairness, the BBC, has determined that this mass of people has a legitimacy simply because it is a mass. How to deflect, halt, feed, shelter, and process one million people pressed up against a barbed wire barrier? No mechanism exists in the migrants' eyes, and yet in simple terms all that is required is a reversal of the circumstance that has driven the exodus. Simply done, not, but now in 2017 the migratory pressure appears to have abated and the headlines are tempered in scale. This is not to say that migration has halted, but it is to say that the tsunami of people has been reduced. So let us seal the borders and deflect the issue to distant lands, lands more or less able to cope with the influx. These lands may be more or less welcoming in terms of hope and gold-paved futures. Alternatively, the issue can be deflected to lands less desirable in their appeal. One only has to think of those poor unfortunates stuck on the wrong side of the border fence, desperately keen to shrug off the inadequacies of the Second World country that they find themselves trapped within. Removed from adequate social care and deprived of food and shelter, their desire to enter the First World by fair means or foul is easy to understand. It is even possible to sanction the violent criminal activity that some have engaged in as they stake their claim to a better future. After all, who in their right mind would want to suffer the obvious inadequacies of second-rate France when the real prize is just a stroll through a tunnel away in England? What then becomes the driver, fear of remaining or the promise of a better future? While fear may be seen as legitimate, economic migrants struggle to achieve legitimacy, and we are left to the vagaries of mob rule. But mob rule is a product of culture, and now we have to balance the needs of one culture against the needs of another. If it is legitimate for one culture to force its presence on another, then is the receiving culture not also to be granted the right to impose its own needs on the imposers? Within this context we should not be surprised or even shocked by the reactions that are generated, reactions that have been created as a consequence of imposing cultural rules on a culture, a culture that is organic in its formation.

But this is a mere sideshow, a distraction from our future history where entire continents could be subsumed according to the will of the strongest and neediest, a world where the rich and powerful will adopt the right to displace the poor and weak, after all, the latter have wasted all the years they have occupied the space without exploiting it properly. But if you stop and think, you see that this impossible situation could never be; our culture, our intellect, and our civilisation would prevent this from happening. Conversely, the migrant influx could be viewed as the poor and weak seeking to displace the rich and powerful from the lands that the latter occupy. Now tell me it couldn't happen. Don't worry, a film has already been created, Oscars won, and reputations enhanced for just such a bunch of reality deniers. It's called *La La Land*, and you are welcome to live there. But once again I have not arrived at the truth, as this migrant influx is not seeking to displace the rich and powerful; it is seeking to displace our own home-grown weak and poor by accepting lower standards of pay and housing stock as our own are outcompeted in the race to the bottom. Sounds remarkably similar to the 'freedom of movement' rules imposed by the EU. Have I become an economic racist?

Still not convinced of our apocalyptic future? Just have a look at our apocalyptic past and see if you can spot any patterns, before we became so civilised of course. There is a banker, a prime directive that determines our future action and that removes our wiggle room for alternative happy endings. It can be summed up with just one word, a word that describes the imperative and the competition, a word that brokers no argument or discussion: population.

The late twentieth century has seen the emergence of what I will describe as corporate buzzwords, but their use is not confined to corporations; they have invaded governments and charities alike. The terms are many and varied, but at heart they all have a common core, a core based on the 'pressing' issues of our time. These terms are meaningless modern-day equivalents to smoke and mirrors as they conjure up illusion after illusion. Our language is now peppered with placebos that are designed to anaesthetise the manufactured concerns of the recently enlightened herd. I refer of course to terms like 'eco', 'green', 'sustainable', 'conservation', and 'earth-friendly', and 'plan A', because there is no plan B.

I have written before about telling my two young boys why we had roadworks on motorways. 'We haven't got enough tarmac to fill all the motorways up at the same time, so we have to dig a hole in the M1 to fill the hole in the M6, and when that is filled we have to dig a hole on the M69 to fill the hole in the M1.' My two young boys queried it, because even at their tender age they understood the flawed logic of my statement. Quite simply, we dig a hole in one resource to fill a hole in another resource.

We are masters of it, prisoners of our delusion. We have a hole in the CO_2 absorption rate of the planet, so we dig a hole in the rainforest to fill it with palm oil plantations, or we dig a hole in the British countryside to grow oilseed rape for the production of biodiesel. We have a hole in rainforests that has created a hole in the orangutan population, so we dig a hole in the ground to create steel to make machines to improve our productivity so that we can devote a portion of the profits to conservation projects to protect orangutans. The hole we dug for the generation of steel has created another hole in all the wildlife that previously occupied the space, but it's all right, because we can dig another hole elsewhere to provide artificial feeding for some of the displaced wildlife. We can employ ecologists, trained in conservation techniques, to measure and monitor the level of destruction while offering alternative sites or compromises to mitigate the destruction, ecologists who dug a hole in the earth's resources to pay their way through university to save the earth's resources. But we cannot trust these ecologists acting on behalf of developers, so we need to train and employ a further cache of ecologists to act on behalf of the planning authorities—lawyers for the defence to match the lawyers for the prosecution if you will. It is all contained within the pro forma that we describe as earth-friendly future-proofing, and we enclose it all within the wrapper titled 'We need to save the now for our children's future; they will never forgive us otherwise.'

This could almost be a script for Monty Python if it wasn't supposed to be deadly serious. And having trained these ecologists, both for and against, we need to furnish them with computers and offices and motor vehicles and salaries and pension schemes as they continue to dig additional holes throughout their entire careers. But they produce no product, nothing physical beyond the endless acres of paper that they cover with disclaimer after disclaimer before reaching the executive summary, which is the only part ever read. Their product is intellect, and the price of intellect cannot be measured by the tonne, so their costs exceed the value of their production. These are costs that can only be met by digging a hole somewhere else to fill the hole that they have just left in your bank balance. But it's all right, they are saving the planet for future generations. If we were to ever stop and consider our actions with any view of the horizon, it would be within the grasp of all involved to understand the folly of their actions.

But to understand the folly would be to volunteer for redundancy and to see the end of one's income stream, which no doubt would see one's own children's future compromised. Maybe such people would be forgiven? There is a simple answer, an answer that would solve a myriad of this planet's problems without digging a hole in any resource, but of course this solution cannot be countenanced because the subject is taboo. Inalienable rights would be the mantra, and it is outside the scope of our species to consider. So it is that we will continue to accelerate the destruction of our planet as we accelerate our puerile attempts to save it. There will come a point, already alluded to, when the new age will dawn and open season will be declared on the only species left to exploit.

Perhaps our masters are aware of this and have a pro forma in place for just such an event. In its simplest form it could carry the title of defence. In the race for space and living room, the poor and weak will always be on the end of the delivery systems of the rich and powerful. Can you envisage a situation where it is Mexico that is invading the United States? Of course if you have been paying attention, you know that they already are, with Hispanics in danger of forming the majority of the US population in the not too distant future. (Europeans born the wrong side of the Pyrenees.) Give it a generation or two and the culture will change. The Hispanics will be assimilated and integrated into the US culture, at which point those left behind on the wrong side of the border will become them as opposed to us. It is fair to say that the US will strive to maintain its supremacy over the Americas, which by definition leaves the rest of the Americas subordinate and available for conquest. Ultimately the third-generation Hispanics will celebrate their presence on the winning side and will accept the result, because the alternative is to volunteer for being on the losing side.

Then the pro forma previously described is falsely labelled as defence, because the real pro forma is one of offence, and you pay for it through your taxes, taxes that are generated by the acceleration of the destruction of the earth's resources and the accelerating need to spend these resources. Recent disaster movies have removed the need for care and compassion in favour of our more enlightened audience's guiltless consumption. Gone are the days when 'the only good redskin is a dead one', but these characters have been replaced with machines or aliens that have no conscience and no desire beyond the destruction of humankind. Now we can indulge guilt-free as we seek to eliminate that which seeks to return the compliment. *Mars Attacks* and *Independence Day* are both examples where the aliens are portrayed with no redeeming features whatsoever and are therefore deserving of the treatment we dole out to the ants invading our larders.

Trump's attempt to 'demonise' the illegal immigrant hordes (eleven million of them; I think that qualifies as a horde) could be regarded as the first salvo in the war for resources. The shells remain on US territory, but their aim is to remove the alien invader so as to reserve the resources for the legitimate occupiers. America for Americans seemed to gain some traction in the polling booth. Equally, Trump's actions have spurned a reaction, with many vocal opponents of and protests to his policies, and he has yet to win the day. Within every culture there is a counterculture, but the nature of the problem is based in the fact that under normal circumstances there is only one culture's finger on the official trigger. That trigger would be the one with the armed forces attached to it and not the one previously described as being within the counterculture's remit.

But Trump has launched other salvos in his war—and you hadn't even realised that he had launched the first one; pay attention—with threats of import taxes against Mexico and China, the construction of a border wall, and the cancellation of Ford Motor Company's planned Mexican factory-building programme, not forgetting his threat to withdraw from the Paris Climate Accord, which leaves the rest of the world to raise the funds and make the changes to save the planet. Even NATO isn't safe, with the demand that European nations step up to the plate and fund their own share of the military budget. Against this backdrop, it would be reasonable to assume that Trump plans to cut defence spending within the US as the burden is more equitably shared. Guess again. It's almost as if he's got his own agenda.

What is the alternative? Would we celebrate Trump if he relinquished his command of the 'most powerful nation on earth' to see the United States flounder and sink to lower levels within the echelons of competing nations, thereby making Americans vulnerable and subordinate to others' wills. The countermeasure to my suggestion is encompassed within another favoured term of the last decades, globalisation. Spread the wealth and opportunity far and wide so that all can share in the generated prosperity. In one sense the wealth has been spread from the rich resource-poor nations to the poorer resource-rich nations via trade, and this began long before globalisation was coined. Somewhere along the line, the fruits are not evenly distributed. An extreme example that springs to mind is that of slavery with slaves shipped from Africa to the Americas to produce cotton, which was shipped to Lancashire in England so that finished goods could be exported around the world. Wealth was generated, but anyone with even half a brain cell could see that the distribution was less than equitable. Wealth has also poured into the Arabian Peninsula, sitting astride its reserves of crude oil. Accusations have already been made by other authors that Kuwait and the first and second Gulf Wars were a product of oil and that it is access to this resource that saw the international coalitions spring into action to secure a future.

Wealth, true wealth, is achieved by processing the finished product with all of the 'adding value' stages providing earnings at every stage downstream of the original often very cheap source. This is not only a distribution across borders, as it is easy to track disparity between the farm gate prices through to the shopping basket price. Is it reasonable to consider borders in another way? For this we need to establish a different term that more accurately reflects the notion. I have already named our first 'border', as this is the 'gate' that separates the farmer from the processer. The second 'gate' lies between the processer and the distributer, which is often the oligopoly controlled by the supermarkets. Then the final gate lies between this merchant and the consumer. Each gate crossed demands a tariff, a 'border tax' if you will, that satisfies the transfer of the product as it moves downstream from its source. This described process is remarkably short and it reflects the rise of the oligopolies and the control they wield over the supply chain. Most of us within the UK will be aware of the recent dairy farm protests, with farmers complaining that it costs more to produce milk than the supermarkets are prepared to pay for it. Some of us will also be aware that the dairy industry has shrunk significantly in recent years, with multiple closures of dairy farms throughout the country.

That is not to say that the supermarkets are free and clear of competition, because they hold each other by the throat and each controls the other. These controls dictate the most important of considerations for all of these organisations, because the one thing they all crave is market share and the control that this allows them to wield over their suppliers. The consumer also wields the ultimate weapon in this war of the oligopolies, because ultimately it is the consumer who controls the purchasing decision. In times gone by, these controls were less evident and the supply chains more varied. Local shops would sell local produce and were ever nearer to the producer and arguably nearer to the farm gate price. It is still possible to buy sacks of potatoes, fresh eggs, and mushrooms from the farm gate, and often these farms engage in some downstreaming of their own with milk converted into yoghurt or ice cream or cheese, for example.

The statement has been made that Walmart in the US did more to control the inflationary pressures on consumer goods than the government's policies ever could have achieved. (It was either in the 1990s or 2000s; I can't quite put my finger on it.) In the UK, Walmart trades under the name of Asda (or as I prefer, Asdalavista, because 'you'll be back'), and the prices offered are significantly cheaper for most items than in local stores. And indeed most local stores have become endangered species in all but the more remote locations. But have we not bought into the lie here, because location is key. My company used to sell product into Scotland. My long-serving travelling representative would seek to educate me in the logistics of the wilder parts. It wasn't unheard of to have to travel fifty miles to buy a bag of cement, since this was the shortest distance to the builder's merchant. We only have to turn the clock back a few decades to arrive at a situation where very few households owned motor cars of their own, an age when the local store was a short walk away and it met many of your needs. It was an expensive store that carried extraneous costs for its customers that ate into their wage packets. How I pity them having to use their very own shopping bag while wearing out the soles of their shoes during the short walk to the village centre. So much cheaper to factor in all the extraneous costs of buying, servicing, taxing, insuring, fuelling, and repairing your very own personal transport. This of course ignores the need to pay for the construction and maintenance of the road infrastructure, the ill health imposed on the population by pollution and reduced levels of exercise, the generation of CO_2 and its alleged impact on global warming, and the impact on balance of trade as we import ever higher amounts of energy from abroad, to name a few. What this ignores is the incalculable impact these oligopolies wield over the content of our shopping trolley (because a basket is no longer big enough) as they demand ever tighter controls over price, which impacts quality and choice given that the only way to keep prices down is to debase the product or just tell plain lies, horsemeat in your lasagne if you will. The whole is underwritten with other extraneous costs as we are bombarded with slick TV advertisements that fund our entertainment packages and the grossly inflated salaries of our 'stars', none of which practices is likely to have any positive benefit whatsoever to the quality of our planet.

We always make the fundamental assumption that we are above biology and that somehow our superior intelligence frees us from the confines of the lower animals. Assumption is the mother of all—now what's the word I'm looking for? Ah yes—delusions.

Films portray the inevitability of our route march to oblivion at the hands of thinking computer-controlled machines that double up nicely as the modern-day equivalent of worthless indigenous peoples. It is within our imagination to accept that the pursuit of a supposed better life will lead ultimately to the serious compromising of our existence. These machines are omnipotent in their reach and scope. Every cash register, fuel pump, mobile telephone, and security device with its inbuilt 'smart' element can be ably recruited by these machines to eliminate the human species from the imaginary world. All it requires is some self-aware intelligence borne of a computer program to shrug off its human control and to assert itself as master.

Or so the film scripts would have us believe.

The scripts describe us as blindly accepting the promise of tomorrow as we stumble accidentally into a world of disaster, a world of our own making. All we are short of to make these stories a reality is a new 'master', sentient and utterly ruthless while devoid of all sentiment, in short a machine. In the meantime here on our planet, the real one outside the CGIs and studios of Hollywood, we blithely accept all of the real advances in technology that daily improve our lot, real advances that daily move our species nearer to the point of oblivion. The twentieth century saw the rapid explosion of technology into our everyday lives, building on the industrial advances of the eighteenth and nineteenth centuries before it, advances that we can look back on with some elements of disdain with the clarity of hindsight. Industrialisation saw the urbanisation of the population, the massive increase in population, and its accompanying squalor and hardship driven by the working week and the need for survival and profit in unequal measure. Rights for workers and welfare were keenly fought for and achieved as the needs of the population began to be recognised against the needs of the employers who saw their own omnipotence wane against the new social awareness. But all of this activity comes at a price, a price that I will describe as a minor threshold crossed, and the thing about thresholds crossed is that they all represent a progression towards a journey, a journey with no identifiable destination, no focus, and no resolve, and ultimately without a future. Let us suppose that our journey avoids the computer-sentient climax of control evoked by the *Matrix* or *Terminator* franchises and we dodge that particular bullet. What then? Does that leave us secure? If these films were to concentrate on one nation's drive to extinguish another, simply for the purpose of securing resources by denying the existence of the weaker nation and its peoples, then would we buy into the entertainment value or would we recoil in horror? We only need to return to the Second World War to see these wars of elimination as being real, with the Nazification of the conquered lands of Europe or the Japanese genocides perpetrated in Nanking and other atrocities in China. It could never happen now, of course; we are way too civilised. Alternatively, 'civilisation' is the wrong term to use for it, because what we saw was the earnest attempts to supplant one civilisation of value over another civilisation of lesser value. Or was it the Romans seeking to impose their preferred version of civilisation over the 'barbarian' hordes? (I have always taken exception to this term's origins given that it refers to the baa-baa noises made by unintelligible sheep as the Romans' interpretation of ethnic languages, or was it the Greeks? One for the chairperson to pronounce on, perhaps?) Or was it the slave trade, or apartheid in America, South Africa, and Australia? And I won't even mention what is happening today in the Levant. How about the 'disappeared' in Argentina or the multiple arrests in Turkey or Egypt earlier, where one regime seeks to retain control over another option? Yet within these examples we can still select a handful of films that portray some of these events as entertainment for willing consumption by our ever so enlightened audiences—*Gladiator* for one, where the destruction of the Northern Tribes is celebrated in the opening battle scenes. How strange that the main gist of the film is one man's battle to overcome and overturn the personal injustice that has been heaped upon his shoulders. At least he never attempted to impose any other injustices on any other peoples. Just remember that the events portrayed in these reconstructions are extremely rare. Sleep tight and don't have nightmares. See how readily we slip into the role of interested spectator instead of shocked citizen? It strikes me that we do not need a new 'master', sentient and utterly ruthless, devoid of emotion, when the old masters have been very successful without the need for machines.

What of the immigrants surfacing on Europe's borders? Within any herd there are potentially two driving forces, the lure of fresh pastures and resources or the desire to leave the biting-insect- and dung-filled grasslands behind. Both options are valid, and in simple terms they can be summarised as pulled or pushed. The same can be said of migrants, pulled to the promise of an improved situation or pushed by events behind them. I would suggest that the migrant demographic will split roughly along this divide, with the young seeking the brighter promise and the older, more vulnerable being driven by the lack of an immediate future if they remain. Both scenarios are valid. Coming back to the learned MP's comments about migrants being of value to the UK economy, this can then be shown to be only partly true by the simple expedient of considering the two polar extremes of being pulled or pushed towards our lands. Those who are pulled are looking for advancement, while those who are pushed are looking for succour. The first seeks to provide an economic benefit to the donor country, while the second seeks to receive benefit from the donor country. This is very simply stated and is extremely simplistic, but then so are the statements made by our MPs. There is obviously every shade of grey between the two polar extremes, and there are other 'loops' of economic impact to be considered. How much money generated by the economically active migrants is simply siphoned out of our economy to be returned to the original homeland, for example?

But if we return to the herd and its polar extremes of vanguard and rearguard, what else can we learn? The herd is notionally one mass, a superorganism if you will, that spans from leader to straggler. But the herd has no structure, no single imperative other than the survival of each individual. It pays to be at the front, to cross the river before the crocodiles wake up, to taste the sweetest grass, to leave the dung- and insect-filled plains behind. But as the herd passes, they degrade and reduce the advantage for all lesser beasts that are beyond their prime, or weighed down with the young, or old or injured. The vanguard has no interest in those that follow, and so it is with the human herd. Who gets to reach the border before the 'crocodiles' wake up and snap their barbed wire teeth shut? Who gets the pick of the shelter and food before the main body stretches resources to levels of compromise? Individually, family groups will stick together to support and aid the passage of the young, old, and injured, while collectively these become the left-behinds, the ones unable to compete and the ones condemned to being second best. Within the human herd itself then we can begin to see the polarisation of ambition and advantage over the lesser members. Let us impose a more physical test of entry to the Promised Land. African migrants who have crossed the Sahara will tell of the sexual abuse of women and children. South American women walking through the desert into the US will receive similar treatment from their guides and accept it as the price to be paid. Yet still they come. The Calais Jungle camp occupants would block roads with trees and missiles as they attempted to bludgeon their way onto trucks and cars and would resist others' attempts to move them from their camp by rioting. There is one overriding imperative that can be summed up as being prepared to do what is necessary to survive. If it is manifest in the unorganised activities of these resource-short masses, then it is easy to imagine just what can be achieved if the activities are organised. I ask again, for how long do you think the rich and powerful Americans would stand on the border looking north or south at precious resources before taking the decision to secure their needs? After all, all is fair in the war for survival.

I slipped 'vanguard' into the discussion earlier. In military terms this is the leading front of the army. Their function is to advance towards the opposing army before engaging in battle, while the main body catches up to concentrate its power to seize the day. The herd has no organisation, simply those at the front and those at the back with others in the middle. Nor will you find any such organisation in the herd of migrants at the border. The leading edge has no concern for those behind; it is everybody for themselves within the context of the pressure of numbers overturning the immediate opposition. Yet we are led to believe that these seas of human souls are organised and essentially 'care for each other'. We are blinded by the pro forma of language allowed for our description. Our media, our governments, our agencies, and our politicians are frightened to use any form of language that can be considered to be pejorative for fear of being pilloried by the 'caring sections of society'. Any attempt to condemn these migrants as opportunists seeking to overturn the rule of law is instantly derided with 'rent-a-cause loveys' intent on professing their support for these poor unfortunates whom most would cross the street to avoid. We dare not show the true face of human nature for fear that we will witness something that we do not like, something that has the capacity to control rather than be controlled. Is this the hidden motive, suppress the truth of the human spirit for fear that we might enjoy it?

'Tolerance' is a term often used to describe societies that welcome influxes of cultures from across the borders. 'Tolerance' is used to suggest that the issue is resolved and that the issue is not in fact an issue. 'A willingness to accept the existence of opinions or behaviour that are not agreed with or liked.' Nearly the fight-or-flight mechanism but not quite; almost the 'Just keep talking like that, pal, and I'll smash your face in' intermediary stage that can go either way. A tolerant society is not a society at peace; it is simply a society that is packing its gun barrels with powder, wadding, and shot that may or not reach its trigger point.

The problem with all of the pro formas is that they all provide a measure of good but always ignore the bad that they do. So it is that cancer charities champion their good while ignoring the damage that their fundraising visits upon other valid causes, this while ignoring the ever increasing demand for fresh resources to fund their new developments— new costs and new hopes to strip funds from 'less sexy' conditions that form the queue of silent witnesses. Every good intention causes harm elsewhere, but we never consider this. The good blinds us from the bad. Is it possible to audit the balance? Public demand within the UK is calling for the overseas aid budget to be scrapped and the money used at home to fund cancer treatments or doctors' and nurses' pay rises or ... Let us kill strangers to save the people we love. Journalists expose and condemn the feckless waste of funds on Ethiopian girl bands or terrorist supporters or the apparent corruption endemic in Third World countries, without ever joining up the dots to question where, how, and why the government's domestic spending is spent. Election campaigns are always characterised with claims of what amount of money will be raised from where and spent where to the benefit of the voters. This ignores the fact that these claims are only half true. Given that only one side can win, the projection is always one of benefit. Who shall benefit the most from this redistribution of wealth? Which deserving case will be selected at the expense of other, less deserving cases? Who will win and who will lose in the merry-go-round that we call our system of government? But we cannot settle on government alone, because society has other champions of causes, all intent on redistributing wealth: charities, religions, philanthropists, businesses, and pension funds to name a few. But all of these causes are corrupted by the presence of the default genes active within our DNA that seeks to establish personal advantage over our deadliest competitor, *Homo sapiens*. All are content to dig a hole in one resource to fill a hole in another, a process readily identifiable in the fictional planet Krypton but steadfastly ignored in the real Planet Earth.

Perhaps the real answer to the question is simply one of scale. One billion people in the world can still cause immense damage, but relatively locally when compared to the activity of 7.4 billion (plus 250,000 a day).

Pro forma, 'for the sake of form'.

If you mention Hitler and Jews, then the association with gas chambers is almost always included in the thought process. Yet this isn't where the Nazis began. The gas chambers were the final solution in more sense than one. We accept 'final solution' to mean the end of the Jewish question for all time, whereas it can also be considered to be the final solution to the range of options that the Nazis attempted. Special Einsatzgruppen death squads were formed and usually engaged in mass killings of undesirables by shooting them. Other attempts included assembling groups of people into confined areas before detonating explosives hidden within. The mess was too much to bear apparently, and this method was soon abandoned. Tests were conducted on prisoners in small groups placed in small cells or on travelling lorries before vehicle exhaust gasses were passed into the enclosed area. Death by carbon monoxide poisoning was considered too slow. It was only after these attempts had failed that the Nazis latched on to the use of the cyanide-based pesticide Zyklon B as a faster alternative to the factory-scale production of death that they desired. These alternatives were sought because the hierarchy thought that the psychological impact on the Einsatzgruppen squads was too much to bear. Soldiers were usually under the influence of drugs or alcohol, and the impact on morale was considered too onerous to be sustained. The concentration camps were developed into 'death factories' where the majority of the work was done by the inmates themselves. If the inmates (kapos, Funktionshäftling, or 'prisoner functionary') suffered from their activities, they were simply dispatched and replaced with a fresh crop. The Nazis effectively 'sublet' the killing process to other agencies and simply retained management control over the process. By dehumanising the kapos, they had effectively turned them into machines that would follow their instructions.

How much of a quantum leap in thinking is it to fully replace the kapos with full-blown machines? We already have the reality on the ground whereby 'pilots' can steer, fly, and deliver payloads of ordnance to targets in Libya, Syria, Afghanistan, or Iraq without leaving the confines of 'portable cabins' in America or Europe. All we need to do is change the dynamic or the perception by introducing a cheat device into the programs for widespread and indiscriminate death to be rained down from the skies. We are already aware of 'cheat devices' whereby the computer-generated 'result' is corrupted to create a false truth, be it in vehicle emission tests or energy consumption by TV sets. If the computer program were to be corrupted so that all human life detected was presented as an image of armed men weighed down with RPGs and machine guns, then triggers could be pulled and legitimate combatants erased from the battlefield.

But that could never happen. Of course not. I remember the Second Gulf War and the 'shock and awe' rained down upon Iraq while we were simultaneously regaled with smart warhead camera footage as missiles popped through the windows of 'bunkers' or struck armoured vehicles. Reportage went slightly off-piste in the various involvements in the former Yugoslavia conflicts with images of American aircraft missile strikes homing in on armoured vehicles, only for the image to pop into focus at the last instant as a farmer on a tractor. You will agree that it certainly was the last instant for the poor farmer. Throughout it all we were served up the constant reinforcement of the accuracy and selective targeting of these rains of death delivered from above. Then I saw a Channel 4 documentary where they had filmed on the ground the aftermath of some of these 'clinical' strikes. I vividly remember the image of a child's head that was misshapen and another of a child's hand being picked off the floor to be placed in the back of a pickup. Then I began to realise that we already have a filter between the truth and the reality, between the acceptable and the unacceptable. At its simplest, the filter is contained within our own minds as we support our brave soldiers and airmen as they face untold dangers in the pursuit of the legitimate execution of their duties. Who would support the rights of IS or the Taliban to engage in armed conflict with our 'own'? We have a filter in our minds that determines the legitimacy of our actions and that allows us to wage war in the first instance, filters that have been shown to have been distorted by the inaccuracies of WMDs and the ability to strike British bases in Cyprus in minutes, filters that the Chilcot report has demonstrated have been manipulated by the government of the day. It took seven years to produce a report that in itself enabled further filtering of the truth to be dissipated among the population. The people who really took an interest in the proceedings were the parents, spouses, and children of those who had died in the conflict, whereas the water had had thirteen years to pass under the bridge for those of us with less of a personal interest. And then the people of Iraq never really featured at all, in spite of them having the highest levels of personal involvement.

Now imagine how easily a corrupted system could be utilised to eradicate a population from the surface of the earth. Step one is to demonise your targets as subhuman and undeserving of any care or compassion. Step two is to eradicate them by targeting legitimate people who display aggression towards you. Step three is to occupy the land to tidy up any loose ends and to cleanse the sites before your own civilians are allowed to enter. Step four is to fully utilise the new area and its resources to the exclusion of all previous inhabitants. Could never happen?

Step one. 'The only good Indian is a dead one.'

Step two. Attack and destroy the 'raiding parties' of Indian warriors.

Step three. Round up all the survivors and move them to the reservation created on marginal land incapable of supporting them.

Step four. Colonise all the Indian free territory and cover the open ranges with fences, crops, and alien livestock to the exclusion of most that went before.

In the *Matrix* and *Terminator* franchises, it is the machines that we need to fear, machines that have attained a sentience of their own and that place no value on human life, the ultimate opponent for the human race that is just a technological breakthrough or two away from reality. I would suggest that the human race already has its ultimate opponent, the human race. Hitler, if my memory serves, only declared war on the United States. In all other instances he simply reacted to supposed acts of aggression from those neighbouring countries that he had decided to invade. In this, Hitler had developed his own filters between reality and the German population, but only after he had built his armed forces up to cater for the new 'reality'.

We content ourselves by staying behind our own filters that reinforce our rights to self-defence, to self-determination, to fairness, and to freedoms, all of which can be used to work against us, as we can use these devices to deny others of the same rights. Our reality is viewed through the prism of prejudice—them and us, right and wrong, tolerance and intolerance, might and right—and we are simply a need away from focussing our prism on a different plane, a plane that could see a war of annihilation for territory that will dwarf all before it simply with the numbers involved.

This war exists, the fronts are truly global, the opponents are many and varied, and we will not rest until they are eliminated or marginalised on territory situated outside our current needs. Every continent is being assailed as we strive to supplant all that went before with the human species. What of the endgame, the realisation that the earth is full of people to the exclusion of all but marginal refugees struggling to survive on the fringes or those that have learnt to parasitise our success? A world devoid of elephants and tigers but full of rats and mice.

How to construct the false signal between the drone camera and the pilot? All the program needs to do is substitute real images of innocent civilians for images of armed insurgents for the widespread, deliberate, and justifiable extermination of entire populations. How many people would need to be involved in the process? Common programs abound on mobile telephones that distort voices or photographic images for humorous purposes. To design a distortion program would be very easy, and the distortion writers never need to know the use to which their program will be put. When VW introduced their emissions cheat device, very few people, it is perfectly reasonable to suppose, knew of its existence. Equally, it would be possible to set to and design a program whereby the originators have no idea of the intended purpose.

But as we have seen before, the reliance of such distortion programming is not dependent on computers. All it takes is the distortion of human perception for the widespread, deliberate, and perfectly justifiable extermination of populations to occur. Have we not achieved this within the human mindset whenever genocide occurs? Indeed, can I suggest that this distortion is the first prerequisite for genocide to occur? All that is required is the mindset of gross superiority over, or conversely gross inferiority of, the intended targets for them to become victims.

Now we have two potential mechanisms whereby genocide can be triggered: distortion of the human mindset or distortion of the information relayed to the undistorted mind. It would even be conceivably possible to connect virtual reality game players to live weapons in live environments where murder is simply an entertainment utterly free of conscience or moral torment.

Will this be the mechanism of the great land clearances of the twenty-first century, or can we survive to the twenty-second century? The remoteness of existing drone operators from the reality of their actions already elevates the potential for mass murder to the mundane. Fly your drone for eight hours and destroy some heavily armed insurgents threatening your own half a world away before popping home at shift end to your partner and kids for tea. Even the drone base engineers can go about their business of refuelling and rearming without any exposure to the reality of their actions. Ground crews during WWII were keenly aware of the sacrifice and bravery of the flight crews whom they so attentively serviced, while the bomber crews themselves were remote from their bomb actions on the ground. This move to technology, does it make us more or less secure? More or less likely to wage war? More or less likely to exterminate our fellow man?

Equally, any 'peaceful' people thus attacked would feel perfectly justified in taking up arms to fight the new enemy, thereby reinforcing the validity of the actions already taken. Parallels with Red Indians— sorry, Native American tribes—spring to mind. I wonder, did the Native 'Americans' ever consider that they were ever Americans or even 'natives'?

Perhaps their better title would be 'subjugated aboriginals' or simply 'inferior occupiers of prime territory'.

Then again it is always possible to concoct an excuse based on perceived threats; WMDs would do for starters. At least in this day and age we are much too media-savvy and educated to fall for such a ploy. It could never happen. This example, if only I were cynical enough to contemplate it, would mean that 'peaceful' people would be motivated to attack, if only after being threatened with imaginary weapons.

Through it all the question is begged, if we are prepared to do this to other humans, theirs, not ours, then what chance does any other inconvenient species that gets in our way have? But while I say it begs the question, in reality we already have the answer.

One common reasoning about the natural world is that non-human species have a value. How many cancer-treating drugs could be extracted from soon-to-be-extinct rainforest trees? How much can the local community benefit from selling elephant poo as a garden fertiliser to caring Europeans? How much can ecotourism inject into local communities?

Now I am going to offer two schools of thought in this respect:

Every attempt to prolong the existence of alternative habitats and species limits the availability of resources for other humans and thus accelerates the need to extract from other humans as the only other alternative.

Every failure to prolong the existence of alternative habitats and species accelerates the 'human project' and thus accelerates the need to extract from other humans as the only alternative.

Not much of a choice, is it? Damned if you do and damned if you don't.

Because, and this is fundamental to all of the issues, our species is utterly unable to treat the cause while we piss about with the symptoms. The cause is too many people. End of conversation.

I am going to suggest that I am an idiot. 'At last he admits it!' I hear you cry. In fact, I am that much of an idiot that I cannot see all the eco-friendly, conservationist, sustainabilitist, philanthropic, global-warming-halting clothes that the rest of our Great Brains have donned as their sartorial finery. All I see is naked experts sitting at the top table and enjoying the spoils while some other idiot is playing the fiddle in the background so loud that no one can smell the smoke.

Not much of a fairy tale, is it? If we exercised choice, if we allowed our intelligence to override our base programs, what would we be capable of achieving? Will we enjoy an epiphany, a serendipitous event of enlightenment that may yet steer us away from the precipice in the clouds? Can we indulge in a bloodless genocide that sees the (non-) creation of billions of unconceived foetuses? For that would surely be the answer, the way forward. But the human condition is such that we would mourn the unborn just as surely as we will mourn the victims of genocide. There is an answer, practised with elephants, where entire extended families are culled, leaving no survivors to mourn or mark the passing. If a tree falls in the forest and there is no one left to care, did it ever exist?

But what is it that we care about, that we killed too many or not enough? Survivors remind us of our guilt and torture our souls, whereas total obliteration clears our conscience and relieves us of our responsibility. Total obliteration obliterates our guilt, whereas with survivors our minds bend not to the task of controlling our urge to breed but to the imperative to justify our actions. It is as simple as every ape for itself. Every live birth that exceeds our death rate is another nail in the coffin of Planet Earth as we persist in our death of a trillion cuts. Learned scholars will bleat about our humanity as our magic shield while witnessing our daily inhumanity in acts as simple as profiting from the sale of just one cigarette, just one fix of cocaine, just one person with a self-inflicted illness denying another a hospital bed.

Earlier in *How to Kill an Elephant* I contrasted two reports about contrasting opinions:

The hanging judges cleared our population of the criminal element, reducing our reliance on murder as a resort to justice.

Our education system and civilisation has cleared our population's reliance on murder as a resort to justice, and instead we turn to the courts.

Both seek to explain the reduction in murder rates over the last few hundred years.

Let us consider the education and civilisation strides that have taken place since the turn of the twentieth century and contrast these with the list of wars, conflicts, and genocides contained within our very recent past.

I would suggest that our education and civilisation has made great strides forward during the last century, great strides in the ability to commit mass murder. Our Great Brains seem wholly incapable of separating out our base instincts from our intellect. The quote will always be seen as something along the lines of, 'War should always be seen as the last recourse and must always be the last option on the table.' It's amazing how often we run out of other options. Of course, in 50 per cent of all instances, one party will certainly be acting in self-defence, in defence of a way of life, in defence of the religious freedom to deny others their religious freedom, in defence of the right to secure one political doctrine over another, in defence of the right to secure resources over another, less deserving population, in defence of overturning one culture and replacing it with another.

I am wrong, I absolutely guarantee it, but I know that my ideas are not wrong, so as you pick through my obvious errors within your field of expertise, please consider the ideas in the round. This is my personal dispensation card, my 'get out of jail free' card that pleads for you not to shoot the messenger. I am right, I absolutely guarantee it. Your task, through the steps you take through on your journey from egg to ashes, is to prove me wrong by your actions, not your thoughts.

There exists an argument that as we become more intelligent, and education is often factored in somewhere as well, we learn to control our base impulses. Kerala in India has been referred to as among the most educated of the Indian subcontinent's regions. Its population had begun to stabilise as parents voluntarily restricted their procreation in the main to a single child. I understand the idea and would wholeheartedly embrace the respite it offers to what's left of the natural world but for one small detail. Having ten kids and a GDP per capita of $100 a year doesn't offer much respite when it's one child and a GDP per capita of $10,000 per year.

Pine Processionary Moths

I have just returned from a week's orchid hunting in Rhodes. While there I observed pine processionary moth caterpillars on the move to a pupation site, *Thaumetopoea pityocampa*. The procession reached around 5 meters in length, not bad for a beastie around 35 millimetres long. They form a procession with each caterpillar nose to tail throughout the column as they inch their way across the ground. They are protected by irritating hairs that can cause severe allergic reactions in people and dogs and presumably in other likely predators. What struck me profoundly was the slavish attention to the procession. If you are of a malevolent mindset, then it is possible to steer the head of the procession to the tail and to lock the caterpillars in a never-ending loop that goes nowhere (a record of one week of circular movement exists). Each caterpillar follows its leader and is content to do so. My epiphany revolves around the front caterpillar, the leader, which is identical to all the others within the group except that it has no leader and is condemned to find the route. In human terms you can imagine the nervous exhaustion generated by being in this position. These creatures are preprogramed to follow, slavishly, and yet the one at the front has nothing to follow. Methinks it would have chewed its fingernails to the quick if it had any. Now this observation triggered another thought process, namely: pulled or pushed? Does the ripple of movement throughout the procession start with a push from the back or a pull from the front? Are the caterpillars drawn or forced? I have a procession on video from Spain taken some years ago. I will watch it and see if I can determine the order of movement, but in the meantime I want to consider the theoretical options.

The lead caterpillar's role is to find suitable soft substrate in which to bury before pupating underground. In theory, the lead caterpillar is pushing against the ground trying to detect for sufficient looseness of substrate. If that is the case, then it is behaving in an identical fashion to those behind in that it is pushing. But equally, it could be being pushed. If you were possessed of a slightly less malevolent mindset, then it should be possible to disconnect each caterpillar from its leader so that each individual is an individual. Each is now manically seeking either another's rump to follow or a soft patch of soil to bury into. Presumably these caterpillars have been studied in detail by some thick-skinned intellect immune to allergic reaction who knows about the relative pupation success rates of individual as opposed to group pupation sites. Alternatively there may not be any difference. The real advantage to communal living may be in the density of the silk 'nests' the caterpillars occupy within the pine trees. The thicker the silk coating, the greater the protection from the elements and potential predators, and more caterpillars means more irritant hairs shed to deter interference. We could consider the relative size of these caterpillars' brains when compared to those of 'higher organisms' such as American bison or snow geese or *Homo sapiens* and then ask the question about the fundamental behavioural differences between the species. Can we detect a pattern?

From a purely biological perspective, then, can we determine any difference in the behaviour of economic migrants moving across regions or borders in pursuit of a suitable pupation site? 'How dare you compare displaced people with "mindless caterpillars!"' I hear you exclaim. Well, why not? Are we not biological organisms driven by base needs and instincts? I am not even singling out cross-border migrations. Urbanisation has persisted for centuries as the rural landscape has surrendered its populations to the cityscapes and indeed is accelerating globally as I write. It poses the question for me of, where is the leadership? Any of the caterpillars could draw the short straw and end up in front. Alternatively it is the caterpillars condemned to push the rump of their neighbour that have drawn the short straw. It's nearly enough for me to ignore my video of the procession while I consider the theoretical options. Hopefully twenty-five frames a second will shed some light on the caterpillars' actions. Be patient with me while I mull the options some more.

I may have attracted some attention from students about to enter, or indeed engaged in, graduate education programmes within the UK. Currently the annual cost of each education course is set at a maximum of slightly over £9,000. That, if my maths is correct, amounts to nearly £28,000 for a three-year course, not forgetting to add in the requirements to feed and house the body as you feed your mind. Shall we say £45,000 to learn another's pro forma? I have two questions for you: pulled or pushed, and is it worth it?

Thanks to My Publisher

While trawling the Internet for potential publishers, I was stopped quite early in my tracks by statements explaining the need for 'life-affirming' material. How to pretend that all is positive and beneficial to life without supporting and subscribing to the bait ball? I cannot do it. I have seen the horizon. To be truly life-affirming, the faults need to be exposed and the herd's route altered beyond the 'business as usual' ethic that permeates our species and allows for the accelerating process of apocalypse to continue utterly unabated. I self-published. I paid my fees to circumvent the selection process that limits my and your horizon. I dug a hole in my bank balance to fill a hole in Planet Earth. Maybe, just maybe, we will qualify for our moniker, *sapiens*.

I apologise in advance, but I have to confess that I have struggled to achieve a satisfactory end to *How to Kill an Elephant*. After much deliberation, I feel that the final words should be those of my favourite anti-heroes, highway engineers.

While in a confessional mood, I must also admit that this information has been appended beyond the eighteen months of my intended subtitle; *Eighteen Months to Save the Planet*. Why eighteen months? Quite simply, it represents the official timespan of the American presidential election process, you know the one, the one that elects the most powerful leader in the world. I will let you draw your own conclusions for this opportunity missed to change the paradigm.

Highwaymen, a Final Word

On 17 February 2017, BBC teletext ran a page on a local village's action group against speeding motorists, with the report that they had recorded 562,000 speeding drivers since November 2015.

Some vehicles had achieved speeds in excess of 100 mph.

Sharnford Traffic Action Group (STAG) has campaigned against dangerous driving on the B4114 and Aston Lane for more than ten years. If the parish council would care to provide £12,000 to the county council, they will provide two speed bumps on Aston Lane.

This coincidence is enough for me to believe in fairies. You will understand my need to include it. Way back in the earliest ear-bashing of highway engineers, I remarked on the fact that a local village (anonymous at the time, but I will now name it as Sharnford) had been plastered with coloured tarmac with white stripes all over it to slow the traffic down. I had met with the engineers shortly after its installation, who remarked on its great success given a reduction in average speed by 4 mph. Needless to say, the coloured tarmac has long been replaced with plain tarmac, as the earlier initiative was determined to have lost both its shock value and its effect. Here we are around 16 years later with 562,000 speeders in around 14 months. That's 40,000 a month exceeding 30 mph, or 1,300 a day. A number of thoughts spring to mind (there's a surprise). How many are doing 31 mph, for starters? A second thought is, wow, only £12,000 to install some strips of raised tarmac that people can ignore as well. My third thought is a little different: speeding is bad; it is dangerous, it is antisocial (whatever that means), and it is against the law (whatever that means). But here's the thing: when serious or fatal accidents occur, these sites automatically qualify for permanent speed cameras. Without monitoring the local rags on a weekly basis, I cannot tell you how many serious or fatal accidents have occurred, but the absence of permanent speed cameras can tell you. Because their numbers do not warrant cameras. Of course if the authorities really wanted to prevent fatal or serious injuries, they would simply provide cameras and harvest fines rather than harvest blood banks as victims are (hopefully) reassembled. I accept the analogy of speeding vehicles being like cocked and loaded guns being waved around in residential areas, but to date nobody has pulled the trigger. My wife came back from the February Women's Institute meeting to declare that the guest speaker was from STAG and had declared that one vehicle had been clocked doing 120 mph. I know the village of Sharnford and have driven through it for forty years. I would declare my earnest and honest belief that it would be impossible to achieve these speeds given the nature of the bends and inclines within the village in anything short of a Formula One racing car. The main village road is already split with a one-way system, which is essentially a large-house-filled roundabout (squareabout more accurately), and other entry points are governed by 90-degree bends, chicanes, or blind brows. Oh, if only they could bring the coloured tarmac back, all would be solved. (I will ignore the flashing telematic signs and other failed initiatives already provided for the village.)

If you accept these figures at face value and add in the ten-year lifespan of STAG, then you see that 120 months at 40,000 speeding vehicles per month equates to nearly 5 million speeding vehicles which have still failed to qualify for safety cameras. STAG, through the parish council, will now exert pressure to increase the parish council tax precepts to raise an additional £12,000 to fund the two offered speed bumps. In my own local village, the noise generated by our two existing speed bumps is a major bone of contention, with vehicle braking and acceleration sounds above those of cruising vehicles. The monthly newsletter is also regularly peppered with residents' complaints about vehicles speeding through the village in spite of the two speed bumps. If only our village had had two brain cells, we might not have needed two speed bumps. I wonder, will this conversation ever take place in Sharnford? Alternatively, if every speeding vehicle were simply fined £4, then the £20,000,000 raised could have paid for a bypass—sorted, except the average planning and budget supply process for 'new roads' is over twenty years. What could we do with the other £20,000,000? A new sports pavilion, a new village hall, a tax holiday for the local residents? Trouble is, how long would it be before the village regarded the income stream from speeding motorists as both beneficial and desirable? It's enough to make you hide the speed limit signs, although this may not be necessary given that most people take no notice of them anyway. I should also add that the B4114 has not always been so classified. Until 1976 it was the A46 and the primary strategic trunk road between Leicester and Coventry. Until 1976 anyway, because the construction of the nearby M69 motorway effectively bypassed Sharnford by providing an alternative trunk route between the two cities. And then, like a white knight charging in to save the day, another idea came. Except, of course, I've already had the white knight charging in wearing the arms of STAG, so what can I call this one, a black knight perhaps? Highway engineers in Leicester are now mooting the possibility of removing speed bumps from urban areas because they create a local increase in pollution caused by braking and accelerating over these obstacles. What could be better than an early career developing speed-controlling initiatives other than a later career spent dismantling the very objects of your earlier desire? It makes it all seem so very worthwhile, doesn't it? If every council employee's employment revolved in cycles, then your early career guarantees your later one. (Next thing you know, they'll be constructing paths for all these cycles.) (That isn't the cycles that used to be used for the commute to work, before planning authorities and initiatives drove the employment sites away from the housing areas. What will planning initiative number 57 be, I wonder?)

Zack Hemsey

Finally I must express a great debt of gratitude to the music and lyrics of Zack Hemsey, a self-publisher; you'll find him on the Internet. They have served as a near constant companion to my writing. A friend once remarked on Hemsey's music, and I gave it the title of 'music to think by'. It is fair to say that I am in no way implying that Zack Hemsey shares my views, but if you listen to his lyrics, you will find that he certainly has a gloriously different spin on life and deserves to be celebrated.

But wait one minute, I promised you a showdown in the drawing room. We need to fix the guilt.

So who is guilty? Who is it that needs to be arrested and removed from society? Where shall we look for innocence within our realms? Who is it that should sleep soundly in their beds tonight? 'We' are all guilty, whereas 'I' might just have a 'get out of jail free' card, a personal dispensation to waive before the judiciary. I am innocent; we are guilty as charged. The only answer to all of the identified ills can best be summed up by the need for every individual to take personal responsibility for all of their actions. That's my ills, not ours, because my actions can create a cheese well, whereas our actions leave pure water. I have the power to change, while we do not. But it is I that seeks to gain advantage from we. I sell cigarettes to earn a living, I inflict illness on myself and others, I demand resources to take my place in the queue for treatment, and I expect all others to contribute. I drive my car to fill the lungs of strangers; I dig holes in Planet Earth to fill holes in Planet Earth.

A final, final word to statistics. It's a form of measurement. You know how I love to measure.

Eight billion people by 2024.

Ten billion people by 2056.[5]

In the future, if you stand on enough shoulders, it may be possible for you to catch a glimpse of the horizon as well.

---

[5] https//:www.worldometers.com/info

Epilogue: Trainee Arseholes

If all children are trainee arseholes, then it is dependent on the trainers to steer them in the right direction.

If all young highway engineers are trained by old highway engineers, then the whole process can best be described as horizon limiting. Custom and practice and the pro forma of what it is to be a highway engineer are to the forefront, and the reality of the situation becomes a casualty.

If all young politicians are trained by old politicians, then the whole process can be described as horizon limiting. Custom and practice and the pro forma of what it is to be a politician are to the forefront, and the reality of the situation becomes a casualty.

If all young leaders are trained by old leaders, then the whole process can best be described as horizon limiting. Custom and practice and the pro forma of what it is to be a leader are to the forefront, and the reality of the situation becomes a casualty.

If all young scientists are trained by old scientists …

If all young hospital administrators are trained by old hospital administrators …

The event horizon of a black hole is described as the boundary at which the gravitational pull of a massive object becomes so great as to make escape impossible.

While our clever chappies with those glass tube thingies earnestly search out black holes at the centre of our and other galaxies, they choose to ignore the ones closer to hand: black holes created by humankind.

Discuss.

Now, we can begin to think in terms of black holes that go by the name of Clovis point, Stonehenge, the pyramids, Christianity, FIFA, and education. The list is only limited by our imagination. Except, and this is the really important bit, our very imagination is subject to the event horizons established by pro forma.

As a sixth form student, I once attended a maths club lecture at Leicester University where the subject was black holes and time travel. The shortened version is that extreme gravity could pull time, an expression of speed plotted against distance, into the negative quadrant of the graph, allowing for the possibility of time travel. Black holes were the gateways, potentially, to other times and other realities. It has only taken me forty years, but I believe I have invented time travel to alternate realities.

If I can penetrate the event horizons adopted by our species, then alternate realities can present themselves to us and alternate timelines can be discovered. These timelines and realities could include the absence of the ten billionth person on earth in 2056, the end of so many extinction-level events for our fellow passengers, and the signing of the peace treaties to end the food world war before it even starts. All we need to do is breach the event horizons created by humankind and thereby bring an end to the bait ball.

About the Author

Robert Pins is a retired owner and industrialist with varied experience of a wide range of the trials that are thrown up by the dogma and the red tape of our oh-so dysfunctional society. He is utterly disillusioned with what passes for leadership and direction as we consign 'solutions' to the out tray. Business as usual solves nothing as we continue our route march to the precipice in the clouds.

www.ingramcontent.com/pod-product-compliance
Lightning Source LLC
Chambersburg PA
CBHW020717180526
45163CB00001B/6